Financial Management in Agriculture

FIFTH EDITION

Financial

Management

PETER J. BARRY, Ph.D.

University of Illinois
at Urbana – Champaign

PAUL N. ELLINGER, Ph.D.

Texas A&M University

in Agriculture

C. B. BAKER, Ph.D.

University of Illinois
at Urbana – Champaign

JOHN A. HOPKIN, Ph.D.

Texas A&M University

Interstate Publishers, Inc. IPP Danville, Illinois

FINANCIAL MANAGEMENT IN AGRICULTURE

Fifth Edition

Prior Editions: 1973, 1979, 1983, 1988

Library of Congress Catalog Card No. 93-80628

ISBN 0-8134-2971-4

2 3
4 5 6
7 8 9

Order from

INTERSTATE PUBLISHERS, INC.

510 North Vermilion Street
P.O. Box 50
Danville, Illinois 61834 – 0050

Phone: (800) 843 – 4774
FAX: (217) 446 – 9706

PREFACE

As with previous editions, the fifth edition of *Financial Management in Agriculture* develops and applies concepts, analytical methods, and descriptive information about agricultural finance. Many of the important managerial problems and opportunities in agriculture involve finance: financing resource control, investment analysis, asset valuation, financial reporting and planning, managing risks and liquidity, establishing credit relationships with lenders, and understanding financial markets.

Most agricultural production firms differ significantly from large corporations in other industries. Business size generally is smaller, and managers must address financial activities as well as production, marketing, and personnel decisions. Risks associated with land, climate, and biological factors are especially important. Special financial institutions and lending programs also contribute to the uniqueness of agricultural finance.

Financial Management in Agriculture focuses on planning, analyzing, and controlling business performance in agriculture and on the related financial markets. The fifth edition contains 21 chapters organized into six sections. Section One introduces the nature and scope of financial management and the major goals that motivate managerial performance. The three chapters in Section Two address financial statements, financial analysis and control, and financial planning.

Section Three develops the concepts of financial leverage, risk, and liquidity and their effects on business performance. Various strategies for managing business and financial risks are considered. Section Four stresses the concepts and tools of asset valuation and long-term financial decision making. Capital

v

budgeting methods are identified under conditions of risk and inflation and under alternative financing arrangements for owning and leasing business assets.

Section Five contains four chapters that describe and evaluate financial intermediaries serving agriculture and the related management and policy issues. This section provides important information for understanding how the national and international financial markets influence financial decision making. Section Six considers legal aspects of finance, legal forms of business organization, and the management of equity capital in agriculture.

The fifth edition has several significant changes. Chapter 1 contains new material on the strategic management process and on financial structure of the agricultural sector. Chapters 3, 4, and 5 reflect substantial changes in the design and analysis of financial statements in order to achieve consistency with the 1991 Recommendations of the Farm Financial Standards Task Force. Chapters 6 and 7 contain added material about capital structure theories and borrower/lender relationships based on new finance concepts associated with asymmetric information and principal-agent relationships.

Chapter 8 extends risk management analysis by generalizing the portfolio theory approach and showing how portfolio analysis can be applied to analyze debt servicing capacity. Chapter 10 describes the modified internal rate-of-return method of capital budgeting, and Chapter 12 extends the bid-price analysis for farm land investments.

Section Five has been changed considerably. Chapter 15 has been recast to focus on recent financial market developments, new types of financial services, and a perspective on competition in financial markets. A new Chapter 16 centers on the management environment of financial institutions, emphasizing asset – liability management, risk management, and loan pricing. Chapter 17 describes the major agricultural lenders in detail, including substantial new developments affecting the Farm Credit System, trade credit, government credit programs, and the Federal Agricultural Mortgage Corporation (Farmer Mac). Finally, the legal aspects of agricultural finance now make up a new Chapter 19.

Other major changes include a summary at the end of each chapter and a glossary of key terms used in agricultural finance at the end of the book. Minor changes and improvements are incorporated within the rest of the chapters as well.

Financial Management in Agriculture is applicable in most countries of the world, although the institutional setting reflects the financial and legal environment of the United States. The book is intended for undergraduate and beginning graduate students, who are assumed to have some acquaintance with microeconomic principles and with statistics. Otherwise, the book represents a starting point in the study of agricultural finance.

We are indebted to many persons for their assistance and support since the book was first published in 1973. Special thanks are due to F. J. Barnard, J. R. Brake, J. H. Clark, T. L. Frey, S. C. Gabriel, C. G. Gustafson, B. Godfrey, K. S. Harris, W. Hayenga, B. L. Jones, W. F. Lee, S. Lence, B. Pflueger, L. J. Robison, and S. T. Sonka. Numerous students, instructors, and book reviewers have made useful suggestions and raised critical questions that have furthered the book's development. Finally, we are grateful to Phyllis Blackford for her tireless, effective, and always cheerful efforts to type and compile the manuscript and to Pat Ward for her excellent editorial work at Interstate Publishers, Inc.

Peter J. Barry
Paul N. Ellinger
John A. Hopkin
C. B. Baker

CONTENTS

SECTION TWO

Financial Analysis, Planning, and Control

SECTION THREE

Capital Structure, Liquidity, and Risk Management

SECTION FOUR

Capital Budgeting and Long-Term Decision Making

SECTION FIVE

Financial Markets
for Agriculture

SECTION SIX

Other Topics

SECTION ONE

Introduction

1

Nature and Scope of
Financial Management

The financial management of a business, an agency, a household, or other economic unit involves the acquisition and use of financial resources and the protection of equity capital from various sources of risk. Thus, the evaluation of new investments, financial planning, liquidity management, and relationships with financial intermediaries are important. Financial management has a strong micro focus, but it also occurs within a setting of financial markets and institutions and public policies. Thus, the study of financial management not only considers the concepts, tools, and methods of analysis that financial managers use but it also addresses the financial market, institutional, and policy issues as well.

In this chapter we introduce the nature and scope of financial management in agriculture. The following sections identify the characteristics of the financial function of agricultural businesses, establish the importance of time and risk in financial management, review the management process and other fundamental tools of financial management, and introduce some of the structural characteristics of agricultural finance.

THE FINANCIAL FUNCTION

Figure 1.1 portrays a flow chart of an agricultural business that shows

3

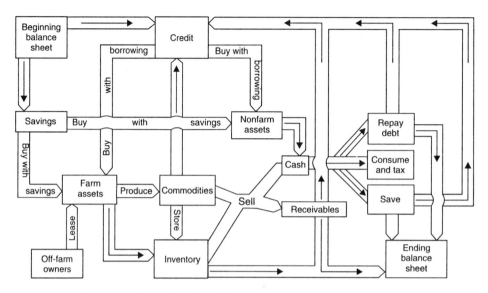

Figure 1.1. Organizational flow chart of an agricultural firm.

how financial activities are related to production, marketing, and consumption. At the start of the accounting period, the firm's beginning balance sheet, in the upper left corner of the figure, is the source of current and noncurrent assets, which are used to produce salable commodities. These assets, along with new purchases, are financed by the firm's equity capital (its "savings") and by borrowed capital. Other assets may be leased or hired from external sources.

Credit plays a central role in providing cash for financing asset acquisitions. Credit also serves as a source of liquidity for meeting financial obligations and countering risks. The income expected from sales influences the cost, the availability, and other terms of loans from various lenders. Credit is also influenced by the firm's asset holdings, past borrowing activities, and financial market conditions. In turn, these credit effects and other factors influencing cash flows determine the firm's liquidity position.

Produced commodities can be sold for cash or for "receivables" involving payment later. Some commodities may be stored in inventory for later sale, thus entering the ending balance sheet. The cash from commodity sales is used to pay debt obligations and to meet tax obligations and outlays for family consumption. The remaining funds go into savings and, along with other assets, enter the ending balance sheet (at the lower right corner of Figure 1.1). The effects of non-business activities on the firm's cash flow and balance sheet also are accounted for. These processes then are repeated as another year begins.

An income statement or a cash flow budget can be used to measure the flow effects of the firm's financial performance over each period of time. And, a balance sheet can be used to measure the stock effects of the firm's financial performance at a point in time. These fundamental financial statements provide a rich and extensive set of indicators for evaluating profitability, liquidity, and solvency.

RISK AND TIME

The financial activities described above occur in a decision environment that is characterized by high degrees of risk and uncertainty. The traditional business risks in agriculture stem mostly from production activities and from resource and commodity markets. These business risks occur as unanticipated variations in agricultural production and in commodity and resource prices, uncertainties about personnel performance, changes in technology, and changes in the firm's legal environment. These business risks combine with financial risks, attributed to borrowing and leasing, to bring strong challenges in risk management for farmers and their lenders.

The impacts of time and risk together are important in financial management. Since business performance occurs over time, trends in key performance measures are important to evaluate. Long planning horizons often are needed to properly analyze capital investments. Observing the "life cycle" effect in a farm business, which also reflects the life cycle of the firm's manager, is common. A farm business typically passes through four stages in its life cycle: establishment, growth, consolidation, and transfer. The manager's objectives and financial activities may differ significantly in these various stages.

The linkage between risk and time is based on uncertainties managers experience in formulating expectations about future events and decision outcomes. Risk and uncertainty reduce the reliability of future plans and tend to shorten planning horizons, thus hampering financial planning and investment analysis. For some decisions, especially routine repetitive ones, the manager can accurately identify the possible outcomes and judge their likelihoods. In other cases the manager may have difficulty delineating the possible outcomes and, even more, their likelihood. However, as time passes, new information becomes available to the firm, learning opportunities arise, and experience is accumulated. All of these factors contribute to effective forward planning and are an important part of risk management.

In general, business and financial risks complicate financial management. They may trigger substantial efforts to manage risks and to protect cash flows and equity capital. Reducing the likelihood of risks, building risk-bearing capacity, and transferring risks to other economic units all warrant careful consideration in financial management.

TOOLS OF FINANCIAL MANAGEMENT

The concepts and tools of financial management are the central focus of this text. In this section we briefly introduce some of the basic tools under the headings of the strategic management process, information flows, budgeting methods, and understanding financial markets and institutions. These tools are used in various ways in the chapters that follow.

The Strategic Management Process

The process of strategic management can be ordered into a systematic set of six interrelated steps. The steps are identified as follows:

1. *Defining and developing a firm's mission*

   This step defines the business and the purposes of the firm. It establishes the future direction of the firm, articulates a mission statement, and helps answer the questions:

   What business or businesses are we in?

   Why are we in business?

   What are the markets?

   What social and ethical responsibilities do we have?

   Where are we headed?

2. *Formulating objectives*

   Objectives and goals guide the decision making and strategy formulation process. One of the purposes of objectives is to translate the mission statement into specific targets. The manager must identify relevant, achievable, clearly stated, and quantifiable objectives. The objectives are often separated into financial and strategic objectives, and further subdivided into short- and long-run objectives.

3. *Assessing the firm and evaluating the environment*

First, the manager should assess the internal strengths and weaknesses of the firm. The current physical and human resources should be evaluated. Weaknesses may indicate vulnerable areas that may need corrective action. The manager should also evaluate the external environment and identify those factors that could have an impact on the firm's performance. The competitive forces within the industry and within the economy as a whole should be evaluated to identify possible opportunities or threats. This assessment is commonly referred to as *SWOT* (strengths, weaknesses, opportunities, and threats) analysis.

4. *Building strategy*

Building strategy involves the managerial tasks of implementing procedures to achieve the desired objectives. Data, facts, and other information are needed to support the strategy-building process. Included are the manager's possible actions, the likelihood of various events, the available resources, legal requirements, the firm's financial position, and information about relevant external factors. The information must be analyzed so that the possible contributions of potential strategic alternatives can be estimated. After considering the relevant alternatives, the manager must decide on a strategic approach and a course of action.

5. *Implementing strategy*

The plan must be administered throughout the business. Budgets need to be developed to support the plan. The plan may require changes in personnel, capital expenditures, or financial restructuring. The plan must have an implementation timetable that will allow any changes to occur in an organized, efficient, and profitable manner.

6. *Evaluating performance and implementing any corrective actions*

As changes occur in technology, laws, economic conditions, markets, and personal situations, prior decisions must be re-evaluated to determine new courses of action. Steps 1 through 5 are never final and need constant evaluation and corrective adjustments.

A business manager typically applies the strategic management process to many problems, both large and small. The need for decision is triggered by changes in technology, economic conditions, markets, and personal situations. Each decision situation may vary in the size of potential gains or losses, frequency of occurrence, urgency of pending action, and flexibility or cost of change following the decision. Managerial effectiveness depends in part on accurate problem definition and efficient use of the management process. Impediments to the process arise from limitations on human abilities to observe, order, and analyze complex information. External changes can create new problems or opportunities quickly enough to reorient the decision process before current issues are resolved. Objectives must be feasible and realistic. The symptoms of a problem must be separated from the underlying cause. Searching for information and problem solutions may be time consuming, costly, and emotionally demanding. In general, superior managers show an orderly approach to the process, strong analytical skills, decisiveness, and the capacity to accept the consequences and to revise their decisions as conditions warrant.

Information Flows

Information about past, present, and expected business performance is essential for the financial manager. It comes in part from a financial accounting system that reports the firm's profitability, liquidity, and solvency positions. An information system aids in financial control, risk management, meeting legal requirements, and financial planning. The system also provides information about credit worthiness and management ability to lenders and other outside parties.

Other information comes from the firm's external environment, including colleges of agriculture, government agencies, communications media, farm organizations, agribusinesses, and investment and marketing services. Efficient access to external information provides a timely information base and a stronger capability for responding to new conditions.

Budgeting Methods

Budgeting methods involve an orderly approach to assembling and analyzing information and choosing between financial alternatives. In financial management, the emphasis is on cash flow budgets and capital budgets. Cash flow budgets are essential for establishing financial programs and managing liquidity in light of seasonal patterns for farm production and marketing. Capital budgets enable the manager to compare new investments under different financing alternatives, using various decision criteria.

Often budgeting can be computerized by using simulation or optimization techniques. Electronic data processing and recording systems provide much of the planning data. A computer can greatly accelerate the budgeting process and increase analytical capacity.

No planning procedure is perfect, however; nor is budgeting error-free. The plans derived from empirical observations, logical deduction, pencil and paper analysis, or computerized techniques are subject to the quality of the available data and the careful application of the evaluation process. Moreover, the best plans are still subject to considerable uncertainty.

Understanding Financial Markets and Institutions

Financial institutions serve as intermediaries for channeling external savings into investment. Both debt and equity capital are involved, although most agricultural production firms do not obtain outside equity capital. Rather, farm businesses primarily rely on their own savings as a source of equity capital and on credit markets as a source of borrowed funds to finance asset acquisitions and to meet liquidity needs.

The major institutional sources of loan funds for U.S. agriculture include commercial banks, the Farm Credit System, life insurance companies, and government agencies. Other important credit sources are merchants, dealers, agribusiness firms, and individuals, especially sellers of farm land.

Each of these groups of agricultural lenders is unique in terms of funding sources, geographic scope, degree of specialization, legal and regulatory environments, organizational structure, types of loans, eligible borrowers, and operating characteristics. Moreover, each type of lender may respond differently to monetary conditions and other forces in national and international financial markets that determine the cost and availability of credit to agriculture. Thus, understanding these characteristics and the intermediaries' responsiveness to financial market conditions is essential to financial managers. Such understanding contributes to evaluating the cost and availability of credit, to organizing the financing of farm businesses, and to formulating strategies for building credit worthiness.

STRUCTURAL CHARACTERISTICS OF AGRICULTURAL FINANCE

Agricultural production units in the United States have traditionally been characterized as relatively small operations with a largely non-corporate form of business organization that concentrates ownership, management, and risk bearing in the hands of individual farmers and farm families. These size and concentration characteristics are in contrast with those in larger scales of operations in other business sectors in which ownership, management, and labor are separate, specialized functions and in which risk-bearing is spread over numerous corporate shareholders. Of course, many exceptions occur to these generalizations, as in the case of cattle feedlots, poultry and egg production, and orchards. Many of these units are larger scale and more industrialized. Similarly, some crop farms, hog production units, and others are very large with complex contractual arrangements for ownership, management, labor, and financing.

The aggregate wealth and income positions of the agricultural production sector are presented using the financial statements and ratios in Tables 1.1 through 1.3 for the time period from 1950 to 1990. Despite the small business orientation, agriculture is still a capital-intensive industry with investments in farm land, buildings, machinery, equipment, and breeding livestock dominating the asset structure of most types of farms. As shown in Table 1.1, the aggregate value of farm real estate has made up more than 70 percent of the total assets since the early 1970s, reaching nearly 80 percent of the total assets in the early 1980s when prices per acre of farm land in the United States reached their peak. Inventories of livestock, machinery, crops, and other non – real estate farm assets generally make up 10 to 15 percent of the total assets. The dominance of farm real estate, together with the relatively small holdings of financial assets, indicates the high capital intensity and low asset liquidity of the agricultural production sector. Finally, consistent with the dominance of farm land, real estate debt consistently comprises about 55 percent of the sector's debt obligations.

Table 1.2 indicates aggregate net farm income and off-farm income earned by farm operators during the 1950 – 1990 period. Interesting features are the volatile nature of net farm income (the buoyant 1970s were characterized by net farm income of $25.5 billion in 1975), the importance of governmental payments in recent years, and the significant role of off-farm income. Off-farm earnings from labor and investments (mostly earned by operators of very small farms) in aggregate began to exceed net farm income in the late 1960s and did exceed net farm income by relatively wide margins in the 1980s.

Table 1.3 reports selected financial indicators calculated from the sector balance sheets and income statements over the 1950 – 1990 period. The general decline in the ratio of net farm income to gross income shows the growing pressures on profit margins in agriculture during this period of time. A major component of the cost – return squeeze has been the increase in interest expenses on debt as a proportion of total expenses. General declines in net cash income relative to total debt are consistent with the profit margin patterns. (Note the significant changes in ratios that can occur in high income years, such as 1990.)

The aggregate debt-to-asset ratio climbed steadily to reach the 15 to 18 percent range in the 1970s, and then rose above 20 percent in the mid-1980s, reflecting the decline of farm real estate values that characterized this time period. Substantial reductions in farm debt levels and modest recoveries of values of farm land and other assets in the late 1980s brought this solvency measure back to the 15 to 18 percent range.

The farm sector debt-to-asset ratio of 15 to 18 percent appears low relative to many other economic sectors; however, this range is consistent with the agricultural sector's heavy reliance on a non-depreciable asset such as farm land in which a significant portion of its economic returns occurs as capital gains or losses on real estate assets. As shown by several studies (e.g., Barry and Robison; Ellinger and Barry), the debt-carrying capacity of non-depreciable assets (e.g., land) is much lower than that of depreciable assets. Thus, farmers who lease much of the land they operate generally can have higher debt-to-asset ratios than owner-operators.

Table 1.4 indicates various structural and financial characteristics of the farm sector. The data are categorized by age of farm operator, along with sector averages at the bottom of the table. Financial characteristics associated with the age of the farmer are important because of the high capital requirements necessary to enter farming, the relatively older age of farmers, and the traditional life cycle of agriculture in which farmers initially rely heavily on leasing, then combine ownership and leasing, and finally focus more on their owned land as they approach retirement. The census data in Table 1.4 largely are consistent with these traditional views.

As Table 1.4 reveals, an estimated 52.8 percent of the 1.88 million U.S. farms reported the use of debt capital in their businesses as of December 31,

Table 1.1. Farm Sector Balance Sheet

	1950		1960		1970	
	- - - - - - - - - - - - - - - - - - *(Billion $)* - - - - - - - - - - - - - - - - - - -					
Farm Assets						
Real estate	88.9	(58.0)	139.7	(66.6)	224.5	(69.2)
Livestock and poultry	17.1	(11.2)	15.6	(7.4)	23.7	(7.3)
Machinery and motor vehicles	14.1	(9.2)	22.2	(10.6)	34.4	(10.6)
Crops	7.1	(4.6)	6.2	(3.0)	8.5	(2.6)
Purchased inputs	NA		NA		NA	
Household equipment and furnishings	9.6	(6.3)	8.7	(4.1)	10.0	(3.1)
Financial assets	16.5	(10.8)	17.5	(8.3)	23.2	(7.2)
Total	153.3	(100.0)	209.9	(100.0)	324.3	(100.0)
Farm Debt						
Real estate	6.1	(50.0)	12.8	(52.0)	30.5	(57.8)
Non – real estate	6.1	(50.0)	12.0	(48.0)	22.3	(42.2)
Total	12.2	(100.0)	24.8	(100.0)	52.8	(100.0)
Equity Capital	140.7		185.2		271.5	

Numbers in parentheses are percentages of totals.

Source: *Economic Indicators of the Farm Sector, National Financial Summary, 1990,* USDA – ERS, ECIFS 10 – 1, November 1991.

1988. Thus, nearly half of U.S. farmers made no use of debt at the time of the census survey. Of course, the year-end figures could understate the true extent of indebtedness if some farmers repaid short-term operating loans within the year. Among regions, the incidence of indebted farmers was the highest in the Midwest, followed by the West, Northeast, and South, respectively. Among individual states, the top three states by proportion of indebted farmers were North Dakota (74.6 percent), South Dakota (70.1 percent), and Iowa (70.0 percent), while the lowest three states were West Virginia (34.9 percent), Tennessee (39.1 percent), and Alabama (39.9 percent). Moreover, incidences of indebtedness are higher in the younger age classes of farmers, and then steadily decline in the age classes beyond 50 years of age. Similarly, average debt-to-asset ratios for farm operators are highest in the younger age classes and decline steadily as age increases.

(Including Operator Households)

1980		1985		1990	
- -		- - - - *(Billion $)* - - - -		- -	
850.1	(78.1)	657.0	(73.6)	702.6	(70.5)
60.6	(5.6)	46.3	(5.2)	69.1	(6.9)
86.9	(8.0)	88.3	(9.9)	91.7	(9.2)
32.8	(3.0)	22.9	(2.6)	22.4	(2.2)
NA		1.2	(0.1)	2.8	(0.3)
19.4	(1.8)	27.8	(3.1)	46.3	(4.6)
39.3	(3.6)	49.3	(5.5)	61.1	(6.1)
1,089.1	(100.0)	892.8	(100.0)	996.0	(100.0)
97.5	(54.6)	105.7	(56.3)	78.4	(54.0)
81.2	(45.4)	82.2	(43.7)	66.7	(46.0)
178.7	(100.0)	187.9	(100.0)	145.1	(100.0)
910.5		704.9		850.9	

Within the agricultural sector, leasing of farm land has been a widespread method of financing that is especially effective for expanding farm size. As shown in Table 1.4, 45.4 percent of the total land in farms in the United States in 1988 was operated by farmers under rental arrangements with the land owners. The remaining acreage was farmed by owner-operators. The declining pattern of leasing as farmers get older is consistent with the life cycle and the debt-carrying capacity effects cited previously. That is, older farmers tend to rely more heavily on land ownership, and farm real estate has a lower debt-carrying capacity than non – real estate farm assets. Other age-related characteristics in Table 1.4 indicate the tendency for farm size and off-farm income to increase with the age of the operator until ages 50 to 60, and then to decline modestly.

Table 1.5 indicates the levels and market shares of loans provided by the major agricultural lenders over the 1950 – 1990 period. Since 1970, more than

Table 1.2. Net Farm Income and Off-Farm Income, Farm Sector

	1950	1960	1970	1980	1985	1990
	- - - - - - - - - - - - - - - - - - - (Billion $) -					
Gross Farm Income						
Gross cash income	28,764	34,958	54,768	143,296	157,854	185,978
(Government payments)	(283)	(703)	(3,717)	(1,285)	(7,705)	(9,298)
Value of inventory adjustment	812	397	6	−6,300	−2,269	2,889
Non-cash income	3,527	3,233	4,044	12,278	5,615	6,256
Total	33,103	38,588	58,818	149,274	161,200	195,123
Expenses						
Cash expenses						
Non-interest	15,832	21,387	33,830	94,825	93,983	111,760
Interest	598	1,347	3,381	16,261	18,613	14,472
Total	16,430	22,734	37,211	111,086	112,596	126,232
Non-cash expenses	3,025	4,642	7,241	22,053	19,836	18,059
Total	19,455	27,376	44,452	133,139	132,433	144,291
Net Farm Income	13,648	11,212	14,366	16,135	28,768	50,832
Off-Farm Income	NA	8,482	17,617	34,694	55,160	66,976

Source: *Economic Indicators of the Farm Sector, National Financial Summary, 1990,* USDA – ERS, ECIFS 10 – 1, November 1991.

half of the total farm loans have been provided by the Farm Credit System (FCS) and by commercial banks. The FCS share, in particular, exhibited strong growth until the 1980s when the system was hard hit by the financial stress in the agricultural sector. Farm loans provided by commercial banks were also curtailed during the early 1980s; however, the banks emphasis on non – real estate lending and greater diversity in their loan portfolio permitted a faster recovery. Included in the recovery was substantial growth in the banks' holdings of farm real estate debt toward the end of the 1980s.[1] U.S. government credit provided by the Farmers Home Administration (FmHA) and the Commodity Credit Corporation (CCC) tends to be low during favorable economic times in agriculture and then increases significantly as financial performance

---

[1]See Chapters 15 and 16 for an extended discussion and data about the major agricultural lenders.

Table 1.3. Selected Financial Indicators of the Farm Sector, 1950 – 1990

Year	Net Farm Income to Gross Income	Interest Expense to Total Production Expense	Net Cash Income to Total Debt	Total Debt to Total Assets	Total Debt to Total Equity
			(%)		
1950	41.2	3.1	101.1	8.0	8.7
1960	29.1	4.9	49.3	11.8	13.3
1970	24.4	7.6	33.3	16.3	19.5
1980	10.8	12.2	18.0	16.4	19.6
1985	15.5	14.1	24.1	21.0	26.7
1990	26.1	10.0	41.2	14.6	17.1

Source: *Economic Indicators of the Farm Sector, National Financial Summary, 1990,* USDA – ERS, ECIFS 10 – 1, November 1991.

in agriculture deteriorates. Finally, relative holdings of total farm debt by life insurance companies and other lenders have been trending downward over the 1950 – 1990 period.

IMPLICATIONS OF RISK AND LIQUIDITY

The agricultural sector's financial statements have indicated a reasonably solvent industry, but one that experiences chronic liquidity problems and cash flow pressures resulting from relatively low but volatile current rates-of-return to production assets. In part, these conditions reflect the dominance of farm real estate among the assets of the agricultural sector and the tendency for much of the sector's total returns to occur as unrealized capital gains or, on occasion, as capital losses. These characteristics make the aggregate debt-servicing capacity and credit worthiness of the sector vulnerable to downward swings in farm income and reductions in land values, as happened through much of the 1980s.

The complex risk environment of agriculture is well illustrated by events of the 1970s and 1980s. Low exchange rates for foreign currencies and low interest rates in the 1970s increased exports, farm income, and land prices, leading to significant growth in farm debt and optimistic expectations about the 1980s. However, a reversal in these patterns — lower exports, declining farm income and land values, and rising interest rates — in the 1980s created

Table 1.4. U.S. Farm Financial Data

Age of Operator	Number of Farms	Percent of Farms	Percent with Debt	Farm Size, Acres
Under 25 years	26,001	1.4	59.7	340.9
25 to 34 years	219,198	11.7	66.2	363.1
35 to 44 years	346,851	18.5	65.2	418.2
45 to 49 years	207,879	11.1	64.3	447.1
50 to 54 years	213,199	11.3	58.7	502.4
55 to 59 years	233,552	12.4	50.9	492.3
60 to 64 years	222,750	11.9	44.9	517.8
65 to 69 years	178,432	9.5	36.6	473.6
70 years and over	231,705	12.3	27.0	448.0
Total	1,879,567	100.0	52.8	453.4

[1]Includes farmers with no debt.

Source: Bureau of the Census, *Agricultural Economics and Land Ownership Survey (1988)*, Vol. 3, 1987, Census of Agriculture, AG 87 – R5 – 2, January 1991.

Table 1.5. Total Farm Debt by Lender

	Farm Credit System		Farmers Home Administration		Life Insurance Companies	
	($ mil.)	(%)	($ mil.)	(%)	($ mil.)	(%)
1950	1,504	11.5	586	4.5	1,353	10.4
1960	4,107	15.7	1,142	4.4	2,975	11.4
1970	12,660	23.2	3,235	5.9	5,610	10.3
1980	56,735	31.1	19,560	10.7	12,928	7.1
1985	59,146	28.8	27,147	13.2	11,836	5.8
1990	36,988	24.8	18,745	12.5	10,186	6.8

[1]Includes credit provided by sellers of farm land; by merchants, dealers, and other agribusiness lenders; and by other lending groups.

Source: *Economic Indicators of the Farm Sector, National Financial Summary, 1990*, USDA – ERS, ECIFS, 10 – 1, November 1991.

Classified by Age of Operator, 1988

Acres Rented, %	Acres Owned, %	Off-Farm Income, $ per Farm	Debt-to-Asset Ratio, Farm Operators[1]
75.2	24.8	16,862	0.269
70.4	29.6	23,463	0.312
56.6	43.4	32,490	0.274
47.7	52.3	37,525	0.232
40.0	60.0	36,785	0.208
41.4	58.6	30,074	0.185
36.0	64.0	25,481	0.145
36.0	64.0	29,586	0.115
33.9	66.1	23,645	0.068
45.4	54.6	29,978	0.193

(Including Farm Households), December 31, 1950 – 1990

Commercial Banks		Individuals and Others[1]		Commodity Credit Corporation		Total
($ mil.)	(%)	($ mil.)	(%)	($ mil.)	(%)	($ mil.)
3,510	26.9	5,286	40.5	812	6.2	13,051
6,583	25.1	9,982	38.1	1,390	5.3	26,179
14,874	27.3	16,228	29.8	1,876	3.4	54,483
40,135	22.0	47,901	26.2	5,292	2.9	182,551
46,898	22.8	42,578	20.7	17,905	8.7	205,510
50,140	33.6	29,000	19.4	4,382	2.9	149,441

significant financial problems for heavily indebted farmers and for many of their lenders as well. The financial crises of the 1980s are so recent that they have significantly influenced financial decision making by agricultural producers in the 1990s.

In general, the combined effects of business and financial risks are high for most types of farms; thus, there is a high premium on effective financial management.

Summary

Important highlights of Chapter 1 can be summarized as follows:

- Financial management involves the acquisition and use of financial resources by economic units and the protection of the units' equity capital from business and financial risks.

- Financial activities combine with the production, marketing, and administrative functions of a business unit and may be viewed as a flow of activities over time or as stock of effects at a point in time.

- Long-term planning horizons, lengthy payoff periods, seasonality, and uncertainties about future events and decision outcomes make the effects of risk and time very important in financial management.

- Significant tools of financial management include the management process, information flows, cash flow and capital budgeting methods, and a clear understanding of the workings of financial markets.

- Major structural characteristics of the agricultural production sector include:

  - a capital-intensive industry dominated by farm real estate

  - relatively small business sizes

  - low asset liquidity

  - volatile farm income

  - sizable capital gains and occasional losses in farm assets

  - significant off-farm income earned mostly by operators of small farms

  - increasing debt loads over time

  - strong reliance on leasing of farm land, especially by younger producers

  - life cycle effects of farm ownership, size, and debt use

–   more than half of farm debt provided by commercial banks and the Farm Credit System

–   a reasonably solvent industry that experiences chronic liquidity problems and cash flow pressures

–   a complex risk environment, including both business risks and financial risks

These issues and others are further developed in the chapters that follow.

Topics for Discussion

1.   Identify some of the interrelationships between decision making in finance, production, and marketing. Why are they important?

2.   Identify the steps of a strategic management process. Where do impediments lie? How are risk and uncertainty involved?

3.   How are the effects of risk and time important to financial management? Illustrate some of the financial methods for managing risks.

4.   Identify the information flows that are important to financial management. How are they related to alternative budgeting methods?

5.   Why is understanding financial markets important to financial management?

6.   What significant financial characteristics distinguish agricultural production firms from large corporate businesses?

7.   How does the capital intensity of agriculture affect the sector's liquidity position?

8.   Describe major structural changes in the agricultural production sector over time. How are they important to financial management?

9.   What is the life cycle effect in agriculture? Identify its important financial implications.

References

1.   Barry, P. J., ed., *Risk Management in Agriculture*, Iowa State University Press, Ames, 1984.

2.   Barry, P. J., and L. J. Robison, "Economic Versus Accounting Rates

of Return on Farmland," *Land Economics*, 62:4(November 1986): 388 – 401.

3. Brake, J. R., and E. O. Melichar, "Agricultural Finance and Capital Markets," *A Survey of Agricultural Economics Literature*, Vol. I., University of Minnesota Press, Minneapolis, 1977.

4. Ellinger, P. N., and P. J. Barry, "The Effects of Tenure Position on Farm Profitability and Solvency: An Application to Illinois Farms," *Agricultural Finance Review*, 47(1987):106 – 118.

5. Hallam, A., ed., *Determinants of Size and Structure in American Agriculture*, Westview Press, Boulder, Colorado, 1993.

6. Harl, N. E., *The Farm Debt Crisis of the 1980s*, Iowa State University Press, Ames, 1990.

7. Hughes, D. W., S. C. Gabriel, P. J. Barry, and M. D. Boehlje, *Financing the Agricultural Sector*, Westview Press, Boulder, Colorado, 1986.

8. Lee, W. F., et al., *Agricultural Finance*, 8th ed., Iowa State University Press, Ames, 1988.

9. Peoples, K. L., D. Freshwater, G. D. Hanson, P. T. Prentice, and E. P. Thor, *Anatomy of an American Agricultural Credit Crisis*, Rowman and Littlefield Publishers, Inc., Lanham, Maryland, 1992.

10. *Technology, Public Policy and the Changing Structure of American Agriculture*, Office of Technology Assessment, U.S. Congress, Washington, D.C., 1986.

11. *Transition in Agriculture: A Strategic Assessment of Agricultural Banking*, American Bankers Association, Washington, D.C., 1986.

12. Van Horne, J. C., *Financial Management and Policy*, 8th ed., Prentice-Hall, Inc., Englewood Cliffs, New Jersey, 1989.

13. Weston, J. F., and E. F. Brigham, *Essentials of Managerial Finance*, 10th ed., Dryden Press, Hinsdale, Illinois, 1993.

14. Weston, J. F., and T. Copeland, *Managerial Finance*, 9th ed., Dryden Press, Hinsdale, Illinois, 1992.

2

Managerial Goals

Before the principles and tools of financial management can be developed, the goals that will guide managerial behavior and that will be used in evaluating the financial performance and managerial decisions of agricultural businesses should be established. These goals are expressed as a goal or a utility function containing three elements: **profitability, risk,** and **liquidity.** We assume that these three criteria, taken together, determine the overall well-being or utility of the firm's decision makers.

Nearly all the material in the chapters that follow relates to these criteria. Information systems are designed to report business performance for analysis in terms of profitability, risk, and liquidity. Decisions on investments, financing, growth rates, and capital structure are also evaluated according to these criteria. So are the performance and functions of financial markets and institutions. Because these criteria are important, their behavioral implications, origins, and interrelationships should be considered.

GOAL IDENTIFICATION

It is nearly impossible to evaluate business performance meaningfully without having a knowledge of the goals or objectives associated with that performance. **Goal identification** leads, in turn, to the development of **measures** of goal attainment, of **targets** for these measures, and of **analytical procedures** for relating plans and expectations to the measures and to the targets.

The results of goal identification are an improved understanding of a business's past performance and better plans for its future.

These procedures sound straightforward. However, **goal identification is not an easy task.** Much more than economics and finance is involved. Social scientists agree that individuals have many goals, both economic and social-psychological, which they are simultaneously striving to achieve. Many of these goals may be organized into hierarchies in which one goal must be achieved before the next goal can be focused on. As an example, one social scientist (Maslow) has identified five ordered goals. An individual's most basic need is to provide for physical necessities such as food, clothing, and shelter. Then comes the need for security, followed by social belonging, recognition and esteem, and self-actualization. Clearly, individuals at different stages of this ordering exhibit different kinds of behavior.

Maslow's list seems remote from economics and financial management. But it helps to indicate the scope of individual goals and the difficulties an analyst has in designing performance measures that capture the unique characteristics and aspirations of every business and every financial manager. By the same token, these multiple hierarchies of goals also imply that using a single goal or measure (e.g., annual income) to evaluate the well-being of many individuals may not be satisfactory.

Instead, a middle ground that portrays several meaningful financial characteristics of business performance and that is suited for measurement and analysis is needed. For our purposes, this middle ground is reflected by the multiple goals of **increasing profitability, reducing risk,** and **providing liquidity.** The following sections discuss these criteria and their use in financial management.

Profitability

Economists generally analyze supply, demand, and prices by making precise assumptions about the objectives of consumers and business firms and by initially assuming away most of the problems of risk and time. Rational consumers are assumed to maximize the utility of consuming goods or services. Rational businesspersons are assumed to maximize the profits obtained from the production of goods and services. The consumers' search for utility maximization underlies the demand for goods and services, while the businesses' search for profits gives rise to supplies of goods and services. Taken together, demand and supply determine the prices of goods and services.

The well-being of consumers can be evaluated by their utility attainment, and the well-being of businesses can be evaluated by their profits. Consider utility first. Why should utility be treated along with profits when business performance is evaluated? One reason is that all businesspersons are also consumers, and if utility maximization explains their consumption behavior, then it may also influence their business behavior. This influence may be especially strong in agriculture because of the close relationship between the business and the household in most farms and ranches. Another reason is that the framework for understanding the responses to risk by firm managers and investors is best developed by using an expected utility approach.

What is **utility?** It is not bankable, tradeable, or even easily measurable. It is simply a concept. Economists say that **utility** *reflects the level of preference or enjoyment attached to a good or a service.* The higher the preference or enjoyment, the higher the utility. Profits and wealth have utility, and more profits and wealth mean more utility. Similarly, the physical and financial assets that generate or protect profits have utility for their owners. Thus, much like the temperature scale on a thermometer, utility serves as an index for overall goal attainment.

The quest of a business for profits is a more tangible concept. **Profits** can indeed be measured. They *are the difference between the value of goods and services produced by the firm and the costs of resources used in their production.* However, there are numerous accounting approaches to measuring profits, and careful judgment is needed in choosing a particular measure.

In addition, the economic rules for organizing a business's inputs and enterprises so as to maximize profits are clear. A firm with a production function characterized by diminishing marginal productivity and operating in perfectly competitive markets will achieve profit-maximizing organizations of inputs and enterprises by producing so that marginal value products (the added value produced from added inputs) are equal in all enterprises and, in turn, are equal to marginal input costs. Or, when viewed from a cost–size standpoint, profit maximization guides the firm to a level of production where marginal revenue from added output equals marginal cost of added output. Moreover, in the long run, competition will cause this condition to occur at the firm's lowest point on its long-run average-cost curve.

These economic principles have served well in explaining market performance and in predicting production responses to changes in prices of a firm's inputs and outputs. But, they assume away the principal effects of time and uncertainty, both of which are important determinants of real-world financial behavior.

Effects of Time

Consider first how the effects of time modify profit maximization. Most firms plan to operate over long periods of time. They have long planning horizons and are concerned with the level and timing of their profits during the planning horizon. Most firms prefer that their profits and net worth grow over time. A growth objective may mean sacrificing some profits in the near future in order to expand profits in the more distant future. How much sacrifice should occur to maximize the well-being or utility of the firm? Agricultural investments such as orchards, range improvement, and herd expansion also build up returns slowly over time and may therefore "pay off" more slowly. Yet, with more time, their total profits may exceed those from other investments. Financing terms on loans and tax considerations also influence the level and timing of returns.

These effects of time mean that static measures of profit are often inadequate. Instead, concepts and measures are needed to account for the level and timing of profits over a firm's planning horizon. We will show that the economic values of a firm's investment opportunities are directly determined by the level and timing of the profits, measured as net cash flows, projected during the firm's planning horizon. Such an economic value is called **present value;** it is found by summing the annual projected net cash flows, appropriately *discounted* for time.

The derivation of present value models, their data requirements, and their uses will be developed in later chapters. For now, we emphasize that our concept of "profits" is expressed by the value of the firm. In the absence of risk, we assume that investors desire to maximize the economic value of their claims on the firm's assets. More specifically, they are assumed to maximize the value of the firm's equity capital.

Risk

Profit maximization as it is portrayed in the preceding section ignores risks. Yet, risks occurring as unanticipated changes are pervasive, and most individuals are averse to risks. Hence, the assumption of profit maximization must be modified to reflect the presence of risk and various attitudes toward risk.

Risk Aversion and Risk Premium

The risk attitudes of investors generally are classified as **risk averse, risk neutral,** and **risk preferring,** although most of the emphasis is on risk averse. Risk aversion does not mean that individuals are unwilling to take risks. Rather, **risk aversion** means *that individuals must be compensated for taking risks and that the required compensation must increase as the risks and / or the levels of risk aversion increase.* The levels of risk aversion may vary considerably. That is, some individuals have low levels of risk aversion, while others exhibit medium or high levels of risk aversion. Moreover, the levels of risk aversion may change over time as changes occur in an individual's wealth, experience, etc. While levels of risk aversion are difficult to measure, they are important determinants of financial behavior.

In comparing a risky investment to a risk-free investment, the risk averter expresses the need for compensation by requiring a higher expected return on the risky investment than on the risk-free investment in order to be indifferent between them. This increase in required returns is called a **risk premium,** or price of risk, *that compensates the individual for bearing the risk.*

As an example, consider an investment in a U.S. government bond, which is essentially risk-free, and an investment in a beef-cow enterprise, which is known to be risky. Risk in this case means that declining cattle prices or death losses may result in economic losses for the investor. The bond is known to yield a 5 percent annual interest rate, while the beef investment has an expected annual rate-of-return of 10 percent. The investor decides to be indifferent between the two investments. The 5 percent difference is the risk premium required by the investor to make the two investments comparable. The risk premium compensates the investor for the added risk. Moreover, the risk premium increases as the level of risk or risk aversion increases.

The problems with using profit maximization to evaluate the behavior of a risk averter should now be clear. If the beef investment is expected to yield 10 percent while the bond yields 5 percent, then a profit maximizer would choose the beef investment. However, we have just seen that the investor's judgment of investment risk, together with the level of risk aversion, leads to indifference between the two investments. Moreover, if the expected profit rate of the beef investment were 9 percent instead of 10 percent, it would be unacceptable to this risk-averse investor, even though its expected profits would be greater than the bond investment.

Market behavior produces the interesting and realistic observation that more profitable ventures are also more risky, with the differences in expected profits generally being just enough to compensate the average investor for the greater risk. Thus, an investment choice under conditions of risk involves a trade-off between the expected level and the riskiness of the investments' returns.

Risk Dominance

The trade-off between risk and expected profits is illustrated in Figure 2.1, where the risks of several investments are plotted against their expected profits. Higher risk is reflected by movement along the horizontal axis, while higher profits are measured on the vertical axis. Suppose that the investor is considering 10 investments, with estimates of risk and expected profits reflected by points A through H in Figure 2.1. Even without knowing the investor's risk-return preference, some of these investment choices can be eliminated. For example, investments A and B both yield the same expected profit (E_{AB}); however, the risk of $A(V_A)$ is lower than the risk of $B(V_B)$. Hence, B is eliminated as an inferior or dominated investment. In a similar fashion, A dominates C because A yields higher expected profit for the same level of risk.

Now, compare investments A and D. Investment D has both higher expected profits and higher risk than A. Neither investment dominates the other. Instead, the preferred investment is based on the risk premium required by the investor.

An important pattern is evident in Figure 2.1. Prospective investments are divided into two sets — a **dominated set** and a **dominating set.** The dominating set is also called the **risk-efficient set.** It *contains those investments or combinations of investments that provide minimum risk for alternative levels of expected profits.* In Figure 2.1, the dominating set contains investments A, D, G, H, and J, which are aligned along the risk-profit curve at the upper boundary of all the investment choices. The investor's optimal, or expected utility-maximizing, choice will occur from among this dominating set.

Several characteristics of the risk-profit curve are important.

- Investments with higher expected profits also have higher risks.

- The increase in risk is faster than the increase in profits.

- The investment with maximum expected profit is at the top of the dominating set (investment J in Figure 2.1).

- The profit-maximizing investment is also the most risky of the domi-

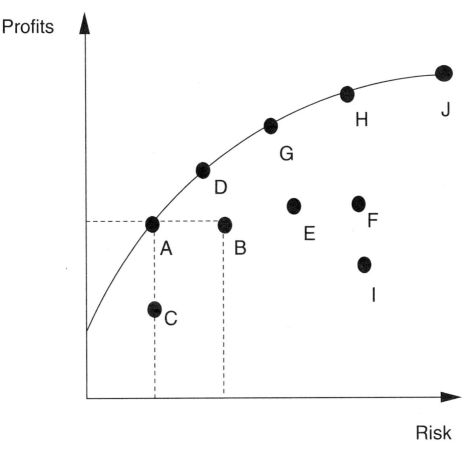

Figure 2.1. Trade-offs in risk and expected profits.

nating investments. However, it is not necessarily the most risky of all investments.

- If enough investments are considered, the dominating set becomes quite large and appears as a risk-profit curve or efficient "frontier," like the curved line, ADGHJ.

- The optimal, or expected utility-maximizing, choice cannot be predicted without knowing the investor's risk-return preferences — that is, the investor's level of risk aversion.

Risk Measures

The next task is to establish procedures for measuring risks. To be meaningful, **a risk measure should reflect how risks are perceived by an investor.**

While there is no sure way to model risk perceptions, most analysts assume that risk perception is expressed as the likelihood of several possible outcomes. Some of these outcomes may cause adversity or losses for the investor.

Hence, rather than use a single projection, it may be more realistic to describe anticipated outcomes from an investment in terms of a range or distribution of possible events, each associated with a degree of likelihood, or probability. The probability distribution essentially reflects how an investor formulates expectations about the outcomes of risky events.

This concept of risk focuses on the random, or unanticipated, variability of outcomes, some of which are favorable for the investor, while others may cause losses or adversity. An alternative concept of risk focuses on the probability of loss or failure to meet a financial obligation. In the latter case, the probability distribution can be used to estimate the probability of loss.

The idea of a probability distribution is not difficult to grasp. Suppose that an investor is considering both a beef-cow investment and a hog investment. The investor has made forecasts of rates-of-return and probabilities of return for each investment under three possible states of nature: optimistic, most likely, and pessimistic. The respective probabilities are given in Table 2.1. The discrete probability distribution for the beef investment indicates a 30 percent probability of a 9 percent profit rate, a 40 percent probability of a 12 percent profit rate, and a 30 percent probability of a 15 percent profit rate. The beef and hog distributions are graphed in Figure 2.2, with probabilities measured along the vertical axis and profit rates along the horizontal axis.

Table 2.1. Probability Distributions for Beef Cattle and Hog Investments

Forecast	Beef Cattle		Hogs	
	Profit Rate	Probability	Profit Rate	Probability
Optimistic	0.15	0.30	0.21	0.10
Most likely	0.12	0.40	0.15	0.60
Pessimistic	0.09	0.30	0.08	0.30

These distributions are used to estimate three statistical measures that aid in decision making under risk: **the expected value, the standard deviation, and the coefficient of variation. The expected value,** or **mean,** of the probability distribution is *the weighted average of the profit rates for the projected outcomes, with the probabilities used as weights.* Expected values are found by summing the products of each possible outcome multiplied by its probability. This process is expressed in an equation as:

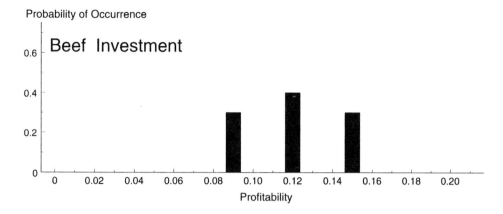

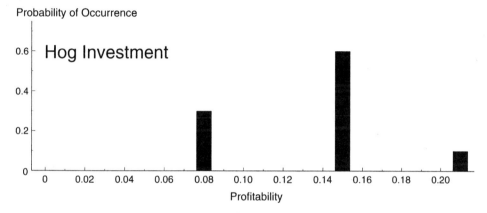

Figure 2.2. Probability distributions for investment alternatives.

$$E(\overline{V}) = \Sigma_i P_i V_I \tag{2.1}$$

where $E(\overline{V})$ is the expected value, Σ is a summation sign, P_i is the probability for each forecast i, and V_i is the profit projected for each forecast.

The expected values of the two investment choices in Table 2.1 are:

$$E(\overline{V})_1 = (0.30)(0.15) + (0.40)(0.12) + (0.30)(0.09)$$

$$= 0.12, \text{ or } 12\%$$

$$E(\overline{V})_2 = (0.10)(0.21) + (0.60)(0.15) + (0.30)(0.08)$$

$$= 0.135, \text{ or } 13.5\%$$

The expected value of the hog investment (13.5 percent) is larger than the expected value of the beef investment (12 percent). Thus, its expected profitability is higher.

The standard deviation (σ) is *a statistical measure of the amount of dispersion or variation of the projected outcomes about the expected value.* Thus, standard deviation serves as an absolute measure of the amount of risk, when risk is synonymous with variability. The standard deviation is found with this equation:

$$\sigma_T = \sqrt{\Sigma_i(P_i)(V_i - \overline{V})^2} \tag{2.2a}$$

or

$$\sigma_T = \left[\Sigma_i(P_i)\,(V_i - \overline{V})^2\right]^{\frac{1}{2}} \tag{2.2b}$$

Equation 2.2a or 2.2b indicates the following procedure for finding the standard deviation. First, find the difference between the expected value ($\overline{V}$) of the distribution and each projected outcome. Square each difference. Then multiply each squared difference by its probability. Add up these products (the sum is called the variance of the distribution). Finally, take the square root to find the standard deviation.

Using equation 2.2b, determine the standard deviations of the beef and hog investments as follows:

$$\sigma_1 = [(0.30)(0.15 - 0.12)^2 + (0.40)(0.12 - 0.12)^2 + (0.30)(0.09 - 0.12)^2]^{\frac{1}{2}}$$
$$= 0.023, \text{ or } 2.3\%$$

$$\sigma_2 = [(0.10)(0.21 - 0.135)^2 + (0.60)(0.15 - 0.135)^2 + (0.30)(0.08 - 0.135)^2]^{\frac{1}{2}}$$
$$= 0.040, \text{ or } 4.0\%$$

Thus, investment 2 has the greater amount of absolute risk because it has the higher standard deviation.

Note that both the expected value and the standard deviation of investment 2 are greater than those of investment 1. Thus, on the average, investment 2 is both more profitable and more risky than investment 1. How, then, can these investments be compared? One useful statistical measure is the coefficient of variation (CV), which is the standard deviation divided by the expected value:

$$CV_1 = \frac{\sigma_1}{E(\overline{V})_1} = \frac{2.3\%}{12.0\%} = 0.192 \tag{2.3}$$

$$CV_2 = \frac{\sigma_2}{E(\overline{V})_2} = \frac{4.0\%}{13.5\%} = 0.296 \tag{2.4}$$

The **coefficient of variation** provides a relative measure of an investment's degree of variability. Thus, it *serves as an indicator of the amount of risk relative to the amount of expected return*. Since the CV for investment 2 is larger, we may deduce that it is the more risky venture. While its expected value is greater, the expected variation increases even more. Thus, an investor's choice between the two investments depends on his or her risk-return preferences — that is, his or her level of risk aversion.

Estimating Probabilities

Estimating probabilities is important in risk analysis because these probabilities are used to assess levels of risk. There are two schools of thought about the estimation of probabilities. The more traditional view contends that probabilities cannot be effectively used to describe the outcomes of *specific* investments, or other events, for several reasons. First, the classical view of probability theory holds that no statement can be made about the probability of a *single* event — only about the *frequency* of occurrence of events that are repeatedly observed under identical conditions. Second, historical data on costs, revenues, output, etc., needed to estimate probability distributions may not be sufficiently available or may be generated by conditions (technical, economic, managerial) that are likely to change in the future. Thus, using historic data to measure levels of variability for use in the future would be invalid.

Another more recently developed view of estimating probability has gained widespread acceptance. This view holds that investors can formulate "subjective" probabilities of specific events that may be based on, but not be limited to, historical frequencies. **Subjective probabilities** are *those that the individual assigns to the possible outcomes based on his or her judgment, experience, and level of available information*. Thus, subjective probabilities are the same as personal probabilities.

Because individuals differ according to judgment, experience, and information, their subjective probabilities may differ as well. Moreover, changes in these factors over time will lead to changes in subjective probabilities. Often, subjective probabilities may be based on historic data; however, if an investor feels that historic conditions are no longer relevant for future analyses, he or she may ignore them. In fact, future expectations based on combinations of historical data and subjective judgments may be the most effective approach.

Much theoretical and empirical work has centered on the use of probability measures in decision making. The findings have indicated that both schools of thought about probabilities are applicable in different situations. The frequency concept is more valid to use when events have occurred frequently and when conditions underlying the events have been constant. Subjective probabilities apply in other cases when frequency data are not available or when other data and considerations are driving the decision situation. In general, most individuals can express their expectations in probabilistic terms. As examples, consider the quantitative estimates of weather forecasters about the chance of rain, doctors' predictions of success in surgical procedures, and the odds on athletic events quoted by the sports media.

To illustrate further, consider a university student trying to estimate his or her likely grade from an agricultural finance course. The possible events and probabilities are arranged as follows:

Grade	Probability (P_i)
A	_____
B	_____
C	_____
D	_____
E	_____

The student must be sure that the probabilities for each grade are non-negative $(P_i \geq 0)$ and sum to 1 $(\Sigma_i P_i = 1)$.

What sources of information might be used? The student might conclude that the five events are equally likely and assign to each a 20 percent probability. Or, the student might observe the relative frequency of grades from an historic time series of similar classes taught by the same instructor, finding a distribution as follows: A = 25 percent, B = 45 percent, C = 20 percent, D = 8 percent, and E = 2 percent. Finally, the student could take into account his or her own major, course background, grade performance, and interest in the course and then determine that a personal probability distribution of A = 95 percent and B = 5 percent is probable.

Try this procedure in evaluating your own grade expectations in probabilistic terms. Are the results realistic? The best possible? Do they help explain your selection of the class and your actions in the class? The same concepts of risk and measures of variability will apply to managers of agricultural firms who must consider the effects of business risk and financial risk on their decision making. Once the probability distribution is obtained, then the various statistical measures can be determined to show the amount of risk. The result is a richer information base for making decisions under conditions of risk.

Liquidity

Besides profitability and risk, financial managers must also be concerned with liquidity. **Liquidity** refers to *the ability to generate cash in order to meet cash demands as they occur and to provide for both anticipated and unanticipated events.* Cash is needed to pay for the usual business and personal transactions: paying operating expenses, paying for capital items, meeting scheduled debt payments, and satisfying family needs. Unanticipated events could include new investment opportunities or an adversity that causes economic losses and makes it difficult to meet cash demands. Failure to meet these cash demands because of inadequate liquidity could seriously jeopardize business survival and credit worthiness.

The firm's need for liquidity is closely tied to its risk position; the occurrence of risks is one of the factors causing the need for liquidity. Also, increased liquidity can lower the manager's risk premium. Hence, one way a risk-averse investor can display risk aversion is through effective liquidity management.

Considering the firm's sources of liquidity will clearly show the linkage between risk and liquidity. Liquidity is generally provided by holding highly liquid, salable assets or by having the capacity to borrow additional funds. However, the occurrence of serious risks may make liquidating assets difficult, especially if they have declined in value, as well as making borrowing additional funds difficult. For some kinds of events, commercial insurance provides another source of liquidity. Liquidity can be measured in various ways, as discussed in Chapters 4, 5, and 7.

LIFE CYCLE OF THE FIRM

The close household – business relationship of most farms and ranches closely links the life cycle of the firm to the life cycle of the manager. Moreover, the manager's objectives may change over the life cycle.

The life stages are indicated in Figure 2.3. The beginning manager focuses on becoming established in agriculture with a viable organization and acquiring control of sufficient resources for future growth. Wealth and liquidity are generally low; risk is high, because of unproven management ability and possible use of borrowed funds to finance asset control. Family assistance has always been important, including parent – children business arrangements, gifts, marriage, and family loans or sales. Leasing of assets is also important. These strategies for entering agriculture may significantly influence cash flow, profits, and growth potential for many years.

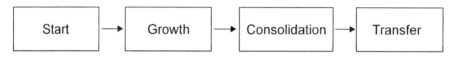

Figure 2.3. Stages of a firm's life cycle.

Once established in a business, the typical manager seeks to grow to more fully utilize management capacity, to gain economic security, and/or to maintain a competitive position in the industry. Beyond the growth stage, farm businesses enter the stage of consolidation of gains and then the stage of transfer to new management when their operators reach retirement age.

Objectives will change during these stages. In the consolidation stage, capital accumulation and growth may have lower value. The manager now has control of the business and has a stronger financial position. However, continued growth may be needed to maintain the financial position. Thus, growth may continue into the consolidation stage.

In the transfer stage, the manager is long on experience and capital but short on energy and length of planning horizon. The manager controls resources but may let others assume their use. Expansion of wealth and income-generating capacity may become less important, and investments with faster payback may be preferred. Attention turns to assuring a stable source of retirement income that will provide liquidity for emergencies and protection against inflation. As the business is transferred to new management, management continuity and production efficiency are primary concerns, as is minimizing the cost of transferring assets to heirs in an equitable fashion. In large firms with several specialized managers, the effects of the transfer stage are less noticeable.

These life stages will vary among individuals in length, in style, and even in behavior. Some beginners are highly risk averse from the start. Others strive mightily for growth. Still others, who might benefit from an orderly transfer to new managers, struggle to retain their control well beyond the point of rational transfer. Understanding these characteristics provides valuable insight about the relationships between changes in managerial goals and the profitability, risk, and liquidity position of farm businesses.

Summary

An understanding of financial goals is essential to financial analysis because goal-driven behavior has a major influence on financial performance. The key elements of a financial goal system for agricultural firms are **profitability, risk,** and **liquidity.** The relative importance of these three elements can change over time depending upon the life cycle of the agricultural firm. The stages of the household – business life cycle are **start, growth, consolidation,** and **transfer.**

Other highlights are as follows:

- In financial analysis, a profitability goal generally is expressed over time to include discounted present values of projected earnings and trends in rates-of-return on assets and equity capital.

- A risk goal must jointly account for sources of risk, levels of risk, risk attitudes of investors, and methods of managing risk.

- A risk-averse investor requires compensation (a **risk premium)** for taking risks, a risk-neutral investor requires no compensation for taking risks, and a risk-preferring investor is willing to pay a premium for taking risks.

- A condition of **risk dominance** exists when *one investment has a higher expected profit and the same or lower risk than another.*

- A **risk-dominating** or **risk-efficient set** contains *those investments or combinations of investments that provide minimum risk for alternative levels of expected profits.*

- **Risk** often is described by probability distributions for various levels of random outcomes. **The expected value** is *the mean of the probability distribution,* **the standard deviation** *measures its degree of dispersion,* and **the coefficient of variation** *measures the standard deviation relative to the expected value.*

- **Objective probabilities** are based on *a frequency concept of variation,* while **subjective probabilities** are *personal assessments of outcomes based on an individual's judgment, experience, and information.*

- **Liquidity** refers to *the ability to generate cash in order to meet demands for cash as they occur and to provide for both anticipated and unanticipated events.*

Topics for Discussion

1. Why is goal identification important in explaining managerial behavior?

2. Explain the nature of utility and utility maximization. How do these concepts differ from profits and profit maximization?

3. Briefly explain how the profit concept is modified to account for the effects of time.

4. Define **risk premium.** What is it used for? Why are more profitable ventures also more risky? How is the risk premium related to risk aversion?

5. Suppose a risk-averse investor is choosing between two investments: Investment A promises a 10 percent expected rate-of-return with a standard deviation of 5 percent, while investment B also promises a 10 percent expected rate-of-return with a standard deviation of 3 percent. Which investment is preferred? What would be the investor's preference with the addition of a third investment, investment C, promising a standard deviation of 3 percent and an expected rate-of-return of 12 percent?

6. Explain the relationships between an investor's expectations and the use of probability distributions. How do probability distributions reflect investment risks? How can estimates of probability distributions be obtained?

7. Explain the meaning of **liquidity,** and indicate why it is an important element of a manager's goal function.

8. Explain the linkages between liquidity and investment risk.

9. Explain how the manager's goals may differ during the various stages of his or her life cycle. How does the role of financial management change over the life cycle?

References

1. Anderson, J. R., J. L. Dillon, and J. B. Hardaker, *Agricultural Decision Analysis*, Iowa State University Press, Ames, 1977.

2. Barry, P. J., ed., *Risk Management in Agriculture*, Iowa State University Press, Ames, 1984.

3. Bourdeaux, K. L., and H. W. Long, *The Basic Theory of Corporate Finance*, Prentice-Hall, Inc., Englewood Cliffs, New Jersey, 1977.

4. Halter, A. N., and G. W. Dean, *Decisions Under Uncertainty with Research Application*, South-Western Publishing Company, Cincinnati, 1971.

5. Levy, H., and M. Sarnat, *Capital Investment and Financial Decisions*, 3rd ed., Prentice-Hall, Inc., Englewood Cliffs, New Jersey, 1986.

6. Luce, R. D., and H. Raiffa, *Games and Decisions: Introduction and Critical Survey*, John Wiley & Sons, Inc., New York, 1957.

7. Maslow, A. H., "A Theory of Human Motivation," *Psychological Review*, 50(1943):370.

8. Robison, L. J., and P. J. Barry, *The Competitive Firm's Response to Risk*, Macmillan Publishing Co., Inc., New York, 1986.

SECTION TWO

Financial Analysis,
Planning,
and Control

3

Financial Statements and Information Flows

An essential activity in financial management is the development of a timely, comprehensive information system that provides a means for the decision maker to measure, evaluate, control, and improve the financial performance of the farm business. The information system is based on a coordinated set of financial statements that organize the data needed to analyze profitability, risk, liquidity, and other relevant performance criteria. The primary users of the information include managers, owners, lessors, short- and long-term lenders, and in some cases auditors. Furthermore, the information system needs to provide specific information to satisfy legal obligations such as taxation and corporate disclosure.

The information system should encompass the separate analytical objectives of the various uses and users of the information. The system should be designed to *report* historical business performance to managers, owners, lenders, and lessors and to satisfy legal obligations. It can also be used to *forecast* future financial conditions and performance. Finally, the system should provide the data needed for *managerial decision making* in specific types of businesses.

In this chapter we emphasize the design of the financial reporting system that is based on the 1991 Recommendations of the Farm Financial Standards Task Force (FFSTF). We review the important financial statements used in measuring business performance and present convenient forms for their implementation in agriculture. These financial statements furnish the information needed for various types of financial analysis presented in later chapters. We also outline procedures followed in accounting for income tax obligations in the United States because these obligations strongly influence accounting and financial management. Financial and tax accounting are closely related, yet separate issues.

AN AGRICULTURAL
ACCOUNTING PERSPECTIVE

The traditional practices in agricultural accounting have largely been informal, simplistic, and in many cases different from the generally accepted accounting principles (GAAP) of the accounting profession. In contrast, larger corporate organizations must have audited financial statements prepared by certified public accountants. Historically, most farms have been relatively small operations, organized as sole-proprietorships, partnerships, or family corporations. In many cases, the farm operator had to maintain the accounts, along with the other management and operation duties. Business and personal assets were closely associated.

A primary motivation for maintaining accounting records was to provide enough information for income tax purposes. Thus, farm operators relied heavily on cash accounting methods rather than on accrual accounting (see "Cash vs. Accrual Accounting"), in part due to the privilege of reporting taxable farm income on a cash basis. In keeping records, farmers often used "single-entry" rather than "double-entry" accounting. Consequently, evaluation of business performance from these records was severely limited.

These accounting practices in agriculture are changing rapidly as farms become more commercialized, relying more heavily on borrowed funds and leased capital to finance their operations. In addition to the needs of lenders and investors, managers of these growing and more complex operations have also recognized the benefits of more complete records. The development of computerized accounting systems has automated the record keeping process and made the preparation of reports less burdensome. Consequently, the needs for uniform methods of financial reporting have increased, as have the demands for effective financial management in agriculture.

These needs were amplified during the farm financial stress of the 1980s. In conjunction with the 1987 Agricultural Credit Act, a National Commission on Agricultural Finance was appointed by President Reagan to examine the financial outlook of agriculture. One finding of the Commission was the need for greater standardization of financial reports and analysis of the farm industry. Consequently, the Farm Financial Standards Task Force (FFSTF), composed of representatives from the lending institutions, academic institutions, regulatory agencies, computer software firms, farm management consulting industry, and accounting industry, was formed by the Agricultural Bankers Division of the American Bankers Association (ABA). This task force provided guidelines regarding the reporting of financial information for farm businesses. The format of the financial statements outlined in the following sections conforms to the recommended guidelines.

COORDINATED FINANCIAL STATEMENTS

Four financial statements are needed to effectively assess and monitor the financial position and progress of the farm business. They are (1) **balance sheet,** (2) **income statement,** (3) **statement of owner equity,** and (4) **statement of cash flows.** In addition to pro forma representations of each of these four statements (see "Pro Forma Financial Statements"), a projected cash flow budget may be used to forecast future performance.

The basic structure and purposes of each financial statement, the required data, and the kinds of information conveyed are presented in this chapter. The various linkages between these statements also are shown, and some common farm accounting issues are outlined. However, the discussion here is not intended to be a substitute for a course or a text in accounting, which is itself an important discipline in business administration. Emphasis here is on understanding and interpreting the financial statements, rather than on preparing them.

The Balance Sheet

The balance sheet is *a systematic listing of all that the business owns (its assets) and all that it owes (its liabilities) at a specific moment in time.* It is a static picture of the firm's financial condition as of that date. It is also called the net worth statement, the statement of financial condition, or the statement of financial position.

The classic accounting identity underlying the balance sheet is **Assets equals Liabilities plus Net Worth.** Liabilities are the claims on the firm's assets by lenders and other creditors. Net worth represents the claims of owners on those assets. The total claims of creditors and owners cannot exceed the total value of the assets.

Classification of Assets

For analytical purposes, assets are classified according to their liquidity. **Liquidity** refers to *the firm's capacity to generate cash quickly and efficiently to meet its financial commitments as they fall due.* Those assets that can quickly be converted to cash, with little or no delay or loss in the net value of the firm and without disrupting the operations of the firm, are considered highly liquid. In contrast, highly nonliquid assets cannot be readily converted to cash without a substantial loss in value to the firm.

Two broad classes of assets that are commonly used to express the degree of liquidity are **current** and **noncurrent** (fixed).[1] Current assets are represented by cash and near-cash items whose values will likely be realized in cash or used up during the normal course of business. Besides cash, current assets also include interest-bearing transactions accounts at financial institutions and other highly liquid financial assets; notes and accounts receivable to be paid during the year; hedging accounts with commodity brokers; prepaid expenses; market livestock, livestock products, and crops on hand (including those growing in the field) that will be converted to cash during the operating cycle of the business; and inventories of supplies such as feed, fertilizer, seed, and other farm supplies used in production. Marketable stocks, bonds, and the cash surrender value of life insurance also are listed as current assets because each could readily be converted to cash if needed without interrupting the farm business.

Noncurrent assets yield services to a business over several years. Many noncurrent assets eventually are fully depreciated and are replaced, or they are liquidated. Depreciable assets include machinery and equipment, breeding livestock, as well as buildings and some other types of real estate. Other noncurrent assets are notes receivable with maturities greater than one year, capital leases, retirement accounts, remainder interests in trusts and life estates, royalties, non-marketable equities in cooperatives, and other types of financial

---

[1]Historically, three classes of assets were used — current, intermediate, and fixed. The Farm Financial Standards Task Force suggested the combination of the intermediate and fixed categories into a noncurrent category. This classification of current and noncurrent is also more consistent with general accounting standards.

assets such as participation stock in a unit of the cooperative Farm Credit System (FCS).

Longer-term noncurrent assets include various types of farm real estate — land, buildings, water-handling facilities (irrigation, tile), a residence, and various improvements, such as establishing orchards and vineyards, controlling brush, and building and repairing fences, roads, and ponds. Mineral and hunting rights associated with real estate may be accounted for separately. Except for land, the other real estate assets depreciate over long periods of time. Liquidation of noncurrent assets is more disruptive and costly than liquidation of current assets; consequently, noncurrent assets have lower liquidity than current assets.

Sole-proprietor family-operated farms are still very common in the agricultural sector. Many of these operations do not separate business-related assets and liabilities from personal accounts. Separate asset and liability accounts should be established for personal assets and liabilities so that the financial performance of the farm business may be adequately assessed and distinguished from personal business and household transactions.

Valuation of Assets

The valuation of farm assets has received considerable attention. In general, farmers have used the current market valuation, while accountants have used cost-basis valuation. With market valuation, each asset is represented at the current fair market value, while with cost-basis valuation, assets are valued at their original cost less any accumulated depreciation. Market valuation is beneficial with security valuation and current wealth measurement, while cost-basis valuation is useful for maintaining consistency among financial statements over time and for measuring business contributions to growth in net worth.

Some accountants formulate balance sheets with two asset columns, one based on the book value (original cost less accumulated depreciation) of capital and financial assets and the other showing current market values. This approach has merit, especially in more sophisticated accounting systems. It yields net worth figures for both the cost concept and the current market value concept of valuation. Another approach, utilized in examples formulated here, reports capital assets at current market values. In addition, the book values for specific capital assets are also noted on the balance sheet, and the net worth section is separated into equity retained through profits and equity resulting from valuation changes.

When the balance sheet is formulated while production is underway, determining the value of growing crops or livestock in production poses valuation problems. An accepted and generally conservative approach is to use the accumulated cost of production to value these items. Thus, the value of the items in production is reflected by the costs of the inputs used in their production. This approach provides a reasonable safety margin against risk of yield loss and changes in product prices and at least for crops is more appropriate when producers use crop insurance to protect against weather loss and when crop prices are supported to the extent of production costs by federal government programs. An alternative approach is to use estimated market values of finished products with an appropriate discount to reflect the product's progress through the production cycle. Of course, this approach is subject to the risks of production losses and price changes before the product sales eventually occur.

Another valuation problem relates to valuation of raised breeding stock. The preferred method is to accumulate costs associated with raising breeding animals. These accumulated costs are then capitalized and depreciated over the useful life of the animals.

Valuation of capital (financial) leases has also become a larger concern because the degree of leasing has increased over the past decade. Capital leases are obligations that usually require the lessee to make a series of payments over a future period of time while the lessee effectively gains ownership in the asset. The debate over capital leases basically centers on whether a specific lease is effectively a purchase (capital lease) or a rental contract. If it is a capital lease, the value of the capital item is reported on the balance sheet. The leased asset is then depreciated. The same method that is used to depreciate other owned capital assets is used to depreciate the leased asset.[2]

Classification of Liabilities

Liabilities are also classified as being current and noncurrent. Current liabilities are listed first. They include existing obligations that are payable within the next year. Several examples of current liabilities are accounts payable to merchants and supply companies and notes payable to lending institutions. Also included as current liabilities are accrued expenses, as well as the current portions of noncurrent liabilities that are payable during the next year. **Accrued expenses** are *expenses that have been incurred but as of the statement date*

---

[2]See Chapter 13 for a discussion and illustration of the procedures for accounting for the effects of capital leases on financial statements.

have not been paid. Accrued expenses include interest, tax, rent, and lease payments. For example, income tax payments for income in the current year are typically due the following March or April. Thus, accrued income tax represents the unpaid obligation to the government based on income from the current accounting period.

With the increased reliance on debt financing, accrued interest is an item that needs specific attention. Accrued interest on a specific loan is the amount of interest that has accrued since the last payment. If the loan were to be paid off, accrued interest would be the amount of interest due. A simple method to estimate accrued interest is to take the total outstanding balance times the annual interest rate paid on the loan times the number of months since the last payment and then divide by 12.

Noncurrent liabilities are obligations having a maturity greater than one year. They primarily involve the noncurrent portion of notes payable on non – real estate and real estate assets. A subclassification of noncurrent liabilities by loan maturities can also be useful in analyzing liquidity in addition to assessing the maturity structure of the liability section and how it corresponds to the asset section. Obligations on capital leases also constitute noncurrent liabilities.

If the market values of assets are reported on the balance sheet, deferred taxes associated with the possible sale of assets also need consideration. If, for example, a current asset is liquidated at the value reported on the balance sheet, the estimated taxes on the income are considered deferred taxes. If a noncurrent asset is valued at a value equal to the original cost less accumulated depreciation (taxable basis), then no tax liabilities should occur. But if an asset is sold at current market values, involving substantial capital gains, then tax obligations may arise if the asset is sold. Considerable judgment is needed in determining the extent of the tax obligation.

Deferred taxes are separated into current and noncurrent portions. The current portion consists of the taxable obligations that will result from the sale of current assets, such as grain and livestock inventory, while the noncurrent portion relates to the tax obligation that will result from the sale of noncurrent assets.

Classification of Net Worth

Net worth or owner equity is calculated by subtracting total liabilities from total assets. The Farm Financial Standards Task Force recommends the net worth or owner equity section contain at least two components. The first component should be a **valuation component** that represents the difference between the cost value of the assets and the current market value adjusted for deferred taxes. This component changes each period, based on market valuation changes. The second component of the owner equity section is a **retained earnings / contributed capital component.** This component represents the capital contributed by owners plus all the accumulated retained earnings of the firm. Retained earnings for a period are the residual profits for the period after family living expenditures, gifts, dividend payments, and / or other capital distributions and contributions have been accounted for.

Sample Illustration

The balance sheet is illustrated in Table 3.1. Joseph and Julie Farmer are farmers who produce wheat, cotton, and forage; have a modest but growing beef-cow herd; and purchase additional stocker cattle for grazing during the winter months. Assets are reported at the current market value for 1993 and 1994. The book values for machinery, equipment, and farm real estate are noted at the bottom of the statement. The total value of assets on December 31, 1994, is $870,562. In this example, the current market value of farm land at year-end 1994 is $340,680. The land was originally purchased in 1972 at $350,000, which is the book value reported at the bottom of the statement.

Current liabilities include an operating line of credit and the current portions of real estate and machinery loans. Other current liabilities for Joseph and Julie Farmer include accounts payable, accrued expenses, and the current portion of deferred taxes. On December 31, 1994, Joseph and Julie Farmer also owe $750 from two credit cards used for personal purposes.

Joseph and Julie Farmer have three years remaining on a five-year machinery installment loan. Their annual payments are based on a 10 percent interest rate. The remaining balance of the loan is $51,970, with $12,060 principal due in 1995. Payments on the installment loan are normally made in March and September. Thus, approximately three months of interest or $1,299 ($51,970 × 3 ÷ 12 × 0.10) has accrued on the installment loan. In other words, if Joseph and Julie Farmer had to settle their installment loan on December 31, 1994, the amount of interest that would be due on this note would be $1,299.

Joseph and Julie Farmer also have a 20-year real estate mortgage. The fixed-rate mortgage carries an interest rate of 10 percent and has 15 years to maturity. The remaining balance of the loan is $172,960, with $5,280 due in 1995. Since payments are due in March and September, three months of interest or $4,324 (172,960 × 0.10 × 3 ÷ 12) has been accrued as of December 31, 1994.

The accrued interest balances for the two term loans ($1,299 + $4,324 = $5,623) is added to the accrued interest for the operating loan ($1,937) and reported as accrued interest in the current liability section ($7,560). The principal amounts for the two loans that are due in 1994 ($12,060 and $5,280) are summed and reported as the current portion of all term debt ($17,340). The remaining portions of the installment loan and real estate mortgage are reported in the noncurrent liability section.

The calculation of the current portion of deferred taxes for 1994 is reported in Table 3.2. The current portion of deferred taxes is estimated by subtracting the taxable basis from the market value of current assets and then adjusting the difference by the amount of current liabilities that will result in deductions for tax purposes when they are paid. For Joseph and Julie Farmer, the difference between the market value and the taxable basis for 1994 is $169,124. Deductions total $8,310, resulting in the current portion of deferred tax income equal to $160,814. The estimated federal and state income tax rate for Joseph and Julie Farmer is 20 percent, while the self-employment tax rate for them is 13 percent of their taxable income up to a maximum payment of $6,250. The resulting current portion of deferred tax expense for Joseph and Julie is $38,413.

The calculation for the noncurrent portion of deferred taxes is shown in Table 3.3. The sum of the differences between market values and taxable basis for noncurrent assets is $39,040. The deferred tax expense of noncurrent assets is $7,808 when an estimated tax rate of 20 percent is used ($39,040 × 0.20).

As shown in the owner equity section of Table 3.1, Joseph and Julie Farmer contributed $90,000 of their own capital when they established their farming operation in 1970. Since that time, the accumulated retained earnings generated by the farm business have been $366,067. This figure suggests that the owner has reinvested $366,067 worth of accumulated profits generated by the business. The owner equity or net worth of the firm has also increased $54,194 because of increases in the market value of assets held in the farm business.

Table 3.4 reports a summary of the valuation / personal asset equity for Joseph and Julie Farmer in 1994. The calculation is the difference between the market values and the book values for selected capital and financial assets. The market values of these assets are adjusted to account for deferred taxes. This adjustment reflects the income taxes that would be due if these assets were sold at the reported market values. In Joseph and Julie's case, the valuation equity of farm land is negative. The current market value of farm land is $340,680, while the book value is $350,000, thus reducing the overall equity by $9,320. However, the machinery and equipment market value is $196,810, while the book value is $170,050, thus providing $26,750 of equity.

Table 3.1. Consolidated Balance Sheet for Joseph and Julie Farmer,

As of December 31		BALANCE SHEET
Assets	**1993**	**1994**
Current Assets		
Cash	$ 3,270	$ 8,620
Savings and short-term time deposits	10,050	20,540
Marketable securities (at market)	8,900	8,900
Accounts receivable	12,570	16,260
Inventories		
Livestock	90,100	72,350
Crops	57,352	70,922
Feed	1,400	1,400
Supplies and other	900	900
Cash investment in growing crops	0	0
Prepaid expenses	8,000	6,080
Other current assets — farm	900	900
Other current assets — personal	0	0
Total Current Assets	**$193,442**	**$206,872**
Noncurrent Assets		
Breeding livestock (at market)	$ 42,410	$ 46,000
Machinery and equipment (at market)	206,100	196,810
Investments in capital leased assets	0	0
Investments in cooperatives	12,500	12,500
Investments in other entities	0	0
Retirement accounts	18,800	18,800
Cash value life insurance	0	0
Long-term financial assets and nonmarketable securities	0	0
Farm real estate (at market)		
Land	349,200	340,680
Buildings	20,100	18,900
Other noncurrent assets — farm (at market)	0	0
Other noncurrent assets — personal (at market)	30,000	30,000
Total Noncurrent Assets	**$679,110**	**$663,690**
Total Assets	**$872,552**	**$870,562**
Book values (original cost less accumulated depreciation)		
Breeding livestock	$ 35,000	$ 38,590
Machinery and equipment	175,000	175,140
Real estate — land	350,000	350,000
Real estate — buildings	10,000	8,800

End of Year 1993 and 1994

for Joseph and Julie Farmer

Liabilities and Owner Equity	1993	1994
Current Liabilities		
Accounts payable	$ 9,380	$ 420
Notes payable within one year	82,350	69,420
Current portion of all term debt	16,040	17,340
Accrued interest	8,990	7,560
Accrued expenses		
Income taxes	9,000	11,000
Accrued rents	0	0
Other accrued items	0	0
Current portion – deferred taxes	36,828	38,413
Other current liabilities – farm	0	0
Other current liabilities – personal	750	750
Total Current Liabilities	**$163,339**	**$144,903**
Noncurrent Liabilities		
Noncurrent portion of term farm debt:		
Non – real estate farm debt:		
Notes with original maturity less than 10 years	$ 34,920	$ 39,910
Notes with original maturity greater than 10 years	0	0
Farm real estate debt		
Notes with original maturity less than 10 years	0	0
Notes with original maturity greater than 10 years	174,470	167,680
Noncurrent portion – deferred taxes	11,398	7,808
Other noncurrent liabilities – farm	0	0
Other noncurrent liabilities –personal	0	0
Total Noncurrent Liabilities	**$220,788**	**$215,398**
Total Liabilities	**$384,127**	**$360,301**
Owner Equity		
Contributed capital	$ 90,000	$ 90,000
Retained earnings	329,872	366,067
Valuation/personal asset equity	68,554	54,194
Total Owner Equity	**$488,426**	**$510,261**
Total Liabilities and Owner Equity	**$872,552**	**$870,562**

Table 3.2. Calculation of Current Portion of Deferred Taxes for Joseph and Julie Farmer, 1994

Current Portion	Market Value	Tax Basis	Difference
Marketable securities	$ 8,900	$7,688	$ 1,212
Inventories	145,572	0	145,572
Accounts receivable	16,260	0	16,260
Prepaid expenses	6,080	0	6,080
Cash investment in growing crops	_____	0	0
1. Excess of market value over tax basis of current assets . . .	$176,812 –	$7,688 =	$169,124
Accounts payable			420
Accrued interest			7,560
Other accrued expenses			0
State and local income taxes payable			330
2. Total deductions			$ 8,310
3. Current portion of deferred taxable income [1 – 2]			$160,814
4. Estimated federal and state income tax rate			20%
5. Deferred federal and state tax expense [3 × 4]			$ 32,163
6. Estimated self-employment tax			13%
7. Maximum self-employment payments			$ 6,250
8. Deferred self-employment tax [the minimum of 3 × 6 or 7]			$ 6,250
9. Total current portion of deferred income taxes [5 + 7] . . .			$ 38,413

Table 3.3. Calculation of Noncurrent Portion of Deferred Taxes for Joseph and Julie Farmer, 1994

Noncurrent Portion	Market Value	Tax Basis	Difference
Breeding livestock	$ 46,000	$ 44,400	$ 1,600
Machinery and equipment	196,810	170,050	26,760
Farm real estate and improvements	359,580	356,400	3,180
Personal assets	30,000	22,500	7,500
1. Noncurrent portion of deferred taxable income			$39,040
2. Estimated federal and state income tax rate			20%
3. Total noncurrent portion of deferred taxes [2 × 3]			$ 7,808

**Table 3.4. Calculation of Valuation/Personal Asset Equity
for Joseph and Julie Farmer, 1994**

	Market Value	Cost Basis	Difference
Marketable securities	$ 8,900	$ 7,688	$ 1,212
Breeding livestock.	46,000	44,400	1,600
Machinery and equipment	196,810	170,050	26,760
Farm real estate — land.	340,680	350,000	(9,320)
Farm real estate — buildings	18,900	6,400	12,500
Personal assets.	30,000		30,000
Less liability items			
Personal liabilities	750		(750)
Noncurrent portion deferred taxes	7,808		(7,808)
Total valuation/personal asset equity			$54,194

The balance sheet in Table 3.1 presented the balance sheets for 1993 and 1994. Livestock inventories declined in 1994, while crop inventories increased. Retained earnings increased approximately $36,195, while valuation equity decreased by about $14,360. Thus, the overall value of net worth increased from 1993 to 1994. However, the sources of the increase in net worth cannot be identified from these two static (balance sheet) pictures. The income statement is needed for this purpose.

The Income Statement

The income statement is *a summary of the revenue (receipts or income) and expenditures (costs) of the business over a specified period of time.* The usual accounting period is one year, although monthly, quarterly, and semi-annual profit summaries are important in some businesses. The income statement is also known as the operating statement or the profit and loss (P & L) statement. Contrary to the balance sheet, which is a static picture, the income statement covers the business actions of the firm over a specified period of time. The usual format for the income statement has four parts: (1) **farm revenue,** (2) **farm expenses,** (3) **nonfarm adjustments,** and (4) **income taxes.**

Cash vs. Accrual Accounting

A distinction must be made at this point between the **cash** and **accrual** basis of accounting. On a cash basis, revenue is composed of the cash actually received during the year, irrespective of when the goods were produced or when they were sold. Similarly, operating costs are included as expenses only in the period when they are paid, irrespective of when they were incurred. On an accrual basis, revenue represents the value of items the business produced during the year, whether or not the products were sold and cash received, and expenses represent all costs incurred by the business in producing those products, whether or not the costs were paid for during the period. The distinction between cash and accrual accounting is necessary because the U.S. Internal Revenue Service (IRS) permits most farmers and ranchers to compute their income on a cash basis for tax purposes. However, accrual accounting provides a more accurate assessment of profitability; thus, the following discussion and forms are on an accrual basis.

Farm Revenue

The farm business can generate revenue in several ways, such as cash receipts from the sale of production items, from a government payment received under a price support or land diversion program, or from an increase in the value of items produced that were not sold. This last category is important and should not be ignored. Assume, for example, that Joseph and Julie Farmer produced $80,000 of wheat, which remained in storage at the end of the year. Obviously, no cash receipts were generated. However, a valuable product was created, and the accrual basis of accounting reflects its contribution to income during the period in which it was produced.

A convenient format for the Farmer's 1994 income statement is presented in Table 3.5. So that the value of everything produced by the farm business during the year can be properly identified, the following items are included under revenue: (1) All cash received from the sale of farm products, government payments, and other farm cash receipts; (2) all *changes* in the value of inventories of farm products on hand at the beginning and end of the year; and (3) changes in accounts receivable between the beginning and the end of the year. The latter two items are determined from beginning and ending balance sheets. In Table 3.5, categories are established for crops and feed, market livestock and poultry, livestock products, and other revenues. The Farmer's gross revenue in 1994 was $369,450. Purchased feed and livestock are subtracted from this value to compute the value of farm production (VFP).

Table 3.5. Consolidated Income Statement
for Joseph and Julie Farmer, 1994

INCOME STATEMENT
for Joseph and Julie Farmer

For the period January 1, 1994, to December 31, 1994

Item	Subtotal	Total
Farm Revenue		
Crops and feed		
Cash receipts	$212,430	
Inventory adjustments	13,570	
Market livestock and poultry		
Cash receipts	131,030	
Inventory adjustments	(17,750)	
Cash sales of other livestock products	0	
Change in value due to raised breeding stock	0	
Gain/loss culled breeding stock sales	0	
Government program payments	19,020	
Other farm income	7,460	
Change in accounts receivable	3,690	
Gross revenue	**$369,450**	
Less purchases of market livestock	− 92,740	
Less cost of purchased feed/grain	− 9,920	
Value of farm production	**$266,790**	**$266,790**
Farm Expenses		
Farm operating expenses		
Cash operating expenses	$154,480	
Noncash adjustments to expenses		
Change in accounts payable	+ (8,960)	
Change in prepaid expenses	− (1,920)	
Change in unused supplies	− 0	
Change in cash in growing crops	− 0	
Change in other accrued operating expenses	+ 0	
Depreciation	+ 26,960	
Total farm operating expenses	**$174,400**	
Interest expense on farm loans		
Cash interest paid	$ 29,960	
Change in accrued interest	(1,430)	
Total interest expense	**$ 28,530**	
Total farm expenses	**$202,930**	**− $202,930**

(Continued)

Table 3.5 (Continued)

Item	Subtotal	Total
Net income from operations	$ 63,860	$ 63,860
Gains/losses on sales of farm capital assets	0	0
Net farm income before taxes	$ 63,860	$ 63,860
Nonfarm Adjustments		
Wages	$ 19,280	
Interest and dividend income	900	
Other nonfarm income net of expenses	0	
Gains/losses on sales of nonfarm assets	0	
Total nonfarm income	$ 20,180	+ $ 20,180
Income before taxes and extraordinary items	$ 84,040	$ 84,040
Income Taxes		
Cash income tax expense	$ 16,490	
Change in income tax accruals	2,000	
Change in current portion of deferred taxes	1,585	
Total income tax expense	$ 20,075	− $ 20,075
Income before extraordinary items	$ 63,965	$ 63,965
Extraordinary items (after tax)	$ 0	$ 0
Net income	$ 63,965	$ 63,965

Considerable controversy has evolved regarding whether to report gross revenue or value of farm production. The format followed in this example allows the representations of both gross revenue and value of farm production. The value of farm production is a value-added concept that allows for more useful comparisons among different farm operations than does gross revenue. For example, two similar livestock operations feed a substantial amount of corn. The only difference between the operations is one operation produces all its own corn and the other sells the corn and purchases all corn used for feed when needed. The values of farm production for these two farms are quite similar. However, the gross revenues could differ substantially. Because many performance measures are based on either gross revenue or value of farm production, we chose to format the income statement that incorporated the VFP concept to allow for more useful comparisons among firms. In Chapter 4 we will discuss comparisons of specific financial measures with peer groups in more depth.

Farm Expenses

Farm operating expenses include outlays for all inputs purchased or hired that are used up in production during the year. Examples include seed, fertilizer, machinery repair, and rent. As illustrated in Table 3.5, the expenses are grouped into cash operating expenses and noncash adjustments to expenses.

The noncash adjustments to expenses shown in Table 3.5 insure that the expense section accurately reflects the costs incurred in this year's production. From the beginning balance sheet to the closing balance sheet for the particular accounting period, besides depreciation, changes in the value of selected items, such as prepaid expenses, supplies, accounts payable, and accrued expenses, must also be included. For example, Joseph and Julie Farmer began the year with $8,000 of prepaid expenses. This figure included a prepayment for seed expense for the 1994 crop. During the year they purchased additional seed, the cost of which is included under cash operating expenses. At the end of the year, they made a prepayment of $6,080 for seed to be used in 1995. During the year they used up $8,000 – $6,080 = $1,920 more seed than was purchased. Thus, an additional $1,920 is added to noncash expenses to reflect the change in prepaid expenses.

Furthermore, at the beginning of the year, Joseph and Julie Farmer owed $9,380 to the farm implement dealer for repair work done in 1993 (accounts payable). During the year, they paid the dealer the outstanding balance and also had other repair costs that are included under cash operating expenses. However, at the end of the year, Joseph and Julie owed $420 for repair work done in 1994. Thus, the cash repair costs are adjusted by the change in accounts payable to reflect the expense incurred during 1994. The adjustment subtracts the difference of the beginning and ending accounts payable balances ($9,380 – $420 = $8,960) from cash operating expenses to reflect accrual costs.

Interest paid or accrued on borrowed funds is also a farm expense. As stated earlier, accrued interest adjustments can be a significant component of interest expense. Capital expenditures for machinery and equipment, breeding livestock or dairy cattle are not included as farm expenses. They are capital investments that must be charged against farm income over several accounting periods as depreciation (see "Computing Depreciation").

As illustrated in Table 3.5, farm expenses are conveniently grouped into three categories: (1) **operating expenses,** (2) **noncash adjustments to expenses,** and (3) **interest expense on farm loans.** Subtracting the total farm expenses from the value of farm production gives the **net income from farm operations.**

For Joseph and Julie Farmer the corresponding data are $266,790 – $202,930 = $63,860 (see Table 3.5).

The sale of capital items (machinery, purchased breeding livestock, etc.) is separated from farm revenue because only the gain (loss) from the sale of capital assets should affect income. The gain (loss) represents the difference between the sale price of the asset and its taxable basis, which is its purchase price less depreciation. Moreover, income from capital sales does not always reflect proficiency of management or the value of the firm's production, which is an important objective of the income statement. Because Joseph and Julie Farmer had no gain or loss from the sale of farm capital assets, the *net farm income before taxes* also is $63,860.

Nonfarm Adjustments and Income Taxes

Nonfarm revenue is added to determine Joseph and Julie Farmer's taxable income and to identify the total income available for use by both the business and the family. Joseph and Julie had nonfarm wages of $19,280 and interest income of $900. Thus, their *income before taxes and extraordinary items* was $84,040. Their accrued income tax expense was $20,075, leaving a *net income after taxes* of $63,965.

Computing Depreciation

Depreciation is *an accounting procedure by which the purchase cost of a depreciable asset is prorated over its projected economic life.* Depreciation should reflect the anticipated decline in an asset's production value over time. However, determining depreciation is complicated because different assets may depreciate at different rates because of differences in use, care, and obsolescence. Moreover, the accepted methods of computing depreciation at best can only approximate the actual decline in an asset's value over time. It should be noted that **land is an asset that does not depreciate.** Theoretically, land is not used up, nor does it become obsolete. Buildings, fences, and other improvements that are attached to land may depreciate, but land does not.

Depreciation charges reflect the funds that need to be set aside and accumulated from gross revenue in order to replace the depreciating asset. However, depreciation is not a cash transaction. It reduces profit, but not cash. In fact, changing depreciation may have a positive effect on cash in the short run because depreciation is a tax-deductible expense.

The actual process of computing depreciation is relatively simple. At the time a depreciable asset is purchased, the manager projects the length of its useful life and estimates its salvage value, if any, at the end of its life. A separate account is maintained for each asset showing its original cost, accumulated depreciation, and "book" value or depreciated value at any given time. At the end of each accounting period, the depreciation for the preceding period is calculated and recorded. This procedure reduces the book value and increases the accumulated depreciation. The depreciation for all assets is totaled and entered under the depreciation heading as a noncash adjustment to operating expense on the income statement. An example of a depreciation record for a single machine is illustrated in Table 3.6.

Table 3.6. Record of Depreciation

Asset:. . . . .   Tractor

Year purchased:. . . . .   January 1993

Method of depreciation used:. . . . .   Double declining balance

Original cost:. $30,000

Estimated useful life:.   10 years

Estimated salvage value:. $2,000

Year	Depreciation Expense for Year	Accumulated Depreciation	Book Value, End of Year
1993	$6,000	$ 6,000	$24,000
1994	4,800	10,800	19,200
1995	3,840	14,640	15,360
1996	3,072	17,712	12,288
1997	2,458	20,170	9,830
1998	1,966	22,136	7,864
1999	1,573	23,709	6,291
2000	1,258	24,967	5,033
2001	1,007	25,974	4,026
2002	2,026	28,000	2,000

Various methods of depreciation are used in accounting. **Straight-line, declining balance,** and **sum-of-the-year's digits** are among the more common methods. The straight-line method provides the same amount of depreciation during each year of the asset's life. It is determined by:

$$D_{sl} = \frac{OC - SV}{N}$$

OC is the original cost of the asset, SV is the projected salvage value, and N is the asset's expected life. Consider, for example, a tractor costing $30,000, with a useful life of 10 years and a projected salvage value of $2,000. The annual depreciation charge would be:

($30,000 – $2,000) ÷ 10 = $2,800

The declining balance method provides an accelerated charge-off for depreciation. It can be implemented at different rates. The double declining balance method (or 200 percent declining balance method) is a common example. It is determined by:

$$D_{ddb} = \frac{2}{N} \times R$$

R is the remaining book value of the asset at the beginning of the year, and N is the expected economic life. This procedure is followed until the book value reaches the salvage value. Using the $30,000 tractor as an example, the first year's depreciation is $6,000, or 20 percent of the tractor's book value. The second year's depreciation is 20 percent of the remaining $24,000, or $4,800. The salvage value is not quite reached in year 10. The operator could continue to depreciate the asset in subsequent years until the salvage value is reached, provided the machine is still used. If the machine is salvaged at the end of year 10, the unused depreciation could be added to the depreciation charge for the tenth year, as shown in Table 3.6. Different rates of depreciation for the declining balance methods are obtained by varying the numerator (e.g., 1.5, 2, 3) in the equation.

The sum-of-the-year's-digits method adds the successive years of the depreciable life of the asset to determine the denominator. In the preceding case, the denominator is:

1 + 2 + 3 + 4 + 5 + 6 + 7 + 8 + 9 + 10 = 55

A simple formula for this calculation is:

(N) (N + 1) ÷ 2, so that (10) (10 + 1) ÷ 2 = 55

The numerator for the first year is the remaining years of depreciable life for the asset at the beginning of the year, or, in this case, 10. For each successive year the numerator is reduced by 1. Thus, depreciation for the first year is 10/55 of the total depreciable charge (original cost less salvage value). The first year's charge is 10/55 of $28,000, or $5,091; the second year's charge is 9/55 of $28,000, or $4,582; and so on. As shown in Table 3.7, the method of computing depreciation strongly influences the level of depreciation in each year.

Table 3.7. Depreciation Schedules for Three Methods of Computation[1]

Year	Straight-Line	Double Declining Balance	Sum-of-the-Year's Digits
1993	$2,800	$6,000	5,091
1994	2,800	4,800	4,582
1995	2,800	3,840	4,073
1996	2,800	3,072	3,564
1997	2,800	2,458	3,055
1998	2,800	1,966	2,545
1999	2,800	1,573	2,036
2000	2,800	1,258	1,527
2001	2,800	1,007	1,018
2002	2,800	2,026[2]	509

[1]Original cost, $30,000; salvage value, $2,000; economic life, 10 years.

[2]Represents the difference between projected salvage value and book value at the beginning of the year, assuming the machine was salvaged at year end.

A difficult problem relating to depreciation is the impact of inflation on asset values over time. Appealing arguments can be made for charging depreciation according to the costs of replacing an asset, rather than according to its original cost, or otherwise adjusting the depreciation charge to reflect the effects of inflation. In preparing financial statements, accountants have traditionally adhered to the assumption of a stable monetary unit and to the principle of historic costs. The basic problem lies in changes in the purchasing power of the dollar and the effects of these changes on the amount of periodic income, when it is measured in the conventional manner. If the general price level has risen in the period between the incurrence of a cost (asset purchase) and its recognition as an expense (depreciation), the expense is reported in terms of dollars having greater purchasing power than current dollars. On the other hand, current revenue is reported in terms of dollars with current purchasing power. Thus, conventionally measured business income is overstated in periods of rising prices.

The Statement of Owner Equity

With a comprehensive information system, the financial statements are each linked to one another. **The statement of owner equity** *reconciles the change of net worth from one period to the next.* The statement of owner equity has two major components. The first component reconciles the change in owner equity that results from retained earnings and changes in contributed capital. The second component summarizes changes caused by fluctuations in market valuation. Table 3.8 presents the statement of owner equity for Joseph and Julie Farmer. Net income is typically the primary source of retained earnings, while family withdrawals reduce the amount of earnings retained in the business. Joseph and Julie Farmer's net income for 1994 was $63,965, while family withdrawals were $27,770, resulting in an addition of $36,195 to retained earnings.

The second component of the statement of owner equity is a reconciliation of the changes in valuation for capital and financial assets. Farm real estate values declined $8,520, while machinery and equipment values decreased $9,430, resulting in a decrease in asset valuation of $17,950. The change in the noncurrent portion of deferred taxes must be accounted for in order for the total change in valuation equity to be estimated. The noncurrent portion of deferred taxes declined $3,590; thus, the total decrease in valuation equity was $14,360. Combining the increase in retained earnings of $36,195 with the decline of valuation equity of $14,360 yielded a $21,835 increase in owner equity for Joseph and Julie Farmer in 1994.

The Statement of Cash Flows

Cash is indisputably the most liquid of all assets. Cash gives a manager flexibility to invest and direct resources to profitable activities that can then be used to generate additional cash. Thus, cash is a primary measure at the beginning and end of the production or operating cycle of a business. The balance sheet, income statement, and statement of owner equity give explicit information on profitability and net worth of an operation. However, these

Table 3.8. Consolidated Statement of Owner Equity
for Joseph and Julie Farmer, 1994

STATEMENT OF OWNER EQUITY
for Joseph and Julie Farmer

For the period January 1, 1994, to December 31, 1994

Item	Subtotal	Total
Beginning of period owner equity	**$488,426**	**$488,426**
Changes in contributed capital and retained earnings		
Net income	$ 63,965	
Withdrawals for family living	– 27,770	
Capital contributions		
Debt forgiveness.	+	
Gifts.	+	
Inheritances	+	
Capital distributions.		
Dividends	–	
Gifts.	–	
Inheritances	–	
Total change in contributed capital and retained earnings	**$ 36,195**	**$ 36,195**
Changes in market valuation (market value over cost/basis) of		
Marketable securities.	0	
Breeding livestock	0	
Machinery and equipment.	(9,430)	
Farm real estate.	(8,520)	
Buildings	0	
Other farm capital assets.	0	
Personal assets	0	
Personal liabilities.	0	
Total change in market valuation.	**($ 17,950)**	
Change in noncurrent portion of deferred taxes.	**(3,590)**	
Total change in valuation equity	**(14,360)**	**($ 14,360)**
Total change in owner equity.		**21,835**
End of period owner equity		**510,261**

statements only provide limited information regarding sources and uses of cash during the accounting period.

Furthermore, **cash rather than net income** is the mechanism used to provide family withdrawals, repay loans, and invest in new capital assets. Highly successful businesses need to have the ability to generate cash as well as profits to remain viable. In addition to knowing what the cash demands of their firms are, managers need to be aware of the ability of their firms to generate cash.

The statement of cash flows is *a summary of the cash inflows and outflows over a specified period of time.* It separates the cash inflows and outflows according to operating, financing, and investment activities of the firm. Thus, this statement can provide information on the operating performance, short-term liquidity, long-term solvency, and investment analysis of the firm. A statement of cash flows for Joseph and Julie Farmer is provided in Table 3.9.

Table 3.9. Consolidated Statement of Cash Flows for Joseph and Julie Farmer, 1994

STATEMENT OF CASH FLOWS
for Joseph and Julie Farmer

For the period January 1, 1994, to December 31, 1994

Item	Subtotal	Total
Cash Flows from Operating Activities		
Inflows		
Cash received from farm production and government payments	$369,940	
Cash received from net nonfarm income	20,180	
Total cash inflows from operating activities		**$390,120**
Outflows		
Cash paid for operating expenses	$154,480	
Cash paid for operating and short-term loan interest	7,970	
Cash paid for term loan interest	21,991	
Cash paid for feed and market livestock	102,660	
Cash paid for other items purchased for resale	0	
Cash paid for income and social security taxes	16,490	
Cash withdrawals for family living	27,770	
Total cash outflows from operating activities		**− $331,360**
Net cash flows provided by operating activities		**$ 58,760**

(Continued)

Table 3.9 (Continued)

Item	Subtotal	Total
Cash Flows from Investing Activities		
Inflows		
Cash received on sale of machinery/equipment/real estate	$ 0	
Cash received from sale of breeding livestock	0	
Cash received from withdrawals of savings	0	
Cash received from sale of personal assets/retirement accounts	0	
Total cash inflows from investing activities		$ 0
Outflows		
Cash paid to purchase machinery/equipment/real estate	$ 25,900	
Cash paid for purchase of breeding livestock	3,590	
Cash paid for deposits to savings accounts	10,490	
Cash paid to purchase marketable securities	0	
Cash paid to purchase personal assets/retirement accounts	0	
Total cash outflows from investing activities		– $ 39,980
Net cash flows provided by investing activities		($ 39,980)
Cash Flows from Financing Activities		
Inflows		
Proceeds from operating loans and short-term notes	$ 77,740	
Proceeds from term debt financing	15,540	
Cash received from capital contributions	0	
Total cash inflows from financing activities		$ 93,280
Outflows		
Cash repayment of operating and short-term loans	90,670	
Cash repayment of term debt scheduled	16,040	
Cash repayment of term debt unscheduled	0	
Cash repayment of capital leases	0	
Cash payments of dividends and other capital distributions	0	
Total cash outflows from financing activities		– $106,710
Net cash flows provided by financing activities		($ 13,430)
Net increase (decrease) in cash flows		$ 5,350
Beginning of year cash balance		$ 3,270
End of year cash balance		$ 8,620

The inflows and outflows for each of the three activities are presented for Joseph and Julie Farmer. The cash inflows from operating activities include cash received from farm production and government payments along with cash generated by operating activities. The cash outflows include cash paid for farm expenses, income taxes, and family living withdrawals. Subtracting the outflows from the inflows gives the *net cash flows provided by operating activities.* For Joseph and Julie Farmer, the sources or inflows provided by operating activities amounted to $390,120, while the uses of cash or outflows from operating activities amounted to $331,360, resulting in a net cash flow from operating activities of $58,760.

The next section presents the inflows and outflows resulting from investing activities. The inflows include items such as capital sales, withdrawals of savings deposits, and sales of personal assets. The outflows include items such as capital purchases, deposits to savings accounts, and purchases of personal assets. Joseph and Julie Farmer had no inflows of cash from investing activities, while their outflows included $25,900 to purchase machinery, $3,590 to purchase additional breeding animals, and $10,490 to deposit in their savings account. Thus, the net outflow from investing activities was $39,980.

The cash flows from financing activities are also provided in the next section. Inflows from financing activities include proceeds from loans and cash from capital contributions. Outflows include payments of debt, capital leases, and dividends. Joseph and Julie Farmer borrowed $93,280 in 1994 — $77,740 to pay operating expenses and $15,540 to pay for machinery items. However, Joseph and Julie made operating loan principal payments of $90,670 and term loan principal payments of $16,040. Because they repaid in principal more than they borrowed, the net cash flow from financing activities is –$13,430.

The overall change in cash flows is calculated by summing the net cash flows from the three activities. Joseph and Julie Farmer had a $58,760 increase from operating activities, a $39,980 decrease from investing activities, and a $13,430 decrease from financing activities, resulting in a net cash increase of $5,350. Because Joseph and Julie had a beginning cash balance of $3,270, the resulting ending cash balance was $8,620. Of course, this ending balance should reconcile with the cash balances reported on the balance sheet.

PROJECTING FINANCIAL PERFORMANCE

Each of the four financial statements discussed provides a means of reporting historical accounting information. However, managers need to set goals and objectives regarding the future outlook and potential performance of their particular firms. Creditors must project the cash demands and debt-carrying capacity of their farms, while potential investors are probably interested in long-term viability and profitability. Reviewing historical accounting reports does not provide this information. The task of the managers, creditors, and investors is to use the historical information along with projections on the future economic outlook to estimate the future financial performance of their farm businesses.

Historically, the cash flow budget was the only projected statement prepared. The primary motivation for producing this report was to satisfy short-term lenders. However, the ability to generate a positive net cash flow is not necessarily associated with profitability. Thus, it is essential that the financial projections provide financial information regarding future profitability, liquidity, and solvency as well as projections on cash flow.

Cash Flow Budget

The cash flow budget is *a projection of all the cash transactions relating to the business that occur during the accounting period, usually one year.* Noncash items are not included. All cash inflows and the cash on hand at the beginning of the period constitute the *sources* of cash for the business. The cash outflows and the cash on hand at the end of the period make up the *uses* of cash, which, of course, must always equal the sources for accounting purposes. The cash outflows can be used to pay operating expenses, make capital investments, reduce debt, support family withdrawals (including income taxes), accumulate savings, or they can remain as cash on hand at the end of the period.

The cash flow budget could also be formulated on an historical basis to provide an accounting check. However, we are primarily concerned with its use as a financial planning tool in evaluating the effects of production, marketing, investment, and financing plans on the firm's liquidity and loan repayability.

Format for the Cash Flow Budget

Considerable flexibility can be used in designing a cash flow budget, depending on the intended uses of the information and the amount of detail desired. For example, for most farm businesses, monthly summaries are appropriate, although quarterly summaries may be sufficient for a moderate-size dairy farm where cash flows are stable throughout the year. Moreover, some managers may prefer to have a detailed breakdown of expenses in their cash flow summary, while others prefer only a single entry each month for all cash operating expenses, relying on other records for the needed details for monitoring and controlling expenses.

The monthly format illustrated in Table 3.10 has four major components: (1) cash available, (2) cash required, (3) new borrowings and savings withdrawals, and (4) repayment of operating loan and savings deposits. It also indicates the outstanding balances for operating loans, term loans, and accrued interest due under the operating line of credit (short-term loan).

Cash Available

The initial source of cash is the cash on hand at the beginning of the year. For Joseph and Julie Farmer, the cash on hand is $8,620 (see Table 3.10). This information is taken directly from the current portion of the balance sheet for December 31, 1994, shown in Table 3.1. The cash received from the sale of farm products during the various months of the year is added to the beginning cash on hand. This information comes from the Farmers' production and marketing plans. Joseph and Julie Farmer project the sale of crops in each month, totaling $242,360, and livestock sales of $131,750 in March. Joseph and Julie expect government payments and other farm receipts of $26,800. Nonfarm income of $17,320 is entered on line 5, and planned futures market withdrawals are entered on line 6. Any sales of capital assets such as machinery, vehicles, breeding assets, and personal assets are listed on lines 7 through 9. Planned capital contributions are entered on line 10. Line 11 summarizes the amount of cash available from these sources by months and on an annual basis. The high variability from month to month in cash available is typical of many farm businesses.

Cash Required

An obvious need for cash is to meet the monthly farm operating and interest expenses, projected on lines 12 through 15. In projecting these expenses, the manager might start with the records of previous years and then modify them based on planned changes in the farming operation, input prices, and/or interest rates.

An important cash requirement for most farm businesses is the cash to make the payments due on term debt. These obligations are usually set by contract for a specified amount to be paid by specific dates. These obligations should always be included in the cash flow budget because cash must be available to pay them. Payments are broken down into interest and principal, because only the interest portion is charged as an operating expense. Joseph and Julie Farmer must make a semi-annual payment of $11,025 on their farm mortgage note in March, with $8,385 for interest and $2,640 on the principal. Their semi-annual payment on their machinery note payment of $1,995 interest and $6,030 on principal is also due in March. The sum of the interest payments ($10,380 = $8,385 + $1,995) and the sum of the principal payments ($8,670 = $2,640 + $6,030) for March are shown on lines 15 and 16 of the cash flow budget.

Capital expenditures for the purchase of assets with a normal life expectancy that exceeds one year are listed on line 17. Joseph and Julie Farmer plan to invest $10,150 in a planter in February and $20,310 in a grain bin in July. They have budgeted approximately $2,340 per month for family living during the year for a total of $28,050. Line 19 includes withdrawals from the business for family and personal purposes. Other cash expenditures include nonfarm business expenses, income taxes, deposits to futures accounts, and purchases of marketable securities (lines 20 through 23, respectively).

Total cash required, on line 24, also varies from month to month from a low of $12,250 in June to a high of $107,210 in October when stocker cattle are purchased. These monthly cash requirements are compared with the monthly cash available to identify periods of cash shortages. This comparison is done on line 25. In January, for instance, Joseph and Julie Farmer obtain a cash surplus of $27,380.

Joseph and Julie wish to maintain a minimum balance of $8,500 in their checking account and $20,000 in their savings account. They also wish to repay their operating note before they deposit additional money in checking or savings. Because they want to keep $8,500 in their checking account, $18,880 ($27,380 – $8,500) is available to repay the operating loan.

Line 25 shows the projected net cash position of the business for each month. A cash deficit of $5,510 occurs in February. Only $540 can be transferred from savings, since Joseph and Julie want to maintain a minimum balance of $20,000. They must meet the remaining cash deficit with new borrowing. Borrowing must be great enough to cover the deficit plus the $8,500 needed in the bank account at the end of the month (shown on lines 26 and 27). Joseph and Julie plan to borrow $6,093 for the planter purchase. Hence, they must borrow $7,377 from their operating line of credit for February. Whether the cash deficits are financed with the annual operating line of credit each month

Table 3.10. Projected Cash Flow Budget

CASH FLOW BUDGET
For the period January 1, 1995,

Item	January	February	March	April
Cash Available				
1. Beginning cash balance	$ 8,620	$ 8,500	$ 8,500	$ 14,193
2. Crop receipts	35,100	11,900	25,100	87,600
3. Livestock receipts	0	0	131,750	0
4. Government payments and other farm production receipts	3,900	1,300	2,800	9,700
5. Nonfarm income	1,400	1,400	1,400	1,400
6. Futures market withdrawals				
7. Sales of machinery/equipment/real estate				
8. Sales of breeding livestock				
9. Sales of personal assets/retirement accounts				
10. Capital contributions				
11. Total Cash Available	**$ 49,020**	**$ 23,100**	**$169,550**	**$112,893**
Cash Required				
12. Farm operating expenses	$ 17,000	$ 13,800	$ 14,900	$ 13,300
13. Feed and market livestock purchases	2,300	2,320	1,150	1,550
14. Interest payments on operating loans				3,982
15. Interest payments on term loans and capital leases			10,380	
16. Principal payments on term loans and capital leases			8,670	
17. Machinery/equipment/real estate purchases		10,150		
18. Breeding livestock purchases				
19. Family living expenditures	2,340	2,340	2,340	2,340
20. Nonfarm business expenses				
21. Income taxes				16,490
22. Deposits to futures accounts				
23. Purchase of marketable securities				
24. Total Cash Required	**$ 21,640**	**$ 28,610**	**$ 37,440**	**$ 37,662**
25. Net Cash Position	**$ 27,380**	**($ 5,510)**	**$132,110**	**$ 75,231**
New Borrowings and Savings Withdrawals				
26. Term debt financing		6,093		
27. Borrowed from operating loan	0	7,377	0	0
28. Withdrawals from savings accounts	0	540	0	0
Repayment of Operating Loan and Savings Deposits				
29. Additions to savings accounts	0	0	60,000	60,000
30. Payments on operating loan	18,880	0	57,917	0
End of Month Balances				
31. Cash	8,500	8,500	14,193	15,231
32. Savings	20,540	20,000	80,000	140,000
33. Operating loan	50,540	57,917	0	0
34. All term debt	224,930	231,023	222,353	222,353
35. Accrued interest on operating loan	3,500	3,982	3,982	0

for Joseph and Julie Farmer, 1995

for Joseph and Julie Farmer
to December 31, 1995

May	June	July	August	September	October	November	December	Year
$ 15,231	$ 9,491	$ 12,241	$ 12,756	$ 15,756	$ 20,144	$ 8,500	$ 8,500	$ 8,620
6,200	12,200	11,000	14,900	8,800	5,900	17,000	6,660	242,360
0	0	0	0	0	0	0	0	131,750
700	1,400	1,200	1,600	1,000	600	1,900	700	26,800
1,400	1,400	1,400	1,400	1,400	1,400	1,400	1,920	17,320
								0
								0
								0
								0
								0
$ 23,531	$ 24,491	$ 25,841	$ 30,656	$ 26,956	$ 28,044	$ 28,800	$ 17,780	$426,850
$ 13,300	$ 9,800	$ 10,200	$ 13,300	$ 12,000	$ 8,300	$ 11,500	$ 9,480	$146,880
400	110	2,100	260	2,470	93,590	790	3,060	110,100
					0			3,982
				11,638				22,018
				8,670				17,340
		20,310						30,460
					2,980			2,980
2,340	2,340	2,340	2,340	2,340	2,340	2,340	2,310	28,050
								0
								16,490
								0
								0
$ 16,040	$ 12,250	$ 34,950	$ 15,900	$ 37,118	$107,210	$ 14,630	$ 14,850	$378,300
$ 7,491	$ 12,241	($ 9,109)	$ 14,756	($ 10,162)	($ 79,166)	$ 14,170	$ 2,930	$ 48,550
		12,187						18,280
0	0	0	0	0	10,650	0	5,570	23,597
2,000	0	9,678	1,000	30,306	77,016	0	0	120,540
0	0	0	0	0	0	0	0	120,000
0	0	0	0	0	0	5,670	0	82,467
9,491	12,241	12,756	15,756	20,144	8,500	8,500	8,500	8,500
138,000	138,000	128,322	127,322	97,016	20,000	20,000	20,000	20,000
0	0	0	0	0	10,650	4,980	10,550	10,550
222,353	222,353	234,540	234,540	225,870	225,870	225,870	225,870	225,870
0	0	0	0	0	89	130	218	218

or with term debt depends on the items being financed and the arrangements with the lender(s).

Lines 31 through 35 provide a convenient listing of the cash, savings, operating loan, term debt, and accrued interest balances. The cash, savings, and loan balances are based on the cash flow position each month, while the accrued interest balance is based on the balance of the operating loan. It is computed by multiplying the average loan balance outstanding for the previous month by the monthly interest rate. Joseph and Julie Farmer's annual interest rate is 10 percent; thus, their monthly rate is 0.833 percent.

Annual totals are provided in the right hand column of Table 3.10. The total cash available for Joseph and Julie Farmer in 1995 is projected at $426,850, compared to the projected cash required of $378,300, with a total net cash position of $48,550 for the year. However, the annual totals can be misleading. During the year Joseph and Julie Farmer will experience four months of cash deficits (February, July, September, and October) that will require borrowing to meet necessary cash flows. In the other months, cash surpluses are projected.

The projected monthly cash flow position (line 25) provides useful information for successful liquidity management. With this information, Joseph and Julie Farmer and their lender can design an operating line of credit that will provide the cash needed each month to meet the required cash flows. They can also identify the timing and sources of cash for repaying the loan and the timing of surplus cash for other purposes.

Information from the cash flow projection can help Joseph and Julie in their liquidity management. From line 27, you can readily see how much money must be borrowed each month under the operating line of credit in order to meet the projected demands on cash. Both Joseph and Julie and their lender can see when funds will be available to repay these advances. Moreover, they will be alerted to any periods of temporary cash surpluses and can decide on how best to invest the funds and still provide the needed liquidity.

Joseph and Julie Farmer and their lender will likely want to prepare alternative cash flow budgets. Because considerable uncertainty exists about the future economic environment, changes in the cash flow budget reflecting different assumptions about commodity prices, yields, production costs, capital purchases, and / or expenditures would indicate the sensitivity of the Farmers' operation to alternative economic scenarios.

Preparing the Cash Flow Budget

The cash flow budget is straightforward to prepare. It consists of identifying the timing and the magnitude of expected cash flows into and out of the business. After an individual has gone through the steps and has used the cash flow budget in financial management, the process becomes routine, especially if it is accomplished with the aid of a computerized spreadsheet.

The information needed for the "Cash Required" section is provided by the basic farm production plans that specify the anticipated acreage for each crop, the seed, fertilizer, chemicals, fuel, and other resource requirements. Then the timing and the magnitude of each of these expenses are projected, using the previous year's cash record as a base with modifications to reflect anticipated changes in the production program. Information about *cash available* comes directly from the marketing plans for the coming year.

The process of preparing and modifying cash flow projections lends itself to the use of a computer. In fact, usable software for computerizing the preparation of cash flow budgets, balance sheets, and income statements is available through the agricultural extension service in many states and from commercial vendors as well.

Pro Forma Financial Statements

With the information provided by the cash flow budget and related data, the manager can prepare the pro forma balance sheet and the income statement. **Pro forma** refers to *setting up accounting information in advance.* It answers the question: "If events occur as implied in the cash flow budget, what will be their impacts on the profitability (income statement), as well as on the liquidity and solvency (balance sheet) of the business?"

Table 3.11 shows the pro forma balance sheet for Joseph and Julie Farmer for December 31, 1995. The ending cash balance of $8,500 comes from line 31 of the cash flow projection (Table 3.10). Other details from the cash flow budget needed in preparing the pro forma balance sheet are capital expenditures (line 17), projected balances from the operating loan (line 33), and term loans (line 34).

The current portion of term loans and the accrued interest for each are estimated from information in the files for these debts. Projected income taxes must come from the pro forma income statement. Thus, these two pro forma statements are mutually supportive and should be prepared jointly. A comparison of the beginning balance sheet for December 31, 1994, and the pro forma balance sheet for December 31, 1995 (Table 3.11), indicates that Joseph

Table 3.11. Pro Forma Balance Sheet

As of December 31	PRO FORMA BALANCE SHEET	
Assets	**Actual 1994**	**Pro Forma 1995**
Current Assets		
Cash	$ 8,620	$ 8,500
Savings and short-term time deposits	20,540	20,000
Marketable securities (at market)	8,900	8,900
Accounts receivable	16,260	9,620
Inventories		
Livestock	72,350	76,950
Crops	70,922	50,062
Feed	1,400	1,400
Supplies and other	900	900
Cash investment in growing crops	0	0
Prepaid expenses	6,080	3,020
Other current assets — farm	900	900
Other current assets — personal	0	0
Total Current Assets	**$206,872**	**$180,252**
Noncurrent Assets		
Breeding livestock (at market)	$ 46,000	$ 48,980
Machinery and equipment (at market)	196,810	202,920
Investments in capital leased assets	0	0
Investments in cooperatives	12,500	12,500
Investments in other entities	0	0
Retirement accounts	18,800	18,800
Cash value life insurance	0	0
Long term financial assets and nonmarketable securities	0	0
Farm real estate (at market)		
Land	340,680	330,360
Buildings	18,900	17,700
Other noncurrent assets — farm (at market)	0	0
Other noncurrent assets — personal (at market)	30,000	30,000
Total Noncurrent Assets	**$663,690**	**$661,260**
Total Assets	**$870,562**	**$841,512**
Book values (original cost less accumulated depreciation)		
Breeding livestock	$ 38,590	$ 41,570
Machinery and equipment	175,140	181,000
Real estate — land	350,000	350,000
Real estate — buildings	8,800	7,600

for Joseph and Julie Farmer, 1995

for Joseph and Julie Farmer

Liabilities and Owner Equity	Actual 1994	Pro Forma 1995
Current Liabilities		
Accounts payable	$ 420	$ 2,980
Notes payable within one year	69,420	10,550
Current portion of all term debt	17,340	19,000
Accrued interest	7,560	6,320
Accrued expenses		
Income taxes	11,000	13,000
Accrued rents	0	0
Other accrued items	0	0
Current portion – deferred taxes	38,413	32,944
Other current liabilities – farm	0	0
Other current liabilities – personal	750	750
Total Current Liabilities	**$144,903**	**$ 85,544**
Noncurrent Liabilities		
Noncurrent portion of term farm debt		
Non – real estate farm debt		
Notes with original maturity less than 10 years	$ 39,910	$ 46,130
Notes with original maturity greater than 10 years	0	0
Farm real estate debt		
Notes with original maturity less than 10 years	0	0
Notes with original maturity greater than 10 years	167,680	160,740
Noncurrent portion – deferred taxes	7,808	5,794
Other noncurrent liabilities – farm	0	0
Other noncurrent liabilities – personal	0	0
Total Noncurrent Liabilities	**$215,398**	**$212,664**
Total Liabilities	**$360,301**	**$298,208**
Owner Equity		
Contributed capital	$ 90,000	$ 90,000
Retained earnings	366,067	407,166
Valuation/personal asset equity	54,194	46,138
Total Owner Equity	**$510,261**	**$543,304**
Total Liabilities and Owner Equity	**$870,562**	**$841,512**

and Julie Farmer's net worth (or owner equity) is projected to increase by $33,043 during the year (from $510,261 to $543,304).

Information supplied by the cash flow budget and changes in selected items between the beginning balance sheet and the pro forma ending balance sheet are incorporated into the pro forma income statement. Projected cash receipts (lines 2 through 4), cash operating expenses (line 12), and interest expenses (lines 14 and 15) come from the cash flow budget. Changes in the value of crops and salable livestock, breeding stock, supplies, accounts receivable, accounts payable, accrued expenses, and accrued interest all come from comparisons of the beginning and ending balance sheets for these items. Depreciation used in the income statement must equal the total depreciation charged in determining the value for depreciable assets in the pro forma balance sheet.

The pro forma income statement for 1995 for Joseph and Julie Farmer is shown in Table 3.12. The projected net farm income before taxes is $64,850,

Table 3.12. Pro Forma Income Statement for Joseph and Julie Farmer, 1995

PRO FORMA INCOME STATEMENT
for Joseph and Julie Farmer
For the period January 1, 1995, to December 31, 1995

Item	Subtotal	Total
Farm Revenue		
Crops and feed		
Cash receipts	$242,360	
Inventory adjustments	(20,860)	
Market livestock and poultry		
Cash receipts	131,750	
Inventory adjustments	4,600	
Cash sales of other livestock products	0	
Change in value due to raised breeding stock	0	
Gain/loss culled breeding stock sales	0	
Government program payments	19,710	
Other farm income	7,090	
Change in accounts receivable	(6,640)	
Gross revenue	**$378,010**	
Less purchases of market livestock	– 99,910	
Less cost of purchased feed/grain	– 10,190	
Value of farm production	**$267,910**	**$267,910**

(Continued)

Table 3.12 (Continued)

Item	Subtotal	Total
Farm Expenses		
Farm operating expenses		
Cash operating expenses	$146,880	
Noncash adjustments to expenses		
Change in accounts payable	+ 2,560	
Change in prepaid expenses	– (3,060)	
Change in unused supplies	– 0	
Change in cash in growing crops	– 0	
Change in other accrued operating expenses	+ 0	
Depreciation	+ 25,800	
Total farm operating expenses	**$178,300**	
Interest expense on farm loans		
Cash interest paid	$ 26,000	
Change in accrued interest	(1,240)	
Total interest expense	**$ 24,760**	
Total farm expenses	**$203,060**	**– $203,060**
Net income from operations	**$ 64,850**	**$ 64,850**
Gains/losses on sales of farm capital assets	**0**	**0**
Net farm income before taxes	**$ 64,850**	**$ 64,850**
Nonfarm Adjustments		
Wages	$ 16,470	
Interest and dividend income	850	
Other nonfarm income net of expenses	0	
Gains/losses on sales of nonfarm assets	0	
Total nonfarm income	**$ 17,320**	**+ $ 17,320**
Income before taxes and extraordinary items	**$ 82,170**	**$ 82,170**
Income Taxes		
Cash income tax expense	$ 16,490	
Change in income tax accruals	2,000	
Change in current portion of deferred taxes	(5,469)	
Total income tax expense	**$ 13,021**	**$ 13,021**
Income before extraordinary items	**$ 69,149**	**$ 69,149**
Extraordinary items (after tax)	**$ 0**	**$ 0**
Net income	**$ 69,149**	**$ 69,149**

compared to $63,860 for 1994. The projected net income after adjustments are made for income taxes and nonfarm revenue is $69,149. Table 3.13 reveals that because family withdrawals are projected to be $28,050, retained earnings are

Table 3.13. Pro Forma Statement of Owner Equity for Joseph and Julie Farmer, 1995

PRO FORMA STATEMENT OF OWNER EQUITY
for Joseph and Julie Farmer
For the period January 1, 1995, to December 31, 1995

Item	Subtotal	Total
Beginning of period owner equity	**$510,261**	**$510,261**
Changes in contributed capital and retained earnings		
Net income	$ 69,149	
Withdrawals for family living	− 28,050	
Capital contributions		
Debt forgiveness	+	
Gifts	+	
Inheritances	+	
Capital distributions		
Dividends	−	
Gifts	−	
Inheritances	−	
Total change in contributed capital and retained earnings	**$ 41,099**	**$ 41,099**
Changes in market valuation (market value over cost/basis) of		
Marketable securities	0	
Breeding livestock	0	
Machinery and equipment	250	
Farm real estate	(10,320)	
Buildings	0	
Other farm capital assets	0	
Personal assets	0	
Personal liabilities	− 0	
Total change in market valuation	**($ 10,070)**	
Change in noncurrent portion of deferred taxes	**− (2,014)**	
Total change in valuation equity	**(8,056)**	**(8,056)**
Total change in owner equity		**$ 33,043**
End of period owner equity		**$543,304**

projected to increase by $41,099. An expected decline in land values and only a small increase in machinery values cause a decrease in valuation equity. However, the decrease in valuation equity of $8,056 is less than the increase in retained earnings of $41,099; thus, the total projected increase in equity for Joseph and Julie Farmer is $33,043.

The pro forma statement of cash flows is reported in Table 3.14. This statement reorganizes the annual cash flow budget items into operating, financing, and investment activities. The projected value for net cash flows provided by operations is $90,710, while the net cash flows from investing and financing activities are –$32,900 and –$57,930, respectively.

Table 3.14. Pro Forma Statement of Cash Flows for Joseph and Julie Farmer, 1994

PRO FORMA STATEMENT OF CASH FLOWS
for Joseph and Julie Farmer

For the period January 1, 1994, to December 31, 1994

Item	Subtotal	Total
Cash Flows from Operating Activities		
Inflows		
Cash received from farm production and government payments	$400,910	
Cash received from net nonfarm income	17,320	
Total cash inflows from operating activities		**$418,230**
Outflows		
Cash paid for operating expenses	$146,880	
Cash paid for operating and short-term loan interest	4,200	
Cash paid for term loan interest	21,800	
Cash paid for feed and market livestock	110,100	
Cash paid for other items purchased for resale	0	
Cash paid for income and social security taxes	16,490	
Cash withdrawals for family living	28,050	
Total cash outflows from operating activities		**– $327,520**
Net cash flows provided by operating activities		**$90,710**

(Continued)

Table 3.14 (Continued)

Item	Subtotal	Total
Cash Flows from Investing Activities		
Inflows		
Cash received on sale of machinery/equipment/real estate	0	
Cash received from sale of breeding livestock	0	
Cash received from withdrawals of savings	120,540	
Cash received from sale of personal assets/retirement accounts	0	
Total cash inflows from investing activities		$120,540
Outflows		
Cash paid to purchase machinery/equipment/real estate	30,460	
Cash paid for purchase of breeding livestock	2,980	
Cash paid for deposits to savings accounts	120,000	
Cash paid to purchase marketable securities	0	
Cash paid to purchase personal assets/retirement accounts	0	
Total cash outflows from investing activities		– $153,440
Net cash flows provided by investing activities		($ 32,900)
Cash Flows from Financing Activities		
Inflows		
Proceeds from operating loans and short-term notes	$ 23,597	
Proceeds from term debt financing	18,280	
Cash received from capital contributions	0	
Total cash inflows from financing activities		$ 41,877
Outflows		
Cash repayment of operating and short-term loans	$ 82,467	
Cash repayment of term debt — scheduled	17,340	
Cash repayment of term debt — unscheduled	0	
Cash repayment of capital leases	0	
Cash payments of dividends and other capital distributions	0	
Total cash outflows from financing activities		– $ 99,807
Net cash flows provided by financing activities		($ 57,930)
Net increase (decrease) in cash flows		($ 120)
Beginning of year cash balance		$ 8,620
End of year cash balance		$ 8,500

Relationships Between the Statements

The financial statements — the cash flow budget, the balance sheet at the beginning and end (pro forma) of year, the pro forma income statement, the pro forma statement of owner equity, and the pro forma statement of cash flows — make up a system of financial planning instruments that rely on one another for information, support, and consistency checks.

Cash Flow Budget and Pro Forma Income Statement

Several important differences between the cash flow budget and the pro forma income statement should be clearly understood. Depreciation and the changes in inventory values for many items (crops and livestock, accounts receivable, accounts payable, accrued expenses, etc.) are included in the income statement but are not entered in the cash flow budget. On the other hand, the cash flow budget contains several items that are not entered in the income statement, such as principal payments on debts and capital expenditures.

The pro forma income statement includes all projected expenses (whether paid or accrued) and the value of items produced (whether or not converted to cash). In contrast, the cash flow budget identifies the amount and timing of all projected inflows and outflows of cash. Neither summary is sufficient by itself for financial planning and management. *The pro forma income statement projects profitability but does not indicate liquidity or loan repayability. The cash flow budget assesses loan repayability and provides information vital for liquidity management, but it does not measure profitability.*

Cash Flow Statement and Balance Sheets

The cash on hand at the beginning of the year, as shown in the beginning balance sheet, provides the starting cash position for the cash flow budget. Similarly, the ending cash position for the last period in the cash flow budget is the cash on hand for the projected year-end balance sheet. Moreover, projected capital sales and capital purchases from the cash flow budget supply the necessary information for determining the ending balance sheet values of capital assets. Additionally, information on loans outstanding at year end for the operating line of credit and term loans on the pro forma balance sheet come directly from lines 26, 27, and 28 of the cash flow budget, respectively.

Income Statement and Balance Sheets

Five important relationships exist between the income statement and the beginning and ending balance sheets. They are:

1. Changes in values from the beginning to the ending balance sheet for crops and crop products, market livestock and livestock products, breeding livestock raised, and accounts receivable are entered directly into the revenue component of the income statement.

2. Changes from the beginning to the ending balance sheet for supplies, prepaid expenses, accrued expenses, and accounts payable are entered appropriately in the expenses' component of the income statement.

3. The charge for depreciation in the expenses' component of the income statement is fully accounted for in the changes in the value of depreciable assets from the beginning to the ending balance sheet, net of any purchases and / or sales and valuation changes of these assets during the year.

4. Income taxes due on the closing balance sheet should come directly from that year's income statement.

5. If depreciable and fixed assets are evaluated on the basis of cost **less** depreciation, the change in retained earnings from the beginning to the ending balance sheet should equal retained earnings.

COMPUTER TECHNOLOGY

The technology for formulating cash flow budgets and the other financial statements described in the preceding sections has been significantly advanced through the development of computerized accounting and spreadsheet programs. Many of the accounting programs are structured to produce the historical accounting reports and to provide information to produce cash flow budgets and pro forma statements.

Spreadsheet programs also have been formulated to assist in producing financial statements and projections. Lotus 1-2-3® and Excel® are examples of spreadsheet programs. These spreadsheets make it easy to input the necessary data and to perform the error-free calculations needed to summarize cash flows, project the timing of financing needs, and formulate pro forma income statements, balance sheets, and other financial statements. Changes in input data can easily be made in order to test the sensitivity of the budgeted outcomes to variations in farm business conditions. The budgets can be stored in the computer system for quick recall and revision. Graphics can be developed along with the printed output to depict various financial characteristics and performance trends over time.

These computer developments clearly have enhanced the effectiveness of communication between borrowers and lenders, accelerated the financial planning process, reduced the likelihood of clerical errors (although computer program "bugs" need careful attention), and contributed to the development of financial management skills. Moreover, these developments also can accommodate many of the other concepts and tools of financial management that are treated in other chapters. In particular, the methods of capital budgeting can be linked with those of cash flow budgeting to provide an integrated analysis of a firm's investment and financing alternatives.

ACCOUNTING FOR
INCOME TAX OBLIGATIONS

Income tax obligations are a strong incentive for farmers to maintain appropriate records and to search for ways to maximize after-tax income. While tax accounting is, in general, similar to financial accounting, the rules are specified by law and by the guidelines of the U.S. Internal Revenue Service. In addition, some provisions apply specifically to farmers. Because taxes represent an important withdrawal of funds from a business, the taxpayer's principal objective is to make the tax obligation as low as possible, on the average, from a given income level.

In this section we present some basic information about U.S. farmers' federal income tax obligations. Other income tax obligations are levied by state governments and by the federal government in the form of social security taxes. However, the focus here is on federal income taxes in the United States.[3] The tax rules are voluminous, detailed, and subject to frequent and substantial change. Thus, the tax issues cannot be treated here in any detail or with precision.

---

[3]A summary of the U.S. tax code and how it compares to the Canadian tax code is provided in Table 3.15.

Table 3.15. Comparison of Tax Code Between the United States and Canada, 1991

Provision	United States	Canada
Number of tax schedules for individuals	4	1
Number of individual tax rates	3	3
Maximum individual tax rate	31%	29%
Preferential treatment of capital gains	Maximum rate of 28%	75% of regular rate
Deductions and personal exemptions	$2,150 per exemption; deduct cost of itemized deductions if they exceed the base standard deduction	Deduct 17% of medical expenses and tuition from taxes; charitable expenses greater than $213 are deducted at 29% value; tax exemptions vary by family size; $489 tax credit per child when income less than $20,400
First-year acquisition depreciation allowance	One-half annual depreciation	One-half annual depreciation

Source: Perry, Nixon, and Bunnage, "Taxes, Farm Programs and Competitive Advantage for U.S. and Canadian Farmers: A Case Study," *American Journal of Agricultural Economics*, May 1992.

Cash vs. Accrual Accounting

In determining "gross income" and "business deductions," farmers may select either the cash or the accrual method of tax accounting. Cash accounting facilitates tax management by allowing farmers to shift sales and expenditures from one year to another in response to swings in farm income and to defer tax obligations to later years in a growing business. However, concerns over taking deductions for expenses as they occur but postponing income by accumulating inventories have prompted occasional changes in cash accounting rules. Following the 1986 legislation, the deductibility of expenses for most variable inputs, such as feed, seed, and fertilizer, prior to their actual use, is limited to half of the total farm expenses. This change was intended to limit sharply the use of cash accounting to defer taxes. In general, many successful farmers use accrual accounting for business monitoring and analysis, while they continue to file income taxes on a cash basis.

Depreciation

The determination of depreciation charges allowable for tax accounting in the United States was simplified considerably in the Economic Recovery Act of 1981 and modified further in the significant Tax Reform Act of 1986. Current legislation allows taxpayers a choice between using straight-line depreciation or a Modified Accelerated Cost Recovery System (MACRS). Most taxpayers prefer the accelerated system because of its faster write-off of depreciation charges.

Depreciation charges are computed by writing off the cost (i.e., basis) of an asset over an appropriate recovery period, using the applicable depreciation method and the applicable convention for treating the first and last years of the recovery period. The applicable depreciation method and the recovery period depend on the class life of the asset. These classes are 3-year property, 5-year property, 7-year property, 10-year property, 15-year property, 20-year property, residential real estate property, and non- residential real estate. Most farm machinery and equipment, grain bins, and fences fall in the 7-year class. The 5-year class includes breeding and dairy cattle, sheep, goats, autos, light trucks, and computers. Breeding hogs fall in the 3-year class, and farm buildings and real estate improvements are in the 20-year class.

The 150 percent declining balance method (the MACRS system) is used in depreciating each of the classes. When the depreciation charges need to be maximized, the depreciation method is switched to straight line. The switch occurs for the year in which the straight-line method, when applied to the asset's adjusted basis at the beginning of that year, will yield a larger deduction.

Generally, a taxpayer is required to use the mid-year (half-year) convention for starting and ending depreciation on property other than real estate improvements (some exceptions to the half-year convention are provided). This requirement means that all acquisitions are treated as if placed in service in the middle of the year and are entitled to one-half of the first year's depreciation. The recovery period then starts at that point and runs for the appropriate number of years, so that one-half year of depreciation is also allowed in the year when the asset is disposed of or taken out of service. Thus, each of the classes of property life is stretched out for an additional year. Salvage value is ignored so that any realized salvage value is considered ordinary income.

While the law no longer contains statutorily prescribed percentages for calculating depreciation, such percentages can be calculated for taxpayers who follow the preceding procedures. Table 3.16 indicates the annual percentages that would be applied against an asset's original cost (basis) under the MACRS system.

Table 3.16. Calculations of Annual Depreciation Rates
Under MACRS: 150% Declining Balance Switch to Straight-Line
Applicable Recovery Periods: 3, 5, 7, 10, 15, 20 Years
Applicable Convention: Half Year

	Recovery Year					
	3-Year	5-Year	7-Year	10-Year	15-Year	20-Year
	(depreciation rate in percent)					
1	25.00	15.00	10.71	7.50	5.00	3.75
2	37.50	25.50	19.13	13.88	9.50	7.22
3	25.00	17.85	15.03	11.79	8.55	6.68
4	12.50	16.66	12.25	10.02	7.70	6.18
5		16.66	12.25	8.74	6.93	5.71
6		8.33	12.25	8.74	6.23	5.28
7			12.25	8.74	5.90	4.89
8			6.13	8.74	5.90	4.52
9				8.74	5.90	4.46
10				8.74	5.90	4.46
11				4.37	5.90	4.46
12					5.90	4.46
13					5.90	4.46
14					5.90	4.46
15					5.90	4.46
16					2.95	4.46
17						4.46
18						4.46
19						4.46
20						4.46
21						2.23

The law also allows married taxpayers filing jointly to treat up to $10,000 ($5,000 for single taxpayers) of the cost of acquired depreciable property as a current expense rather than as a capital expenditure. This provision, called **expensing,** is intended for those with relatively low capital expenditures. Thus, the $10,000 allowance is reduced by $1 for each dollar of investment in excess of $200,000 in any one year. Various recapture and carryforward provisions are given as well.

Other Tax Issues

As of 1994, a capital gain from the sale or the exchange of property is included in income in its entirety and taxed at ordinary income tax rates. Some preferential tax treatment still exists however because the capital gains are taxed at a maximum rate of 28 percent compared to 39.6 percent (1994 rate) for ordinary income. In determining the capital gain, a taxpayer must net all of his or her long-term capital gains with long-term capital losses and net short-term capital gains with short-term capital losses. The excess, if any, of capital gains over capital losses is subject to tax. If a capital loss occurs, it may offset up to $3,000 of ordinary income with a carryover of the balance of the loss to future years. Prior to 1987, a portion of the net long-term capital gain was excluded from taxation, thus yielding a lower tax rate on capital gain. However, the 1986 Act repealed this exclusion.

The 1986 Act also repealed the investment tax credit, making it no longer applicable to purchases of depreciable property. The investment tax credit was a fiscal policy tool of the U.S. government used periodically over the years to stimulate investment. Under the old tax code, investments in most farm assets qualified for a 6 to 10 percent tax credit. While subject to possible reinstatement in the future, the tax credit is not available at the present time.

The tax law also contains various other provisions unique to agriculture — the treatment of development expenditures for establishing orchards and raising livestock to maturity, expenses for land clearing, the use of losses in agricultural income to shelter other income, the discharge of farm indebtedness, and others. The details on these provisions are available in the tax literature, which should be referred to for all current tax provisions.

Summary

This chapter has emphasized the development of a coordinated set of financial statements that report various aspects of financial performance to those parties having legal or financial interests in a farm business. The format of the statements is consistent with the recommended guidelines of the Farm Financial Standards Task Force and is applicable to farms of any type or structural characteristic. Specific highlights of the chapter are as follows:

- Financial statements organize the data needed to analyze **risk, profitability, liquidity, efficiency,** and other criteria.

- Financial statements are part of an information system that is used to *report* historical performance, *forecast* future performance, and provide data for *managerial decision making.*

- Four major financial statements are needed to report and assess a farm's position. They are (1) **balance sheet,** (2) **income statement,** (3) **statement of owner equity,** and (4) **statement of class flows.**

- **The balance sheet** is *a systematic listing of the firm's financial status at one point in time.* The main theme of the balance sheet is **Assets equals Liabilities plus Net Worth.**

- Assets are classified as **current** and **noncurrent** (fixed). Assets can be valued at either book or market value, as long as they are consistently valued by the same method.

- Liabilities are also classified as current and noncurrent. **Current liabilities** include *those obligations that are payable within the next year.* **Noncurrent liabilities** are *those that have maturities beyond one year.*

- **Net worth** or **owner equity** should have two components. The first is the *difference between the cost value and the market value of assets.* The second is the *retained earnings of the firm plus the capital contributed by the owners.*

- **The income statement** is *a summary of the receipts and expenditures over a period of time.* There are four parts to the income statement: (1) **farm revenue,** (2) **farm expenses,** (3) **nonfarm adjustments,** and (4) **income taxes.**

- Most income statements and other financial forms are based on the accrual basis of accounting, rather than the cash basis, which is widely used to calculate income for tax purposes.

- Most farm assets can be depreciated. **Depreciation** *allocates the cost of the asset over the useful life of the asset.* Because land is assumed to have an indefinite life, it cannot be depreciated, but improvements and buildings on the land can be.

- **The statement of owner equity** *reconciles changes in the value of retained earnings, contributed capital, and market valuation of assets.*

- **The statement of cash flows** *summarizes the cash inflows and outflows from operating, investing, and financing activities of the firm.*

- **The cash flow budget** *estimates cash inflows and outflows through the year.* It is a good tool for planning and evaluating the effects of decisions on the firm's liquidity.

- **Pro forma financial statements** *represent probable future financial position(s), given that events occur as provided for in the cash flow budget(s).*

Topics for Discussion

1. Explain the concept of "coordinated" financial statements. In what ways are the income statement, balance sheet, statement of owner equity, and statement of cash flows coordinated?

2. Distinguish between operating receipts and capital sales. Which would the sale of a tractor be for a farmer — an operating receipt or a capital sale? For a machinery dealer?

3. Explain the different methods of depreciation. What might influence the choice of depreciation method for a particular asset?

4. Discuss the accounting problems associated with changing values of farm land. What are some alternatives for dealing with these problems?

5. Distinguish between accounts payable and prepaid expenses. How do changes in each impact the income statement?

6. A farmer reports the following financial information for last year's operation:

   Current assets 100,000
   Net worth, end of year 100,000
   Depreciation 10,000
   Operating expenses 30,000
   Ratio of total liabilities to net worth 3
   Proprietor withdrawals 25,000
   Change in inventory 15,000
   Capital purchases 20,000
   Operating receipts 100,000
   Noncurrent assets 200,000
   Current liabilities 50,000

   a. Use this information to formulate an annual income statement.

   b. Complete the year-end balance sheet.

Assets		*Liabilities & Net Worth*	
Current	_____	Current	_____
Noncurrent	_____	Noncurrent	_____
		Net worth	_____
Total	_____	Total	_____

   c. Indicate the beginning-of-year net worth.

7. A farmer contends that the value of growing crops on a balance sheet should be $100,000 to reflect their anticipated market value. The lender argues for a $60,000 value to reflect the accumulated costs of production. Which is the proper figure to use? Why?

8. Discuss the advantages of using fair market value asset valuation compared to cost-basis valuation.

9. Distinguish between accrued income taxes and deferred taxes.

10. Distinguish between cash vs. accrual accounting in formulating an income statement.

11. Discuss the rationale for separating the statement of cash flows into three components.

12. Characterize the differences between retained earnings and valuation equity.

13. Describe the procedures involved in formulating a cash flow budget.

References

1. Benjamin, J. J., A. J. Francia, and R. Strawser, *Principles of Accounting*, 6th ed., Dame Publications, Inc., Houston, 1990.

2. Bernstein, L. A., *Financial Statement Analysis: Theory, Application, and Interpretation*, 5th ed., Richard D. Irwin, Inc., Homewood, Illinois, 1992.

3. Frey, T. L., and D. A. Klinefelter, *Coordinated Financial Statements for Agriculture*, 3rd ed., Century Communications, Inc., Skokie, Illinois, 1989.

4. Frey, T. L., and R. Behrens, *Lending to Agricultural Enterprises*, American Banking Institute, New York, 1982.

5. Helfert, E. A., *Techniques of Financial Analysis*, 7th ed., Richard D. Irwin, Inc., Homewood, Illinois, 1991.

6. James, S. C., and E. Stoneberg, *Farm Accounting and Business Analysis*, 3rd ed., Iowa State University Press, Ames, 1986.

7. Lee, W. F., et al., *Agricultural Finance*, 8th ed., Iowa State University Press, Ames, 1988.

8. Recommendations of the Farm Financial Standards Task Force: Financial Guidelines for Agricultural Producers, Farm Financial Standards Task Force, American Bankers Association, Washington, D.C., 1991.

4

Financial Analysis
and Control

The financial accounting system established in the preceding chapter provides an information base for evaluating the profitability, liquidity, efficiency, and risk of a farm business. Procedures for analysis and control according to these performance criteria are developed in this chapter. The emphasis here is on **analyzing the firm's past and projected performance in order to identify strengths and weaknesses and to develop more feasible financial plans for the future.** This chapter provides part of the analytical base for the methods of financial planning that are discussed in later chapters. In the following sections, we establish the steps of a systematic financial control process; identify various methods for evaluating profitability, liquidity, efficiency, and risk; and apply them to case situations.

THE FINANCIAL CONTROL PROCESS

Financial control is facilitated by the process of measuring and monitoring the performance of a business over time in order to maintain desired standards of performance. The process is a dynamic one; it involves the passage of time and the use of new information that is fed back to the decision-making unit for processing, analysis, and response. The control process provides an orderly framework for responding to an uncertain environment in which various signals caused by events trigger the need for control and response. Hence, an important part of the control process is the design of information systems and strategies for responding to risk.

The control process is systematically expressed by the following steps:

1. *Identifying goals*

   The identification of goals, or performance criteria, is the first step. The mix of goals, their ordering, and weights are important. Profitability, liquidity, efficiency, and risk reduction represent one set of goals for evaluating farm businesses.

2. *Developing measures for the goals*

   Step 2 involves the selection of indices, indicators, or proxies to measure the attainment of goals.

3. *Determining norms for the measures*

   The reality of goal attainment involves norms, targets, or standards for evaluating the degree of firm performance. Often "maximization" or "satisfaction" is too abstract or impractical. Moreover, some goals involve trade-offs in achievement levels; for example, attaining higher expected profits usually means accepting higher risk and lower liquidity. Specific norms provide a tangible basis for analyzing multiple-goal attainment.

4. *Setting tolerance limits on norms*

   Under risk and uncertainty, the norms for the various goal measures will seldom be exactly attained. Setting tolerance limits on deviations from norms allows for reasonable variations in performance measures before corrective actions are needed.

5. *Developing an information system*

   Periodic reports on the performance measures, based on a financial accounting system, keep the decision maker informed of the firm's progress and help to identify corrective actions when tolerance limits on norms are exceeded.

6. *Selecting and administering the analytical and diagnostic tools*

> The appropriate analytical and diagnostic tools need to be selected and adapted to process the financial information to the firm. These tools are used to estimate the performance measures and to provide signals for potential changes in performance.

7. *Identifying and implementing corrective actions*

> When events cause one or more of the performance measures to exceed the tolerance limits for the respective norms, appropriate actions are needed to restore performance to acceptable levels.

This fundamental process of control is essentially the same whether it relates to the health and growth of organisms, an athletic contest, a space flight, an agricultural finance class, or the management of a farm business. To illustrate, consider the task of controlling the health and comfort of the human body. A healthy human body maintains a body temperature of about 98.6°F, a pulse rate and blood pressure that are considered normal for the person's age, and a specified metabolic process and chemical balance. The medical profession has developed measurements to determine deviations from established norms as indicators of health problems. Moreover, a sick body sends its own signals as well. After appropriate analysis and diagnostics determine the cause of the problem, corrective actions are prescribed and monitored to try to resolve it.

We follow a similar process in regulating our environment. One environmental goal is comfort. At work or at home, we specify a temperature norm or target on a thermostat that controls the temperature. The tolerable ranges of deviation from the norm are specified and set on the thermostat and a feedback system keeps the mechanism informed of its present condition. When the tolerance limit is exceeded, a signal is sent to the decision-making center calling for corrective action. If the temperature is too hot, the air conditioner comes on until the temperature declines to the specified norm; if the room is too cold, the furnace comes on to restore the desired temperature.

Clearly, technology can play a major role in the control process. Thermostats have engineering designs to automate temperature control. Computers may be involved in the information feedback activity. They may initiate corrective actions too, as in the case of space technology. Moreover, computers are playing an increasing role in information processing for managing farm businesses.

The process of business control in a farming operation follows the same steps just outlined. However, the process is less intense than its application in space flights, health, and other types of activities, and it involves substantial human input, especially in identifying and implementing corrective actions. In the discussion to follow, we focus primarily on steps 2, 3, 4, and 6 of the financial control process. The goals in step 1 are already established as profitability, liquidity, efficiency, and risk reduction. The information system in step 5 mainly involves the use of financial statements identified in Chapter 3, although sources of information external to the firm are important too. The corrective actions indicated in step 7 are considered more fully in other chapters; they are usually unique to the particular decision situation.

CHARACTERISTICS OF GOAL MEASURES

In selecting measures to reflect the firm's profitability, liquidity, efficiency, and risk, the decision maker should choose measures that are both **meaningful** and **manageable. Meaningfulness** refers to *how well the measure actually reflects the stipulated goals.* The match should be as close as possible. For example, if profitability is the goal, then crop yields per acre or milk production per cow would not be very meaningful, since other factors also influence profitability. A dollar measure of profitability would be preferred.

Manageability refers to *the ease of computation, the ease of comprehension, and the number of measures involved.* Measures that are easier to compute and to comprehend are preferred. The number of measures should be numerous enough to provide for comprehensive analysis but still few enough to work with efficiently. No set number is stipulated, although three or four measures for each goal are common in practice. Indeed, dozens of measures could be formulated from the various financial statements, although only a few are needed for effective analysis.

Several types of comparisons can occur with the measures of business performance. Measures can be compared with norms and tolerance limits to evaluate the degree of goal attainment. Measures can be compared over time to identify important patterns or trends that may indicate needed corrective actions. Finally, measures can also be compared with those of other firms to assess peer performance.

It is important to understand that **financial measures are only indicators or signals of a firm's performance.** They may indicate the symptoms of a problem, but they do not necessarily identify the underlying cause or the solution. Further analysis is needed to identify why changes occur in measured performance. Moreover, the **causal factors** rather than the symptoms or indicators should be the object of corrective actions taken to resolve a properly defined problem. The relationships between the symptomatic role of financial measures and the underlying causes are discussed in later sections.

THE TOOLS OF FINANCIAL ANALYSIS

The financial analyst has many tools available to use in measuring and monitoring the financial performance of a farm business. The choice of tools depends on the objectives of the users of the information and the goals of the analysis. The tools provide the methods of estimating the financial performance measures and diagnostic indicators of financial performance. Furthermore, the tools can divide the financial measures into components. Changes in the components provide explanations for historical movements in financial performance measures as well as signalling future changes in financial performance.

The primary tools of the analysis are (1) **comparative financial statements,** (2) **index-number trend series,** (3) **common-size financial statements,** and (4) **specific financial performance measures.** The historical and pro forma financial accounting statements for Joseph and Julie Farmer are summarized in the appendix to this chapter. Using information about Joseph and Julie Farmer, we will describe each of these tools in the following sections.

Comparative Financial Statements

The values contained within the financial statements provide the basis for financial analysis. In Chapter 3 we discussed the foundations for preparing a coordinated set of financial statements. **Comparative financial statements** are *simply representations of two or more years of accounting statements.* Each particular measure or value for a given year is conveniently placed in juxtaposition with its counterparts in other years. Comparative financial statements can also involve comparing an individual farm to a similar farm, norm, or goal. However, in the following sections we will discuss reports and measures that are more useful for inter-firm comparative purposes. Appendix Tables A4.1

through A4.3 show a time series of comparative balance sheets, income statements, and statements of cash flows for Joseph and Julie Farmer. Values from
these statements are used to generate all the measures outlined in this chapter.

Financial analysis is typically initiated with a comparison of current financial statements with historical statements of one or more years. The next step
in preparing comparative statements involves the preparation of pro forma
statements. Observance of the direction and magnitude of *trend* is the primary
motivation for the preparation of comparative financial statements. Table 4.1
reports a comparative income statement summary for Joseph and Julie Farmer
for 1994 and 1995. The value of farm production is projected to decrease $1,120,
or 0.4 percent, while operating expenses will increase $3,900, or 2.2 percent.
This representation of financial statements can *signal* performance trends of
the farm business and *outline* potential strengths and weaknesses.

Table 4.1. Summary of Annual Year-to-Year Income Statement Changes for Joseph and Julie Farmer

Item	1994	1995	Change $ Amount	Change Percent
Value of farm production	$266,790	$267,910	1,120	0.4
Operating expenses	174,400	178,300	3,900	2.2
Interest expense	28,530	24,760	(3,770)	−13.2
Net farm income from operations	63,860	64,850	990	1.6
Gain (loss) on sale of capital items	0	0	0	0.0
Net farm income before taxes	63,860	64,850	990	1.6
Nonfarm income	20,180	17,320	(2,860)	−14.2
Income before taxes	84,040	82,170	(1,870)	2.2
Income taxes	20,075	13,021	(7,053)	−35.1
Net farm income after taxes	63,965	69,149	5,183	8.1

Index-Number Trend Series

Representing year-to-year changes with financial statements over more
than two years can be a cumbersome process. A simplified approach to represent longer-term trends is the **index-number trend series.** Implementing this
technique involves selecting a base year for specific financial measures. Each
of the other years is expressed relative to the base year. Table 4.2 reports
index-number trend series for some selected balance sheet and income state-

ment items, with 1992 used as the base year for Joseph and Julie Farmer. The calculation of an index for a specific year is:

$$\frac{\text{Value in specific year}}{\text{Value in base year}} \times 100$$

For example, the valuation/personal asset equity values are $65,602 for 1992 and $46,138 for 1995 (Table A4.1). The index for 1995 is:

$$\frac{\$46,138}{\$65,602} \times 100 = 70$$

Thus, the valuation/personal asset equity balance in 1995 is only 70 percent of the same account in 1992. In contrast, the amount of contributed capital and retained earnings is projected to be 115 in 1995, and the total owner equity is estimated to be 109 in 1995.

Table 4.2. Index-Number Trend Series for Joseph and Julie Farmer

	1992	1993	1994	1995
Assets				
Current assets.	100	89	95	83
Noncurrent assets	100	101	99	98
Total Assets	**100**	**98**	**98**	**95**
Liabilities				
Current liabilities	100	95	84	50
Noncurrent liabilities	100	100	98	97
Total Liabilities	**100**	**98**	**92**	**76**
Owner Equity				
Contributed capital and retained earnings	100	97	105	115
Valuation/personal asset equity	100	104	83	70
Total Owner Equity	**100**	**98**	**102**	**109**

Common-Size Financial Statements

Common-size financial statements represent each item on a financial statement relative to a total value for a category of similar items. Thus, the approach utilizes an indexing procedure. For example, on a common-size balance sheet, all items are expressed relative to total assets, and on common-size income statements, all items can be expressed relative to gross revenues or value of farm production.

Common-size financial statements provide two primary functions. First, the **structural components** and **mixture of items** on the financial statements *can more easily be assessed and compared.* If common-size financial statements are prepared over time, changes in the distribution of assets or the maturity structure of liabilities can be determined. Furthermore, the components of the income statement can be separated, and the proportion of revenues consumed by the various costs can be determined.

The second function of common-size financial statements is *to allow for more appropriate comparisons to norms or other firms.* The absolute values on the financial statements cannot be used in making comparisons of firms of different sizes or asset structures. However, with the common-size financial statements, inter-firm comparisons are more appropriate because values are expressed as a proportion of assets or revenue. These inter-firm comparisons can signal potential differences relative to other firms. The manager, creditor, and / or financial analyst can then determine how any of these differences could affect the financial performance of the firm.

A common-size balance sheet for Joseph and Julie Farmer is shown in Table 4.3. Each item of the balance sheet is divided by total assets. Cash and equivalents have increased from 3 to 4 percent of the total assets from 1992 to 1995. The proportion of operating and short-term notes has decreased from 10 percent in 1992 to a projected 1 percent in 1995. Total debt as a proportion of assets has decreased from 44 percent in 1992 to a projected 35 percent in 1995.

The common-size income statement displays each item as a proportion of value of farm production (Table 4.4). Joseph and Julie Farmer's non-depreciation operating expenses are projected to absorb a higher proportion of total revenue in 1995 (57 percent) than in 1994 (55 percent). Consequently, a lower proportion of value of farm production will be available for depreciation, interest, taxes, and net income.

Specific Financial Performance Measures

The analytical tools discussed in the previous three sections report alternative representations of the basic financial statements. Comparative financial

Table 4.3. Common-Size Balance Sheet for Joseph and Julie Farmer

	1992	1993	1994	1995
	-------- (Historical) --------			(Pro Forma)
Assets	------------------ (%) ------------------			
Current assets				
Cash, savings, and marketable securities	3	3	4	4
Crops, feed, and supplies	10	7	8	6
Market livestock	10	10	7	9
All other current farm assets	2	2	3	2
All personal assets – current				
Total current assets	**25**	**22**	**22**	**21**
Noncurrent assets				
Breeding livestock	4	4	4	5
Machinery and equipment	22	24	23	24
Farm real estate	42	42	41	41
Other noncurrent assets	2	2	2	2
All personal assets – noncurrent	5	6	6	6
Total noncurrent assets	**75**	**78**	**76**	**79**
Total Assets	**100**	**100**	**100**	**100**
Liabilities				
Current liabilities				
Operating, short-term notes	10	10	8	1
Current maturities	2	2	2	2
Accounts payable	1	1	0	1
All other current liabilities	7	6	7	6
All personal liabilities – current				
Total current liabilities	**20**	**19**	**17**	**10**
Noncurrent liabilities				
Non – real estate farm debt	3	4	5	6
Real estate farm debt	20	20	20	19
All other noncurrent liabilities	1	1	1	1
All personal liabilities – noncurrent	0	0	0	0
Total noncurrent liabilities	**24**	**25**	**26**	**26**
Total Liabilities	**44**	**44**	**42**	**36**
Owner Equity				
Contributed capital and retained earnings	49	48	52	59
Valuation/personal asset equity	7	8	6	5
Total Owner Equity	**56**	**56**	**58**	**64**
Total Liabilities and Owner Equity	**100**	**100**	**100**	**100**

Table 4.4. Common-Size Income Statement for Joseph and Julie Farmer

	1992	1993	1994	1995
	-------- (Historical) --------			(Pro Forma)
	-------------------- (%) --------------------			
Value of farm production	100	100	100	100
Non-depreciation operating expenses	55	68	55	57
Depreciation expense	9	13	10	10
Interest expense	13	15	11	9
Net farm income from operations.	23	4	24	24
Gain (loss) on sale of capital items	0	0	0	0
Net farm income before taxes	23	4	24	24
Nonfarm income	8	10	8	6
Income before taxes.	31	14	32	31
Income taxes	6	7	8	5
Net farm income after taxes.	25	7	24	26

statements involve presenting the financial statements over time and analyzing annual changes. Index-number trend series represent items relative to a base time period, while common-size statements report individual items relative to the total items on the financial statement. Another tool commonly used is the presentation of specific performance measures and comparisons over time, against norms and among firms.

The performance measures can include absolute measures and ratios. **Absolute measures** refer to *the monetary or physical levels of the measured item.* Absolute measures have limited generality and are primarily useful in evaluating and monitoring individual businesses over time. Inter-firm comparisons of absolute measures are usually inappropriate due to the size differences of the firms. Ratios mathematically express as fractions, decimals, or percentages one item or group of items in relation to another item or group of items. Ratios typically have greater generality in various types of comparisons than do absolute measures.

Financial analysts have developed a commonly followed set of ratios that they compute and compare in determining business progress. Financial ratios for various industries are published by firms such as Dun and Bradstreet and Robert Morris and Associates. Some trade associations and credit agencies compute ratios for their members. Furthermore, the Farm Financial Standards Task Force has made recommendations regarding the calculation and presentation of financial measures. The presentation here will include these measures in addition to other performance measures. Specific measures are classified by **profitability, liquidity, solvency, repayment capacity,** and **financial efficiency.** The calculation of each measure is outlined, and examples for Joseph and Julie Farmer are presented.

Profitability Measures

Profitability can be measured in several ways. The level of **net farm income** reflected in the income statement is a meaningful absolute measure with which to monitor profitability of the business from year to year. It is not very useful, however, for comparing the profitability of one firm to another or to some established standard because each firm represents a unique set and volume of resources. Moreover, the change in profit position should be considered in relationship to other changes in the firm. For example, an annual increase in profits of 2 percent might be unsatisfactory if the firm's total assets increased 20 percent. Under these types of circumstances, ratio measures of profitability provide better general performance indicators than do absolute measures.

The three ratios most often used are **rate-of-return on farm assets (ROFA), rate-of-return on farm equity (ROFE),** and **operating profit margin ratio.** In addition, we will also consider a fourth measure called the **cost of farm debt (COFD)** in order to show the relationships between the ROFA, ROFE, and COFD measures through the structure of the firm's debt and equity capital.

The **rate-of-return on farm assets (ROFA)** is the *rate-of-return on the firm's average total investment in farm assets for the accounting period.* It is given by:

$$\text{ROFA} = \frac{\cdot \, R}{A} \tag{4.1}$$

in which,

$$R = \text{NFIO} + I_d - \text{LC} + \text{OT} \tag{4.2}$$

where NFIO represents net farm income from operations, I_d is farm interest expense, LC is the value of operator and unpaid operator labor, OT is other net income on farm assets, and A represents average farm assets over the accounting period.[1]

---

[1]The inclusion of other income from farm assets (OT) is a deviation from the 1991 *Recommendations of the Farm Financial Standards Task Force.*

Since the numerator represents the return over a period of time, the denominator should represent the investment over the entire period. Thus, an average value for farm assets is used. Unless more detailed data are available, the average of the beginning and end-of-period farm asset values are commonly used to determine the average farm assets.

This measure is not influenced by the way in which the assets are financed. It only assesses the returns to assets, not specific claimants on these assets. Interest expense is added back because it represents the return to the suppliers of debt capital. Thus, rate-of-return on farm assets provides a measure of the profitability of the production and marketing activities of the business that is separate from the financing function. However, this is also a weakness of the measure, since the method and cost of financing the business assets often have a strong influence on profits.

A problem in agricultural accounting for the typical proprietary farm arises because net farm income often includes returns to the farm manager for unpaid labor and management as well as a return to equity capital. In larger corporate firms, the labor, management, and ownership functions are more easily separated and accounted for. Ideally, in estimating the dollar return to capital, each farmer could subtract from net farm income for his or her operations a return to his or her labor and management valued at what could be earned in an alternative employment. This figure might differ from the family withdrawals shown on the statement of cash flows. Precise estimates of these alternative earnings are difficult to obtain. In this analysis we use family living withdrawals as a proxy for the value of operator and unpaid family labor and management.

Another problem in using and interpreting agricultural accounting statements involves the inclusion and valuation of personal assets. As discussed in Chapter 3, many agricultural operations are sole-proprietor, family-owned businesses. Separate financial statements are not typically prepared for personal and business purposes. Furthermore, creditors may use personal items such as marketable securities and nonfarm income as a means of providing debt repayment. However, the return measure used in the numerator should be associated with the assets used to generate the return in the denominator in order to provide an appropriate rate-of-return estimate.

Distinguishing between farm and personal assets can be difficult. Personal residence, personal vehicles, household items, cash value of life insurance, and personal debts are clearly personal items. However, some financial assets along with other assets and liabilities are not easy to classify. In the examples presented here, all short-term financial assets are included as farm assets. However, since the interest or dividend income from these assets is not included in net farm income from operations, an adjustment (OT) is made to asset returns in the numerator of ROFA.

The calculation of rate-of-return on farm assets for Joseph and Julie Farmer for 1994 is shown below. The values in brackets are the respective line numbers from Tables A4.1 through A4.3 for each of the variables included in the calculation of the ratio.

$$\text{ROFA} = \frac{\$63,860\,[41] + \$28,530\,[40] - \$27,770\,[58] + \$425\,[48]}{(\$823,752 + 821,762\,)[34] \div 2}$$

$$= 0.0791, \text{ or } 7.91\%$$

The **rate-of-return on farm equity (ROFE)** measures *the rate-of-return on the equity capital that the owners have invested in the farm business.* It is given by:

$$\text{ROFE} = \frac{I_e}{E} \tag{4.3}$$

where

$$I_e = \text{NFIO} - \text{LC} + \text{OT} \tag{4.4}$$

where E is the average farm equity over the period and NFIO, LC, and OT are defined previously in the ROFA calculation.

The **rate-of-return on farm equity** represents *the return to farm equity employed in the farm business.* The cost of financing has been accounted for in calculating in this measure. The calculation of the rate-of-return on farm equity for Joseph and Julie Farmer for 1994 is shown as follows:

$$\text{ROFE} = \frac{\$63,860\,[41] - \$27,770\,[58] + \$425\,[48]}{(\$440,376 + \$462,211\,)[36] \div 2}$$

$$\text{ROFE} = 0.0809, \text{ or } 8.09\%$$

The **cost of farm debt (COFD)** measure is defined as:

$$\text{COFD} = \frac{I_d}{D} \quad\quad\quad (4.5)$$

where I_d is the dollar cost (interest paid and accrued) on debt capital and D is the firm's average farm indebtedness during the accounting period. The calculation of the cost of farm debt for Joseph and Julie Farmer for 1994 is shown as follows:

$$\text{COFD} = \frac{\$28{,}530\,^{[40]}}{(\$383{,}377 + \$359{,}551\,)^{[35]} \div 2}$$

$$\text{COFD} = 0.0768, \text{ or } 7.68\%$$

In order to show the relationship between these three measures, you should understand that the rate-of-return on farm assets is the sum of the cost of debt plus the rate-of-return on farm equity.

$$R = I_d + I_e \quad\quad\quad (4.6)$$

Moreover, the level of assets is the sum of the levels of debt and equity.

$$A = D + E \qu\quad\quad (4.7)$$

Thus, the rate-of-return on farm assets can be expressed as a weighted average of the rate-of-return on farm equity and the cost of farm debt when the weights are the proportional claims of equity and debt on total assets.

$$\text{ROFA} = \text{ROFE} \times \frac{E}{A} + \text{COFD} \times \frac{D}{A} \qu\quad\quad (4.8)$$

The calculation for Joseph and Julie Farmer is:

$$\text{ROFA} = 0.0809 \times \frac{(\$440{,}376 + \$462{,}211\,)^{[36]} \div 2}{(\$821{,}762 + \$821{,}762\,)^{[34]} \div 2} + 0.0768$$

$$\times \frac{(\$383{,}377 + \$359{,}551\,)^{[35]} \div 2}{(\$823{,}752 + \$821{,}762\,)^{[34]} \div 2}$$

ROFA = 0.0791, or 7.91%

The weighted average concept provides a consistency check on the profitability measures obtained from a set of coordinated financial statements. Moreover, a logical ordering of these three measures is based on the relative risk positions of the firm's debt and equity holders. The higher risk position of equity capital means that the returns (or costs) on equity should, on the average and over time, exceed the costs of debt. Under favorable economic circumstances, returns on assets should fall between the returns on equity and the debt because of the weighted average concept. Thus, under favorable conditions, the logical ordering of these profitability measures is **ROFE > ROFA > COFD.**[2]

To illustrate, the performance measures projected for Joseph and Julie Farmer for 1995 are considered in the following:

$$R = \$64{,}850 + \$24{,}760 - \$28{,}050 + \$410 = \$61{,}970$$

$$I_e = \$64{,}850 - \$28{,}050 + \$410 = \$37{,}210$$

$$I_d = \$24{,}760$$

$$A = \frac{\$821{,}762 + \$792{,}712}{2} = \$807{,}237$$

$$D = \frac{\$359{,}551 + \$297{,}458}{2} = \$328{,}504$$

$$E = \frac{\$462{,}211 + \$495{,}254}{2} = \$478{,}733$$

$$ROFA = \frac{\$61{,}970}{\$807{,}237} = 0.0768, \text{ or } 7.68\%$$

$$ROFE = \frac{\$37{,}210}{\$478{,}733} = 0.0777, \text{ or } 7.77\%$$

$$COFD = \frac{\$24{,}760}{\$328{,}504} = 0.0754, \text{ or } 7.54\%$$

---

[2]The cost of farm debt has not been expressed on an after-tax basis (i.e., net of reduced income tax obligations), although this is a common practice in financial structure analysis. The after-tax specification will be introduced in Chapter 6 where the relationships between the costs of financial capital and financial leverage are developed more fully.

$$\frac{E}{A} = \frac{\$478{,}733}{\$807{,}237} = 0.5931$$

$$\frac{D}{A} = \frac{\$328{,}504}{\$807{,}237} = 0.4069$$

To check returns on farm assets as a weighted average of returns on farm equity and costs of farm debt, insert the appropriate values in equation 4.6.

$$\text{ROFA} = (0.0777) \times (0.5931) + (0.0754) \times (0.4069)$$

$$= 0.0768, \text{ or } 7.68\%$$

In summary, during 1995, Joseph and Julie Farmer are projected to achieve a return on farm assets of 7.68 percent and a return on equity of 7.77 percent and to pay an average cost of debt of 7.54 percent. These ratios are consistent with an equity-to-asset ratio of 0.5931 and a debt-to-asset ratio of 0.4069. Although slightly lower than 1994, the profitability measures are still favorable under projected plans for 1995. As an aid to understanding where to find and analyze the information for assessing profitability, the data in the appendix to this chapter can be used to confirm that the projected rate-of-return on farm assets and the rate-of-return on farm equity for Joseph and Julie Farmer for 1993 are 1.44 percent and – 3.92 percent respectively.

The **operating profit margin ratio** is another commonly used profitability measure. This measure is calculated as:

$$\frac{\text{NFIO} + I_d - \text{LC}}{\text{Value of farm production}} \tag{4.9}$$

The **operating profit margin ratio** measures the *profitability of the firm as a proportion of the volume of production*. This ratio demonstrates how a firm can increase its profitability. A firm can increase its profitability by increasing the operating profit margin ratio while maintaining the same size of operation and the same interest obligations. Or, a farm can boost its profitability by decreasing its interest obligations while maintaining the same profit margin ratio. Furthermore, a farm can increase its profits by raising its volume of production while maintaining the same profit margin ratio and interest obligations. The 1994 operating profit margin ratio for Joseph and Julie Farmer is calculated as:

$$\text{Operating profit margin ratio} = \frac{\$63,860^{[41]} + \$28,530^{[40]} - \$27,770^{[58]}}{\$266,790^{[37]}}$$

$$= 0.2422, \text{ or } 24.22\%$$

For comparative purposes, profitability ratios can also be expressed on a per unit basis, when the units represent the important enterprises in the business. In cash-crop farming areas, for example, profit per acre is often used in comparing the profitability of one farm with another or in evaluating performance over several years. In dairy regions, profit per cow is more applicable. Many other similar profit ratios could be developed.

Liquidity Measures

Liquidity refers to *the firm's capacity to generate sufficient cash to meet its financial commitments as they become due.* Whenever a firm cannot satisfy these financial claims, it is bankrupt even though the firm might be profitable and have a favorable net worth. The part of net worth composed of claims on fixed assets is inaccessible without partially or wholly liquidating the firm.

If the timing and magnitude of cash inflows and outflows could be projected with certainty, the future liquidity of the farm business could be accurately assessed. However, since the future cannot be projected with certainty, liquid reserves are needed to provide flexibility in meeting uncertain cash outflows. These liquid reserves can be considered in terms of both the balance sheet of the business and its cash flows.

BALANCE SHEET MEASURES

The balance sheet is *an important source of information about the various ways in which liquidity is provided at a particular point in time.* Recall from Chapter 3 that the firm's assets and liabilities are classified according to their liquidity into two major categories: current and noncurrent. Liquidity analysis primarily focuses on the current portion of the balance sheet, although on occasion, the noncurrent portion can be a relevant indicator of liquidity as well. Thus, the balance sheet approach to liquidity analysis emphasizes the relationship between current assets and current liabilities and then works down the balance sheet to also consider the noncurrent portions.

In Chapter 3, **current assets** were classified as *those that are converted to cash in the normal operation of the business without significant price discount and* *without interrupting normal business operations.* Thus, current assets can be used to generate cash to meet the known demands on cash, as evidenced by the current liabilities, plus unknown demands that might arise during the year.

An absolute measure of liquidity based on balance sheet data is **working capital,** defined as *current farm assets minus current farm liabilities.* Adjustments for personal assets and liabilities should be made in the estimation of working capital. Working capital should be a positive value to signify a liquid situation. The working capital for Joseph and Julie Farmer in 1994 is calculated as:

$$\text{Working capital} = (\$206{,}872^{[6]} - \$0^{[5]}) - (\$144{,}903^{[19]} - \$750^{[18]})$$

$$= \$62{,}720$$

A more general measure of liquidity, and the one most widely used by financial analysts, is the **current ratio** — *current farm assets divided by current farm liabilities.* It specifies the dollars of current assets that are available for every dollar of current liabilities when the balance sheet is made. The higher the ratio, the higher the firm's liquidity. A current ratio less than 1 signals nonliquidity. The current ratio for Joseph and Julie Farmer is:

$$\text{Current ratio} = \frac{\$206{,}872^{[6]} - 0^{[5]}}{\$144{,}903^{[19]} - \$750^{[18]}}$$

$$= 1.44$$

Joseph and Julie Farmer had \$1.44 in current assets available for every \$1 of current liabilities, confirming that on December 31, 1994, their financial position was reasonably liquid.

The current ratio usually is expected to exceed 1 by a relatively wide margin.

Another commonly used measure of liquidity is **working capital** as a proportion of value of farm production (VFP). Working capital by itself is not very useful for comparing the liquidity of one firm to another or to some established standard because each firm represents a unique set and volume of resources. Furthermore, the current ratio may be misleading in cases when current liabilities are small or the absolute levels of current assets and liabilities are small relative to the overall size of the operation. The working capital to VFP ratio for Joseph and Julie Farmer in 1994 is:

$$\frac{(\$206,872^{[6]} - \$0^{[5]}) - (\$144,903^{[19]} - \$750^{[18]})}{\$266,790^{[37]}}$$

$$= 0.24$$

The relationship between working capital and value of farm production is likely to differ across farm types. As with the other ratios, inter-firm comparisons of this performance measure should be evaluated for similar farm types.

CASH FLOW LIQUIDITY MEASURES

Regardless of the completeness and preciseness of the balance sheet information, the liquidity of a business cannot adequately be assessed based on these data alone. This is especially true for farm and ranch businesses. An important consideration is the time of the year when the balance sheet is taken. The balance sheet of a farm business is usually taken early in the year (January, February) when current assets and current liabilities are at a minimum for many types of farms. Most crop and livestock products have already been sold. Feeder livestock is a notable exception, as are crops stored subject to commodity loans. Even so, cash to be generated from the sales of products produced later in the year usually far exceeds cash that can be generated from the sales of current inventory for most farms. Similarly, most of the firm's cash outflow will be incurred after the start of the year and is not reflected in current liabilities at the beginning of the year. Thus, liquidity assessments based on balance sheet analysis should be combined with cash flow analysis in order to assess fully the liquidity of the farm business.

The cash flow budget provides a month by month assessment of the projected cash inflows and cash outflows, indicating the expected liquidity position of the business each month. In assessing the firm's liquidity from the cash flow projection, first examine the monthly cash position. Reflected on line 25 of Table 3.10, the cash deficits and cash surpluses are indicated for each month. The primary function of the operating line of credit is to provide the needed liquidity in months of cash deficits, with the expectation that the loan will be repaid during months of cash surpluses. Also verify that the end-of-month cash balances are positive. Finally, monitor the amount outstanding under the operating line of credit as the business proceeds through the year. This loan balance is expected to reduce to zero at some period during the year. If not, the business likely has a shortage of working capital and a liquidity problem.

Solvency and Financial Risk Measures

Financial risk relates to the *firm's total structure of assets, liabilities, and equity capital.* It deals primarily with the firm's ability to meet total claims; hence, it is closely related to solvency. A farm business is insolvent, in the final analysis, if the sale of all assets fails to generate sufficient cash to pay off all liabilities. Consider the following example:

Current assets	$ 30,000	Current liabilities	$ 10,000
Fixed assets	200,000	Fixed liabilities	190,000
		Net worth	30,000
		Total liabilities	
Total assets	$230,000	and net worth	$230,000

Despite this firm's high liquidity, it is barely solvent, with only a narrow margin of equity ($30,000) relative to other liabilities. If a sudden emergency compelled the manager to quickly sell the fixed assets at "85 cents on the dollar," the business would be insolvent. From the lender's point of view, measures of solvency indicate loan recovery problems if the business fails. The smaller the safety margin of equity, the greater the financial risk.

The financial analyst is concerned with *trends in capital structure.* Selected measures of solvency should indicate the firm's financial progress in generating new funds, making capital expenditures, retiring debt, and coping with business and financial risks. Information about solvency may come from both the balance sheet and the income statement. The balance sheet measures apply to a specific point in time, while income statement measures reflect an interval of time. The balance sheet measures are typically called **solvency ratios,** while income statement measures are called **coverage ratios** of various kinds. Coverage ratios are treated in the following section.

An important solvency measure is the ratio of total farm debt to farm equity. Called the **leverage ratio,** the ratio of total farm debt to farm equity measures *the firm's total obligations to creditors (lenders and lessors) as a percent of the equity capital provided by the owners.* Again, for Joseph and Julie Farmer the leverage ratio is:

$$\frac{\text{Total farm liabilities}}{\text{Total farm equity}} \tag{4.10}$$

$$= \frac{\$359{,}551 \, [35]}{\$462{,}211 \, [36]}$$

$$= 0.78$$

This ratio indicates that the firm is using $0.78 of debt capital for every dollar of equity capital. Other ratios reflecting exactly the same information are the **debt-to-asset ratio** and the **equity-to-asset ratio.** Joseph and Julie Farmer's debt-to-asset ratio for December 31, 1994, was 43.75 percent and their equity-to-asset ratio was 56.25 percent.

Many farm lenders prefer that borrowers have at least as much investment in their own businesses as their lenders do. Therefore, a standard rule of thumb for the maximum leverage ratio is 1. This standard would be consistent with a debt-to-asset ratio of 0.50 or lower or an equity-to-asset ratio of 0.50 or higher. However, the leverage norm varies substantially among farm businesses and from one type of business to another. It is commonly accepted, for example, that larger farms and farms with higher tenancy positions (i.e., more leased land) can carry relatively greater debt loads. Thus, their leverage ratios tend to be higher. Firms in other industries often have higher leverage ratios. For most farm-related businesses, however, which experience highly variable prices and costs, a much lower leverage is advisable. The critical issue relating to leverage is the firm's ability to generate the cash to meet all expenses and service the debt with an acceptable margin of safety.

Solvency ratios do not indicate an optimal level of leverage for a firm. Each manager must decide this crucial question based on a careful analysis of projected profitability, liquidity, and risk. Selecting the desired level of leverage is a complex decision that is addressed in Section Three. Once the decision has been made, however, progress toward that goal will be indicated by changes over time in the leverage ratio or in other solvency ratios.

Coverage and Repayment Capacity Measures

Coverage ratios taken from the income statement account for the relative claims of debt and equity holders on the returns to farm assets during a period of time. Thus, **coverage ratios** essentially *relate a firm's financial charges to its ability to serve them.* They serve as a counterpart of the solvency ratios that hold for a specific point in time. The **interest coverage ratio** *expresses the ratio of a firm's returns to assets (R) to the amount of interest charges (I_d) for a period.* It is:

$$\frac{R}{I_d} \tag{4.11}$$

$$= \frac{\$63,860 \, [41] + \$28,530 \, [40] - \$27,770 \, [58] + \$425 \, [48]}{\$28,530 \, [40]}$$

$$= 2.28$$

This figure indicates that $2.28 of returns on assets is available for every dollar of interest commitment. Or, 43.8 percent (1 ÷ 2.28) of returns on assets is required to meet interest commitments in the budgeted year. Notice that the interest coverage ratio is analogous to the inverse of the debt-to-asset ratio (net capital ratio) taken from a firm's balance sheet. The coverage ratio relates asset return to debt obligations for a period of time; the net capital ratio (A ÷ D) relates total assets to total debt at a point in time. Generally, the interest coverage ratio is subject to greater volatility than solvency ratios because of the year-to-year swings in farm income and interest rates.

A shortcoming of the interest coverage ratio is that a firm's ability to service debt is related to both interest and principal payments. Moreover, these payments must come from the firm's cash flow rather than from accrual income.

Repayment capacity measures relate the cash generated by the operation to cash obligations of the firm. These measures more precisely *assess the ability of the firm to service debt and / or make alternative investments.* However, measures of repayment capacity require more information and more complex estimation procedures than the other measures.

The Farm Financial Standards Task Force recommends two repayment capacity measures, one an absolute measure and the other a ratio. The first measure, **capital replacement and term debt repayment margin,** is computed as:

Net farm income from operations

+ Total nonfarm income

+ Depreciation

− Total income tax expense

− Withdrawals for family living

= **Capital replacement and term debt repayment capacity**

– Payment on unpaid operating debt from previous period

– Principal payments on current portions of term debt and capital leases

– Annual payments on personal liabilities not included in family living withdrawals

= **Capital replacement and term debt margin**

(4.12)

The second measure, **the term debt and capital lease ratio,** is calculated as:

Capital replacement and term debt replacement capacity

+ Interest on term debt and capital leases

÷ Annual scheduled principal and interest payments on term debt and capital leases

(4.13)

The measures of repayment capacity evaluate the ability of the firm to service debt and generate funds for alternative investments. Each of the measures incorporates the cash generated by farm and nonfarm sources and considers all term-debt obligations, income tax obligations, and family living expenditures. For each measure, the greater the measure, the greater the ability of the firm to service debt.

The calculations of capital replacement and term debt repayment margin and the term debt and capital lease ratio for Joseph and Julie Farmer in 1994 are outlined as follows:

Capital replacement and term debt
replacement capacity (RC) = $63,860[41] + $20,180[44] + $26,960 [39] – $20,075[46] – $27,770[58]

= $63,155

Capital replacement and term debt
replacement margin = $63,155 – $16,040[75]

= $47,115

Term debt and capital lease ratio $= \dfrac{\$63,155\,^{[RC]} + \$21,991\,^{[54]}}{\$21,991\,^{[54]} + \$16,040\,^{[75]}}$

$= 2.24$

A positive capital replacement and term debt replacement margin and a term debt and capital lease ratio greater than 1 indicate that Joseph and Julie Farmer have a moderate margin for capital replacement and debt service.

Efficiency

Efficiency ratios, which represent the *ratio of outputs to selected inputs,* are also useful in monitoring the performance and profitability of the farm. They portray the efficiency of the production process, which is an important component of profitability. In many instances, management problems are reflected in deteriorating production ratios long before they are evident in profit ratios. Hence, monitoring a few critical efficiency ratios can provide an early warning of possible impending income problems.

A decline in the number of eggs per hen per day, for example, may signal a developing health or other management problem that, if not corrected, might seriously affect profits for the month ahead. For a swine farmer, the number of swine sold per litter, the feed cost per 100 pounds of hog, and the pounds of hog sold per $100 of expenses are relevant indexes of operating efficiency. Large commercial feedlot operators closely monitor feed conversion ratios and daily nonfeed costs per head. Each farm business should develop the physical production ratios most relevant for that particular business.

The **asset turnover ratio** is also used in business analysis as one measure of efficiency. It is defined as *the ratio of the value of farm production to the average total farm assets.* The higher the ratio, the more rapid the turnover of assets into income. Also, the higher the ratio, the greater the opportunity for profits, provided profit margins are positive. The turnover ratio for Joseph and Julie Farmer is:

$\dfrac{\$266,790\,^{[37]}}{\$821,762\,^{[34]}}$

$= 0.32$

Four operational ratios are used to analyze the components of value of farm production. The four components are **non-depreciation operating ex-**

penses, depreciation expenses, interest expenses, and net farm income from operations. Each component is divided by the value of farm production, and the sum of the four ratios is 100 percent. These items are equivalent to the elements of the common-size income statements discussed earlier and displayed for Joseph and Julie Farmer in Table 4.4.

The most important financial ratios and measures are summarized for Joseph and Julie Farmer in Table 4.5. These measures and ratios are listed under the five headings of "Profitability," "Liquidity," "Solvency," "Coverage and Repayment Capacity," and "Financial Efficiency." The changes and trends of the firm over time can readily be detected and assessed with this type of representation. Moreover, the performance and trends of the firm can be compared with those of similar firms or with standards for the industry when such information is available.

One of the most important standards against which the actual and projected performance should be tested, however, is the "goals" which the farmer (often in counsel with the lender) has set as achievable targets. These goals should be realistic, but challenging. Because of uncertainty, acceptable ranges should also be established. If the measure falls within the stipulated range, the performance is considered acceptable. If the performance drops outside the acceptable range, then careful study of the causal factors and possible corrective actions are needed. With corrective actions, unacceptable performance should trend toward the acceptable range over time.

As revealed in Table 4.5, the profitability measures for Joseph and Julie Farmer were strong in each year with the exception of 1993. The logical orderings of the profitability measures (ROFE > ROFA > COFD) were consistent in 1992, 1994, and 1995. However, in 1993, the cost of debt was 7.61 percent, while the rates-of-return on assets and on equity were only 1.44 percent and − 3.92 percent respectively. The liquidity measures for Joseph and Julie Farmer exhibit constant improvement with a slight decrease in 1993. The projected current ratio is expected to exceed 2 in 1995. The solvency measures for Joseph and Julie Farmer are also improving. The debt-to-equity ratio declined from 0.87 in 1992 to a projected 0.60 in 1995. The coverage and repayment capacity measures are also strong for 1994 and 1995. The cushion available to repay interest and principal is adequate in these years, although the measures for 1993 exhibit the sensitivity of the coverage and repayment capacity measures in unprofitable years. Finally, the efficiency measures for Joseph and Julie Farmer are also improving. Net income from operations as a percentage of value of farm production has increased from 23.3 percent in 1992 to a projected 24.2 percent in 1995.

Table 4.5. Performance Measures for

RATIO ANALYSIS

Ratio	1992	1993
	--------- (Historical) ---------	
Profitability		
Rate-of-return on farm assets	7.81%	1.44%
Rate-of-return on farm equity	7.76%	−3.92%
Cost of farm debt	7.86%	7.61%
Operating profit margin ratio	25.22%	5.74%
Net farm income before taxes	$60,550	$ 8,902
Liquidity		
Current ratio	1.27	1.19
Working capital	$46,774	$30,854
Working capital to value of farm production ratio	0.18	0.15
Solvency		
Debt-to-equity ratio	0.87	0.87
Debt-to-asset ratio	46.49%	46.54%
Equity-to-asset ratio	53.51%	53.46%
Coverage and Repayment Capacity		
Interest coverage ratio	2.01	0.41
Term debt and capital lease coverage ratio	2.32	0.9
Capital replacement and term debt repayment margin	$47,306	($ 550)
Financial Efficiency		
Asset turnover ratio	0.31	0.24
Operating expense ratio	54.74%	68.18%
Depreciation expense ratio	9.36%	12.78%
Interest expense ratio	12.61%	14.62%
Net farm income from operations ratio	23.29%	4.42%

Joseph and Julie Farmer

for Joseph and Julie Farmer

1994 (Historical)	1995 (Pro Forma)	Goal	Range
7.91%	7.68%	8.00%	6.50% – 9.50%
8.09%	7.77%	9.00%	7.50% – 10.50%
7.68%	7.54%	7.00%	5.50% – 8.50%
24.22%	22.98%	25.00%	23.00% – 27.00%
$63,860	$64,850	$ 70,000	$65,000 – 75,000
1.44	2.12	2.25	1.75 – 2.75
$62,719	$94,558	$100,000	$80,000 – 120,000
0.24	0.35	0.35	0.20 – 0.50
0.78	0.60	0.54	0.43 – 0.67
43.75%	37.52%	35.00%	0.30% – 0.40%
56.25%	62.48%	65.00%	0.70% – 0.60%
2.28	2.50	2.00	1.50 – 3.00
2.24	2.27	2.50	2.00 – 3.00
$47,115	$49,559	$ 50,000	$40,000 – 60,000
0.32	0.33	0.35	0.32 – 0.38
55.26%	56.92%	55.00%	52.00% – 58.00%
10.11%	9.63%	10.00%	8.00% – 12.00%
10.69%	9.24%	10.00%	8.00% – 12.00%
23.94%	24.21%	25.00%	22.00% – 28.00%

EXPLAINING CHANGES IN RATIOS

As indicated earlier, financial ratios and other performance measures are only indicators or signals of a firm's performance. Judgment and the experience of analysts, as well as theoretical relationships and more detailed analysis, play important roles in explaining changes in ratios and in identifying problems calling for corrective actions. Setting reasonable tolerance limits to allow for anticipated variations in ratios due to random events is an important part of the control process. Moreover, the norms that are established for the important performance measures may themselves change as changes occur in farm business conditions, government policy, the national economy, and international markets.

We will illustrate the issues involved in diagnosing changes in several key performance measures that represent a farm business's profitability, liquidity, and solvency. No measure by itself can describe the financial performance of a business in a dynamic environment. As discussed earlier, many of the ratios are inter-related and should be treated as a system. A common technique called the **Dupont analysis** *uses the interrelationships between many of the performance measures to analyze the components and linkages of a business.*

Dupont Analysis

As shown in equation 4.1, ROFA is calculated as:

$$\text{ROFA} = \frac{R}{A}$$

Also shown is that net profit is related to asset turnover and the profit margin. Thus, equation 4.1 may be restated as:

$$\text{ROFA} = \frac{R}{VFP} \times \frac{VFP}{A} \qquad (4.14)$$

and further restated as:

$$\text{ROFA} = \frac{(\text{VFP} - \text{expenses}) + I_d - \text{LC}}{\text{Units produced} \times \text{sales price of units} + \text{other adjustments}}$$

$$\times \frac{\text{Units produced} \times \text{sales price of units} + \text{other adjustments}}{\text{Current} + \text{noncurrent farm assets}} \qquad (4.15)$$

These relationships demonstrate the linkages that management can control to enhance the rate-of-return on farm assets. For example, asset management or expense control could be a potential weakness within the firm. The key focus points and potential weaknesses may be identified more readily by expressing the individual components rather than simply interpreting the ROFA measure. A similar analysis can be performed for rate-of-return on farm equity where:

$$\text{ROFE} = \left[\text{ROFA} - \frac{I_d}{A} \right] \times \frac{A}{E} \qquad (4.16)$$

In equation 4.16, the elements of profitability include the profitability of assets as well as the level of leverage. The same components of value of farm production, net profits, and farm assets could be separately analyzed. The additional component in the ROFE analysis is leverage. Thus, the manager can determine the impact of rate-of-return on farm equity on changes in leverage as well as on other operation changes.

To illustrate, consider the performance measures for Joseph and Julie Farmer for 1992 – 1995. Table 4.6 reveals the individual components in the Dupont analysis for Joseph and Julie Farmer. A slight decline in the rate-of-return on assets is projected for 1995 (0.2% = 7.9% – 7.7%). The Dupont analysis provides a more detailed assessment outlining the decline. Asset turnover actually increases, while the asset returns to sales exhibit a slight decline. Thus, the decline in the rates-of-returns on assets is more likely a result of income generation and/or expense control problems rather than asset utilization.

The rate-of-return on farm equity also is projected to decline from 1994 to 1995. A portion of the decline is a result of the rate-of-return on assets decreasing. However, leverage is also projected to decrease; thus, the rate-of-return on farm equity will also decline. Chapter 6 discusses the linkages between returns, leverage, and firm growth in greater detail.

Table 4.6. Dupont Analysis for Joseph and Julie Farmer

	1992	1993	1994	1995
1. R ÷ VFP	0.254	0.060	0.244	0.231
2. VFP ÷ A (Asset turnover)	0.308	0.242	0.324	0.332
3. Rate-of-return on farm assets (ROFA) [1 × 2]	0.078	0.014	0.079	0.077
4. I_d ÷ A	0.037	0.035	0.034	0.028
5. A ÷ E	1.869	1.871	1.778	1.601
6. Rate-of-return on farm equity (ROFE) [3 − 4] × 5	0.078	−0.039	0.081	0.078

Other Linkages Between Performance Measures

It is critical to evaluate these operational and leverage components to assess properly changes in performance measures over time. First, consider a business that has experienced an increase in its ratio of operating cost to value of farm production from 0.60 to 0.70. On the surface, this change suggests a decline in profitability. It should alert the manager to the possibility that production costs are getting out of control. But, several other factors could be involved too. The ratio could have increased because product prices or production levels declined so that the value of total farm production also declined, while total costs remained about the same.

The higher ratio of costs to value of farm production could also affect the firm's adjustment in output levels toward greater productivity and higher profits. Profit maximization occurs at an output level where marginal returns equal marginal costs. This condition may differ from the level of output that results in minimum average costs of production. This situation indicates a fundamental limitation of ratio analysis; that is, it emphasizes average relationships rather than marginal relationships. Marginal analysis is preferred in planning and decision making; but in financial analysis, the available data are usually best suited for deriving ratios based on average relationships.

Next, consider the possible reasons for an observed decrease in a firm's current ratio, signifying a reduction in liquidity. The ratio is current assets divided by current liabilities. A decrease in the ratio could be attributed to (1) current assets staying constant and current liabilities increasing; (2) current assets declining and current liabilities staying constant; (3) current assets declining more than current liabilities decline; (4) current assets declining and current liabilities increasing; and (5) both current assets and current liabilities increasing, but current liabilities increasing more so.

Beyond these possible combinations, the changes in assets, liabilities, or both could be attributed to a price change, a quantity change, or both price and quantity changes. If assets changed, which ones and why? Could the changes in assets be due to losses, random fluctuations, a change in marketing and inventory policies, a change in the size of the firm, or a change in enterprise mix? If current liabilities changed, which ones and why? Could the changes be the result of changes in interest rates, carryover financing due to losses, growth and capital investment, new inventory financing, larger household demands, other reasons?

Finally, consider the possible reasons for a decline in a firm's debt-to-equity (leverage) ratio from 0.50 to 0.40 during the past year. On the surface, it seems that the firm has made substantial progress in reducing financial risk. However, a careful examination of the balance sheet could indicate that this business sold many of its assets at a substantial loss in order to meet its financial obligations. By selling these assets, the firm paid off a large volume of debt, thereby decreasing its leverage ratio. In the process, however, the firm's net worth was reduced considerably. Similarly, an increase in leverage could be due to a decline in asset values for a given level of debt rather than to additional borrowing or to a revision in the firm's leverage policy.

Each financial manager must determine a preferred or target leverage ratio that is appropriate for his or her particular firm. Many farmers and lenders prefer low leverage ratios and interpret most movements toward lower leverage as indicating financial progress. However, as shown in Chapter 6, higher leverage ratios, under proper circumstances, generate more rapid rates of firm growth. Moreover, as leverage increases, the average cost of financial capital usually decreases to some point, then moves up sharply. Thus, increasing the leverage ratio could be both a managerial goal and an indicator of financial progress.

Summary

In this chapter the procedures for analysis and control are emphasized. The steps of the financial control process are outlined, and the principal analytical tools are described and applied to the sample case. Specific highlights of the chapter are as follows:

- The financial control process provides an orderly framework for responding to signals in an uncertain environment. There are seven steps in the process: (1) identifying goals, (2) developing measures for the goals, (3) determining norms for the measures, (4) setting tolerance limits on norms, (5) developing an information system, (6) selecting and administering the analytical and diagnostic tools, and (7) identifying and implementing corrective actions.

- Measures selected to reflect the status of the firm must be both **meaningful** and **manageable**.

- The primary tools used in financial analysis are (1) **comparative financial statements,** (2) **index-number trend series,** (3) **common-size financial statements,** and (4) **specific financial performance measures.**

- **Comparative financial statements** compare two or more years of accounting statements or compare one firm to another firm or industry norm. Changes in the direction of trends can signal performance strengths and weaknesses of a firm.

- An **index-number trend series** uses a base year to compare changes in financial statements for periods longer than two years.

- **Common-size financial statements** allow comparisons between firms of different sizes. Each item on the financial statement is displayed as a proportion of the total item, such as total assets or sales.

- Profitability measures allow the firm to assess changes in profits relative to the condition of other financial measures. The two most used profitability ratios are **rate-of-return on farm assets** and **rate-of-return on farm equity.**

- **Liquidity measures** tell how well a firm is able to generate cash to meet financial commitments as they come due. Three common liquidity measures are (1) **working capital,** (2) **current ratio,** and (3) **working capital to value of farm production ratio.**

- **Solvency** and **coverage and repayment capacity** refer to the *firm's ability to meet total claims against the business.* Four important solvency and coverage ratios are (1) **debt-to-equity ratio,** (2) **interest coverage ratio,** (3) **capital replacement and term debt repayment margin,** and (4) **term debt and capital lease ratio.**

- Efficiency ratios can provide early warning signals of income problems. The **asset turnover ratio** is an important measure of efficiency in business. A higher ratio signals rapid asset turnover and greater opportunity for profits.

- The **Dupont analysis** uses the interrelationships between many performance measures to analyze a business.

Topics for Discussion

1. Identify the steps of the financial control process. How are they related to time and uncertainty? Identify some possible data sources for developing norms and tolerance ranges for the various measures of business performance.

2. Contrast the use of absolute measures versus ratio measures in analyzing a firm's financial performance. How are both of them useful in various types of comparisons?

3. How is the rate-of-return on farm assets distinguished from the rate-of-return on farm equity? Which is the more valid measure? Why?

4. Suppose a farm business indicates an average cost of farm debt of 12 percent, a rate-of-return on farm assets of 15 percent, and a debt-to-equity ratio of 1. What is the firm's rate-of-return on farm equity? What would be its rate-of-return on farm equity if the debt-to-equity ratio were 2?

5. Suppose a firm experiences a decline in its current ratio from 1.8 to 1.2. Is this a favorable or an unfavorable change? Why? What factors might explain the decline in the ratio?

6. Explain the concepts of solvency and risk. How are they related to the firm's leverage position? What are typical ranges of leverage for farm businesses?

7. Distinguish between solvency and coverage ratios. How are they related? Which financial statements are involved?

8. Why are average values used in calculating profitability ratios?

9. Describe the Dupont analysis and discuss the interpretive benefits of using the analysis.

References

1. American Bankers Association, *Credit Analysis for Agricultural Lending*, Washington, D.C., 1985.

2. Benjamin, J. J., A. J. Rancia, and R. Strawser, *Principles of Accounting*, 6th ed., Dame Publications, Inc., Houston, 1990.

3. Berstein, L. A., *Financial Statement Analysis: Theory, Application, and Interpretation*, 5th ed., Richard D. Irwin, Inc., Homewood, Illinois, 1992.

4. Fraser, L., *Understanding Financial Statements*, Reston Publishing Company, Inc., Reston, Virginia, 1985.

5. Frey, T. L., and R. Behrens, *Lending to Agricultural Enterprises*, American Banking Institute, New York, 1982.

6. Lee, W. F., et al., *Agricultural Finance*, 8th ed., Iowa State University Press, Ames, 1988.

7. *Recommendations of the Farm Financial Standards Task Force: Financial Guidelines for Agricultural Producers*, Farm Financial Standards Task Force, American Bankers Association, Washington, D.C., 1991.

8. Van Horne, J. C., *Financial Management and Policy*, 8th ed., Prentice-Hall, Inc., Englewood Cliffs, New Jersey, 1989.

Table A4.1. Balance Sheet Summary for Joseph and Julie Farmer

		1991	1992	1993	1994	1995
		----------- (Historical) -----------				(Pro Forma)
Assets						
	Current assets					
1.	Cash, savings, and marketable securities	$ 24,900	$ 22,900	$ 22,220	$ 38,060	$ 37,400
2.	Crops, feed, and supplies	80,600	91,030	59,652	73,222	52,362
3.	Market livestock	106,900	85,590	90,100	72,350	76,950
4.	All other current farm assets	17,800	18,260	21,470	23,240	13,540
5.	All personal assets – current	0	0	0	0	0
6.	**Total current assets**	**$230,200**	**$217,780**	**$193,442**	**$206,872**	**$180,252**
	Noncurrent assets					
7.	Breeding livestock	$ 35,000	$ 38,480	$ 40,810	$ 38,590	$ 41,570
8.	Machinery and equipment	185,000	196,300	206,100	196,810	202,920
9.	Farm real estate	383,500	374,090	369,300	359,580	348,060
10.	Other noncurrent assets	14,100	14,100	14,100	19,910	19,910
11.	All personal assets – noncurrent	48,800	48,800	48,800	48,800	48,800
12.	**Total noncurrent assets**	**$666,400**	**$671,770**	**$679,110**	**$663,690**	**$661,260**
13.	**Total Assets**	**$896,600**	**$889,550**	**$872,552**	**$870,562**	**$841,512**
Liabilities						
	Current liabilities					
14.	Operating, short-term notes	$140,000	$ 88,220	$ 82,350	$ 69,420	$ 10,550
15.	Current maturities	13,490	13,820	16,040	17,340	19,000
16.	Accounts payable	6,500	10,440	9,380	420	2,980
17.	All other current liabilities	60,722	58,526	54,818	56,973	52,264
18.	All personal liabilities – current	750	750	750	750	750
19.	**Total current liabilities**	**$221,462**	**$171,756**	**$163,339**	**$144,903**	**$ 85,544**
	Noncurrent liabilities					
20.	Non – real estate farm debt	25,000	26,760	34,920	39,910	46,130
21.	Real estate farm debt	188,700	182,430	174,470	167,680	160,740
22.	All other noncurrent liabilities	8,520	10,660	11,398	7,808	5,794
23.	All personal liabilities – noncurrent	0	0	0	0	0
24.	**Total noncurrent liabilities**	**$222,220**	**$219,850**	**$220,788**	**$215,398**	**$212,664**
25.	**Total Liabilities**	**$443,682**	**$391,606**	**$384,127**	**$360,301**	**$298,208**

(Continued)

Table A4.1 (Continued)

		1991	1992	1993	1994	1995
		----------- (Historical) -----------				(Pro Forma)
Owner Equity						
26.	Contributed capital and retained earnings	$395,876	$432,342	$419,872	$456,067	$497,166
27.	Valuation/personal asset equity	57,042	65,602	68,554	54,194	46,138
28.	**Total Owner Equity**	**$452,918**	**$497,944**	**$488,426**	**$510,261**	**$543,304**
29.	**Total Liabilities and Owner Equity**	**$896,600**	**$889,550**	**$872,552**	**$870,562**	**$841,512**
Balance Sheet Notes						
Book values (original cost less accumulated depreciation)						
30.	Breeding livestock	$ 38,480	$ 38,480	$ 40,810	$ 38,590	$ 41,570
31.	Machinery and equipment	167,390	167,390	169,910	175,140	181,000
32.	Real estate — land	350,000	350,000	350,000	350,000	350,000
33.	Real estate — buildings	8,800	8,800	7,600	8,800	7,600
34.	Farm assets	847,800	840,750	823,752	821,762	792,712
35.	Farm liabilities	442,932	390,856	383,377	359,551	297,458
36.	Farm equity	404,868	449,894	440,376	462,211	495,254

Table A4.2. Income Statement Summary for Joseph and Julie Farmer

		1992	1993	1994	1995
		----------- (Historical) -----------			(Pro Forma)
37.	Value of farm production	$259,940	$201,412	$266,790	$267,910
38.	Non-depreciation operating expenses	142,280	137,320	147,440	152,500
39.	Depreciation expense	24,330	25,740	26,960	25,800
40.	Interest expense	32,780	29,450	28,530	24,760
41.	Net farm income from operations	60,550	8,902	63,860	64,850
42.	Gain (loss) on sale of capital items	0	0	0	0
43.	Net farm income before taxes	60,550	8,902	63,860	64,850
44.	Nonfarm income	19,620	19,680	20,180	17,320
45.	Income before taxes	80,170	28,582	84,040	82,170
46.	Income taxes	15,934	14,252	20,075	13,021
47.	Net farm income after taxes	$ 64,236	$ 14,330	$ 63,965	$ 69,149
Income Statement Notes					
48.	Interest and dividend income received on farm assets	$ 380	$ 450	$ 425	$ 410

Table A4.3. Statement of Cash Flow Summary for Joseph and Julie Farmer

	1992	1993	1994	1995
	------- (Historical) -------			(Pro Forma)
Cash Flows from Operating Activities				
Inflows				
49. Cash received from farm production and government payments	$376,160	$328,840	$369,940	$400,910
50. Cash received from net nonfarm income	19,620	19,680	20,180	17,320
51. **Total cash inflows from operating activities**	**$395,780**	**$348,520**	**$390,120**	**$418,230**
Outflows				
52. Cash paid for operating expenses	$140,830	$136,790	$154,480	$146,880
53. Cash paid for interest on operating loans	11,981	8,957	7,970	4,200
54. Cash paid for interest on term loans	22,440	21,964	21,991	21,800
55. Cash paid for feed and market livestock	103,310	105,360	102,660	110,100
56. Cash paid for other items purchased for resale	0	0	0	0
57. Cash paid for income and social security taxes	16,490	16,490	16,490	16,490
58. Cash withdrawals for family living	27,770	26,800	27,770	28,050
59. **Total cash outflows from operating activities**	**$322,820**	**$316,360**	**$331,360**	**$327,520**
60. **Net cash flows provided by operating activities**	**$ 72,960**	**$ 32,160**	**$ 58,760**	**$ 90,710**
Cash Flows from Investing Activities				
Inflows				
61. Cash received on sale of machinery/equipment/real estate	$ 0	$ 0	$ 0	$ 0
62. Cash received from sale of breeding livestock	0	0	0	0
63. Cash received from withdrawals of savings, marketable securities, personal	38,000	48,003	0	120,540
64. **Total cash inflows from investing activities**	**$ 38,000**	**$ 48,003**	**$ 0**	**$120,540**
Outflows				
65. Cash paid to purchase machinery/equipment/real estate	15,520	27,060	25,900	30,460
66. Cash paid for purchase of breeding livestock	3,480	2,330	3,590	2,980
67. Cash paid for deposits to savings accounts, marketable securities, personal	36,000	46,053	10,490	120,000
68. **Total cash outflows from investing activities**	**$ 55,000**	**$ 75,443**	**$ 39,980**	**$153,440**
69. **Net cash flows provided by investing activities**	**($ 17,000)**	**($ 27,440)**	**($ 39,980)**	**($ 32,900)**

(Continued)

Table A4.3 (Continued)

		1992	1993	1994	1995
		- - - - - - - - (Historical) - - - - - - - -			(Pro Forma)
	Cash Flows from Financing Activities				
	Inflows				
70.	Proceeds from operating loans and short-term notes	$ 95,666	$ 95,207	$ 77,740	$ 23,597
71.	Proceeds from term debt financing	9,310	16,240	15,540	18,280
72.	Cash received from capital contributions	0	0	0	0
73.	**Total cash inflows from financing activities**	**$104,976**	**$111,447**	**$ 93,280**	**$ 41,877**
	Outflows				
74.	Cash repayment of operating and short-term loans	$147,446	$101,077	$ 90,670	$ 82,467
75.	Cash repayment of term debt and capital leases — scheduled	13,490	13,820	16,040	17,340
76.	Cash repayment of term debt and capital leases — unscheduled	0	0	0	0
77.	Cash payments of dividends and other capital distributions	0	0	0	0
78.	**Total cash outflows from financing activities**	**$160,936**	**$114,897**	**$106,710**	**$ 99,807**
79.	**Net cash flows provided by financing activities**	**($ 55,960)**	**($ 3,450)**	**($ 13,430)**	**($ 57,930)**
80.	Net increase (decrease) in cash flows	$ 0	$ 1,270	$ 5,350	($ 120)
81.	Beginning of year cash balance	2,000	2,000	3,270	8,620
82.	End of year cash balance	2,000	3,270	8,620	8,500

5

Financial Planning
and Feasibility Analysis

In this chapter we apply the knowledge of the financial statements and related analyses gained from Chapters 3 and 4 to the design of appropriate financing programs and to an assessment of overall financial feasibility. The financing needs of the various agricultural assets and the financing tools for meeting these needs are considered first. Emphasis is placed on the design of financing programs to support the firm's production, marketing, and capital investments. The use of cash flow budgeting and pro forma projections to test program alternatives is reviewed and applied to the analysis of financial feasibility.

AN OVERVIEW OF
FINANCIAL PLANNING

The development of business plans for farm firms occurs within the three organizational areas of finance, production, and marketing. The focus here is on the financial organization, although it is important to recognize that financial decisions interact with production and marketing decisions.

The relationship between cash flows, loan repayment obligations, and the firm's financial performance must be considered in designing an acceptable financing program for a farm business. The expenditures and cash flows associated with the acquisition and use of resources need careful coordination. Each lending program sets in motion a particular pattern of future cash flows. Each of these flows must be considered together with other cash flows arising from production and marketing activities. The cash flow requirements of the family must also be taken into account. All too frequently a sound long-range financing program is placed in jeopardy by the addition of short-term or non – real estate loans that are not anticipated or that require repayment faster than cash can be generated.

In developing an **optimal loan program,** a borrower would prefer (1) a maturity structure of debt that matches the length of the payoff periods for the assets being financed and (2) a repayment pattern over time that matches the assets' earnings pattern. The result is a set of **self-liquidating loans,** *with the various assets pledged as collateral to secure the loan.* A self-liquidating loan is made for the purpose of generating sufficient income to repay the loan within the maturity period. Two conditions are required for a self-liquidating loan. First, the asset or project being financed must generate more cash returns over its life than the size of the loan plus the interest obligation. Second, the maturity of the note and the schedule of repayments must be such that the payments can be met from cash generated by the investment.

An important advantage of a self-liquidating loan is that it does not absorb the firm's liquidity. Loans for the purchase of current assets, such as fertilizer or feeder cattle, are generally self-liquidating because the loan payment is usually scheduled to occur upon the sale of the crop or livestock being produced. The same principles hold for financing noncurrent assets, although longer-term maturities are needed and more flexible repayment schemes are advisable.

If the loan maturities are too short or if they are not well-coordinated with the time pattern of earnings, then liquidity problems and cash flow pressures may arise. Moreover, when business and financial risks occur, financing arrangements must be flexible so that loan maturities and repayment obligations may be adjusted in response to unanticipated variability in cash flows. Hence, credit reserves with the lender must be large enough to provide additional loans to meet the borrower's financial obligations in periods of adversity and flexible enough to restructure excessive indebtedness for more orderly payoffs. Then, the financing program should be designed to restore a more acceptable debt structure when adverse conditions have passed.

The coordination of cash flows is simplified if all the financing (short-term and noncurrent loans) is provided by one lender. Historically, farmers have generally used different lenders for financing non – real and real estate assets because few lenders could meet all of the farmer's financial needs. However, this condition is changing. The cooperative Farm Credit System has restructured many of its lending associations in order to offer full-service financing to agricultural borrowers from one office location. Similarly, many commercial banks have designed real estate financing programs so that they, too, can meet the total financing needs of farmers.

Despite the clear advantages of coordinating the borrower's total financing needs into a single "package" of loans carefully tailored to the projected cash flows, a separate analysis of loans for financing assets with distinctly different time dimensions and with separate sources of funds for loan repayment has merit. The different loan maturities permit more precise analysis of repayability for each financing need, thus providing borrowers with a wider range of choices.

FINANCING CURRENT ASSETS

Financing current assets in farm businesses involves the acquisition of annual operating inputs. However, considerable financing of grain inventories and other stored commodities occurs as well. The characteristics of operating inputs, the analysis of operating loans, and the types of operating loans are considered in this section.

Characteristics of Operating Inputs

Fertilizer, chemicals, seed, irrigation water, feed, feeder and stocker livestock, veterinary services, labor services, machine fuel and repairs, and utilities are examples of annual operating inputs. These inputs have several unique characteristics that distinguish them from other productive assets. First, as their name implies, they are generally "used up" or converted into other forms during production. Hence, as a rule, they are not carried from one year to another. For example, feed is consumed by animals and converted into livestock production, while fertilizer is converted into crop products. In both cases, the use of the asset in production means that it cannot be recovered in its original form. Similar results apply to water in irrigated crop production and to fuel and labor services.

These conversion features rule out the possibility of rental or leasing for most operating inputs, as occurs with land, tractors, or even breeding livestock.

Moreover, these assets make less effective loan collateral because they do not generate easily reclaimable assets. Generally, the lender must rely on a security interest in other fixed assets as well as in the growing crops and livestock to provide the necessary collateral. Feeder livestock are an important and interesting exception to this generalization, as is a warehoused inventory such as grain.

Operating inputs also tend to have a high marginal value product (MVP) associated with the first few units of use. The payoff from the use of sufficient chemicals to control a serious disease, for example, is very great. Similarly, the marginal returns from hiring labor for harvesting crops are substantial if no other harvest labor is available. Up to some point, each operating input has a sufficiently high marginal value product that, in its absence, a farmer might justifiably pay a very high price for a limited supply. However, the marginal value product of annual operating inputs tends to fall sharply as their use increases. Moreover, capital tied up in excess quantities of operating inputs may generate low or even negative returns. Thus, the amount of an operating input a farmer should use is not significantly influenced by changes in interest rates.

The high payoff associated with annual operating inputs creates a high-priority claim on a farmer's financial resources. The lack of financing for operating inputs may terminate production and jeopardize the business. In contrast, borrowing to finance firm growth or to replace old equipment can likely be postponed without seriously jeopardizing business productivity or profitability, at least for a time.

These conditions indicate the farmer's urgent need to (1) secure sufficient financing to acquire the annual inputs for profitable production, (2) obtain appropriate financing terms, and (3) hold adequate reserves of credit or other financial assets to cope with unexpected needs for additional operating inputs — the need to replant a crop after flooding or a frost, for example. The more critical financing terms for operating inputs are loan maturities and repayment schedules, although collateral requirements and interest rates are also important.

In order for the operating loan to be self-liquidating, loan repayments should be scheduled to come from income generated by the sales of the products being financed. Thus, loans to finance the purchase of fertilizer in the spring of the year should be scheduled for repayment with funds arising from the sales of the products grown with that fertilizer. Such sales generally occur at harvest in the fall, although many producers may utilize storage or contract arrangements that modify the timing of crop sales and cash receipts. In these cases, the financing terms should be modified to reflect the producer's commodity marketing policies, assuming, of course, that these policies are acceptable to the lender.

Another characteristic of operating inputs arises from the seasonality of farm production, which influences the timing of flows of funds and short-term financing needs. Because the patterns of seasonality differ, the timing of acquiring operating inputs may also differ. Crop production is influenced greatly by climatological factors. Livestock production can also exhibit seasonality, although it is often reduced by farmers' efforts to achieve more continuous marketing. Moreover, both crops and livestock have rather lengthy production periods.

The high degree of divisibility provided by operating inputs is an advantage in acquiring, managing, and financing these assets. In contrast to the bulkiness or indivisibility of investments in land and most depreciable assets, operating inputs can be purchased and used in very small-sized units. However, quantity discounts are often available to large producers, and prepayments of operating expenses can sometimes lower income taxes.

Analysis of Operating Loans

The acquisition of operating inputs influences the firm's profitability, risk, and liquidity in much the same fashion as investments in durable assets. However, these influences all generally take place within one production period. Hence, the need for evaluation procedures that directly account for the time value of money is reduced. This does not mean that the time value of money is no longer relevant. Time is definitely important, as evidenced by crop or livestock producers who must acquire their operating inputs several months prior to the sale of their products. These producers must still consider the opportunity costs of funds committed to the acquisition of operating inputs. However, because the product sales occur within a year of the cash outflows for input acquisition, the opportunity costs are smaller and easier to account for. Moreover, if operating inputs are financed by borrowing, the money costs of borrowing are directly reflected in interest charges.

Because an operating loan generally is self-liquidating, when operating expenses are charged against farm sales on the income statement, the borrower, in effect, is setting aside the funds to repay the loan. Operating expenses are usually paid for with cash either as items are purchased or on a monthly billing process, often with cash obtained from an operating loan. If the operating loan is used only to pay for operating inputs, the loan will be self-liquidating provided: (1) the note does not mature before income from the sale of farm products is received and (2) farm sales exceed operating expenses. Thus, the source of repayment for an operating loan is from funds set aside when cash operating expenses are subtracted from cash revenues.

Types of Operating Loans

Three type of loans are used to finance the purchase of operating inputs: **the regular or standard operating loan, a nonrevolving line of credit,** and **a revolving line of credit.** The **standard operating loan** consists of *separate borrowing transactions and notes each time financing is needed.* In each transaction, the borrower and the lender agree on the loan purpose, loan size, interest rate, and repayment date. The borrower signs a note specifying these terms, and the lender provides the loan funds.

This process is cumbersome and inefficient when the borrower needs several operating loans over a short period of time. For example, the crop producer who must borrow money in four consecutive months in the spring would have at least four different notes to repay under the standard operating loan. This approach essentially implies that no cash flow budgeting was used to forecast borrowing needs and loan repayments.

Under a **nonrevolving line of credit,** *the lender agrees at the beginning of the planning period to supply the operating loan funds at the times and in the amounts indicated in a cash flow budget.* Generally, the budget is formulated on a monthly basis, although any number or any length of time periods may be used. The size of the line of credit is the total amount of borrowing projected for the coming year. However, the funds are advanced only as specified in the budget, and repayments occur when cash inflows exceed cash outflows. This approach minimizes the borrower's cost of using the borrowed funds but still assures their availability throughout the year. These concepts are clearly illustrated in the cash flow budget for Joseph and Julie Farmer outlined in Table 3.10. For each month in which a surplus cash position is projected, payment of interest and principal on the operating loan is expected.

The outstanding loan balance at any one time may not necessarily reach the total line of credit because of repayments that occur before all borrowing is completed. The loan balance is monitored each month for Joseph and Julie Farmer on line 33 of Table 3.10. The loan agreement or master note usually specifies the size of the credit line and the maximum loan balance at any one time. It also specifies that funds are available as budgeted in the cash flow projection, although some flexibility is provided to account for reasonable deviations in timing and size of individual transactions. Other loan terms and performance conditions may also be stipulated in the loan agreement.

The nonrevolving line of credit provides several advantages in liquidity management and financial control for both the borrower and the lender. These advantages are: (1) it indicates the uses of the borrowed funds and their sources of repayment, (2) it indicates the expected timing of borrowing and repayments based on the projected cash flows, and (3) the cash flow budget provides an instrument of financial analysis and control for both the borrower and the lender. The size and timing of actual transactions can be checked against the budget. Deviations can be calculated and explanations provided for those that appear extraordinary. The borrower can assess the quality of financial planning, while the lender can assess repayment performance. In this fashion the cash flow budget jointly serves the purposes of financial planning, analysis, and control.

A **revolving line of credit** provides additional freedom to the borrower in using the loan funds. *The lender and the borrower agree on a maximum line of credit for the next planning period* (e.g., three months, six months, one year) as determined from the results of a cash flow budget and from analysis of the firm's profitability, risk, and liquidity. The borrower can then borrow freely up to this limit, but not above it.

A revolving line of credit is especially useful when the borrower has recurring needs for and repayments of large amounts of financing in a short period of time. Cattle feedlots and broiler operations with large inventories of feed and animals that turn over rapidly are examples of businesses needing this kind of financing. The lender must be assured that the budgeted funds will be used frequently, effectively, and for the purposes requested. Generally this type of financing is limited to borrowers who provide accurate, detailed, and reliable cash flow budgets and who exhibit superior financial performance.

Interest on these three types of operating loans is generally charged at the contractual rate on the outstanding loan balance. Hence, interest obligations only occur for the time during which the funds are borrowed.

To illustrate the interest charge, suppose that a crop farmer borrows $100,000 on April 1 at a 9 percent annual rate with interest charged daily on the outstanding or remaining loan balance. The cash flow budget projects $75,000 from crop sales available for loan payment on October 1 and $50,000 from added crop sales available for repayment on November 1. The total interest obligation is found as follows:

$$I = P \times i \times T \tag{5.1}$$

where I is the interest paid, i is the interest rate, and T is the fraction of the year the loan (P) is outstanding.

October 1 $\quad I = \$100,000 \times 0.09 \times \left(\dfrac{183}{365}\right) = \$4,512$

November 1 $\quad I = \$25,000 \times 0.09 \times \left(\dfrac{31}{365}\right) = \$191$

$\qquad\qquad$ Total = \$4,703

The \$100,000 outstanding for 183 days from April 1 to October 1 incurs \$4,512 interest, which is payable along with the \$75,000 principal payment on October 1. The \$25,000 outstanding for 31 days in October incurs \$191 interest, which is paid on November 1 along with the final \$25,000 loan payment, making total interest and principal payments during the year of \$104,703.

Some lenders require that interest always be paid whenever principal payments are made, although most let interest on the outstanding balance accumulate throughout the year, with payment required when the operating loan is fully repaid. In the latter case, it is to the borrower's advantage to repay the loan principal first, then interest, rather than to pay interest along with each principal payment.

FINANCING DEPRECIABLE ASSETS

Financing depreciable assets in farm businesses primarily involves the purchase of machinery, equipment, and breeding livestock. These assets are used over several years, and their capital cost is charged against the business as depreciation over their years of economic life. They contribute significantly to the income-generating capacity and growth of the business. New technology and economies of scale are often involved in the growth process.

Compared to operating inputs, depreciable assets come in larger units, and most of them may be either purchased or leased. Moreover, unless replacement of worn-out assets is urgent, depreciable assets offer more flexibility in the timing of acquisition than operating inputs. Hence, their acquisition can generally be postponed in periods of liquidity crisis or high financing costs. Finally, most depreciable assets provide effective loan collateral. As such, they are very important to a firm's credit worthiness.

Following the principles of self-liquidating loans, a proper financing of a depreciable asset should synchronize the loan maturity and repayment schedule to the cash generated by the asset. In practice, it is difficult to precisely identify the timing and the magnitude of the contributions to profits made by a particular piece of machinery or other depreciable asset. However, when depreciation of the asset being financed is charged against revenue on the income statement, cash essentially is set aside for capital replacement. This cash provides an appropriate source for repaying the loan. In this way, loans made to finance the asset can be repaid at the same pace the asset is paying for itself in the business. The cash flows for earnings and debt servicing then are coordinated.

Depreciable assets with an economic life of five to seven years should be financed with loans of similar maturity. Loan maturities shorter than the depreciation schedule of the asset require a diversion of funds from working capital and thereby aggravate, rather than alleviate, the firm's liquidity problems.

Consider, for example, a farmer who acquires a $42,000 machine for use in the business over a seven-year period, after which the machine has no remaining value. Assuming uniform use over the seven years, which implies straight-line depreciation, the machine should pay for itself at an annual rate of $6,000 = $42,000 ÷ 7 years. Suppose the lender requires a 25 percent downpayment to provide an adequate safety margin of the farmer's equity capital in the investment. A $31,500 loan payable in seven equal annual payments of $4,500 (plus interest) would be required to make the loan self-liquidating, while still satisfying the lender's desired safety margin. If, however, the lender provides only a three-year loan, the required principal payments would be $10,500 each year. The added $4,500 over the $6,000 attributed to depreciation would have to come from working capital or net income, thereby reducing the funds available for family living.

Many industries in which depreciable assets dominate the firms' assets, such as airlines, railroads, and trucking companies, design the financing of their depreciable assets so that the loans can be repaid with funds set aside for depreciation. Some non – real estate lenders, however, prefer to finance farmers' capital acquisitions solely through short-term loans, carrying unpaid portions of the debt over to following years until the loans are fully repaid. For the lender, this practice provides greater financial control. However, this practice leaves the borrower in a nonliquid position, fully dependent on the lender for continued financing, and it provides little capacity for the borrower to adjust his or her financial program to changing business conditions. The general principle in financing depreciable assets is **to match the term of the loan with the anticipated length of the assets used in the business.** A shorter term will save interest payments but may cause liquidity problems because of the larger principal payments.

FINANCING FIXED ASSETS

The primary fixed asset in agriculture is land, although long-lived but depreciable assets such as buildings, fences, water-handling facilities, and other types of real estate improvements are also included. For most farmers, land is the dominant capital resource requirement, although land can be leased as well as purchased.

Financing programs for land are subject to the characteristic that farm land is not used up in the production process. In principle, then, land could be financed with permanent debt, requiring no repayments of loan principal until the property is sold. In practice, of course, farm real estate lenders do require loan repayments, although many farm real estate loans are for long (20 to 30 years) maturities. Even so, as will be seen in Chapter 12, if the economic return from the land is assumed to be the cash rent equivalent (less property taxes and maintenance), most farm real estate loans have not been self-liquidating. That is, the annual rent has not been sufficient to service the loan. This condition arises because an important consideration in determining the market price of farm land is the long-term growth in its earnings due to inflation and other factors. However, the appreciation in land value resulting from the growth in earnings does not provide annual income over the years to help service the debt (see Chapter 12 for further treatment of the financial feasibility of land investments).

The primary source of income for repaying a farm real estate debt is from retained earnings (profits after taxes and after family withdrawals). In years when retained earnings are insufficient to make the loan payments, funds must come from reduced family income, the farm's working capital, or the sale of noncurrent assets. Hence, the decision to purchase farm land often sets in motion future demands for cash that place the firm in a tight liquidity position.

To help alleviate this problem, many borrowers periodically refinance real estate loans in order to restructure or stretch out repayment obligations. This procedure utilizes the high collateral quality of farm real estate, as long as land values are increasing over time.

REPAYMENT PLANS

Financing programs for noncurrent assets are generally based on installment loans in which the loans are **amortized** with regularly scheduled payments. Two alternative payment schedules are used to amortize farm loans: (1) **constant payment,** *in which the total payment in each period remains constant over the term of the loan,* with varying proportions allocated to interest and principal as each payment occurs; and (2) **constant payment on principal,** *in which an equal payment on principal plus interest occurs each period.* In both cases, interest is calculated on the remaining balance.

Figure 5.1 compares the pattern of annual payments on a $200,000 loan

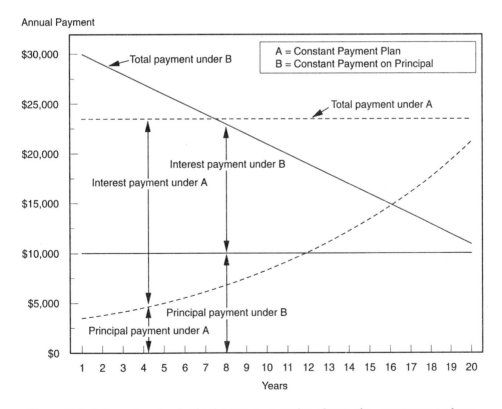

Figure 5.1. Interest and principal payments under alternative repayment plans.

amortized over 20 years for these two methods. The total payments (interest plus principal) are lower for the constant payment method during the early years of the loan. Eventually, however, the payments become lower for the constant payment on principal method. Curvilinear and linear relationships are indicated in Figure 5.1. If payments occur at discrete intervals (annually, monthly), the relationships would have discrete steps.[1]

Although the length of time required to repay the loan is the same under both methods, the outstanding loan balance for any year is greater for the constant payment method. Thus, the total interest paid during the loan is greater for the constant payment method. This method may seem less attractive to some borrowers. However, the higher payment obligations during the early years of the constant payment on principal method reduces the amount of cash flow available for other uses. In particular, it reduces the cash available for servicing non – real estate loans.

A similar situation arises for shorter versus longer maturities on long-term loans. The total interest costs will be less for the shorter-term loans. However, the cash flow required each year to repay principal and interest will be higher, thus adversely affecting the investor's liquidity position. Hence, an incentive exists for longer-term real estate loans.

Another common type of real estate loan is a **balloon payment loan. Balloon payment loans** are *relatively short-term loans (e.g., five years). At the end of the period, the entire unpaid balance of the loan is due.* The principal balance of the loan must be paid in full, or new loan terms must be established. The initial payments are usually based on a longer amortization period (e.g., 10 to 30 years) under the assumption that the loan will be paid off, renewed, or refinanced at maturity. If interest rates fall and credit conditions improve, a borrower could negotiate more favorable loan terms at renewal. However, if interest rates rise or credit tightens, the loan terms may become less favorable. In addition, the borrower's risk is considerably higher because the lender could decide not to renew the loan at maturity.

Table 5.1 illustrates a balloon payment loan. The first four payments are identical to a 20-year fixed payment loan. After the fourth payment, an outstanding balance of $183,794 remains on the loan. This amount must be refinanced at the terms prevailing in year 5 or be paid in full.

A potential disadvantage of all types of farm mortgage loans is the fixed commitment represented by the amortization payment. If possible, this payment should be scheduled within each year to coincide with the seasonal

---

[1]The methods of calculating amortization payments are developed in Chapter 9, and the effects of these repayment schedules on the costs of debt capital are considered in Chapter 14.

Table 5.1. Payment Pattern of a $200,000 Balloon Payment Loan

Year	Beginning Balance	Principal Payment	Interest Payment	Total Payment[1]	Ending Balance
1	$200,000	$ 3,492	$20,000	$ 23,492	$196,508
2	196,508	3,841	19,651	23,492	192,667
3	192,667	4,225	19,267	23,492	188,442
4	188,442	4,648	18,844	23,492	183,794
5	183,794	183,794	18,379	202,173	0

[1]Total payment based on a 20-year amortization, 10% interest rate.

patterns of the borrower's cash flow. Scheduling the payments when cash is available from crop or livestock sales is an example. On occasion, however, payments on all term debt are met from the firm's operating line of credit, which is then repaid later in the year when the crops or livestock are sold.

SOURCES FOR LOAN REPAYMENT: AN EXAMPLE

In the preceding sections we identified the following sources of cash for repaying different types of loans: (1) **from cash set aside to pay expenses** to repay an annual operating line of credit, (2) **from cash set aside for depreciation** to repay a term loan to finance purchase of depreciable assets, and (3) **from retained earnings** (profits less income taxes and family withdrawals) to repay long-term real estate loans.

To illustrate the identification of these cash flows and their use in specific loan payments, consider the simplified case described as follows. Farmer Jones borrowed the following from his lender at the beginning of the year:

1. $437,000 under a nonrevolving operating line of credit to be used as follows:

   a. Pay operating expenses totaling $415,000
   Purchase feeder cattle $260,000
   Purchase feed....................................... 75,000
   Purchase fertilizer and chemicals 25,000
   Pay other operating expenses (other than interest) 55,000
   415,000

b. For family living expenses 22,000

Total advances under the operating line $437,000

2. $40,000 to purchase new equipment. This equipment will be depreciated over five years using straight-line depreciation. Thus, annual depreciation charge-offs are increased by $8,000 per year. Accordingly, the loan principal payments also are set at $8,000 per year. (He already had an installment loan outstanding with annual payments of $7,500 over the next four years.) Total depreciation charges are projected at $20,000 per year.

Farmer Jones also has an existing farm mortgage loan that requires an annual payment of $12,000 on the principal.

The sole source of income for Farmer Jones is from the sale of fat cattle for market, scheduled for December. His pro forma income statement is shown in Table 5.2.

Table 5.2. Pro Forma Income Statement Summary for Farmer Jones

Gross Revenue (sale of fed steers)	$530,000	
Less livestock purchases	− 260,000	
Less feed purchases	− 75,000	
Value of Farm Production		$195,000
Farm Expenses		
Fertilizer, chemicals, and medicines	25,000	
Other noninterest operating expenses	55,000	
Depreciation	20,000	
Total farm operating expenses	100,000	
Total interest expense	40,000	
Total Farm Expenses		$140,000
Net farm income from operations		$ 55,000
Income taxes payable		− $ 9,000
Net Farm Income After Taxes		$ 46,000
Withdrawals for family living		$ 22,000
Change in retained earnings		$ 24,000

Assuming that Farmer Jones performs as projected, all of the loans are self-liquidating, as follows:

- The operating loan of $437,000 is repaid with (1) cash set aside when noninterest farm operating expenses of $415,000 are subtracted from revenue to get farm operating income, plus (2) the $22,000 subtracted for family withdrawals from profits after taxes. Since the actual expenses for feeder cattle, feed, etc., as well as the expenses for family living, are made during the year with funds supplied by the lender under the operating line, the monies set aside from revenue for operating expenses *and* family withdrawals are available to repay the operating loan.

  As long as gross sales of cattle exceed $477,000 (the sum of livestock and feed purchases, total expenses except depreciation, plus family living), Farmer Jones can repay the operating line plus pay the estimated total interest cost (on all loans) of $40,000. A margin of safety is evident, since sales are conservatively projected at $530,000, although part of this safety margin must be used to service the term loans.

- The $7,500 principal payment due on the existing installment loan plus the $8,000 principal payment on the new farm machinery loan are met out of the $20,000 set aside to service depreciation. The cash balance of $4,500 can serve as a margin of safety for long-term loans, or it could be used as a downpayment on new capital purchases, if needed.

- The $12,000 principal payment on his farm mortgage note is paid from retained earnings. Fortunately, with the projected change in retained earnings of $24,000, a safety margin of $12,000 occurs here as well.

The overall assessment of the repayability of Farmer Jones' three separate loans is favorable. Although the basic source of revenue is from the sale of market steers, the funds needed to repay each type of loan are found in different places on the pro forma income statement. Notice that only principal payments on the farm real estate loan are repaid out of profits after taxes. All other debt servicing (including interest) is made with "before-tax" dollars, as operating expenses, livestock and feed purchases, or depreciation.

ANALYZING DEBT STRUCTURE

Determining the capacity of the business to service current and noncurrent debt, as explained in the previous section, provides a framework for assessing the firm's debt structure. The allocation of total debt to current (operating line) and noncurrent debt should be such that:

1. The operating line can be repaid from funds set aside to repay cash operating expenses (plus family living expenses if funds are borrowed for this purpose).

2. The non – real estate loans for depreciable assets can be repaid on schedule from funds set aside to service depreciation.

3. The real estate debt can be repaid from retained earnings.

Failure to meet one or more of the preceding conditions indicates that either the debt burden has not been appropriately allocated or that the total debt burden is too great.

In many cases it is to the benefit of both the borrower and the lender to restructure the borrower's debt. For example, suppose a borrower has not been able to pay off his or her operating line of credit for a number of years. The liquidity of the farm could usually be enhanced by restructuring the debt into a longer-term loan. However, both the lender and the borrower should analyze the financial feasibility and debt repayment capacity of the restructured plan.

ANALYZING FINANCIAL FEASIBILITY

A plan is considered financially feasible if it satisfies the financing terms and performance criteria that are agreed upon by both the borrower and the lender. The major performance criteria are profitability, risk, and liquidity; however, the borrower and the lender may each have a different view of these criteria. Borrowers are concerned with the profitability, risk, and liquidity, of the farm business primarily for the sake of income, survival, and financial growth. Lenders are more concerned with loan repayment capacity (a liquidity characteristic) and loan security (a risk characteristic) for their own benefit. Thus, loan risk analysis has two components: **repayment risk** and **collateral risk.** These two components are closely related because all loan repayment ultimately comes from the sale of the firm's assets, especially current assets; however, they are separated here for consistency with the practice of credit analysis.

Loan Repayment Risk

Repayment capacity refers to *the firm's ability to generate sufficient cash from product sales to repay the loan plus interest according to the contracted financing terms.* An assessment of repayment risks primarily is done with the pro forma statements including the cash flow budget. Although past performance can be an indicator of future performance, the future economic conditions of the industry can significantly impact the future performance of any firm. Thus, future projections of the financial performance of the firm are the best indicators of loan performance.

Evaluating repayment capacity and loan performance necessitates evaluating the firm under alternative economic scenarios. Considerable uncertainty exists about the impacts of future economic and weather conditions on farm performance. The lender and the borrower must evaluate the conditions that permit loan repayment as well as adverse conditions that hamper loan repayments. Understanding the sensitivity of an operation to adverse conditions is essential in determining financial feasibility and repayment capacity. The lender may require a borrower to reduce the impact of the adverse conditions through insurance, hedging, or specific cropping procedures, for example.

The cash flow budget is the initial basis of the loan repayment analysis. However, information from the pro forma income statement and balance sheets provides additional details regarding the profitability, liquidity, and solvency of the farm. These statements should be used as a system to determine loan repayment. Using only one of the statements could result in misinterpretation of an individual situation. For example, a cash flow budget could indicate adequate loan repayment along with an overall reduction of operating debt. However, crop and / or livestock inventories may have been liquidated, or expenses may have been postponed to provide the cash required to make the payments. Thus, the longer-term viability of the farm was jeopardized in order to meet the current loan payments. The pro forma balance sheet and pro forma income statements in addition to the cash flow budget should be used to adequately measure loan repayment ability and projected business performance.

The operating loan balance row of the cash flow budget (line 33, Table 3.10) can be used in assessing and monitoring the repayment ability of the operating line of credit. This will show if the line of credit will be entirely paid off at some time during the year or if any loan carryovers from the preceding year are significantly reduced. If either or both of these conditions are projected, then the operating line of credit is considered to have achieved repayability, provided events occur as projected in the cash flow budget.

Regarding the repayment capacity of term loans, Table 5.3 shows components of the capital replacement and term debt repayment margin, which was introduced and defined in Chapter 4. Items 1 through 10 are selected from the income statement, while the remaining items are taken from the cash flow budget. The income statement measures incorporate accrual adjustments such as changes in inventory and accrual expense adjustments. Items 1 through 6 are included to allow for sensitivity analysis of the debt repayment margin to changes in revenues and expenses.

The analysis summarized in Table 5.3 indicates that the projected capital replacement and term debt repayment margin (CRDRM) of the farm for 1995

Table 5.3. Capital Replacement and Term Debt Repayment Margin Components for Joseph and Julie Farmer

	Components	1995 (Pro Forma)
1.	Crops and feed revenue	$221,500
2.	Livestock and livestock product revenue	+ 136,350
3.	Other revenue and adjustments	+ (89,940)
4.	Value of farm production	267,910
5.	Operating expenses	− 178,300
6.	Interest expense	− 24,760
7.	Net farm income from operations	64,850
8.	+ Total nonfarm income	17,320
9.	+ Depreciation expense	25,800
10.	− Income tax expense	13,021
11.	− Withdrawals for family living	28,050
12.	**= Capital Replacement and Term Debt Repayment Capacity**	**66,899**
13.	− Principal payments on unpaid operating debts	0
14.	− Principal payments on current portions of debt and leases	17,340
15.	**= Capital Replacement and Term Debt Repayment Margin (CRDRM)**	**$ 49,559**
16.	Sensitivity Analysis	
16.1	Total nonfarm income as a percent of CRDRM	35%
16.2	Percent crop revenue can decline and still maintain positive CRDRM	22%
16.3	Percent livestock revenue can decline and still maintain positive CRDRM	36%
16.4	Percent farm operating expenses can increase and still maintain positive CRDRM	28%
16.5	Percent interest expense can increase and still maintain positive CRDRM	200%

is $49,559 (line 15). This measure reflects the risk or safety margin that could be used for capital replacement or additional debt service. The positive $49,559 figure indicates that loan repayability is assured if events occur as projected.

Further analyses shown in lines 16.1 to 16.5 of Table 5.3 indicate, however, that Joseph and Julie Farmer's term loan repayability is vulnerable to adverse events. A crop revenue decline of 22 percent (line 16.2) or a livestock revenue decrease of 36 percent (line 16.3) would eliminate the entire risk margin. Furthermore, an increase in operating expenses of 28 percent (line 16.4) would negate any risk margin. Combinations of the adverse situations would also reduce the risk margin. A negative risk margin would indicate that loan repayment would have to come from alternative sources such as savings accounts, marketable securities, and / or liquidations of other assets.

These figures imply a relatively high degree of vulnerability to fluctuations in various cost and return items. Thus, the relative variabilities of these items due to random or unanticipated events are important to consider. So are the risk attitudes of the borrower and the lender. The greater the risk aversion, the less acceptable are highly sensitive safety margins in cash flows, and vice versa. Reducing leverage and taking other actions to widen these safety margins would have beneficial effects on the firm's repayment risks.

Loan Collateral Risk

Collateral risk refers to *the relationship between the values of assets pledged as security for a loan and the outstanding loan balance.* When a legal security interest in these assets is created, the lender has the right to claim them in the event of loan default by the borrower. Lenders generally require that the value of the collateral exceed the loan balance by an appropriate safety margin, and they periodically inspect the collateral position to assure that this relationship holds. However, possible fluctuations in the value of the collateral serve as the source of collateral risk, since a sharp drop in asset values could drive the safety margins to zero or below.

Operating loans are usually secured by current assets and sometimes longer-term assets as well. Examples of collateral for operating loans are farm supplies, crop and livestock inventories, growing crops, government payments, and deposit accounts. A blanket filing may be used on a line-of-credit financing so that the security agreement applies to essentially all of the current assets, some of the non – real estate assets, and, if stipulated, to property acquired in the future as well. Noncurrent loans are typically secured by the assets being purchased.

The assessment of collateral risk on operating loans should evaluate the effects of seasonality, length of the production process, and marketing policies on the lender's (and the firm's) exposure to collateral risk. As shown in the preceding discussion of repayment risk, these factors can be very important. The monthly cash flow budget can be used to evaluate the seasonality of the operating loan and the cash balances. These balances should then be compared with the projected balances of the pledged collateral to evaluate periods when the collateral risk is higher.

For example, the peak projected operating loan balance for Joseph and Julie Farmer occurs in February ($57,917). However, the current market values of crop and market livestock inventories exceed $140,000; hence, the short-term lender's exposure to collateral risk is not significant. The remaining balance on December 31, 1994, of Joseph and Julie Farmer's non – real estate loan used to purchase a combine is $51,970 ($12,060 current portion and 39,910 deferred portion). The current market value of the combine (pledged collateral) is $89,000; thus, the loan-to-collateral value of the combine loan is 0.58 ($51,970 ÷ $89,000), leaving an adequate safety margin for the non – real estate lender. The remaining balance on December 31, 1994, of Joseph and Julie Farmer's real estate loan is $172,960 ($5,280 current portion and $167,680 deferred portion). This loan was secured by 120 acres with a current market value on December 31, 1994, of $1,800 per acre, a decrease of $400 per acre since the land was purchased. Hence, the current market value of the land secured by the real estate loan is $216,000, resulting in a current loan-to-collateral value of 0.80.

An additional decrease in land prices of 20 percent would reduce the collateral safety margin on this loan to zero. The lender and the borrower(s) (the Farmers) may wish to review this position and evaluate the alternatives the lender may require to improve the collateral position. As demonstrated with Joseph and Julie Farmer, the collateral risk needs to be determined for each loan, and seasonality, along with the projected balances of the pledged collateral, should also be considered.

Feasibility Assessment

Final decisions about overall financial feasibility depend on the lender and the borrower. With Joseph and Julie Farmer's recent favorable trends, their present program seems feasible. Clearly, the diversity provided by the mix of livestock and crops helps to moderate the seasonal swings in financial risk and liquidity for each enterprise. Except for 1993, Joseph and Julie Farmer have maintained adequate profitability levels. The liquidity and solvency positions of Joseph and Julie are strong and projected to improve. Furthermore, cash flow projections indicate the ability to meet all debt obligations with an adequate risk margin.

In general, the impact of seasonality on financial risk is greater for crop producers than for livestock producers. This greater impact occurs because crop production is more seasonal, the effects of weather risks on crops are greater, and growing livestock generally are more marketable than growing crops. Of course, the availability of crop insurance and other risk management practices are offsetting factors. Nonetheless, the seasonality of financial risks and liquidity should be considered for all types of agricultural production units.

If the financial feasibility analysis indicates that the firm's credit worthiness is sufficient to finance the proposed operation and still assure adequate coverage of risk and liquidity, then the plan is feasible. Besides the monthly cash flow analysis, pro forma statements should be conducted for the next three to five years to assure that term debt can be adequately serviced.

If, on the other hand, credit is not judged to be sufficient to finance the projected demands for cash, the plan is deemed financially infeasible. Various alternatives to achieve feasibility then can be considered. These alternatives include one or more of the following, each of which influences profitability, liquidity, and risk to some degree.

1. *Changing the production plans:* changing enterprise combinations, substituting labor for capital, using less fertilizer, etc.

2. *Changing the marketing plans:* holding less inventory, using forward contracting, and gaining price premiums on high-quality products or large- volume transactions.

3. *Improving cost control:* conserving fuel, more fully utilizing labor and machinery, substituting used for new equipment, and gaining price discounts on volume purchases.

4. *Improving consumption control:* producing more home-grown food, postponing household expenditures, and conserving the family living budget.

5. *Refinancing:* refinancing existing term debt or an operating debt carryover for more orderly payoff and improved liquidity.

6. *Instituting other financing sources and methods:* lower downpay-ments, longer loan maturities, lower interest rates, government loan and guarantee programs.

7. *Considering leasing instead of purchasing new capital items.*

8. *Postponing expenditures:* postponing new investments — including asset replacements — to future years when liquidity and risk are improved.

9. *Introducing new equity capital:* bringing new equity capital into the business to finance expansion or to retire existing debt.

10. *Seeking off-farm income:* the farm operator and/or family members trying to secure employment in nearby communities.

11. *Selling highly liquid financial assets and inventories* to generate cash for debt repayment.

12. *Downsizing the scale of operations* through sales of capital assets including machinery, breeding livestock, and land — either with or without a leaseback of some of the assets involved.

Summary

This chapter has shown how information from financial statements can be used to design financing programs and to assess financial feasibility. Financing programs for current, depreciable, and noncurrent assets are reviewed. Specific highlights of the chapter are as follows:

- An **optimal loan program** would include *a maturity structure or repayment pattern that matches the payoff or earnings of the assets being financed.*

- A **self-liquidating loan** is *a loan that finances a project that has returns to cover the loan principal plus interest, and payments are scheduled such that they can be made with cash generated from the project.*

- Cash flow coordination is simplified when all financing needs are pro-vided by a single lender. The lender typically has three options to choose from in offering short-term credit to the farmer: **standard operating loan, nonrevolving credit,** and **revolving credit.**

- The **cash flow budget** is the primary determinant of the type of financing a farmer receives. Matching loan and project maturities and cash inflows and repayment dates reduces the risk for both the borrower and the lender.

- A **nonrevolving line of credit** *allows the borrower to have access to a prede-termined maximum amount of credit*, but funds are received only when they are needed, not all at once. Payments are made on the credit line only when cash inflows are expected to exceed cash outflows.

- A **revolving line of credit** *allows the borrower freedom to borrow up to a maximum amount within a specified time period* based upon detailed and reliable cash flow budgets.

- Installment loans on noncurrent assets can have one of two types of payment schedules: (1) **constant payment** — *where all payments are equal throughout the loan period*, and (2) **constant payment on principal** — *where payments on principal are equal, but payments on interest decrease as the remaining balance decreases.*

- **Loan repayment risk** refers to the *potential failure of the borrower to generate sufficient cash from product sales to repay the loan plus interest.*

- **Loan collateral risk** refers to the *risk lenders take so that the value of the collateral will not decrease below the outstanding loan balance.*

Topics for Discussion

1. Explain the concept of **a self-liquidating loan.** Indicate how the time patterns of earnings for current and noncurrent assets are taken into account in formulating self-liquidating loans.

2. Contrast the various types of operating loans: **standard, nonrevolving,** and **revolving.** Which is best suited for a cash grain farm? A dairy farm? A large cattle feedlot? Why?

3. Calculate the total interest charge on a $200,000 loan on March 1 at a 10 percent annual interest rate with repayments projected as $100,000 on October 1, $50,000 on November 1, and $50,000 on December 1.

4. Design an appropriate financing program for a farm to purchase an $80,000 tractor and related equipment. The machine will be used for six years, and the lender requires the farmer to make a 25 percent downpayment as an equity margin.

5. Explain the role of the cash flow budget in financial feasibility analysis.

6. Explain how seasonality in cash flows in a farming operation influences financial risk and liquidity.

7. If the results of a cash flow budget indicate excessive financial risk, what are some changes that could be considered to make the plans financially feasible?

8. Explain the liquidity of a farm business as indicated by (a) its balance sheet, (b) its income statement, and (c) its cash flow budget. Why are these important?

9. Distinguish between **loan repayment risk** and **loan collateral risk** in evaluating financial feasibility.

References

1. Baker, C. B., "Credit in the Production Organization of the Firm," *American Journal of Agricultural Economics*, 50(1968):507 – 520.

2. Fraser, L., *Understanding Financial Statements*, Reston Publishing Company, Inc., Reston, Virginia, 1985.

3. Frey, T. L., and D. A. Klinefelter, *Coordinated Financial Statements for Agriculture*, 3rd ed., Century Communications, Inc., Skokie, Illinois, 1989.

4. Frey, T. L., and R. Behrens, *Lending to Agricultural Enterprises*, American Banking Institute, New York, 1982.

5. Hanson, G. D., and J. L. Thompson, "A Simulation Study of Maximum Feasible Farm Debt Burdens by Farm Types," *American Journal of Agricultural Economics*, 62(1980):727 – 733.

6. Lee, W. F., et al., *Agricultural Finance*, 8th ed., Iowa State University Press, Ames, 1988.

7. Van Horne, J. C., *Financial Management and Policy*, 8th ed., Prentice-Hall, Inc., Englewood Cliffs, New Jersey, 1989.

8. Weston, J. F., and E. F. Brigham, *Essentials of Managerial Finance*, 10th ed., Dryden Press, Hinsdale, Illinois, 1993.

SECTION THREE

Capital Structure, Liquidity, and Risk Management

6

Capital Structure, Leverage, and Financial Risk

In previous chapters, we assumed that a firm's desired mix of debt and equity capital was already established. In financial management, however, decisions about leverage are of major importance and have profound effects on expected profitability, risk, and liquidity. In this chapter, we treat financial leverage as a decision variable in financial planning and develop concepts and measures for evaluating how leverage influences business performance.

The following section reviews some of the key theoretical relationships between financial leverage and the firm's costs of debt and equity capital. Then, conceptual models are developed that relate a firm's profitability and risk to leverage and other business characteristics. Numerical examples demonstrate that under proper conditions higher leverage can accelerate the growth rate of equity capital (a profitability measure), but it can also increase the risk of loss of equity. This risk – return trade-off means that the optimal level of leverage will differ between farm businesses, depending on the decision makers' risk – return attitudes and expectations about returns and costs. The final sections of the chapter consider the relationships between leverage and other performance factors.

CAPITAL COSTS AND LEVERAGE

Costs of capital are the costs a firm incurs for its financial capital. **Financial capital,** in turn, refers to the debt and equity claims making up the liabilities side of the firm's balance sheet. The combination of debt and equity reflects the firm's financial leverage or capital structure. Together, debt and equity provide the means for financing the firm's total assets.

Both debt and equity command payments — interest for debt and profit shares for equity — that constitute costs to the firm. Thus, one conceptual approach to specifying optimal leverage is to seek the firm's least-cost combination of debt and equity capital. Once the minimum cost leverage is found, the firm continues to finance its assets with that combination of debt and equity.

Income tax obligations are important in financial structure analysis and must be considered at the outset. The major point is that the costs of both debt and equity capital are expressed on an after-tax basis. As indicated in Chapter 3, the returns to equity and thus the costs of equity capital are determined after payments of income tax obligations have occurred. Thus, these equity measures are already expressed on an after-tax basis. In contrast, the interest costs of debt capital must be converted to an after-tax basis, since an interest obligation is a tax-deductible business expense. If, for example, the firm's cost of debt capital is i_d, and its income tax rate is t, then the after-tax cost of debt capital is $i_d(1 - t)$.[1]

In the remainder of this section, we will consider two conceptual approaches developed in the finance literature to explain the capital structure of businesses. These conceptual approaches are called the **trade-off** or **equilibrium theory** approach and the **pecking order theory** (Myers). Both approaches were developed in corporate finance, but the insight they offer is applicable to agricultural finance as well. At this point in time, one approach is not necessarily better or more valid than the other. Financial economists have vigorously debated capital structure theory, and more study is needed. Moreover, as shown later in this chapter, a third approach applicable to the non-corporate proprietary firm is based on the risk – return trade-off of financial leverage and the risk attitude of the decision maker.

---

[1]To illustrate, suppose that the interest expense for a farm business is $12,000 on an average outstanding loan balance of $100,000. The average cost of debt then is $i_d = 0.12$ (12,000 ÷ 100,000). If the farmer's tax rate is 30 percent, i.e., t = 0.30, then the tax savings generated by the interest expense is 30 percent of $12,000, or $3,600, leaving $8,400 as the after-tax expense. Alternatively, one can say that the after-tax cost of debt capital is $i_d(1 - t) = (0.12)(1 - 0.3) = 0.084$, or 8.4 percent, which is the same as $8,400 ÷ $100,000 = 0.084, or 8.4 percent.

Trade-off or Equilibrium Theory

The original equilibrium theory proposed by Modigliani and Miller in 1958 set forth the argument that capital structure did not matter to the wealth maximization of a corporate firm because any changes in risk at the firm level, caused by changes in the firm's financial leverage, could be offset by individual investors through changes in the composition of their own portfolios. However, this line of reasoning ignored the tax deductibility of interest payments on debt, administrative and other agency costs incurred by lenders, and costs of potential financial distress and bankruptcy associated with higher financial leverage.

In response, the finance literature has been full of studies exploring the capital structure implications of including these factors. Tax deductibility of interest favors the use of debt, while administrative, agency, and financial distress costs constrain the use of debt. Thus, a trade-off occurs between factors that favor and those that constrain the use of debt. It is generally believed, however, that as leverage increases, the increases in administrative, agency, and distress costs eventually outweigh the cost-reducing tax effects, thus leading to an equilibrium or value-maximizing combination of debt and equity capital.

To evaluate the relationships between capital costs and leverage, we begin with the following propositions about the costs of equity and debt. First, the cost of debt should be less than the cost of equity because of the equity holder's higher financial risk. Second, we accept the traditional view in finance theory that the costs of both debt and equity will eventually increase as leverage increases. Equity costs increase because of greater financial risk associated with higher financial leverage. Debt costs increase because of the greater likelihood of repayment problems, reduced liquidity, and other lending risks associated with higher leverage. With these propositions we can determine how the firm's average cost of capital responds to changing financial leverage.

These relationships between capital costs and leverage are illustrated in Figure 6.1. Suppose, for example, that the cost of equity capital at zero leverage ($D \div E = 0$) is 10 percent and that as leverage increases, the cost of equity also increases, as shown in line i_e in Figure 6.1. Let the cost of after-tax debt at zero leverage be 8 percent, and let it also increase with leverage as shown in line i_d. Line i_a indicates the weighted average cost of debt and equity as leverage increases.

At zero leverage, where no debt is used, the average cost equals the cost of equity. As leverage increases, the substitution of lower-cost debt for higher-

Cost of Capital

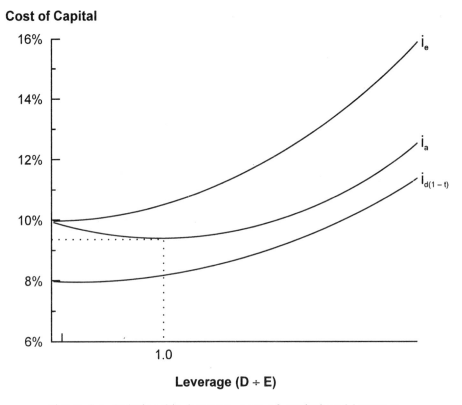

Figure 6.1. Relationship between cost of capital and leverage.

cost equity lowers the average cost until higher costs of both debt and equity cause the weighted average to increase.

The derivation of a weighted average cost of debt and equity is illustrated in Table 6.1, based on seven leverage positions and using the weighted average cost of capital formula. In Chapter 4, the focus was on partitioning the returns on farm assets into returns on farm equity and costs of farm debt based on measured returns to assets, debt, and equity and on the weights of debt and equity relative to total assets. The equation was expressed as:

$$ROFA = (COFD)(D \div A) + (ROFE)(E \div A)$$

Here, we relabel the weighted average equation in order to achieve consistency with the cost concepts identified above and introduce the after-tax specification for debt. The relabeled equation is:

$$i_a = (i_d)(1 - t)(D \div A) + (i_e)(E \div A) \tag{6.1}$$

where $i_d (1 - t)$ is the after-tax cost of debt capital, i_e is the cost of equity capital, and i_a is the weighted average cost of debt and equity, using $D \div A$ and $E \div A$ as the weights. This approach assumes that the costs of debt and equity are known values that serve as the basis for establishing leverage policies.

Table 6.1. Weighted Average Cost of Debt and Equity Capital for Alternative Leverage Positions

Leverage (D ÷ E)	Debt Capital		Equity Capital		Weighted Average (i_a)
	Cost [i_d(1 − t)]	Weight (D ÷ A)	Cost (i_e)	Weight (E ÷ A)	
0.00	0.08	0.00	0.10	1.00	0.100
0.50	0.08	0.33	0.10	0.67	0.093
1.00	0.08	0.50	0.10	0.50	0.090
1.50	0.10	0.60	0.12	0.40	0.108
2.00	0.12	0.67	0.14	0.33	0.127
2.50	0.14	0.71	0.16	0.29	0.146
3.50	0.16	0.75	0.21	0.25	0.173

This equation is used to estimate the weighted average costs of capital for the seven leverage positions in Table 6.1. For $D \div E = 1.0$, for example, the weights are $D \div A = 0.5$ and $E \div A = 0.5$, and the costs of after-tax debt and equity are $i_d = 0.08$ and $i_e = 0.10$. The average cost of capital is then:

$$i_a = (0.08)(0.5) + (0.10)(0.5)$$
$$= 0.09, \text{ or } 9\%$$

Average costs for other levels of leverage are found in a similar fashion. Through a debt-to-equity ratio of 1.0, the costs of equity and debt remain constant, thereby lowering the average cost as leverage increases. For leverage positions of 1.5 and greater, the average cost increases because of higher costs of both debt and equity capital. These leverage positions represent seven of the points on the average cost of capital line, i_a, in Figure 6.1.

The optimal combination of debt and equity occurs at the minimum average cost of capital — a debt-to-equity ratio of about 1.0 in Figure 6.1. In concept, the firm should finance its growth in assets by using the least-cost combination of debt and equity. Hence, each new dollar of assets should be financed with 50 cents of debt and 50 cents of equity.

The idea of a minimum average cost of capital is appealing and instructive. It explicitly identifies the factors associated with leverage decisions and expresses the risks associated with higher leverage through the respective costs of debt and equity capital. In practice, however, the applicability of the trade-off model is hampered by several factors. First, costs of debt and equity capital may be difficult to measure, especially when leverage is changing, and for smaller firms or non-corporate firms that lack direct access to equity capital markets.[2] The smaller-scale firms in agriculture are mostly organized as **farm proprietorships, partnerships,** and **family corporations.** Their main source of equity capital is retained earnings whose costs are difficult to measure, whose amounts are subject to high uncertainty, and whose rates of growth may not be fast enough to finance large investments. As a result, agriculture has relied heavily on borrowing and leasing, as well as on retained earnings and unrealized capital gains on farm assets, to finance its growth.

Second, the costs of adjusting in response to changes in the costs of debt and / or equity may cause lags in the adjustments of capital structures over time. Third, the weighted average cost curve may be relatively flat over a wide range, so that relatively large differences in capital structure could occur without significantly affecting a firm's value. Thus, the trade-off approach has not been completely successful in explaining differences between firms and industries in their actual capital structures. Alternative theories that may have greater explanatory power have been sought.

Pecking Order Theory

The pecking order theory sets forth the view that the capital structure of a business reflects a preferred hierarchy or "pecking order" to various sources of financial capital. This theory is attributed to Myers (1984) and to Myers and Majlauf who formulated the underlying concepts based on earlier observations by Donaldson (1961) that firms exhibit a clear ordering of their financing sources. Such an ordering reflects a primary preference for retained earnings, then debt, and finally reliance on external equity.

To be more specific, and incorporating some of the financing characteristics of agricultural firms, the pecking order theory suggests the following:

- Firms prefer to rely on retained earnings as a source of equity capital to finance assets and prospective investments.

---

[2]Procedures for measuring the costs of debt and equity capital are considered in Chapter 14.

- Firms set their dividend or withdrawal policies in response to anticipated investment opportunities over the long run so that capital expenditures under normal conditions can be met by retained earnings plus depreciation. However, dividends and other withdrawals are sticky and adjust primarily in an upward direction over time.

- If retained earnings are insufficient to meet investment needs, especially when earnings and investment opportunities are unpredictable, the firm first draws down its holdings of cash and financial assets and then turns to external sources of funds.

- When external sources of funds are contemplated, the firm first considers debt or leasing, then debt that is convertible to equity, and lastly new issues of equity securities.

A major implication of the pecking order theory is that a firm has no preferred capital structure or target level of leverage. Rather, its holdings of debt relative to equity at any time reflect the accumulated effects of past financing decisions. Thus, the holding of debt reserves or unused borrowing capacity is an important element of a financial strategy that bases the use of debt capital on unanticipated depletions of internal sources of funds.

Two major factors motivate the pecking order theory. One is the transaction costs of using various sources of financial capital. Consistent with the ordering indicated previously, the inferences would be that retained earnings have the lowest transaction costs, debt has an intermediate level of costs, and new issues of common stock incur the highest costs. Most evaluations of actual transaction costs are consistent with this ordering, although some of the costs of using various sources of external capital are difficult to estimate.

A second motivation to the pecking order theory involves differences in information (i.e., asymmetric information) between a firm's manager and its suppliers of debt and external equity capital. Especially applicable to corporate finance, this motivation basically reflects the view that the manager of a firm knows more than lenders and outside investors about the firm's prospects and performance. Thus, the lender's costs of monitoring a borrower's performance and assuring compliance with the terms of a loan contract increase the costs of debt capital relative to retained earnings. Similarly, sales of common stock are interpreted by potential investors as bad news about the firm's financing needs, thus raising the costs of external equity to high levels.

The pecking order theory is intuitively appealing and seems plausible in terms of informal observations of financing methods in agriculture. However, no empirical studies have yet been conducted to validate its applicability to farm businesses.

FIRM GROWTH
IN AGRICULTURE

The growth of agricultural firms is a long-standing phenomenon, as evidenced by the increased sizes of agricultural operations and the declines in farm numbers. Even though the significance of firm growth was diminished in the 1980s because of the widespread financial stress in agriculture, the accumulation and protection of equity capital remains a major objective of most farm businesses. Thus, it is appropriate to address the issue of firm growth in agriculture. Two important features of firm growth are **measures of growth** and the **economic environment for growth.**

Measures of Growth

The term *growth*, as used in financial management, refers to *increases in business size. Rates of growth* refer to *how fast changes in size occur over time.* Hence, measures of size and growth are closely related.

Several types of growth measures are possible. The measures may be financial or physical and may focus on inputs or outputs. Examples of physical measures of growth for inputs and outputs are acres, litters of hogs, cattle fed, cows milked, and broilers raised. Everyday discussions of farm size often are of physical measures — for example, a 1,000-acre grain farm, a 100-cow dairy herd, and a feedlot capacity of 20,000 head. However, these physical measures of size have limited usefulness in comparing different types of businesses. Acreage may be used as a measure in comparing the size of two wheat farms. But acreage is meaningless in broiler operations or cattle feedlots where the birds and cattle are raised in confinement with purchased feed.

Financial measures of growth are easier to compare. Gross income, for example, may be satisfactory for comparing the growth in the size of businesses of similar type and capital intensity. However, gross income may not be appropriate as a financial measure when farm types with different ratios of cash expenses to gross receipts (dairy cattle versus feeder cattle, for example). Total assets and level of net worth are widely used financial measures of business size, especially in loan analysis by lenders.

Financial measures provide for more useful comparisons among farms with different operating characteristics. Rate of growth of equity or of net income flows reflects the firm's profitability. As indicated earlier, and developed further below, we will use the percentage change in equity capital as the growth measure. It is synonymous with the expected rate of growth of equity.

Economic Environment for Growth

Conditions for growth exist when the firm has underutilized resources, less than optimal resource allocation, and / or savings from disposable income to be invested. Moreover, two types of incentives for economic growth can stimulate the manager. One is a "push" incentive, in which the manager seeks growth as a means of fully exercising management ability. The manager is seeking greater control and stronger challenges from the environment; expanding the business is one means of accomplishing these goals. The second is a "pull" incentive, in which the business grows in order to operate more efficiently, achieve economies of size, and enhance its income-generating capacity. These two incentives are not independent; elements of both generally guide growth-oriented decisions. Moreover, their relative importance depends upon the economic environment of farm businesses.

The pull incentive is attributed to the combined effects of several structural characteristics of U.S. agriculture. The farm sector is largely comprised of relatively small firms, despite having relatively large capital investments per person employed. Low price and income elasticities for many farm commodities, subject to weather and other uncontrollable events, have led to wide swings in commodity prices. Moreover, individual farmers in general have had little capacity to influence either resource or commodity prices. These factors have traditionally destabilized farm income and placed downward pressure on profit margins per unit of farm production.

The 1930s through the 1960s witnessed a long, steady period of structural change in the numbers and sizes of farms in the United States and in their financial and market characteristics in response to these income and resource problems. Farm numbers declined, and many people left agriculture. The remaining farms experienced substantial mechanization, modernization, and growth in size as farmers responded to the cost efficiencies associated with new technologies and the quest for greater income-generating capacity. Farmers also became more dependent on markets for acquiring farm resources.

In response to these events, the per capita income for farm families became quite comparable to that of nonfarmers during the late 1970s, particularly when nonfarm income to farmers is included. This was a period of high inflation, substantial growth in farm sizes, and rapidly growing use of farm debt. Moreover, farmers experienced increased wealth through unrealized capital gains, which helped generate high payoffs from financial leverage.

Declines in farm prices beginning about 1979, combined with high real interest rates, depressed farm income during the 1980s. Farm lenders experienced increased farm loan delinquencies and farm foreclosures. The total value of farm assets began to decline, followed soon after by declines in farm debt. Financial stress in agriculture became widespread and adversely influenced farm lenders as well. Financial conditions in agriculture eventually stabilized toward the end of the 1980s and into the 1990s.

Prospects for the future suggest continued incentives for firm growth in agriculture, based primarily on the impact of new technology. Studies by the Office of Technology Assessment and by the American Bankers Association, among others, have projected a sharply declining number of mid-sized family farms; a steady to increasing number of part-time farms; and increased numbers of large, commercial farms having the scale, management skills, and financial strengths to exploit new technologies. This last group will seek economic growth in a complex risk environment, which will include greater reliance on contractual relationships with input suppliers and product handlers in the various stages of the food and fiber system. Thus, the development of appropriate leverage policies will be challenging.

FINANCIAL LEVERAGE
AND FIRM GROWTH

Financial leverage influences a firm's growth through its effects on the expected returns to equity capital. In general, increasing financial leverage will boost the growth in equity as long as the marginal returns from the use of a loan exceed the costs of borrowing. To illustrate, consider the financial organizations of two farm businesses that are similar in all respects except for their levels of financial leverage. Both Farmer C and Farmer D expect returns on assets of 16 percent and interest rates on debt of 12 percent. Each of them has an income tax rate of 20 percent and a consumption rate of 50 percent from net after-tax income. Each of them has $100,000 of equity capital; however, Farmer C's financial structure indicates a debt-to-equity ratio of 0.5, while Farmer D's debt-to-equity ratio is 1.0.

Table 6.2 indicates the balance sheets for Farmers C and D, along with the set of calculations needed to project their rates of growth in equity capital. As the calculation procedures show, the returns to assets (line 5) are reduced by the interest obligation on debt (line 6) to yield taxable income (line 7). Payments for taxes (line 8) and consumption (line 10) according to the specified rates yield the retained earnings (line 11). Dividing retained earnings by the level of equity capital gives the rate of growth of equity — 7.2 percent for Farmer C and 8.0 percent for Farmer D. Thus, Farmer D's higher leverage yields a higher expected growth rate for equity capital, with all other variables equal.

The calculation process in Table 6.2 is generalized by developing a growth model that expresses the rate of growth of equity capital as a function of the rate-of-return on assets, the interest rate on debt, the rates of taxation and consumption, and the level of financial leverage. The model is formulated as:

$$g = (rP_a - iP_d)(1 - t)(1 - c) \qquad (6.2)$$

or, more briefly, as:

$$g = (rP_a - iP_d)k \qquad (6.3)$$

where

$$k = (1 - t)(1 - c)$$

and

g = the rate of growth of equity capital

r = the average net rate-of-return, except for interest (i) and taxes (t), on total assets owned by the firm

i = the average interest rate paid on debt

t = the average rate of income taxation

c = the average rate of withdrawals for family consumption, dividends, and other non-business flows

P_a = the ratio (or proportion) of assets to equity, and

P_d = the ratio (or proportion) of debt to equity, which is called the *leverage ratio*.

For greater generality, leverage is modeled in proportional terms as the weights of assets (P_a) and debts (P_d) per dollar of equity capital.[3] To illustrate, Farmer D's balance sheet in Table 6.2 shows \$2 of assets and \$1 of debt for every \$1 of equity capital. Thus, the weights (or ratios) of assets and debts relative to equity for Farmer D are $P_a = 2.0$ and $P_d = 1.0$, and the value of P_d represents the leverage ratio. Following the balance sheet identity, in which assets minus debt equals equity, these weights must satisfy the relationship $P_a - P_d = 1.0$.

Using equation 6.2 and the preceding data, we can project Farmer D's rate of growth of equity to be:

$$g = [(0.16)(2.0) - (0.12)(1.0)](1 - 0.2)(1 - 0.5)$$

$$= 0.080, \text{ or } 8.0\%$$

Similarly, the projected growth rate for Farmer C is:

$$g = [(0.16)(1.5) - (0.12)(0.5)](1 - 0.2)(1 - 0.5)$$

$$= 0.072, \text{ or } 7.2\%$$

---

[3]An alternative formulation of the growth model is expressed by the dollar values of assets (A), debt (D), and equity (E) as follows:

$$g = \left[\frac{rA - iD}{E}\right](1 - t)(1 - c) \tag{6.2a}$$

To show the impact of leverage on the growth in equity, equation 6.2a can be modified to directly include the debt-to-equity ratio. Substitute $D + E = A$ into:

$$\frac{rA - iD}{E}$$

to give

$$\frac{r(D + E) - iD}{E}$$

and simplify to

$$\frac{D}{E}(r - i) + r$$

Then,

$$g = \left[\frac{D}{E}(r - i) + r\right](1 - t)(1 - c) \tag{6.2b}$$

Table 6.2. Rate of Growth of Equity for Different Leverage Ratios

Item	Farmer C	Farmer D
Balance Sheet		
1. Assets	$150,000	$200,000
2. Debt	$ 50,000	$100,000
3. Equity	$100,000	$100,000
4. Leverage (D ÷ E)	0.5	1.0
Income Statement		
5. Return to assets (16% of line 1)	$ 24,000	$ 32,000
6. Interest on debt (12% of line 2)	– $ 6,000	– $ 12,000
7. Taxable income	$ 18,000	$ 20,000
8. Taxes (20% of line 7)	– $ 3,600	– $ 4,000
9. After-tax income	$ 14,400	$ 16,000
10. Consumption (50% of line 9)	– $ 7,200	– $ 8,000
11. Retained earnings	$ 7,200	$ 8,000
12. Rate of growth of equity (line 11 ÷ line 3)	7.2%	8.0%

Notice that the weights of assets and debt relative to equity for Farmer C are $P_a = 1.5$ and $P_d = 0.5$, reflecting the composition of Farmer C's balance sheet. These weights satisfy the requirement of $1.5 – 0.5 = 1.0$.

While this model presents a simplified functional form, it includes the major variables affecting a firm's rate of growth of equity capital. It combines balance sheet (P_a, P_d) and income (r, i) elements as well as specifying external cash withdrawals (c, t). Thus, this model can be used to measure the effects on growth of each component.

Growth rates determined by leverage ratios ranging from 0.0 to 5.0 and rates-of-return ranging from 8 to 30 percent are indicated in Table 6.3 when rates of interest, taxes, and consumption are 0.12, 0.20, and 0.50, respectively. Row 1 shows the growth rates for a debt-free situation at each rate-of-return in the column headings. These growth rates are calculated by using equation 6.4, which is the reduced version of equation 6.3 when $P_a = 1.0$ and $P_d = 0.0$:

$$g = rk \qquad (6.4)$$

Thus, for an earnings rate of 12 percent, the rate of growth is:

$g = (0.12)(0.80)(0.50) = 0.048$, or 4.8%

Notice that when $P_d = 0.0$, the growth rates are wholly determined by the returns on assets (r) and the k factor representing the combined effects of taxation and consumption. The growth rate (g) is less than r because of the dampening influence of consumption and tax payments. Notice also, from Table 6.3, that equity growth is unaffected by leverage when the interest rate equals the net rate-of-return on assets. When both i and r equal 12 percent, all earnings from additional borrowing are used to pay interest. Moreover, if r is less than i, then the growth rate declines with higher leverage and eventually becomes negative.

Table 6.3. Growth Rates for Alternative Returns to Assets and Leverage[1]

Leverage (D ÷ E)	Returns to Assets (r)					
	0.04	0.08	0.12	0.16	0.20	0.25
	(%)					
0.00	1.60	3.20	4.80	6.40	8.00	10.00
0.50	0.00	2.40	4.80	7.20	9.60	12.60
1.00	− 1.60	1.60	4.80	8.00	11.20	15.20
2.00	− 4.80	0.00	4.80	9.60	14.40	20.40
3.00	− 8.00	−1.60	4.80	11.20	17.60	25.60
4.00	−11.20	−3.20	4.80	12.80	20.80	30.80
5.00	−24.00	−4.80	4.80	14.40	24.00	36.00

[1]Rates of consumption, taxation, and interest are 50%, 20%, and 12%, respectively.

The remainder of the table illustrates the close relationship between leverage (P_d) and the rate-of-return on assets (r), and their effects on growth in equity, given the values of i, t, and c. If the rate-of-return exceeds the interest rate and other variables remain constant, increases in financial leverage will increase the rate of growth, and thus the firm's profitability. Alternatively, if the rate-of-return on assets is less than the cost of borrowing and other variables remain constant, then increases in financial leverage will reduce the rate of equity capital growth, with the growth rate eventually becoming negative.

The effects on the growth of changes in consumption, taxation, and interest rates are evaluated by holding other factors constant. The reader may explore these effects by changing these variables. The growth rates are increased by higher rates-of-return, lower costs of debt capital, and lower rates of consumption and taxation. In contrast, growth rates are reduced by the opposite changes in these variables.

This analysis suggests that financial leverage can have the effects of a double-edged sword. In favorable economic times, higher leverage can stimulate firm growth. In unfavorable economic times, it can cause business performance to deteriorate rapidly. Given the behavior of agricultural lenders, it is unusual to find agricultural firms with leverage ratios that exceed the 1 to 2 range. But, the risk behavior of agricultural producers also limits leverage. We need to consider how these factors reduce the incentive for higher leverage.

LEVERAGE AND RISK

One factor working against higher financial leverage is the increase in the firm's financial risk. The effects of farm business risks are magnified considerably by the increase in financial risk when leverage increases.

Financial risks arise in several forms. As financial leverage increases, the potential loss of equity capital increases, the variation of expected returns to equity increases, and liquidity provided by credit reserves lessens. These effects are important because as leverage increases, unfavorable events have greater impacts on the firm than do favorable events. These greater variations arise from the fixed contractual obligations associated with interest payments and liabilities. These obligations remain fixed even though the firm's returns on assets may fluctuate considerably.

Moreover, volatile interest rates and greater use of floating-rate loans by most farm lenders have created considerable uncertainty about the costs of borrowing. Many loan contracts now contain variable (or floating) interest rates, thus providing another source of financial risk to be considered along with leverage and business risk.

Leverage and Business Risk

To illustrate the risk effects of financial leverage, consider how a 10 percent reduction in both asset values and returns to assets influences the level of equity and growth rates for Farmers C and D. Table 6.4 indicates the changes in equity and growth rates as asset values (A_0) decline by 10 percent to A_1, and as asset returns (R_0) decline by 10 percent to R_1, with levels of debt, interest, taxes, and consumption remaining the same.

Table 6.4. Effects of Financial Risk for Alternative Levels of Leverage

Item	Farmer C	Farmer D
Balance Sheet		
1. Asset, A_0	$150,000	$200,000
2. Asset, A_1	$135,000	$180,000
3. Debt	$ 50,000	$100,000
4. Equity, E_0	$100,000	$100,000
5. Equity, E_1	$ 85,000	$ 80,000
6. Percentage change in E	15%	20%
Income Statement		
7. Returns to assets, R_0	$ 24,000	$ 32,000
8. Returns to assets, R_1	$ 21,600	$ 28,800
9. Interest	$ 6,000	$ 12,000
10. Taxes	$ 3,600	$ 4,000
11. Consumption	$ 7,200	$ 8,000
12. Retained earnings, I_0	$ 7,200	$ 8,000
13. Retained earnings, I_1	$ 4,800	$ 4,800
14. Growth rate, g_0	7.2%	8.0%
15. Growth rate, g_1	4.8%	4.8%
16. Percentage change in g	33%	40%

For Farmer C, a 10 percent decline in asset values causes a 15 percent decline in the level of equity and a 33 percent decline in the projected growth rate, given Farmer C's leverage ratio of 0.5. For Farmer D, the same 10 percent decline in asset values causes a 20 percent decline in the level of equity and a 40 percent decline in the projected growth rate. The relative decline is greater for Farmer D, who has a higher leverage ratio, and the fixed obligations associated with debt and interest as asset values and returns on assets fluctuate. Even with adjustments for the rates of taxation and consumption, the relative decline would still be greater for higher leverage.

As shown, the relationship between financial leverage and risk can be generalized by extending the growth model to account for both the expected rate of growth and its variability, as measured by the standard deviation. Suppose that the rate-of-return on assets (r) is viewed as a normally distributed random variable with an average expectation of $\bar{r} = 0.16$, or 16 percent, and a standard deviation of $\sigma_r = 0.06$, or 6 percent. Recall from Chapter 2 that the standard deviation is a measure of the amount of variability of returns, and thus an indicator of risk. Using the standard deviation as the measure of variability means, for example, that asset returns should fall within 6 percentage points of the expected value about 66 percent of the time and within 12 percentage points of the expected value about 95 percent of the time.

Suppose also that the interest cost of borrowing is fixed at $i = 0.12$, or 12 percent, over the term of the loan. Hence, for now the standard deviation of the interest rate is zero ($\sigma_i = 0$).

The expected rate of growth under risk is still:

$$\bar{g} = (\bar{r}P_a - iP_d)k \tag{6.5}$$

The standard deviation of the growth rate is:

$$\sigma_g = \sigma_r P_a k \tag{6.6}$$

which is the weighted standard deviation of the risky assets, adjusted by the withdrawals (k) for taxation and consumption.

The relative risk for a given level of leverage is shown by the coefficient of variation (CV), which is the standard deviation of growth divided by the expected rate of growth (see Chapter 2).

$$CV = \frac{\sigma_g}{\bar{g}} \tag{6.7}$$

By using equation 6.5, one would expect that Farmer C, whose financial data are shown in Table 6.2 and whose leverage ratio is 0.5, would have the following growth rate:

$$\bar{g} = [(0.16)(1.5) - (0.12)(0.5)](0.8)(0.5)$$

$$= 0.072, \text{ or } 7.2\%$$

and the standard deviation is:

$$\sigma_g = \sigma_r P_a k$$

$$\sigma_g = (0.06)(1.5)(0.4)$$

$$\sigma_g = 0.036, \text{ or } 3.6\%$$

Thus, the coefficient of variation would be:

$$CV = \frac{\sigma_g}{\overline{g}} = \frac{0.036}{0.072} = 0.50$$

Farmer D's risk position should be greater than Farmer C's because Farmer D has greater leverage. The expected rate of growth for Farmer D is:

$$\overline{g} = [(0.16)(2.0) - (0.12)(1.0)](0.8)(0.5)$$
$$= 0.080, \text{ or } 8.0\%$$

The standard deviation for D is:

$$\sigma_g = (0.06)(2.0)(0.4)$$
$$= 0.048, \text{ or } 4.8\%$$

and the coefficient of variation is:

$$CV = \frac{4.80}{8.00} = 0.60$$

The higher coefficient of variation for Farmer D indicates the greater financial risk from higher leverage. That is, the amount of variability relative to the expected rate of growth is greater for the higher leverage position. Greater variability, in turn, means a greater risk of loss in equity capital.

Table 6.5 (Section A) indicates expected values, standard deviations, and coefficients of variation of growth rates for several leverage ratios. Higher leverage increases the expected rates of growth, given the values of other growth-related variables. However, higher leverage increases the variability of growth even faster. Thus, the total risk of a business relative to expected growth increases with greater leverage, as indicated by the increasing coefficients of variation.

Leverage and Interest Rate Risk

Now, suppose that interest rates on debt are also subject to unanticipated

Table 6.5. Expected Values and Risk Measures for Growth Rates Under Different Leverage Positions, Excluding and Including Interest Rate Risks

Measure	Leverage (D ÷ E = P_d)						
	0.0	0.5	1.0	2.0	3.0	4.0	5.0
A. No interest rate risk							
Expected growth rate, $\bar{g}$, %	6.40	7.20	8.00	9.60	11.20	12.80	14.40
Standard deviation, σ_g, %	2.40	3.60	4.80	7.20	9.60	12.00	14.40
Coefficient of variation, $\sigma_g / \bar{g}$	0.38	0.50	0.60	0.75	0.86	0.94	1.00
B. With interest rate risk							
Expected growth rate, $\bar{g}$, %	6.40	7.20	8.00	9.60	11.20	12.80	14.40
Standard deviation, σ_g, %	2.40	3.69	5.05	7.88	10.73	13.60	16.47
Coefficient of variation, $\sigma_g / \bar{g}$	0.38	0.51	0.63	0.82	0.96	1.06	1.14

variation so that the expected rate is $\bar{i} = 0.12$, or 12 percent, and the standard deviation of interest rates is $\sigma_i = 0.04$, or 4 percent. This measure of variability means that interest rates should fall within 4 percentage points of the expected rate (a range of 8 to 16 percent) about two-thirds of the time. In general, we would expect that interest rate risk would add to the firm's total risk, thus diminishing the incentive for higher leverage. This, indeed, is the case, although the complexity of the growth model increases relative to the situation in which interest rates are known with certainty.

The expected rates of growth, $\bar{g}$, for different levels of leverage are the same as before; however, the variability of expected growth increases with higher leverage to reflect the presence of interest rate risks. The standard deviation of growth is now expressed as:[4, 5]

$$\sigma_g = \sqrt{\left(\sigma_r^2 P_a^2 + \sigma_i^2 P_d^2\right) k^2} \tag{6.8a}$$

or

$$\sigma_g = \left[\left(\sigma_r^2 P_a^2 + \sigma_i^2 P_d^2\right) k^2\right]^{1/2} \tag{6.8b}$$

---

[4]These equations first find the statistical variance of the growth rate and then take the square root to yield the standard deviation.

[5]This analysis has assumed that the rates-of-return on assets and interest rates on debt are statistically independent. That is, they exhibit zero correlation. This assumption may be reasonable in practice, since the level of market interest rates and farm income are not strongly related. If, however, these rates are not independent, then the degree of covariation must be included in the risk analysis.

Continuing the numerical examples, the expected rate of growth for Farmer C, with a leverage ratio of 0.5, is still $\bar{g} = 0.072$, or 7.2 percent. Including interest rate risk, the standard deviation of Farmer C's growth rate is now:

$$\sigma_g = \sqrt{[(0.06)^2(1.5)^2 + (0.04)^2(0.5)^2](0.4)^2}$$

$$= 0.037, \text{ or } 3.7\%$$

This result indicates that the standard deviation of growth has increased from 3.6 to 3.7 percent in response to interest rate risk — a minor increase, reflecting Farmer C's relatively low leverage. Farmer C's coefficient of variation increases to $CV = 0.51 = (3.7 \div 7.2)$, compared to $CV = 0.50$ without interest rate risk.

For Farmer D, the presence of interest rate risk increases the standard deviation of growth to:

$$\sigma_g = \sqrt{[(0.06)^2(2)^2 + (0.04)^2(1)^2](0.4)^2}$$

$$= 0.051, \text{ or } 5.1\%$$

compared to a standard deviation of 4.8 percent without interest rate risk. Thus, the magnitude of risk increase is greater for Farmer D than for Farmer C, reflecting Farmer D's higher leverage. Farmer D's coefficient of variation increases to $CV = 0.64 = (5.1 \div 8.0)$, compared to $CV = 0.60$ without interest risk.

The impacts of interest rate risk on total risk (σ_g) and on the coefficients of variation are shown in Section B of Table 6.5 for the other levels of leverage. Again, the coefficients of variation increase as leverage increases and exceed the CV values for the same levels of leverage without interest rate risks, although by rather modest amounts.

THE LEVERAGE CHOICE

Clearly, the choice of a desirable level of leverage is not as clear when risk considerations are introduced. In the absence of risk, higher leverage can accelerate the expected rate of growth of equity if the rate-of-return on assets exceeds the cost of debt. In the presence of risk, however, managers must consider the level of risk relative to expected rates of growth. Since higher rates-of-return come with higher risk, the desired level of leverage from the investor's standpoint depends on his or her risk attitude. Lower levels of risk aversion generally are associated with higher leverage, and higher levels of risk aversion are associated with lower leverage. Of course, the lender's judgment may be important too. The terms of financing may influence the leverage choice as well.

We have focused primarily on the potential losses arising from higher leverage rather than on the potential gains, even when the expectations on each are of equal amount and probability. We have done this because risk aversion, of varying degrees, tends to dominate investor behavior. The leverage incentive declines because an assumption of diminishing marginal utility of money reflects greater utility from avoiding losses than from acquiring gains. Businesspersons have more to fear (bankruptcy) than to hope for (spectacular profits), and different attitudes toward risk are, in effect, different assessments of the utility of expected gains and losses.

PROPRIETOR WITHDRAWALS

Expenditures for family consumption, income tax payments, and other withdrawals of funds can rapidly drain farm income and severely constrain growth. If capital accumulation were the predominant goal, a minimum level of current consumption would be desirable. But, consumption also has value to the manager, who, when making growth decisions, must consider the time value of money and its implications for savings versus consumption.

Although consumption is likely to increase as income increases, it is commonly observed that above some minimum level, consumption increases less rapidly than income. Such a consumption pattern is influenced by several factors: (1) expected increases of income, (2) capital gains, (3) level of wealth, (4) leverage, (5) type of farm, and (6) stage of life cycle. Furthermore, consumption may be related more to average or normal levels of income in order to moderate the levels of consumption from volatile swings in income from year to year. The manager should examine the withdrawal rate (c) for any possible waste. However, the objective is not just to minimize the family consumption rate but to insure that the family members are getting their money's worth from the level of consumption and the rate of firm growth.

Tax management can also play a significant role in farm planning. Growth-conscious managers should examine their business records and tax situation to see if the tax obligation (t) can be reduced through better planning, record keeping, and analysis. However, the objective is **not to minimize taxes but to maximize** *after***-tax profits.**

An interesting effect on growth rates arises from the interaction of tax and consumption rates. As income increases, the tax rate (t) also likely increases because of the progressive structure of marginal tax rates. On the other hand, the rate of family consumption expenditure tends to decrease as income increases. Hence, the rate of savings (k) from higher income is likely to be more stable than it first appears.

RESOURCE PRODUCTIVITY AND GROWTH

How does firm growth affect the rate-of-returns (r) on business assets? Is r expected to increase, remain constant, or decrease? The matter is not so simple. Some factors generate economies of size, and other factors generate diseconomies as firm size increases. At the same time, changes in technology can alter the productivity of resources and perhaps increase the earning potential of new investments that are fully utilized. For example, a more efficiently designed milking parlor and feeding system may reduce average costs of milk production if the system is used to capacity. Nonetheless, diminishing returns (r) generally are expected to occur at some firm size. But how do diminishing returns actually appear?

Part of the answer lies in the adjustment of management to new large-scale investments. As financial growth takes place, lags may occur in the adjustment of management to new technology and increased size of business. These lags in managerial productivity arise from the new production decisions needed for crop, livestock, machinery, and labor management; from new marketing skills needed for buying and selling; and from the need for more effective management of investments, financing, and cash flows. The dairy farmer, for example, who rapidly expands herd size and mechanizes his or her operation has difficulty maintaining a high level of milk production per cow. Eventually, as expansion slows, the manager can exercise more intensive culling, employ more precise feeding and health care, and master the "mechanics" of mechanization. Thus, management experience is essential in attaining high levels of production efficiency.

These managerial lags often are not properly accounted for in business planning. Generally, a constant level of management productivity is used for planning. This approach overlooks any temporary reductions in production efficiency. The lags in management may severely jeopardize the firm's financial position, especially when a large investment reduces the firm's liquidity. Overcoming these lags is important when financial obligations require a rapid investment payoff.

While diminishing productivity of firm resources may slow the rate of growth, other attributes of resources may also hamper their acquisition. The land market may be inactive. High-quality hired labor, seasonal or full-time, may not be available. Many resources are lumpy or indivisible in size and may not be easily matched with one another. Examples include land, buildings, feeding systems, large-sized machinery, and irrigation equipment. Even full-time labor represents a lumpy kind of resource. This lumpiness requires large blocks of financial capital to purchase the assets and may sharply change the firm's leverage, risk, and liquidity.

EXTERNAL FINANCING

Access to external financing, as reflected by the firm's credit capacity, significantly influences its growth rate. Moreover, the terms of external financing can influence both the rate and the direction of firm growth. Our focus in this chapter is on the rate of growth.

The term *external capital* (or *credit*) *rationing* is used when a borrower has exhausted all sources of loanable funds but still finds that the marginal value product of resources acquired with borrowed capital exceeds the marginal cost (interest and noninterest) of borrowing. Under these conditions, a firm's economically desirable rate of growth exceeds its financially supportable rate of growth.

External credit rationing arises from the behavior of lending institutions. In turn, they are influenced by their own loanable-funds supplies, alternatives in lending and investment, and their evaluation of a borrower's credit worthiness. (The properties of credit worthiness will be discussed in detail in Chapter 7.) The typical farm manager will probably have exhausted all available credit, as evaluated by conventional lenders, at a debt level about equal to the equity of the farm: a leverage ratio of 1.0. Leverage ratios up to 2.0 or beyond usually denote either a superior financial manager or a manager who is in pending financial disaster as evidenced by the depletion of assets or excessive debt.

The operator who can secure a land purchase contract may attain a leverage as high as 3.0, provided non – real estate lenders do not curtail credit for financing other assets. Some nonfarm businesses attain much greater leverage.

A number of commercial banks, for example, operate with debt-to-equity ratios exceeding 15.

The response of interest rates (i) as borrowing and leverage increase also must be considered. In theory, interest rates are expected to differ between borrowers, with variations in loan risk, purpose, length, and / or size. More specifically, loan rates are expected to be higher for higher-risk borrowers. In the past, however, considerable evidence suggested that interest rates on agricultural loans from specific lenders did not vary much between borrowers. Instead, interest rates primarily varied over time in response to changes in the lender's costs of acquiring loan funds.

The reliance of agricultural lenders on interest rate adjustments to differences in lending risks between borrowers is increasing. Institutions of the Farm Credit System, in many cases, have adopted risk classes in pricing loans so that borrowers with higher lending risks pay higher interest rates. Commercial banks are also tailoring interest rates to the risk and other characteristics of farm borrowers. Surveys of agricultural banks in the United States indicate that more than half of these banks charge different interest rates to borrowers at any one time. Moreover, about 95 percent of the banks charging farm borrowers different rates do so because of differences in the borrowers' credit risks.[6]

MARKET DEMAND

Market demand is beyond the control of most individual producers. Yet, market factors that influence product prices also affect the rates-of-return to farm business assets. Thus, the growth of a farm business is not solely attributed to the firm's capacity to produce; an expanding business is necessary but not sufficient for financial growth. Growth in demand for farm products must also occur.

The generally low price elasticity of demand for farm products, together with the competitive structure of commodity markets, complicates the situation. Thus, the aggregate implications of firm growth and individual investment decisions must be considered. In fact, market-oriented producers who are considering long-term investments in highly specialized capital equipment are quite concerned with demand conditions for their commodities. They are

---

[6]Other factors associated with differences in interest rates include differences in the borrowers' deposit balances, loan size, loan maturity, and loan purpose. Reviews of the survey results are found in references 5 and 8.

becoming increasingly aware that product prices and farm incomes may be adversely affected when many producers invest in a fashion that increases total production.

Not all agricultural investments increase output in the aggregate. When a producer expands farm size by purchasing or leasing a neighboring tract of land, this agricultural investment will not increase aggregate output if the same cropping patterns are maintained. It will simply add to the expanding farm's resource base and income-generating capacity. On the other hand, if a producer expands by adding to livestock-producing capacity, then aggregate livestock output increases. Similarly, bringing additional farm land into production or farming existing land more intensively will add to total production capacity. If numerous producers follow suit, the impact on aggregate output may be substantial. Given an inelastic demand for most agricultural products, the increased output could result in a relatively larger decline in product price and thus gross returns to its producers.

The effects on producers' net income depend on the cost-reducing impact of the new investment. If costs per unit decline by a relatively greater proportion than prices decline, income may actually increase in the short run. The prospects of higher income may in turn attract further investments by new producers, thus putting renewed pressure on the cost – price margin and bringing additional resource adjustments in the future.

The dynamics of the situation affect different producers in different ways. As an example, the early investors in new technology may benefit substantially from cost reductions and increased volume before price declines from increased output by later investors. The demise of some later (or inefficient early) investors may alleviate the pressure. By then, still newer technologies may come along to continue the process. All these factors influence the rate of earnings over time and, thereby, the growth of the firm.

Summary

The choice of leverage position for an agricultural business depends jointly upon the costs of financial capital, the expected returns on assets, risk, and attitudes toward risk. In turn, the level of leverage can influence the expected rate of growth of equity capital, which also is constrained by proprietor withdrawals, resource productivity, financing terms, and aggregate market conditions.

More specific points are as follows:

- The costs of debt capital increase as leverage increases because of the increased credit risks experienced by lenders.

- The costs of equity capital increase as leverage increases because of greater financial risk for the borrower.

- The weighted average cost of debt and equity capital first declines as leverage increases and then rises because of increased financial and credit risks. In principle, optimal financial leverage occurs at the minimum weighted average cost of capital.

- The "pecking order" theory of capital structure suggests a hierarchy approach to financing decisions in which leverage at any time depends upon the accumulated effects of past financing decisions.

- In general, increasing financial leverage will increase the expected growth in equity as long as the marginal returns from the use of a loan exceed the cost of borrowing, and the marginal rates of consumption and taxation remain constant.

- Increasing financial leverage will increase the variation of expected returns on equity, heighten the potential loss of equity capital, and reduce the size of liquid credit reserves. Variations in interest rates further magnify these financial risks as leverage increases.

- Because greater expected rates-of-return come with greater risks as leverage increases, the preferred level of leverage for an investor depends on his or her risk attitude.

- Leverage decisions also are influenced by consumption and income tax expenditures, economies of size, management ability, credit availability, and aggregate market conditions.

Topics for Discussion

1. Explain the concept of a weighted average cost of capital and a minimum cost level of leverage. Contrast the equilibrium theory and pecking order theory approaches to explaining a firm's capital structure. How applicable are these theories to agricultural finance?

2. Explain why the costs of debt and equity are expected to increase as leverage increases.

3. What is the average cost of capital when the after-tax cost of debt is 12 percent, the cost of equity is 16 percent, and the ratio of equity to assets is 0.40?

4. Contrast and critique the use of physical and financial measures for evaluating the growth of agricultural firms.

5. Given the following information about a farm business:

   Debt-to-equity ratio............................. 2.0
   Expected return on assets........................ 15%
   Expected interest rate on debt................... 10%
   Consumption rate 60%
   Tax rate 20%
   Standard deviation of return on assets 5%
   Standard deviation of interest rate................ 3%

   a. What is the expected rate at which this firm could grow?

   b. What might the manager do to increase the rate of growth?

   c. What level of risk is associated with the rate of growth found in a?

   d. Suppose another lender allows a maximum debt-to-equity ratio of 2.5, with an interest rate of 9 percent and a standard deviation of 4 percent. How do expected growth and risk for these conditions compare with the terms cited above?

6. With a tax rate of 20 percent, a consumption rate of 50 percent, a rate-of-return on assets of 14 percent, and an interest cost of 7 percent:

   a. What leverage ratio must a farmer achieve in order to grow at a 12 percent rate?

   b. What factors should both this farmer and the lender consider when they are deciding on whether the farmer's leverage should be increased to this point?

7. Explain the relationship between business risks as evidenced by variability of prices, yields, asset values, etc., and financial risk associated with borrowing and greater leverage.

8. Discuss how the effects of diminishing returns are modified over time for a firm.

References

1. Baker, C. B., and J. A. Hopkin, "Concepts of Financial Capital for a Capital-Using Agriculture." *American Journal of Agricultural Economics*, 51(1969):1055 – 1065.

2. Barry, P. J., and C. B. Baker, "Financial Responses to Risk," *Risk Management in Agriculture,* Iowa State University Press, Ames, 1983.

3. Barry, P. J., C. B. Baker, and L. R. Sanint, "Farmers' Credit Risks and Liquidity Management," *American Journal of Agricultural Economics,* 63(1981):216 – 227.

4. Brigham, E. F., and L. Gapenski, *Financial Management: Theory and Practice,* 6th ed., Dryden Press, Hinsdale, Illinois, 1991.

5. Calvert, J., and P. J. Barry, *Loan Pricing and Customer Profitability Analysis in Agricultural Banks,* AE – 4533, College of Agriculture, University of Illinois at Urbana – Champaign, 1982.

6. Donaldson, G., *Corporate Debt Capacity,* Boston: Harvard Business School, 1961.

7. Economic Growth of the Agricultural Firm, Tech. Bull. 86, College of Agriculture Research Center, Washington State University, Pullman, 1977.

8. Ellinger, P. N., P. J. Barry, and M. A. Mazzocco, "Farm Real Estate Lending by Commercial Banks," *Agricultural Finance Review,* 50(1990):1 – 15.

9. Fama, E. F., *Foundations of Finance: Portfolio Decisions and Securities Prices,* Basic Books, Inc., Publishers, New York, 1976.

10. Fama, E. F., and M. H. Miller, *Theory of Finance,* Dryden Press, Hinsdale, Illinois, 1972.

11. Gwinn, A. S., P. J. Barry, and P. N. Ellinger, "Farm Financial Structure Under Uncertainty: An Application to Grain Farms," *Agricultural Finance Review,* 52(1992):43 – 56.

12. Langemeier, M. R., and G. F. Patrick, "Farmers' Marginal Propensity to Consume," *American Journal of Agricultural Economics,* 72(1990):309 – 316.

13. Lee, W. F., et al., *Agricultural Finance,* 8th ed., Iowa State University Press, Ames, 1988.

14. Levy, H., and M. Sarnat, *Capital Investment and Financial Decisions,* 3rd ed., Prentice-Hall, Inc., Englewood Cliffs, New Jersey, 1986.

15. Modigliani, F., and M. H. Miller, "Cost of Capital, Corporation Finance and the Theory of Investment," *American Economic Review,* 68(1958):261 – 297.

16. Myers, S. C., "The Capital Structure Puzzle," *Journal of Finance,* (1984):575 – 592.

17. Myers, S. C., and N. Majlauf, "Corporate Financing and Investment Decisions When Firms Have Information That Investors Do Not Have," *Journal of Financial Economics,* 18(Summer 1989):39 – 58.

18. *Technology, Public Policy and the Changing Structure of American Agriculture,* Office of Technology Assessment, U.S. Congress, Washington, D.C., March 1986.

19. *Transition in Agriculture: A Strategic Assessment of Agricultural Banking,* American Bankers Association, Washington, D.C., 1986.

7

Liquidity and Credit

The relationships between a firm's leverage and liquidity positions are very important considerations in financial management. As shown in Chapter 6, higher levels of financial leverage may, under proper conditions, accelerate the growth of the firm's equity capital. However, higher leverage also adds to the firm's financial risk and thus to its total risk. Holding liquidity is an important way to counter risk, but higher leverage depletes the firm's credit reserves, which are valuable sources of liquidity. Moreover, higher leverage creates additional repayment obligations in the future that reduce the margin of liquidity provided by anticipated net cash flows. Thus, a significant trade-off exists between the firm's leverage and liquidity positions. This trade-off highlights the need for the manager to consider decisions about levels of indebtedness along with decisions about the sources and levels of the firm's liquidity.

This chapter focuses on the firm's liquidity management under risk, giving emphasis to the development and management of credit relationships with lenders as a means of providing liquidity for leveraged farm businesses. We begin by establishing the concept of liquidity management — why it is needed and how it is provided. The relative merits of various sources of liquidity are discussed. Then we consider the role of credit as a source of liquidity and the lender's evaluation of a borrower's credit worthiness. The lender's own objectives and operating environment influence the terms by which loan funds are made available and thus may have significant influence on the borrower's managerial decisions.

NEEDS FOR LIQUIDITY

As established in earlier chapters, liquidity involves the firm's capacity to generate cash to meet cash demands as they occur and to provide funds for responding to unanticipated events. *Liquidity should not be confused with solvency.* A firm is solvent when the current market value of its assets exceeds its debt obligations and when it can meet these obligations over a sufficiently long period of time. Liquidity is a shorter-run concept. A firm may be solvent and either liquid or illiquid. Or, it may be insolvent and either liquid or illiquid. Generally, however, problems of insolvency also involve problems of low liquidity. Thus, the two concepts are related to one another, differing primarily in the length of time involved for covering the firm's financial obligations.

Liquidity is needed for three basic purposes. The first is a **transactions demand** for liquidity that involves having sufficient cash available to meet known cash demands arising from various market transactions anticipated to occur in the near future. This need arises only because of known or anticipated differences in the timing of cash receipts and cash expenses. Even if there were no uncertainty, liquidity would be needed for this purpose. The second is a **precautionary,** or **safety, demand** for liquidity in which cash is needed for responding to possible adversities in business or household performance. The third is an **investment demand** for liquidity — sometimes called a **speculative demand** — in which funds are needed for responding to new, unanticipated investment opportunities.

First, consider the transactions demand for liquidity. In the commercial agriculture of the United States, most of the items needed for farm production or for family consumption are acquired with cash outlays. While subject to some uncertainty, the general characteristics of seasonal patterns of cash inflows and outflows can be predicted with relatively high accuracy for most farm businesses. Projections of these patterns can indicate seasonal cash deficits and / or cash surpluses — the deficits must be covered and the surpluses must be managed, perhaps through short-term interest-bearing investments.

These seasonal patterns may differ in systematic ways among various types of farm businesses, creating different demands on liquidity management for transactions purposes. For example, a dairy farm typically has a more uniform flow of cash throughout a year than does a grain farm. Thus, a dairy farmer's approach to managing cash flows is less demanding, but still important. The grain farmer, however, must organize his or her financial reserves to cope with relatively large swings in seasonal deficits and surpluses.

The precautionary demand for liquidity involves having sources of financial reserves available to meet unanticipated (or random) fluctuations in cash flows caused by variable prices, yields, expenses, and so on, and by severe hazards, such as fire, hail, tornados, sickness, and death. Since many of these unanticipated events involve market conditions, the transactions and precautionary demands for liquidity are closely related. For example, a grain farmer may be counting on selling his or her expected crop production at harvest at a targeted price level in order to provide the cash needed to cover loan payments and other financial obligations. If, however, low prices and / or yields cause the revenue from crop sales to fall short of expectations, then a shortfall in cash flow occurs as well. Thus, some other source of liquidity is needed to meet the financial obligations.

The level of reserves to carry for safety purposes is influenced by many factors — e.g., the magnitudes and sources of risk involved; the manager's risk attitudes; the use of various responses to risk, which include commercial insurance; the characteristics of various sources of liquidity; and the extent of fixed cash demands on the business. These fixed cash demands can vary with the operator's personal characteristics, investment patterns, and financing practices. To illustrate, expenditures for family consumption by farm businesses have tended to increase over time and become fixed at the higher levels, despite continued variability in farm income. In the case of investment and financing, acquiring a new machine may improve the firm's productivity, but it likely reduces liquidity and diverts cash to meet the downpayment and the future debt repayment requirements. This reduction in liquidity occurs at a time when greater borrowing and the associated financial risk have increased the firm's need for liquidity.

The investment demand for liquidity involves having the financial capacity to respond to new, seemingly favorable investment opportunities that appear quickly and perhaps unexpectedly over time. In the financial markets, for example, a sharp change in monetary conditions might create new opportunities for capital gains, thus heightening the demand for stocks, bonds, and other financial assets. In a farm community, a nearby tract of land may have unexpectedly become available for sale as a result of the owner's death, retirement, or other departure from farming. Since the land may not become available for purchase again for a long time, neighboring farmers may have considerable interest in bidding on this tract — if they have the financial resources to do so. Closely related to the land example are the investment opportunities arising at various auctions and farm sales in which having the financial capacity to bid on used machinery and other farm assets could prove timely and cost-effective. Clearly, then, the investment demand for liquidity is important, along with the transactions and precautionary demands, thus making liquidity management an important and challenging task.

SOURCES OF LIQUIDITY

In this section we consider sources of liquidity in terms of the firm's assets and liabilities, or sources of financial claims on assets. Assets are a source of liquidity through the funds that can be generated by their sales or liquidation. Credit reserves — unused borrowing capacity — have been a valuable source of liquidity for many types of farm businesses. The following sections focus on the concepts and general characteristics of these sources of liquidity. In Chapter 8 we will further consider the uses of these sources of liquidity in formulating a firm's risk management strategies. In addition, the use of insurance as a contingent source of liquidity will also be introduced in Chapter 8.

Asset Liquidity

The most obvious source of liquidity among a firm's assets is its cash balance (currency, checking accounts, and other transactions accounts). In many respects cash is the most accessible source. Yet, in many firms, borrowing occurs while cash remains on hand. Clearly, even when borrowing, the manager must perceive an advantage to leaving some cash on hand. Nearly everyone has a "reservation price" or required return that must be met to use additional cash for either productive, investment, or consumption purposes. The same concept applies to the holdings of other financial assets and, as shown later, to the holdings of credit reserves.

Other assets also contribute to the firm's liquidity position. The contribution ranges over a wide continuum from cash to near-cash substitutes (money market accounts, time deposits, government securities, etc.), through assets held for sale (grain inventories, market-finished livestock, etc.), through intermediate assets (machinery, breeding animals, etc.), and through real property (land, land improvements, buildings, etc.). As indicated in Chapter 3, these assets are typically classified into current and noncurrent categories, which, in general, reflect their liquidity.

In concept, the degree of liquidity in farm assets depends on an asset's sale value relative to its economic value to the firm. An asset is considered perfectly liquid if its sale generates cash equal to or greater than the reduction in the value of the firm resulting from the sale. Assets become increasingly less liquid as their potential sale reduces the value of the firm by more than their expected sales value.

There are several possible explanations as to why an asset's sale value might be less than its contribution to the firm's economic value. One of these involves the transactions costs needed to complete the sale — the commission charges, transportation costs, interim insurance, installation and assembling costs, and losses in transit. For most financial assets, transportation and handling costs are insignificant. However, these costs are much more important for most noncurrent farm assets and for inventories of grain and livestock.

Another liquidity cost involves characteristics of the markets in which the various types of assets are traded. A market is perfect only if the asset can be bought or sold for the same price at a particular point in time. Hence, a perfect market for an asset implies perfect liquidity; no spread exists between a bid from the buyer and an asking price from the seller. Clearly, transaction costs are a source of market imperfections. Other factors that affect marketability include the quality and availability of market information, volume of trading, number of market participants, location of participants, and availability of secondary markets in the case of financial assets.

Liquidity costs also are associated with contingent tax obligations. If an asset has experienced long-term capital gains, the sale of the asset could create a capital gains tax obligation that reduces the net cash proceeds from the sale. No such tax obligation would occur if the asset were not sold.

Liquidity characteristics also interact with the urgency of cash demands and the speed with which cash must be generated. In general, the greater the urgency, the greater are the liquidation costs, or price discounts, experienced in accomplishing the sale. Indeed, forced sales of assets with relatively inactive, informal markets (e.g., land, forage, growing crops, many kinds of equipment) may result in significant price discounts.

Liquidity risk is another factor to be considered in evaluating an asset's liquidity. Liquidity risk involves the unanticipated variability of the asset's value over time and how these changes in value are correlated with the needs for liquidity. In general, the less variable the asset's market price, the greater its liquidity. Moreover, an asset yielding a high price when the needs for liquidity (i.e., cash demands) are high is liquidity-preferred; one yielding a low price when liquidity needs are high is liquidity-adverse; and one whose return is independent of cash demands is liquidity-neutral.

Finally, the liquidity characteristics of assets are also influenced by their impact on a firm's capital integrity. Capital integrity refers to the importance of an asset's income-generating role in the firm. Selling current assets, such as inventories or goods in production, is a normal part of the firm's operations. The sales of current assets have little, if any, impact on income-generating capacity over time. In contrast, the liquidation of other assets, such as machinery, breeding livestock, and real estate, may reduce the firm's scale of operations as well as the efficiency of operations if the desired balance between assets (e.g., matching machines to land acreage) is adversely affected. Thus, sales of these assets may reduce the firm's economic value by more than each asset's individual sale value. These assets are then illiquid, even though some may have high market availability, low transactions costs, or other liquidity-preferred characteristics.

All these characteristics may differ with regard to the type of asset and within individual firms. In general, no nonfinancial assets found in the balance sheet are perfectly liquid, although some, such as finished feedlot cattle, market weight hogs, and grain inventories, are nearly so. Markets for these commodities are well-established, and with some adjustment for marketing costs, these commodities may be sold quickly at prices known at least a short time in advance.

Other assets are less liquid. Growing crops are not easily sold unless neighboring farmers are the buyers. To sell a breeding or a dairy animal prior to the time of optimal replacement would lower the firm's economic value. The same result holds for selling machinery and equipment. Other assets such as buildings, land, and land improvements are highly illiquid. Marketing costs for these assets generally are high, and few buyers may be available. Finally, the liquidation of many fixed assets is a major step toward dissolving the firm.

Credit Reserves

A **credit reserve** is *a source of liquidity that arises from the liability side of the firm's activities.* Specifically, a firm's credit reserve is represented by its unused borrowing capacity. At any point in time, one or more lenders may be willing to loan additional funds to a firm or an individual borrower in order to finance transactions, risk-related needs for funds, or new investment opportunities. The firm's credit reserve is the difference between the maximum amount of potential borrowing and the amount already borrowed.

A credit reserve may be difficult to measure. Its size and structure differ among lenders and will change over time as changes occur in the firm's financial performance, industry conditions, financial markets, and the lender's own financial performance. Thus, the procedures for measuring credit reserves may be imprecise, with considerable uncertainty about the results. Sometimes the uncertainty may be reduced by formal agreements between the borrower and the lender — a line of credit and a credit commitment are examples. But even then, the agreements may not cover significant adversities affecting the firm's liquidity position.

The data in Table 7.1 illustrate these points. This table presents the maximum credit capacities of a case farm in terms of limits on long-, intermediate- and short-term borrowing of $200,000, $60,000, and $88,500, respectively. When long-, intermediate-, and short-term borrowing actually occur at $150,000, $40,000, and $45,000, respectively, the remaining credit reserves are $50,000, $20,000, and $43,500. This is a highly simplified version of borrowing capacity. Multiple lenders may be involved, and sometimes unused long- and intermediate-term borrowing may be used to support short-term borrowing above its apparent limit. Moreover, as indicated previously, these figures only hold at a point in time.

Table 7.1. Evaluating the Credit Reserve of a Case Farm

Source of Credit	Credit Capacity	Credit Used in Borrowing	Credit Reserve
Long-term assets	$200,000	$150,000	$50,000
Intermediate-term assets	60,000	40,000	20,000
Short-term operating credit	88,500	45,000	43,500

In practice, increased borrowing, carrying over or extending loans, deferring loan payments, refinancing high debt loads, and otherwise utilizing credit during times of financial distress are all means of drawing upon credit reserves to provide liquidity under adversity. The borrowing may be from a firm's primary lenders, from other commercial lenders, or even from government credit programs. Thus, debt management is an integral part of overall credit management.

The advantages and disadvantages of holding credit reserves as a source of liquidity also can be established. In general, credit is considered a highly efficient way of providing liquidity for the following reasons: (1) using credit does not disturb a firm's basic asset structure and production organization; (2) no liquidation of assets occurs; (3) the transactions costs are relatively low; (4) institutional sources of loan funds are generally available in the financial markets; and (5) credit has a high degree of flexibility because loan funds may be channeled to numerous uses. However, the costs of establishing, maintaining, and borrowing from credit reserves must be considered. Holding reserves reduces the returns from investment opportunities that are foregone from further financial leverage. Interest must be paid when borrowing occurs. Non-interest charges such as deposit balances and loan fees are sometimes used to compensate lenders for establishing a line of credit. The future costs and availability of credit may be subject to considerable uncertainty. Financial flexibility is lost as the firm's assets become subject to the collateral claims of lenders. Finally, considerable financial control may shift to the firm's lenders, especially when additional borrowing occurs under stress conditions.

CREDIT CONCEPTS
AND BORROWING BEHAVIOR

Developing a conceptual framework in which the costs and returns of the alternative uses of credit are explicitly considered will help you to understand further the holdings of liquid credit reserves (and other reserves as well) as a response to risk. In developing this framework, the decision maker should recognize that credit, or borrowing capacity, (1) can be used in borrowing and (2) can be held in reserve. The choice presumably depends on the value of credit in each use — that is, on the returns earned from the business by utilizing borrowed funds and on the liquidity value of the credit reserve. Between these two values, the liquidity value is more difficult to measure. It is a subjective value that is strongly influenced by the behavioral characteristics (i.e., risk attitude) of the decision maker. Nonetheless, in developing a formal approach to credit management, the decision maker must consider both types of values jointly.

As a point of departure, let's assume that a farmer has established a business relationship with one major lender (i.e., his or her institutional source of credit) and is engaged in the production of a single farm commodity. The farmer and the lender have reached an agreement about the maximum amount of funds that can be borrowed, based on the farmer's credit worthiness at the present time. However, the farmer prefers not to borrow all these funds but to hold a portion as a liquid reserve. Thus, the key question is: How much of the total credit should be allocated for borrowing and how much should be held in reserve?

The answer to this question is based on the costs and returns of credit in its two possible uses. Optimal credit use reflects equality at the margin between a payoff schedule from using borrowed funds in the business and a cost-of-borrowing schedule that includes both the interest obligation to the lender and the farmer's liquidity premium on the maintained credit reserve.

To show this result, consider the framework in Figure 7.1. The horizontal axis measures the percentage of total credit used in borrowing. Movement to the right along the axis indicates more of credit used for loans. The remainder constitutes the credit reserve. The vertical axis measures the value of credit used in borrowing or in reserve. The body of the figure contains two curves that represent the costs and returns from borrowing.

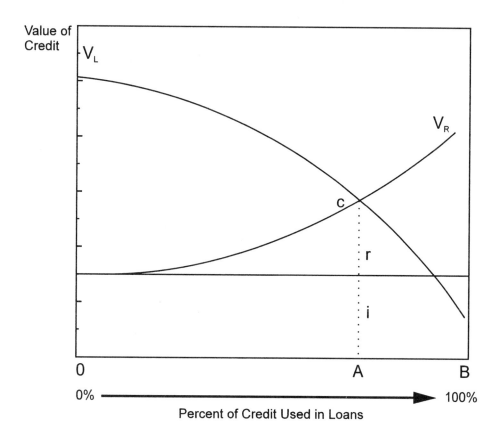

Figure 7.1. Equilibrium in credit allocation.

First, consider the returns from borrowing. The familiar law of diminishing marginal returns indicates that the returns (i.e., marginal value products) from additional units of resources and resource services acquired with borrowed funds (i.e., used credit) will decline at an accelerating rate, as indicated by curve V_L. Thus, curve V_L is a payoff schedule from using borrowed funds in the business; alternatively, it is an opportunity cost schedule of maintaining the credit reserve.

Curve V_R represents the cost-of-borrowing schedule. It has two components: the interest rate (i) charged by the borrower and the liquidity premium (r). The interest rate is established as part of the loan contract; thus, it is observed in the lending market. In Figure 7.1, the interest rate is drawn as a constant rate over the amount borrowed; however, it could have an upward slope as well.

The liquidity premium is determined by the borrower's level of risk aversion. It is similar to a reservation price or a required rate-of-return on the next unit of borrowing. A liquidity premium (r) of 10 cents, for example, indicates that the manager requires a 10 percent net return over and above the interest rate (i) on the next unit of credit used for borrowing. Thus, the next dollar of unused credit has a net value of 10 cents to the manager. If the interest rate on the next dollar borrowed is 12 percent, then the total cost of borrowing $(R = i + r)$, including the liquidity premium, is 22 percent.

The upward slope of the liquidity premium curve V_R shows that the remaining units of unused credit become increasingly valuable as the credit reserve becomes smaller in size. This relationship is logical to expect. Just as a thirst-craved wanderer on a desert would value highly his or her last "drops" of liquid, so a debt-ridden borrower would value highly the last "drops" of liquidity!

The height and shape of the liquidity curve are determined jointly by the manager's risk attitudes, levels of risk, business characteristics, and the practices employed in risk management. The higher (lower) the risk aversion, the higher (lower) the curve, with other factors held constant. Similarly, the greater the risks facing the business, the higher the liquidity premium schedule, with other factors held constant. Finally, the greater the availability and use of other methods of managing risk (see Chapter 8), the lower the needs for liquidity, and the lower the value of holding reserves. An individual's liquidity premium curve may shift and change shape over time as a result of experience, age, attitudinal changes, or other changes that alter the importance of risk.

Given curve V_L as the marginal value product of additional assets acquired with loans and curve V_R as the marginal value of liquidity from credit held in reserve, the optimal level of borrowing is OA, with AB the amount of credit in reserve. At this point, the marginal value product of assets (V_L) equals the marginal cost of borrowing, including both the interest rate (i) and the liquidity value (r). The marginal returns to the manager of using additional credit in borrowing are less than the marginal cost of borrowing (i + r). If r were ignored or considered to be zero, the optimal level of borrowing would be found by equating V_L with i. Since farmers and other types of borrowers seldom borrow to this point, it is logical to conclude that they do reserve credit as a part of their liquidity management.

This framework indicates that the demand for credit is a broader concept than the demand for loans. Another point to note is that conservative lenders may restrict the size of total credit so that the borrowers experience external credit rationing. In this case, the marginal value product curve (V_L) encounters the right-hand margin of the diagram in Figure 7.1 before it crosses V_R, leaving an excess return above the total cost of borrowing. In addition, as noted previously, the interest line (i) may curve upwards, if farmers face higher cost loans as borrowing increases. Finally, the amount of credit (c) may change frequently and be subject to considerable uncertainty. Changes in credit, in turn, due to macroeconomic forces and other factors influence the total cost of borrowing and introduce an additional source of financial risk.

We began this section with the assumption that the borrower utilized a single lender and thus had a single source of credit reserve. However, multiple sources of credit generally are available, and credit reserves can be distinguished with each of them. Moreover, the liquidity values of the credit reserves may differ depending on the lender. For example, non-specialized credit that is available from a commercial bank is likely higher in liquidity value than is credit from a trade firm. The latter generally is available only for acquiring a specific input such as fertilizer or a farm machine. Thus, the value of credit liquidity is limited to this type of transaction.

DETERMINANTS OF CREDIT
AND THE COST
OF BORROWING

The process of determining the amount of available credit — from whom it is available, what its cost is, and how the business and personal characteristics of the borrower influence credit — is a complex one that entails unique measurement problems. Identifying the person who makes the credit decisions is one consideration. A second consideration is knowing the agency costs that are inherent in any lender-borrower relationship. A third consideration involves identifying and understanding the full set of forces that affect the availability and cost of credit at a particular time and for a particular borrowing situation. Understanding the distinction between a lender's use of "price" and "non-price" methods of making credit decisions is a fourth consideration. These considerations are treated in this section.

Who Determines Credit?

Borrowers generally are not likely to evaluate their own credit. Rather, their credit is evaluated by lenders who are relied upon for borrowing. Nevertheless, borrowers who are knowledgeable about the credit evaluation process may be able to anticipate a lender's response, influence it through their managerial decisions, and thus have considerable control over their own credit position.

Lenders are guided by their own objectives for profits, risk, and liquidity as well as by forces operating in the financial markets. Hence, the lender's viewpoint in a loan transaction may differ from that of the borrower on several counts. For example, loans constitute the earning assets for lenders, who prefer a high rate-of-return over a short period of time. Conversely, loans generate liabilities and costs for borrowers. Hence, they prefer low-cost loans for a relatively long period of time. The borrower participates directly in any gain or loss that arises from the financial investment. Lenders do not and are restricted to the returns called for by the debt contract. They only indirectly benefit from the borrower's profits through possible growth in financing needs over time.

Other terms of financing may reflect the needs of the lender as well as the needs of the borrower. For example, to meet their short-term obligations to depositors, commercial banks must maintain a high degree of liquidity in their asset structure, an important part of which is their loan portfolio. Hence, to meet liquidity needs, the lender may prefer a loan length shorter than the economic life expectancy of the asset being financed. Or, the borrower might prefer a longer length of loan. Finally, some lenders are limited by law in the loan maturities they can offer.

An Agency Cost Perspective

The aspects of lender-borrower relationships cited above are reflected in the concepts of agency relationships and agency costs, which have emerged as important components of modern finance theory. An agency relationship (also called a principal-agent relationship) is defined as:

> . . . an explicit or implicit contract in which one or more persons (the principal) engage another person (the agent) to take actions on behalf of the principal. The contract involves the delegation of some decision making authority to the agent [16, p. 39].

In a credit relationship, the lender (the principal) is considered to contract with the borrower (the agent) to productively utilize and repay (with interest) the lender's funds.

During the course of the loan contract, the borrower (the agent) is expected to behave in concert with the objectives of the lender so that these objectives can be optimally attained. However, due to self-interest seeking by the borrower, informational deficiencies (asymmetries) as the terms of the contract are formulated and carried out, and uncertainties about future events, loan contracts tend to be incomplete, and the objectives and activities of the principal and the agent may not completely coincide. Thus, agency costs are incurred in structuring, administering, and enforcing loan contracts in order to better align the goals of the borrower with those of the lender, resolve problems associated with informational deficiencies, and deal with contingencies during the term of the loan contract.

In general terms, agency costs may include (1) expenditures incurred by the lender in monitoring the borrower's activities, (2) bonding expenditures (pledging collateral is an example) incurred by the borrower to insure the lender against adverse actions by the borrower, and (3) any residual loss incurred by the lender arising from incompleteness in the loan contract and failure to fully align the objectives of the principal and the agent (see Jensen and Meckling for the landmark article on agency relationships in finance). Because agency costs are borne by the contracting parties, it is logical to expect efforts to structure contracts so as to minimize agency costs. However, these efforts are economical only to the point where the marginal cost of contract control equals the marginal gain in reducing the residual loss.

Focusing on agency cost control in credit relationships entails asking the following basic questions:

1. Is the borrower a greater risk than believed when the loan contract was established (an adverse selection problem)?

2. Will the borrower take on greater risks during the term of the loan

contract than anticipated by the lender when the contract was established (a moral hazard problem)?

In part, such actions could reflect asymmetric goals held by the two parties — i.e., the borrower focuses on profitability and wealth accumulation, while the lender emphasizes loan repayability and safety. Moreover, asymmetric information is directly involved because the lender may know less than the borrower about the borrower's goals and actions, as well as about the characteristics of the projects being financed. Thus, lenders may utilize extensive financial contracts in order to better align incentives, monitor and control borrower performance, and deal with unforeseen contingencies affecting loan repayment.

Availability and Interest Cost of Credit

The availability and interest cost of credit to individual borrowers is determined by a set of four factors, two of which are found in the financial markets and two of which arise from the industry in which the borrower is engaged — in this case, the agricultural sector. In turn, each of the two factors involves macro and micro conditions, respectively.

First, consider those factors in the financial markets that influence the interest cost and availability of credit. One set of factors is based on the macro or aggregate financial condition of the marketplace. Included here are current monetary and fiscal policies, inflation conditions, structural characteristics of domestic and international financial markets, aggregate economic performance, and other factors influencing the levels of interest rates, money supply, loan demand, and so on. A second set of factors deal with the micro conditions that characterize individual financial institutions available to particular borrowers. Included here are the lender's profit performance, loss rates and loan delinquencies, cost of funds, size of operation, lending policies, legal structure, regulatory environment, and personnel.

These macro and micro factors interact to influence the interest cost and availability of credit, irrespective of the characteristics of individual borrowers. Moreover, these factors sometimes may contrast with each other; for example, the financial markets in general may have low interest rates and abundant funds to loan, while selected lending institutions may have higher interest rates on loans and reduced fund availability because of problems in their own financial performance.

Next, consider those factors in the borrower's business (i.e., agriculture) that influence the interest cost and the availability of his or her own credit. Macro and micro factors can also be distinguished here. Macro factors include the aggregate profit prospects for various farm commodities as determined by their supply and demand conditions, changes in the values of farm land and other resources, international trade conditions, and government farm programs. Micro factors include the business and personal characteristics that determine the credit worthiness of individual farm borrowers. Credit worthiness basically involves the lender's evaluation of the profitability and risks associated with lending to individual borrowers. Credit worthiness is evaluated on the basis of evidence borrowers furnish to assure lenders that lending risks will be minimal and that debt servicing will meet the terms of the loan contract. In the discussion to follow, we focus primarily on the concepts and methods used in evaluating credit worthiness.

Price Versus Non-price Responses

A price response by a lender is characterized by a change in the interest rate charged on the loan. This approach recognizes that the interest is the major "price" involved in a loan transaction. Other prices are fees and service charges. In contrast, a non-price response by a lender refers to changes in the amount of credit that is available and other terms of the loan agreement. These other terms might include changes in collateral requirements, loan maturities, loan documentation requirements, loan supervision, and loan disbursement and repayment control practices.

In practice, lenders use a combination of price and non-price responses in differentiating between borrowers according to differences in their lending risks, lending costs, and other attributes of business performance. Historically, heavier reliance has been placed on non-price responses, although beginning in the late 1980s, the balance shifted toward price responses in which interest rates were tailored more closely to the risk position and other financial characteristics of individual farm borrowers. The future may bring still greater reliance on differential loan pricing between borrowers to reflect their unique characteristics.

THE CREDIT PROFILE

The concepts of credit evaluation can be portrayed in a credit profile or matrix in which a borrower's internal determinants of credit are valued by the firm's lenders. While lenders use various terminology and classifications to describe the factors affecting credit worthiness, these factors basically can be defined in terms of the following four items: (1) assets available for loan security, (2) repayment and income expectations, (3) personal characteristics, and (4) other financial management practices. The other financial management practices include the borrower's attempts to manage risks, liquidity, taxes, and other factors that, in turn, may influence the lender's credit evaluation.

Since farm lenders have differing degrees of specialization, all lenders will not consider all of the farm sources of credit. In addition, one lender's response to a loan request might depend on whether the borrower has a loan from another lender and the loan terms and repayment obligation involved. For example, lending associations of the Farm Credit System are legally required to use land as security for real estate loans. Once they have a lien on the land, then another lender's use of land as security (i.e., as a source of credit) is subject to the lending association's prior lien. Similar cases exist when other lenders have liens on machinery, growing crops, and livestock.

Using the balance sheet data and income expectations shown in Table 7.2, let's apply some typical lender rules of thumb to a case farm situation to illustrate the credit evaluation process. The focus here is on the first two determinants of credit worthiness cited previously. We will assume that a primary real estate lender requires a one-third equity in purchased land. This requirement implies a maximum debt-to-equity ratio of 2. Non – real estate lenders are assumed to allow a maximum debt-to-equity ratio of 1 for non – real estate assets. In addition, operating credit arising from profit projections for a typical farm manager is generated at 70 percent of the expected gross value of crop production and 80 percent of the expected gross value of cattle feeding.

As indicated in Table 7.2, the $100,000 of equity in real estate held by the case farm will support $200,000 of real estate credit, and the $60,000 of equity in non – real estate assets will support $60,000 of non – real estate credit. Thus, the total credit generated by assets is $260,000. The operating credit generated by income expectations is estimated to be $52,500 for 250 crop acres and $36,000 for 150 cattle. Hence, the total operating capacity is $88,500.

Thus, under these particular rules of thumb, the case farmer could expect total credit of $348,500 at this point in time. Changes in asset structure and

Table 7.2. Evaluating the Credit Capacity of a Case Farm

Source	Asset Value (A)	Debt Out-standing (D)	Equity in Assets (E = A − D)	Maximum (D ÷ E)	Credit Capacity [(E)(D ÷ E)]
A. Asset Credit					
Real estate	$100,000	0	$100,000	2	$200,000
Non – real estate assets	60,000	0	60,000	1	60,000
Total	$160,000	0	$160,000		$260,000
	Gross Value per Unit	**Credit Rate**	**Credit per Unit**	**No. of Units**	**Credit Capacity**
B. Income Credit					
Crops	$300 / acre	70%	$210	250 acres	$ 52,500
Cattle	$300 / head	80%	$240	150 head	36,000
Total					$ 88,500

income expectations would alter the composition and amount of credit, as would changes in the lender's rules of thumb. It is important to note that this approach determines the maximum amount of credit that may be borrowed from available lenders. This amount may or may not have been borrowed. The unused portion represents the borrower's credit reserve.

Consider next the other internal determinants of credit. Personal factors, such as honesty, integrity, and reliability, are essential attributes of credit worthiness. The absence of any of these factors could quickly reduce the loan limits to zero. Therefore, personal factors have a high priority for the financial manager.

What financial practices can the borrower utilize to expand or restructure credit? Obviously, growth in equity capital, or improved production efficiency and marketing, which increases income expectations, would generate increased credit. However, such growth often depends on the availability of greater credit. The farm borrower should seek lenders who are knowledgeable about current production techniques and financing requirements of agriculture. The results should be more reliable credit evaluations that are tailored to the borrower's situation. Farmers can also signal superior profitability, financial progress, and repayment ability to lenders through more complete systems of financial reporting and planning. Complete and comprehensive financial statements that document past performance and project future performance are essential tools for building credit worthiness.

The various strategies used to manage risks in production, marketing, and finance also may significantly influence the firm's credit. The producer who fixes the crop price by contracting at planting time for its later sale may increase the credit generation rate — possibly beyond the 70 percent of expected gross value indicated in Table 7.2. Similarly, the cattle feeder who reduces price uncertainty by hedging on the futures market might raise credit to 90 percent or more of the expected gross value of the cattle. The use of insurance may also expand credit.

Credit also depends on the terms of obligation undertaken by the borrower. The amount of borrowing on an open account may be small. It can usually be increased, however, if the borrower is willing to sign a promissory note. It can be increased even more if collateral is pledged to secure the loan repayment, with greater collateral leading to greater credit.

LENDER PREFERENCES
AND RESOURCE ALLOCATION

Consistent with the agency cost relationships cited earlier in this chapter, the rate at which the borrower's credit is used up by borrowing depends on the compatibility between the loan characteristics and the preferences of the lender. Less preferred high-risk loans will use up credit at a rapid rate, if all other terms and provisions of the loan agreement remain constant. The result is relatively low loan limits. Consequently, the farmer's credit decisions are influenced by the loan purpose.

In general, lenders prefer loans that are both self-liquidating and asset-generating.[1] As established in Chapter 5, self-liquidating loans are made for purposes that will generate sufficient income to repay the loans within their maturity periods. Such loans reduce the lender's risk; however, as discussed earlier, loan maturities may often fall short of an asset's economic life, thus requiring a diversion of funds from net income to meet the loan payments. Most loans for feeder livestock, fertilizer, and other crop inputs have maturities that allow for loan repayment upon the sale of the fed livestock or the crops. Machinery, equipment, and buildings tend to rank lower in self-liquidation because the loan maturity often is shorter than the period over which the asset is depreciated.

---

[1]Of course, lenders also prefer making loans to borrowers with high management ability. High management ability may even substitute for other loan preferences. However, in considering the preferences toward self-liquidating and asset-generating loans, we will hold the borrower's management ability constant.

An asset-generating loan provides a valuable source of collateral that the lender can use to secure the loan in the event a default in repayments occurs. Thus, such a loan will be secured by a high degree of reclaimable assets. Based on reclaimability of assets, a loan to purchase feeder cattle is an asset-generating loan. In fact, cattle become more valuable as they grow to market weight. Their high marketability enhances their reclaimability. Loans for machinery also are asset-generating, although the machinery depreciates over time. A loan to purchase fertilizer is not strictly asset-generating because fertilizer is not reclaimable once it is applied. Loans for buildings, permanent fixtures, and drainage do not generate easily reclaimable assets. However, land is a readily reclaimable asset and is one of the most sought-after sources of loan security.

These two loan criteria (self-liquidating and asset-generating) are combined to determine the relative merits of alternative loan purposes for lenders and, in turn, for borrowers. This situation is illustrated in Figure 7.2. Loan purposes ranking high in both self-liquidation and asset reclaimability are most preferred by the lender. Loan purposes ranking high on one criterion but low on another have an intermediate preference from the lender's standpoint. And, loan purposes ranking low on both criteria have a low level of preference.

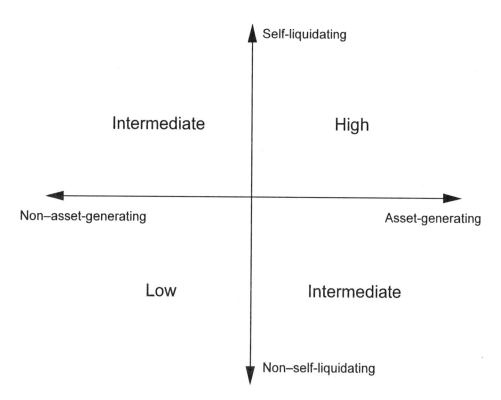

Figure 7.2. Lender preferences for combinations of loan characteristics.

The reader can classify other loan purposes in Figure 7.2. Where, for example, is a loan to purchase breeding livestock or dairy cows located? What maturities do such loans require to be self-liquidating?

In summary, an important element of credit management is recognizing that lender behavior influences, and is influenced by, the farm manager's resource allocation, risk management, and product choices. More loanable funds may be available for some purposes than for others. The effects of these lender responses in turn may influence managerial decisions. It is even possible, for example, for expansion to occur in what appears as a less profitable enterprise simply because more financing is available for this purpose. Under these conditions, the traditional economic theory of the firm, with resource allocation and product choices based on marginal value productivity, must be modified to account for the effects of financing. In fact, resource allocation may be influenced whether or not borrowing actually occurs because credit liquidity is valuable to the firm and because any transformation of assets affects the firm's credit. And, following the leverage propositions in Chapter 6, the borrower's access to credit will also influence the rate of growth of equity capital.[2]

These factors suggest that numerous strategies in borrowing and debt management arise from alternatives in production and marketing, in choice of lender, in the sequence and source of borrowing and repayment, in choice of financing instrument (note, mortgage, contract, lease) and financing terms, and in interactions between credit evaluations by different types of lenders. These factors are important to both borrowers and lenders because each choice modifies those attributes of loan purposes that are desirable to the lender: asset generation and self-liquidation of loans.

CREDIT EVALUATION
AND SCORING PROCEDURES

We will conclude this chapter by considering the practical side of credit evaluations by lenders and illustrating the numerical techniques employed in

---

[2]The influence of financing on the rate of growth and allocation of resources is not unique to agriculture. It permeates all types of production and consumption decisions. Consider, for example, the purchase of a house. The amount of available financing may influence an individual's decision as to what size of house may be purchased. Similarly, the financing terms that reflect the preferences of the lender may influence the type or age of a house that one buys. It is often easier to purchase a new house than an older house when, except for financing, the price of the old house might be more attractive. The financing terms on the new house generally include a lower downpayment, lower interest rate, longer maturity, and lower monthly payments than do the financing terms on the older house. This discrepancy reflects the higher lending risk for the older house.

making credit decisions. In practice, lenders evaluate their borrowers' credit worthiness by using various approaches that range from highly subjective informal methods to scoring techniques based on sophisticated statistical analysis of the borrower's financial data. Whatever the approach, however, the fundamental principle is the same — that is, to appraise the major attributes of the borrowers' credit worthiness in order to determine the credit risk involved in the lending situation. The results of the credit evaluation may serve several purposes: to distinguish between good and poor credit risks; to accept or reject loan applications; to price borrowers' loans; to identify credit situations needing special control, monitoring, and supervision; and to assist in examining the quality of the lender's loan portfolio.

It is important to recognize that credit evaluation cannot be reduced completely to an objective procedure based only on financial data. Rather, numerically based credit-scoring models are tools lenders may use along with their experience, judgment, and other information about the borrower's personal characteristics, management ability, and long-term financial prospects. Many instances arise involving young borrowers with good potential, growth situations, or perhaps temporary business setbacks in which a lender may be willing to take more lending risk in the short term in return for greater loan profitability in the future. The most widespread use of numerical credit-scoring techniques is in consumer lending (a mail-in application for a credit card is an example). Nonetheless, credit-scoring concepts are receiving substantial use in commercial and agricultural lending.

Credit scoring has several basic steps. The first is to identify key variables, factors, or characteristics that best distinguish between the different levels of credit worthiness. These variables are labeled $X_1, X_2, \ldots X_n$. The next step is to assign each variable a proper weight. These weights are labeled $a_1, a_2, \ldots a_n$. Then, the level of each variable should be multiplied (X) times its respective weight (a) and summed to a total (Y) to score each loan. The result is a weighted average credit score:

$$Y = a_1 X_1 + a_2 X_2 + \ldots a_n X_n$$

In principle, the credit worthiness of a population of borrowers ranges along a continuum from the lowest to the highest levels, as represented by the distribution of credit scores. In practice, however, lenders tend to group their borrowers into a few discrete categories for purposes of credit evaluation, loan pricing, monitoring, and so on. Thus, the estimated credit score could be used to assign borrowers to the various categories of credit worthiness, based on predetermined cut-off levels for each category. For example, the simplest classification scheme could involve two categories: acceptable and unacceptable loan applications. A more detailed classification scheme might break the acceptable category into four risk classes (low, intermediate, high, very high) so that risk-adjusted interest rates can be assigned to each class. The hierarchy of rates would reflect the differences in lending risks between the borrowers that make up the various classes. Moreover, in commercial loan programs from large banks, some loan pricing models also let the borrower's interest rate reflect the types and amount of deposits the borrower holds with the bank and the borrower's use of other financial services.

To illustrate the credit-scoring approach, we will consider a case situation in which a lender puts the prospective or existing borrowers in five credit classes, based on the results of the credit-scoring model. A worksheet for the model and the results for two farm borrowers are shown in Table 7.3. The lender has identified five of the borrower's performance characteristics that are especially important to the credit evaluation: X_1 is the profitability of the borrower's business, as measured by the rate-of-return on equity capital; X_2 is liquidity, as measured by the borrower's current ratio; X_3 is solvency, as measured by the borrower's debt-to-equity ratio; X_4 is the collateral position, as measured by the ratio of secured assets to maximum loan balance; and X_5 is a repayment capacity variable, as measured by the borrower's term debt and capital lease coverage ratio.

Each variable is weighted as follows: profitability, 10 percent; liquidity, 20 percent; solvency, 25 percent; collateral, 20 percent; and repayment capacity, 25 percent. Clearly, this lender gives more weight to solvency and repayment capacity than to liquidity, collateral, and profitability. Of course, considerable variation will exist among lenders in their use of particular variables, measures for these variables, weights on the variables, and even in their use of an approach such as this.

As shown in Table 7.3, each variable is scored on a range from 0 to 50 points, based on the location of the variable's value within a set of intervals representing a typical range of values for that variable. The higher the score, the stronger is that portion of the borrower's credit worthiness. To illustrate, the intervals for the profitability measure (return on equity) range from under 0 to over 20 percent, with four intervals in between. Each of the six intervals has a score that falls in the 0 to 50 range, at 10-unit intervals. If, for example, the borrower's rate-of-return on equity were 18 percent, then this measure would fall in the 15 percent to 19 percent interval and receive a score of 40 points. The other variables would be scored in a similar fashion.

The summary score (Y) is the weighted average of the score for each variable, using the lender's weights identified previously. The summary score then is assigned to the appropriate credit class for further analysis. The weighted average score and the credit classes are shown at the bottom of Table 7.3.

This process is illustrated in Table 7.3 for two farm situations. One is that of Joseph and Julie Farmer (the same case farm analyzed in Chapters 3 through 5). Based on the budgeted year data found in Table 4.5, the following performance measures are applicable to this scoring model: an expected rate-of-return on equity capital of 7.77 percent, a current ratio of 2.12, a debt-to-equity ratio of 0.60, a collateral ratio of 2.12, and a term debt and capital lease coverage ratio of 2.27.[3] The evaluation of each of these performance measures in the worksheet yields a summary credit score of 40, which places the borrower in credit class 5, "Low Risk." Thus, based on this numerical analysis alone, Joseph and Julie Farmer would be acceptable borrowers.

In contrast, Robert and Regina Rancher are new borrowers whose financial data indicate an expected rate-of-return on equity of 12 percent, a current ratio of 1.60, a debt-to-equity ratio of 1.20, a collateral ratio of 1.50, and a term debt and capital lease ratio of 1.60. The Ranchers' summary credit score is 30, which places them in credit class 4, "Intermediate Risk." Thus, the credit worthiness of Robert and Regina Rancher is less than that of Joseph and Julie Farmer, at least according to these financial data and the results of this credit-scoring model.

Clearly, changes in credit worthiness can occur over time because of changes in the various components of the credit-scoring model. The borrower's performance measures could be altered, the weights on the variables could be modified, new variables could be added or current ones deleted, and the designation of the credit classes could be changed. Trends in the credit score over time can indicate favorable or unfavorable directions for credit worthiness.

The scoring approach also enables the lender to advise the borrower about the factors that make up the credit score and about trade-offs or changes needed in selected variables to warrant reclassification. If Robert and Regina Rancher want to improve their credit score, they could evaluate changes in ratios that would put them into the "Low-Risk" classification. Other trade-offs could be similarly evaluated.

---

[3]The collateral ratio is measured as the total summed assets divided by the maximum operating loan plus the term loan balance. The total summed assets come from the sum of lines 1, 2, 7, 8, and 9, in Table A4.1. The operating and term loan balances are derived from lines 33 and 34 in Table 3.10.

Table 7.3. Credit-Scoring Worksheet with Applications to Joseph and Julie Farmer and Robert and Regina Rancher

Variable	Measure	Weight	Range	Score	Joseph and Julie Farmer	Robert and Regina Rancher
Profitability (X₁)	Return on equity	10% (A₁)	20% and greater	50		
			15 – 19	40		
			10 – 14	30		X
			5 – 9	20	X	
			0 – 4	10		
			less than zero	0		
Liquidity (X₂)	Current ratio	20% (A₂)	2.50 and greater	50		
			2.00 – 2.49	40	X	
			1.50 – 1.99	30		X
			1.00 – 1.49	20		
			0.50 – 0.99	10		
			less than 0.50	0		
Solvency (X₃)	Debt-to-equity ratio	25% (A₃)	less than 0.50	50		
			0.50 – 0.99	40	X	
			1.00 – 1.49	30		X
			1.50 – 1.99	20		
			2.00 – 2.49	10		
			2.50 and greater	0		
Collateral (X₄)	Ratio of secured assets to maximum loan balance	20% (A₄)	1.80 and greater	50	X	
			1.60 – 1.79	40		
			1.40 – 1.59	30		X
			1.20 – 1.39	20		
			1.00 – 1.19	10		
			less than 1.00	0		
Repayment capacity (X₅)	Term debt and capital lease coverage ratio	25% (A₅)	2.50 and over	50		
			2.00 – 2.49	40	X	
			1.50 – 1.99	30		X
			1.00 – 1.49	20		
			0.50 – 0.99	10		
			less than 0.50	0		

Table 7.3 (Continued)

Summary Score	Variable	Weight	Joseph and Julie Farmer	Robert and Regina Rancher
	Profitability (X_1)	0.10	20	30
	Liquidity (X_2)	0.20	40	30
	Solvency (X_3)	0.25	40	30
	Collateral (X_4)	0.20	50	30
	Repayment capacity (X_5)	0.25	40	30
	Score (Y)		40	30

Credit Class	Scoring Range	Joseph and Julie Farmer	Robert and Regina Rancher
1. Non-acceptable loan	0 – 9.99	_____	_____
2. Very high – risk loan	10 – 19.99	_____	_____
3. High-risk loan	20 – 29.99	_____	_____
4. Intermediate-risk loan	30 – 39.99	_____	X
5. Low-risk loan	40 – 50	X	_____

As indicated earlier, a credit-scoring model serves as a tool to aid lenders in evaluating the credit worthiness of their farm borrowers. The credit-scoring approach provides a systematic, comprehensive way in which to assess the borrower's financial data and, along with the lender's judgment and other relevant information, reach a valid assessment of the borrower's credit worthiness. In turn, the credit evaluation yields important information about the firm's borrowing capacity, credit reserves, and ability to manage liquidity.

Summary

Credit relationships between borrowers and lenders are essential to liquidity management in agriculture. Compared to other sources of liquidity, credit reserves are flexible and efficient but also subject to random variation in credit availability. The size and structure of credit reserves are based on a borrower's credit worthiness, which is determined by lenders, based on formal or informal methods of credit scoring and risk assessment.

More specific points are as follows:

- Liquidity involves the firm's capacity to generate cash to meet cash demands and to provide funds for responding to unanticipated events.

- Cash demands include liquidity needed for transactions, safety, and investment purposes.

- Holdings of various types of assets, especially financial assets, provide an important source of liquidity. Major factors affecting asset liquidity are transactions costs, market characteristics, liquidity risk, and a firm's capital integrity.

- Holdings of credit reserves, distinguished by type of credit and lending institution, provide another valuable source of liquidity, subject to both advantages and disadvantages.

- **Credit,** defined as *borrowing capacity,* is determined by lenders, although borrowers may form expectations on credit terms and influence credit through their financial management and business characteristics.

- The results of credit evaluations are expressed by lenders in both price and non-price terms, although greater reliance on risk-adjusted interest rates in recent years has increased the importance of price responses.

- Differences between borrowers and lenders in incentives and goals may influence the cost and availability of credit, thus affecting managerial decisions by borrowers.

- **Credit scoring** is *a systematic, numerical approach to evaluating credit worthiness* that weighs financial ratios and other measures representing a borrower's profitability, liquidity, solvency, collateral, and repayment ability.

Topics for Discussion

1. Identify and explain the various needs for liquidity.

2. Describe the notion of "credit." Who evaluates credit? How is the evaluation accomplished? What leads to increased credit? What can credit be used for?

3. In what ways can credit evaluation and borrowing influence a manager's resource allocation decisions?

4. Why is liquidity important to a farm manager? What determines an asset's liquidity?

5. What are the major factors that influence a farmer's credit worthiness? Explain how changes in these factors may influence the level of available credit.

6. Contrast the effects on credit for loans to finance (a) feeder cattle, (b) storage facilities, and (c) dairy cows.

7. Identify the factors affecting the liquidity of various assets. How would you rank the liquidity of (a) grain inventories, (b) farm land, (c) farm machinery, and (d) U.S. Treasury bills. Why do they differ in liquidity?

8. Explain the concept of a credit reserve. How is the credit reserve related to actual amounts of borrowing? Why is the credit reserve considered a source of liquidity? How can the value of the credit reserve be expressed? How does the liquidity associated with a credit reserve differ between farm lenders?

9. Distinguish between a lender's price and non-price responses in a credit evaluation. Which is the preferred method? Why?

10. Explain the purposes and procedures of credit scoring. How might the results change over time?

11. Explain the principal-agent relationship as it applies to financing transactions. What are agency costs and how are they controlled?

References

1. Baker, C. B., "Credit in the Production Organization of the Firm," *American Journal of Agricultural Economics*, 50(1968):507 – 520.

2. Baker, C. B., and J. A. Hopkin, "Concepts of Financial Capital for a Capital-Using Agriculture," *American Journal of Agricultural Economics*, 51(1969):1055 – 1065.

3. Barry, P. J., C. B. Baker, and L. R. Sanint, "Farmers' Credit Risks and Liquidity Management," *American Journal of Agricultural Economics*, 63(1981):216 – 227.

4. Barry, P. J., and C. B. Baker, "Management of Financial Structure: Firm Level," *Agricultural Finance Review*, 37(February 1977):50 – 63.

5. Barry, P. J., and C. B. Baker, "Reservation Prices on Credit Use: A Measure of Response to Uncertainty," *American Journal of Agricultural Economics*, 53(1971):222 – 227.

6. Barry, P. J., and D. R. Fraser, "Risk Management in Primary Agricultural Production: Methods, Distribution, Rewards and Structural Implications," *American Journal of Agricultural Economics*, 58(1976):286 – 295.

7. Barry, P. J., and D. R. Willmann, "A Risk-Programming Analysis of Forward Contracting with Credit Constraints," *American Journal of Agricultural Economics*, 58(1976):61 – 70.

8. Boehlje, M. D., and L. D. Trede, "Risk Management in Agriculture," *Journal of the American Society of Farm Managers and Rural Appraisers*, 41(1977):20 – 29.

9. *Economic Growth of the Agricultural Firm*, Tech. Bull. 86, College of Agriculture Research Center, Washington State University, Pullman, 1977.

10. Ellinger, P. N., N. S. Splett, and P. J. Barry, "Consistency of Credit Evaluations at Agricultural Banks," *Agribusiness: An International Journal*, 8(1992):43 – 56.

11. Harris, K. S., and C. B. Baker, "Financing East-Central Illinois Farmers Who Hedge," *Agricultural Finance Review*, 642(1982):9 – 15.

12. Jensen, M., and W. Meckling, "Theory of the Firm: Managerial Behavior, Agency Costs, and Ownership Structure," *Journal of Financial Economics*, 3(1976):305 – 360.

13. Miller, L. H., and E. L. La Due, "Credit Assessment Models for Farm Borrowers: A Logit Analysis," *Agricultural Finance Review*, 49(1989):22 – 36.

14. Pflueger, B. W., and P. J. Barry, "Crop Insurance and Credit: A Farm-Level Simulation Analysis," *Agricultural Finance Review*, 46(1986): 1 – 15.

15. Robison, L. J., and J. R. Brake, "Applications of Portfolio Theory to Farmer and Lender Behavior," *American Journal of Agricultural Economics*, 61(1979)158 – 164.

16. Smith, C., "Agency Costs," *The New Palgrave, A Dictionary of Economics*, eds., Eatwell, J., M. Milgate, and P. Newman, Macmillan Press, London, 1987.

17. Turvey, C., "Credit Scoring for Agricultural Loans: A Review with Applications," *Agricultural Finance Review*, 51(1991):43 – 54.

8

Risk Management

Risks are pervasive in agriculture. Types of business risks for farmers include (1) production and yield risks; (2) market and price risks; (3) losses from severe casualties and disasters; (4) social and legal risks from changes in tax laws, government programs, trade agreements, and so on; (5) human risks in the performance of labor and management; and (6) risks of technological change and obsolescence. Financial risks arise from the financial claims on the firm. The greater the financial leverage, the greater the risks in meeting financial obligations to lenders and lessors. Financial risks are further increased by unanticipated variations in interest rates, credit availability, and other changes in loan terms, as well as in leasing terms.

When combined, business and financial risks magnify potential losses in farmers' equity capital and net income and create inefficiencies in resource use by hampering business planning. Risk management involves the selection of methods for countering business and financial risks in order to meet the decision maker's risk-averting goal. However, lower risk is generally associated with reduction in expected returns. Thus, it is important to account for the risk – return trade-off in designing risk management strategies.

In this chapter we identify the major choices in risk management for managers of agricultural production units, emphasizing financial responses to risk and their relationships to production and marketing responses. The risk management process is established, the portfolio modeling approach is used to present the fundamentals of diversification and to evaluate the capacity to service debt and to meet other financial obligations under risk, and the major methods of responding to risks are presented and analyzed.

A RISK MANAGEMENT PERSPECTIVE

The process of managing risks is based on the steps of financial control outlined in Chapter 4 — setting goals, measures, norms, and tolerance limits; designing an information system; and initiating corrective actions when needed. Control occurs over time and under uncertainty. Within the control process, risk management focuses on the strategies and corrective actions needed to cope with possible adversities in business performance.

The first phase in risk management is designing strategies to cope with risks. These generally are long-range plans for using various risk responses that should hold over a period of years and over a range of uncertain events. The second phase in risk management is the implementation of risk responses and their specific use when adversity occurs. The third phase restores the firm's capacity to implement the risk strategies when distress conditions have passed. Our emphasis here is on phase 1 of the risk management process — the design of appropriate strategies.

Some risk responses may focus on reducing risks within the business. Effective diversification of several types of assets and business enterprises is an example. Other responses may focus on transferring the risks outside the business. Hedging through forward or futures contracts for commodities or farm inputs is an example. Still other risk responses may not reduce the likelihood of occurrence but may better enable the firm to bear risks when they occur. Holding liquid reserves of cash or credit, or using insurance, illustrates the building of risk-bearing capacity within the firm. Finally, some risk responses may exhibit several of these effects, as with purchasing insurance, which builds liquidity and transfers risk acceptance outside the business.

Some methods of managing risks are feasible for all types of farms. Others are only feasible for certain sizes and types of farms, qualities of management, financial structures, and other characteristics. The methods can be categorized in terms of the production, marketing, and financial organizations of farm businesses.

In production, risk responses include enterprise diversification, informal insurance (pesticides, reserve equipment, supplemental irrigation), organization flexibility, multiple production practices (e.g., planting several seed varieties), and avoidance of high-risk enterprises.

In marketing, risk responses include inventory management and forward and futures contracts. Participation in government programs may be a response to production or marketing risks, or both, depending on the program. Vertical integration may also present opportunities for reducing some production and marketing risks. Financial responses to risk reflect the firm's capacity to bear risks in production and marketing and mostly involve the management of leverage and liquidity.

The payoffs associated with these methods of managing risks are often difficult to measure because they involve the protection of the firm's equity from possible losses. Furthermore, the financial responses to risk may involve other risk responses too. The risk responses in production, marketing, and finance generally involve trade-offs. Thus, emphasis on one means of countering uncertainty (e.g., liquidity) may allow an investor to carry greater risks in production or marketing. The reverse may also be true.

FUNDAMENTALS OF DIVERSIFICATION

The fundamentals of diversification are extremely important in risk analysis. It may be possible to reduce the total variability of returns by combining several assets, enterprises, or income-generating activities without unduly sacrificing expected returns. As a rule, higher (lower) risk investments carry higher (lower) expected returns. Thus, reducing risk usually involves lower expected returns as well. Frequently, however, gains in business planning and risk efficiency can occur if the principles of diversification are followed.

To illustrate, consider an investor who is evaluating two farm units, one located in the Corn Belt and the other in the Great Plains area of the United States. A financial analyst reports to the investor that the two farm units have about the same expected profitability and risk. More specifically, each farm is expected to earn a 20 percent return to assets, with a standard deviation of 10 percent. These data indicate that returns from either investment should average 20 percent over time and should fall between 10 and 30 percent about two-thirds of the time and between 0 and 40 percent about 95 percent of the time. Moreover, the chance of returns falling below zero (two standards deviations below the mean) is about 2.5 percent.

Based on these figures, the investor should be indifferent between the two farm units. But, suppose the investor could divide available funds and invest equally in both the Corn Belt and the Great Plains units. What gains could arise from this investment strategy? The expected returns would still be 20 percent, but the standard deviation of returns from the combined investments could decline to 7.07 percent, as we will show in the following sections. The diversified investment would reduce the anticipated variability of returns, and thus the investment's risk, without reducing the expected levels of returns. Hence, the diversified investment would be preferred over investing in only one unit.

Why does risk decline from combining two seemingly comparable investments? The answer is based on the statistical relationship between the two investments. We develop this relationship by using a portfolio model approach.

Portfolio Model

The word *portfolio* refers to *a mix, or combination, of assets, enterprises, or investments.* It is often used to describe holdings of financial assets such as stocks and bonds. However, it can also be applied to holdings of tangible assets such as grain inventories, growing crops, livestock, machines, land, and apartment buildings. Hence, the portfolio of a grain – livestock producer is considered to hold two investments — grain and livestock — although different crops and livestock could further diversify the portfolio. Or, the investor described above could hold two different farm units in the portfolio.

The portfolio model indicates how different combinations of investments may reduce an investor's risk more than having a single investment. *Holding combinations of investments* is called **diversification,** with the potential for risk reduction determined by (l) the number of investments held, (2) correlation (or covariation) between the expected returns of the individual investments, and (3) possible changes in the levels of costs and returns per unit of investment as a result of diversifying. In particular, this chapter will demonstrate how the gains in risk reduction from diversification increase as the correlation between investments declines and as the number of investments in a portfolio increases.

Positive covariation means that high profits in one investment are associated with high profits in another investment. Negative covariation means that high profits in one investment are associated with low profits in another investment. Zero covariation means that there is no statistical association between the variations of returns of these investments. The investments are statistically independent.

Diversification as a risk-reducing strategy becomes more effective as the covariation among investments is lower. These effects are demonstrated with statistical measures of standard deviation, variance, covariance, and correlation. To illustrate, let X_1 and X_2 be two investment alternatives having expected rates-of-return $\bar{r}_1$ and $\bar{r}_2$, respectively. Let P_1 and P_2 be the proportions of total resources invested in X_1 and X_2, respectively, with $P_1 + P_2 = 1$ to assure that all resources are accounted for.

The portfolio's expected return ($\bar{r}_t$) is now:

$$\bar{r}_t = \bar{r}_1 P_1 + \bar{r}_2 P_2 \tag{8.1}$$

Let σ_1 and σ_2 be the standard deviations of X_1 and X_2, and σ_{12} be their covariance. An alternative expression of covariance is $\sigma_{12} = c\sigma_1\sigma_2$, where c is the correlation coefficient between returns r_1 and r_2. The correlation coefficient varies between positive and negative $1(-1 \leq c \leq 1)$, with a positive value indicating positive covariation, a negative value indicating negative covariation, and a zero value indicating no covariation. The advantage of using correlation is that it is an index or relative measure of the strength of association between two investments, while covariance is an absolute measure of the association.

The total variance of the portfolio is the sum of the individual proportional variances plus (or minus) the covariance:

$$\sigma_T^2 = \sigma_1^2 P_1^2 + \sigma_2^2 P_2^2 + 2P_1 P_2 c\sigma_1\sigma_2 \tag{8.2}$$

and the total standard deviation is:

$$\sigma_T = \sqrt{\sigma_T^2} \tag{8.3}$$

Some special cases of the portfolio model are expressed as follows:

If all resources are in X_1, so that $P_1 = 1.0$ and $P_2 = 0.0$, then:

$$\sigma_T^2 = \sigma_1^2$$

For the following correlation values, total variance is:

$c = -1, \sigma_T^2 = \sigma_1^2 P_1^2 + \sigma_2^2 P_2^2 - 2P_1 P_2 \sigma_1 \sigma_2$

$c = 0, \sigma_T^2 = \sigma_1^2 P_1^2 + \sigma_2^2 P_2^2$

$c = 1, \sigma_T^2 = \sigma_1^2 P_1^2 + \sigma_2^2 P_2^2 + 2P_1 P_2 \sigma_1 \sigma_2$

Total variance, or portfolio risk, now depends on the relative proportions (P_1, P_2) of X_1 and X_2 in the portfolio, their variances, and the correlation of their returns. Equation 8.2 clearly shows that the higher the value of c, the higher the portfolio risk for any combination of investments X_1 and X_2. Similarly, the lower the value of c, the lower the portfolio risk for any combination of investments X_1 and X_2. Hence, the lower the correlation, the greater the reduction in risk associated with diversification.

The risk estimates for portfolios of X_1 and X_2 in equation 8.2 can be combined with the portfolio's estimates of expected returns in order to evaluate risk – return trade-offs for different combinations of X_1 and X_2. You can also expand these formulas to include other investments by adding terms representing their expected returns, standard deviations, correlations with other investments, and weights in the overall portfolio.

Portfolio Analysis

To illustrate the gains in risk reduction from diversification, consider again the investor's choice of a farm unit in the Corn Belt (X_1) or one in the Great Plains (X_2). Recall that the expected returns and standard deviations of these farm units are 20 percent and 10 percent, respectively, and assume a zero correlation between their returns. These data are expressed as:

$\bar{r}_1 = 0.20$, $\bar{r}_2 = 0.20$, $\sigma_1 = 0.10$, $\sigma_2 = 0.10$, and $c = 0.00$

Investing in either the Corn Belt unit or the Great Plains unit would yield an expected return ($\bar{r}_t$) of 20 percent and a standard deviation (σ_t) of 10 percent. Moreover, a portfolio composed of equal proportions ($P_1 = P_2 = 0.50$) of the Corn Belt and Great Plains units would yield the following expected returns.

$$\bar{r}_t = (0.20)(0.50) + (0.20)(0.50) = 0.20, \text{ or } 20\% \tag{8.4}$$

However, using equation 8.2, find the total variance of the diversified portfolio, which will be:

$$\sigma_T^2 = (0.10)^2(0.50)^2 + (0.10)^2(0.50)^2 + 2(0.50)(0.50)(0.00)(0.10)(0.10) \tag{8.5}$$

$$= 0.0025 + 0.0025 + 0.0000$$

$$= 0.0050$$

and the standard deviation, which will be:

$$\sigma_T = \sqrt{0.0050} \qquad (8.6)$$

$$= 0.0707, \text{ or } 7.07\%$$

Thus, the diversified portfolio yields the same expected return but with about 30 percent less risk, based on an assumption of zero correlation between the returns of these investments. If the correlation were less than zero, the risk reduction from diversification would be greater.

In this example, an assumption of zero correlation for the returns of these investments may be realistic, since the farm units are far apart geographically and perhaps subject to different economic forces. Suppose, however, that the investor was considering two farm units (X_{1A}, X_{1B}) in the Corn Belt located immediately adjacent to one another. Both farm units have an expected return of 20 percent and a standard deviation of 10 percent. However, because the units grow the same crops and are subject to the same risks, their returns are perfectly positively correlated; that is, $c = 1$.

Are there any gains from diversification now? The total variance of a portfolio with equal proportions of the two investments is:

$$\sigma_T^2 = (0.10)^2(0.50)^2 + (0.10)^2(0.50)^2 + 2(0.50)(0.50)(1)(0.10)(0.10) \qquad (8.7)$$

$$= 0.0025 + 0.0025 + 0.0050$$

$$= 0.0100$$

and the standard deviation is:

$$\sigma_T = \sqrt{0.0100} \qquad (8.8)$$

$$= 0.10, \text{ or } 10\%$$

Thus, no gains in risk reduction occur from diversifying investments with returns that are perfectly positively correlated. This case is an exception, however. Diversifying does reduce risk for less than perfect correlation, with the reduction in risk increasing, the lower the level of correlation.

Next, consider how portfolio risk declines as the number of assets held in the portfolio increases. To illustrate, suppose a third farm unit, (X_3), located in the Pacific region, is added to the portfolio, along with the units in the Corn

Belt and the Great Plains regions. Let the expected return and standard deviation of X_3 also be 20 percent and 10 percent, respectively, and assume independence (zero correlation) among the returns of the three investments. Using an expanded version of equation 8.2, find the expected returns from a portfolio composed of equal proportions ($P_1 = P_2 = P_3 = 0.333$) of the three investments, which will be:

$$\bar{r}_t = (0.20)(0.333) + (0.20)(0.33) + (0.20)(0.333) \tag{8.9}$$

$$= 0.20, \text{ or } 20\%$$

Similarly, the total variance of the portfolio will be:

$$\sigma_T^2 = \sigma_1^2 P_1^2 + \sigma_2^2 P_2^2 + \sigma_3^2 P_3^2 \tag{8.10}$$

$$= (0.10)^2(0.333)^2 + (0.10)^2(0.333)^2 + (0.10)^2(0.333)$$

$$= 0.003327$$

and the standard deviation will be:

$$\sigma_T = \sqrt{0.003327}$$

$$= 0.0577, \text{ or } 5.77\%$$

Comparing the standard deviation of 5.77 percent for the three-investment portfolio to the standard deviation of 7.07 for the two-investment portfolio and to 10 percent for the single-investment portfolio indicates the gains in risk reduction from diversification as the number of investments increases. Recall that this example assumed zero correlation among the investments. In reality, the correlation coefficients among assets that make up large portfolios usually average greater than 0, but much less than 1. As a result, diversification over more investments does indeed reduce portfolio risk, although the gains in risk reduction diminish as the number of investments increases. Evidence from portfolios of financial assets, for example, indicates that holding 15 to 20 common stocks largely exhausts the possible gains in risk reduction.

An Extended Portfolio Model

The portfolio model can be extended to the case of more than three assets in a straightforward fashion. If, for example, 20 assets were available to form a portfolio, then the portfolio's expected rate-of-return would be the weighted sum of the expected rates-of-return of each of the 20 assets. Similarly, the variance of the portfolio would be the weighted sum of the variances of each asset plus the weighted sum of the covariances of all possible pairs of the assets. Writing these equations in detail would be very cumbersome and space-consuming. It is easier to express these equations in general terms, using summation signs and notation for N assets, where N can range from 1 to infinity.

The portfolio's expected rate-of-return then is:

$$\bar{r}_t = \sum_{n=1}^{N} r_n P_n \qquad (8.11)$$

and the portfolio variance is:

$$\sigma_T{}^2 = \sum_{n=1}^{N} P_i^2 \, \sigma_i^2 + \sum_i \sum_j P_i P_j \sigma_{ij} \qquad (8.12)$$

where i and j represent a covariance pairing of assets i and j. The first term to the right of the equals sign in equation 8.12 is the sum of the variances of the N assets and the second term is the sum of the covariances between each and every possible pair of assets. Clearly, the data needed for portfolio analysis increase substantially as more assets are evaluated. Using a computer to perform the calculations becomes essential.

Portfolio Theory, Risk Dominance, and Risk Programming

Portfolio theory also is related to the concept of risk dominance or risk efficiency introduced in Chapter 2. Recall that risk dominance occurs when the variance of one asset, or a portfolio of assets, is lower than the variance of another asset, while the first asset has the same or greater than expected rate-of-return. The dominating (lower variance) asset or portfolio generally is preferred by risk-averse investors to the dominated (higher variance) asset or portfolio.

A risk-efficient set of portfolios is composed of the portfolios of assets that minimize variance for different levels of expected returns. That is, variance in equation 8.12 is minimized for different levels of expected returns in equation 8.11. The result is the risk-efficient set portrayed in Figure 2.1 and repeated as Figure 8.1 in this chapter. The optimal choice from among the risk-efficient set will vary among investors, depending on their level of risk aversion.

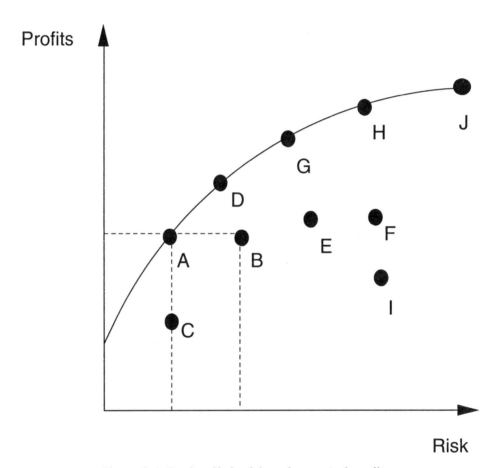

Figure 8.1. Trade-offs in risk and expected profits.

In practice, a computerized mathematical procedure called risk programming [2, Ch. 10] can be used to derive a risk-efficient set. In standard portfolio analyses involving stocks, bonds, and other financial assets, the risk programming procedure finds the combinations of financial assets that minimize portfolio variance for different levels of expected return, subject to the constraint that the proportional holdings of the assets sum to 100 percent. Data needs include the expected rates-of-return, variances, and covariances for all the assets under consideration.

In applications of risk programming to agricultural businesses, the concept of an asset is broadened to include crop and livestock enterprises; acquisition of machinery, buildings, and land; hiring labor; financing alternatives; consumption and tax activities; and investments in financial assets. A broader set of constraints is specified as well, including limitations on resource availability, borrowing capacities, cash flows, and accounting and tax requirements. The risk-efficient solutions to the risk programming model then may be characterized by various forms of diversification, and other risk management methods may be analyzed as well.

DIVERSIFICATION, RISK, AND FINANCIAL SERVICING CAPACITY

Portfolio theory and diversification principles may also be applied to the financial servicing capacity of farm businesses in order to estimate the probabilities of meeting debt and other financial obligations under different leverage positions. To illustrate, consider Robert and Regina Rancher who must decide how to allocate $700,000 among three investment alternatives. Alternative A is a livestock enterprise that has an expected net cash return of 14 percent and a standard deviation of 8 percent. Alternative B is a crop enterprise that has an expected net cash return of 12 percent and a standard deviation of 5 percent. Alternative C is a financial asset expected to yield 7.5 percent with a standard deviation of 1 percent.

The Ranchers are considering three portfolios that represent different combinations of these enterprises. The portfolios are (1) 100 percent invested in the livestock enterprise, (2) 50 percent invested in the crop enterprise and 50 percent in the livestock enterprise, and (3) one-third invested in each of the three enterprises. The correlation coefficients for the enterprises are $c = 0.30$ between livestock and crops, $c = -0.40$ between livestock and the financial asset, and $c = -0.10$ between crops and the financial asset. Under each portfolio, other financial obligations, including family living, total $25,000.

The portfolios will be evaluated under four leverage positions: debt-to-equity ratios of $D/E = 0.15, 0.50, 1.00,$ and 1.50. Table 8.1 summarizes these data. The debt-to-equity ratios are applied to the $700,000 of assets in each portfolio to determine the levels of debt. Annual debt payments then are found

Table 8.1. Portfolio Data for Financial Servicing Analysis Under Risk

	Investment Alternatives		
	A	B	C
Investment parameters			
Expected cash returns per $1.00 invested	$0.14	$0.12	$0.075
Standard deviation of returns	$0.08	$0.05	$0.01
Coefficient of variation	0.57	0.42	0.13

Correlations between investment alternatives

Alternative A with B 0.30

Alternative A with C . . . . −0.40

Alternative B with C . . . . −0.10

	Portfolio Weights of Each Investment Alternative		
	A	B	C
Investment portfolio			
1	100%	0%	0%
2	50%	50%	0%
3	34%	33%	33%

by amortizing the debt over an average payment period of 15 years at an annual interest rate of 10 percent (see Chapters 9 and 14 for a discussion of amortization procedures).

Based on the financial leverage principles discussed in Chapter 6, we would expect a portfolio's financial risk to increase as the debt-to-equity ratio increases. Greater financial risk means lower probabilities of meeting debt repayment requirements. Based on diversification principles, we would expect a farm's business risk to decline as assets or enterprises with lower degrees of correlation are added to the portfolio. Lower business risk means greater probabilities of meeting debt repayment requirements. The net effect on repayment probabilities of increasing financial risk and decreasing business risk will depend on the relative magnitude of the two changes.

These relationships are illustrated in Table 8.2 for the particular numerical specifications introduced above. To understand the calculation procedures, consider the data for portfolio 2 in Table 8.2. The expected net cash return (NCR) of $91,000 results from allocating 50 percent of resources to each of the crop and livestock enterprises.

Table 8.2. Financial Servicing Capacity Under Risk Conditions

	Investment Portfolio		
	1	**2**	**3**
Returns per dollar of investment			
1. Expected cash return	14.0%	13.0%	11.1%
2. Standard deviation of return	8.0%	5.3%	3.4%
Dollar returns from $700,000 investment portfolio			
3. Expected cash return	$98,000	$91,000	$77,385
4. Standard deviation	$56,000	$37,206	$23,850
5. Coefficient of variation [4/3]	0.571	0.409	0.308
6. Other obligations, including financial obligations	$25,000	$25,000	$25,000
Debt-to-Equity Ratios			
D/E = 0.15			
7. Principal and interest obligations	$12,004	$12,004	$12,004
8. Minimum funds needed for withdrawals and debt obligations [7 + 6]	$37,004	$37,004	$37,004
9. Probability of cash returns meeting debt and other financial obligations	86.2%	92.7%	95.5%
D/E = 0.5			
10. Principal and interest obligations	$30,677	$30,677	$30,677
11. Minimum funds needed for obligations [10 + 6]	$55,677	$55,677	$55,677
12. Probability of cash returns meeting debt and other financial obligations	77.5%	82.9%	81.9%
D/E = 1.00			
13. Principal and interest obligations	$46,016	$46,016	$46,016
14. Minimum funds needed for obligations [13 + 6]	$71,016	$71,016	$71,016
15. Probability of cash returns meeting debt and other financial obligations	68.4%	70.4%	60.5%
D/E = 1.50			
16. Principal and interest obligations	$55,219	$55,219	$55,219
17. Minimum funds needed for obligations [16 + 6]	$80,219	$80,219	$80,219
18. Probability of cash returns meeting debt and other financial obligations	62.4%	61.4%	45.3%

$$\text{NCR} = [0.14 \times 0.50 + 0.12 \times 0.50] \times \$700{,}000$$

$$= 0.13 \times \$700{,}000$$

$$= \$91{,}000$$

The standard deviation of the portfolio is found by using equations 8.2 and 8.3 where the weights on the enterprise units are 50 percent.

$$\sigma_t^2 = 0.50^2 \times 0.08^2 + 0.50^2 \times 0.05^2 + 2(0.50)(0.50)(0.3)(0.08)(0.05)$$

$$\sigma_t^2 = 0.00282$$

$$\sigma_t^2 = 0.053$$

Applying the standard deviation of the net cash return to the $700,000 investment yields:

$$\sigma_{\text{NCR}} = 0.053 \times \$700{,}000$$

$$\sigma_{\text{NCR}} = \$37{,}206$$

$$\text{CV} = \frac{\sigma_{\text{NCR}}}{\text{NCR}} = \frac{37{,}206}{91{,}900} = 0.409$$

For a debt-to-equity ratio of 0.15, the $700,000 of assets are financed by $608,696 of net worth and $91,304 of debt, yielding an annual amortized debt payment of $12,004.[1] Thus, debt plus other financial obligations on this farm is $37,004. The difference between the expected cash returns ($91,000) and the required debt payments and other obligations of $37,004 is $53,996. This margin of difference indicates that the obligations of $37,004 are 1.45 standard deviations (Z) below the expected cash returns of $91,000. That is:

$$Z = \frac{91{,}000 - 37{,}004}{37{,}206} = 1.45$$

Using a normal probability distribution table (see Appendix Table I), you can determine the probability of available funds is 92.7. To use the table, follow

---

[1]Amortized debt payment of 12,004 is calculated as:

$$\frac{\$91{,}304}{\text{USPV}_{0.10,\ 15}}$$

the column headed "z" to "1.4" and go across to the column headed "0.05." This value, "0.9265," represents the probability that income will exceed minimum funds needed for debt and other financial obligations. The other data in Table 8.2 are found in a similar fashion.

As shown in the rest of Table 8.2, the probabilities of meeting financial obligations decline as leverage increases for a given portfolio. Moreover, as additional assets are added in portfolios 2 and 3, the coefficients of variation signifying business risk decline. However, the probabilities of meeting the repayment requirements increase only in a couple of cases. The lower expected returns for enterprises 2 and 3, together with the fixed repayment and other obligations, result in lower probabilities of meeting these obligations in most of the cases, even though business risk is declining across the portfolios. These examples illustrate an approach that shows how portfolio analysis can be applied to evaluate debt-servicing capacity under conditions of risk.

ENTERPRISE DIVERSIFICATION IN AGRICULTURE

Diversifying among several types of farm enterprises, and even between farm and nonfarm activities, is a traditional approach to risk management in agriculture. It is premised on the condition that low or negative correlation of returns among some enterprises will stabilize total returns over time. When the income of one enterprise is low, the income from another is high.

Enterprise diversification in farm businesses should be carefully considered, although most farms experience a limited range of available enterprises without sacrificing too much in expected returns. Moreover, both prices and yields of most crops grown in a given area tend to be positively correlated, often highly so. This correlation occurs because in roughly the same location, most crops experience similar weather patterns, use similar resources, and experience similar market factors. Combining livestock and crops is likely the most promising approach. Nonetheless, the average of the correlations will probably still be considerably less than $c = 1.0$.

Examples of correlations among enterprise returns are provided by two studies of farmers' risk management. In a study of Iowa farms [5] over the 1965 – 1974 period, the only negative correlations in enterprise returns occurred between soybeans and various livestock enterprises (hogs, feeder cattle, dairy). Small positive correlations were reported between hogs and dairy (0.179), beef cows and dairy (0.248), and fed cattle and dairy (0.254). Other correlations were higher, as in the case of hogs and fed cattle (0.658), beef cows and corn (0.648), and corn and soybeans (0.481).

A similar study in Georgia [17] focused on a wider range of crops and found negative correlations between net returns of such dissimilar crops as wheat and tobacco (– 0.63), cotton and tobacco (– 0.50), and cotton and peanuts (– 0.16). Correlations were positive between corn and wheat (0.24), corn and grain sorghum (0.95), corn and soybeans (0.68), and between other feed grains.

Several studies have considered the possible benefits from diversification in both farm and nonfarm investments. In general, the results indicate that the correlation coefficients are low and in some cases negative. Young and Barry [23] observed negative correlations ranging from – 0.136 to – 0.408 between returns to grain farms and returns earned by the Standard and Poors 500 Index, small company stocks, treasury bills, certificates of deposit, preferred stocks, and corporate, government, and municipal bonds. Moss, et al., [16] found negative correlations between returns to farm assets and treasury bills, corporate and government bonds, and a small positive correlation (0.144) with small stocks. In a third study, Crisostomo and Featherstone [6] also found negative correlations between returns to treasury bills and returns to assets and equity for crop, dairy, swine, and beef feeding enterprises. Negative correlations with common stocks also were observed for swine farms and general farms, although other stock correlations were modestly positive. In general, then, diversification between farm and nonfarm investments appears promising as a risk management strategy.

A major problem with enterprise diversification is the *losses in efficiencies and returns from specialized production*. These losses, called **diseconomies of scope,** could outweigh the value of any risk reduction from diversification. Consequently, specialization often increases rather than decreases as farms become more commercialized, in order to gain the higher expected returns. The result is greater emphasis on other methods of risk management.

Both economies of scale and economies of scope are involved here. Economies (diseconomies) of scale refer to possible reductions (increases) in costs per unit of production as the scale of operation increases, with the mix of enterprises held constant. Economies (diseconomies) of scope refer to reductions (increases) in costs per unit of production as the number of enterprises increases, with the scale of operation held constant. Both types of cost relationships, along with the effects of diversification on business risk, should be carefully evaluated.

As illustrated in the earlier examples, the geographic scope of a farm business also may represent a form of diversification. Geographic diversification involves spreading production out over a wide space. By diversifying location and seasonality of production, lettuce growers in the western United States have greatly stabilized annual income, expanded their credit, and increased their rate of firm growth. Similar patterns of geographic diversification are developing for other perishable products and could occur for grain production and integrated beef operations, as far as yield risks are concerned. Even farming several tracts of land within a small geographic area (a township or a county) may help the farmer cope with yield uncertainty. The gains from reduced risk must be compared, however, with possible reductions in expected returns from geographic dispersion.

MARKETING ALTERNATIVES

Marketing alternatives provide other methods of risk management for farmers. The grain producer who stores the crop at harvest meets a considerable price risk during the storage period. Without effective inventory management to protect against unfavorable price movements, the producer must maintain a large reserve of liquidity.

Inventory management refers to those production and storage policies that influence the timing and magnitude of market transactions. Inventory policies can reduce risk through greater flexibility and frequency in marketing. Spreading sales over time results in price averaging over the marketing period and reduces the total variability of returns. In effect, spreading sales amounts to diversifying transactions over time rather than over different enterprises. If the principles of diversification in the preceding section are followed, the more frequent the sales, the lower the variability of returns; and the lower the correlation of prices and returns over time, the lower the variability of returns. However, increases in the frequency of sales and reductions in the amounts sold also increase the transactions costs of selling, per unit of sales.

Much investment in on-farm grain storage has occurred, and farmers have begun to consider sequential marketing as a viable strategy. A survey of farmers in Indiana [18] showed, for example, that 70 percent of the surveyed farmers market corn more than once a year, with 31 percent of these farmers selling corn six times or more in any one year. About 58 percent of the farmers spread their sales of soybeans across different time periods, but less so than corn.

Selling agricultural products throughout the year is more common with livestock, where production is less seasonal. Linkages with financial risk are also important — a farmer who invests in storage facilities may have to borrow money to finance the investment and carry the inventory.

Another response to price risk involves hedging. At harvest time, the producer could sell a futures contract timed to be consistent with the firm's storage policy. If, during the storage period, the price increases, the producer gains in terms of stored grain but loses in terms of the futures contract. The reverse could hold if the price movement were reversed. The net result is to hedge most of the price risk associated with storage.

The principles of diversification are involved in hedging too. A perfect hedge occurs only if the cash price and futures price for the commodity are perfectly correlated. In practice, these prices are not perfectly correlated, so that *new risks associated with hedging* — called **basis risks** — must be considered.

A major problem with hedging is the scarcity of opportunities to hedge. For many farm products no futures contracts exist, or the specifications on product quality, sizes of contract, or delivery dates do not coincide well with the characteristics of the producers' crop. History also suggests the need for educational activity to improve the producers' understanding of hedging. On the other hand, hedging may not be warranted if the producers expect the product price to increase, and they have sufficient liquidity to assume the risk. A hedge would lock in the low price and lose the expected gain. Margin requirements by commodity brokers in response to price changes during the hedging period require a flexible credit arrangement. Hedging may be better suited for buyers of farm products (grain elevators, cooperatives, livestock dealers), who by hedging themselves can extend its benefits to farmers through forward contracts.

Forward contracting directly with a buyer is another market response to risk. Forward contracts provide the manager with more certain price expectations and more precise planning. Contracting the sale of output may even increase the farm's credit, since lenders prefer the greater price certainty.

Forward contracting has limitations as well. Procurement contracts will probably specify production performance, harvest conditions, and terms of sale. These specifications reduce the manager's flexibility and freedom of decision making. Forward contracts that stipulate delivery amounts also produce interactions between yield and price risk. Shortfalls in yields may force the producer to purchase from the cash market at harvest to cover the contracted volume of production. Contracts may also reduce price expectations because the buyer should be rewarded for accepting the price risk.

For broiler, turkey, and egg production, as well as many large cattle feedlots and some hog production, contract production, or custom feeding, has become a significant organizational approach. The typical arrangement is for the agricultural producer to provide the buildings, equipment, and labor, and for other parties to provide ownership of the birds and animals, feed, and managerial assistance, along with a payment per unit of production to the producer to cover his or her labor and capital contributions [12, 22]. This organizational approach reduces the capital and financing requirements for producers and limits their exposure to market risks.

The use of commodity options is still another type of market response to risk. By paying a market-determined premium, a farmer may acquire the right, but not the obligation, to sell or buy a commodity at a stipulated price. Similar to forward contracting and hedging, options provide another alternative for dealing with price risk.

Choosing products with markets influenced by government programs may offset some price and production uncertainty. Through production and price specifications, these programs may add stability to the respective markets. Cotton, wheat, rice, feed grains, soybeans, peanuts, tobacco, and dairy products are some of the major farm products that benefit from political action to reduce uncertainties. Yet, new uncertainties arise about future changes in government programs and underlying economic conditions. Over-reliance on foreign markets, for example, to sell farm products may be risky in the long run.

For some commodities, pooling arrangements that may be available through marketing cooperatives enable farmers to transfer the storage, sale, and pricing function to larger organizations. Frequent marketing of pooled commodities helps stabilize farmers' returns. The market firms' knowledge of market conditions, their large size, and their specialized management may add to expected profits as well.

FLEXIBILITY

Greater flexibility in the business organization enables the manager to respond more quickly as new information becomes available to the firm. Flexibility does not directly reduce risk; it provides a means of coping with risk. Cost, resource, and behavioral flexibility should be considered by the manager.

One way to increase flexibility is to reduce fixed costs relative to variable costs. Short-lived assets can be changed more often than long-lived assets. Organizing grain and livestock enterprises to use more labor relative to capital investments increases enterprise flexibility. Consider, for example, a hog producer who can choose between a capital-intensive, specialized confinement system of production and a pasture system that uses more labor. This means of gaining flexibility is declining, however, as labor resources become more scarce and more costly. More flexible organizations often add to the total cost and forego possible size economies. Hence, the manager usually chooses a less flexible and more efficient organization.

Another common way to achieve flexibility is to choose nonspecific resources in place of specific resources. In this strategy, general-purpose buildings and machines are preferred to specialized buildings and machinery. Dual-purpose animals replace beef or dairy animals. General-purpose inputs can respond to changes in market opportunities. However, these general-purpose inputs usually have high costs per unit of output.

Behavioral attributes of flexibility are important too. A manager must recognize when changes are needed and be willing to make the changes as conditions warrant. Searching for new information about market conditions and production techniques is an example. The quality of expectations is improved, the degree of uncertainty is reduced, and the firm can better respond to changing conditions.

The general shortcoming of some types of flexibility is similar to that of diversification. With a flexible organization, the manager loses the benefits of specialization. The higher total costs of flexibility may make its choice infeasible. While flexibility is effective in some cases, it is generally limited as a risk response for modern commercial farms. It is usually cheaper to choose a more specific input and to manage risks in other ways.

FINANCIAL RESPONSES
TO RISK

Financial responses to risk refer to a firm's capacity to bear risks in production, marketing, and financing and to spread these risks among the financial claimants on the firm. In large corporate firms, the wide dispersion of ownership spreads business risks over a large number of stockholders who may themselves be well diversified. The smaller-scale, non-corporate structure of the farm sector reduces the feasibility of this risk response. One exception is in the leasing of farm land where extensive use of share rents allocates expected business risks between the farm operator and the land owner.

Most of the financial responses to risk by farmers involve the management of leverage, liquidity, and formal insurance. These actions affect both the firm's assets and its liabilities and are related to the risk responses cited previously. Some sources of liquidity were presented in Chapter 7; they are briefly reviewed here along with other financial responses to risk.

The Pace of Investment

Maintaining flexibility in the pace of farm investment is an important financial response to risk. Postponing capital expenditures, including asset replacement, is a common financial control mechanism under adversity. It has the multiple effects of postponing large financial outlays, reducing cash outflows, and restraining indebtedness. Generally, capital investments offer more flexibility in timing than do production and marketing activities, which must be carried on each year to sustain the firm's operations. Ultimately, of course, capital assets must be replaced to sustain the firm's productivity.

Control over the pace of investments may be exercised either by the borrower or the lender. The borrower simply decides not to undertake an investment at the present time. If financing is involved, then the lender may exercise control by declining to fund a loan request or by making the financing terms stringent enough that the borrower declines the loan. If lenders indeed intend to discourage borrowing, they generally do so by declining the loan request.

The Pace of Disinvestment and Withdrawals

As discussed in Chapter 7, many assets have important liquidity attributes and provide relatively low-cost access to cash. Willingness to liquidate assets to meet financial obligations is a crucial financial response, especially under crisis conditions. Reducing financial reserves is the first step. However, for younger operators with expanding operations, liquid financial assets are relatively scarce, and the opportunity costs of asset liquidation are relatively high. Growing crops and raising livestock in feedlots that are not ready for sale are also considered illiquid assets. In contrast, harvested crops held in storage are highly liquid assets and contribute significantly to risk management.

Other farm assets that are dominated by land, machinery, breeding livestock, and other fixed assets are costly and disruptive to liquidate. Selling these capital assets is usually a last-resort effort, although depleting livestock herds is a common cyclical behavior. Selling a tract of land is perhaps the most anxiety-producing response, especially in traditional family farms. But, if land acquisition is part of the growth strategy, then land disposal can be part of the risk control strategy as well. In general, when businesses experience heavy stress, asset sales can generate cash in larger amounts and more quickly than most other risk responses.

Control of withdrawals by family members and other business owners for consumption, taxes, and other purposes is also an important risk response. However, farmers today are more dependent on cash purchases and have less capacity than in the past to adjust family living to swings in farm income. Options in meeting tax obligations provide further opportunities for stabilizing income, especially in shifting taxable gains and losses among years.

Credit Reserves

Establishing effective credit relationships with commercial lenders is very important in farming. These relationships may enable borrower to carry over loans, defer payments, refinance high debt loads, or otherwise utilize credit liquidity during times of financial distress. As shown in Chapter 7, the firm's credit position, as evaluated by lenders, is strongly sensitive to factors affecting the farm's credit worthiness, including risk responses in production and marketing. Moreover, borrowers must consider differences between farm lenders as they plan, coordinate, and communicate in establishing credit relationships. Nonetheless, working with commercial lenders is a prominent risk response for many farmers.

Part of the credit relationship involves establishing acceptable targets on a farm's financial leverage. Basing leverage on risk – return considerations, adequate safety margins in collateral, and projected repayment capacity under normal conditions sets the stage for effective financial control when adversity occurs. Cash flow budgets, capital budgets, and pro forma analyses with careful documentation and sensitivity analysis are important tools in managing leverage and liquidity.

Credit reserves provided by public programs are important to farmers as well, although they are subject to their own uncertainties. Price and income support programs administered through the Commodity Credit Corporation of the U.S. Department of Agriculture furnish crop farmers who participate with sophisticated financial control programs. These programs provide (1) inventory financing for farmers when market prices are low, (2) mechanisms for stabilizing commodity prices, (3) added flexibility in marketing crops, and (4) maintenance of farm incomes.

Other public credit programs in the United States administered mainly through the Farmers Home Administration (FmHA) provide direct or guaranteed credit liquidity for high-risk borrowers, including on occasion those who experience natural or economic disasters. FmHA lending to farmers occurs through direct loan programs, guarantees of farm loans made and serviced by commercial lenders, and occasional emergency loan programs that often have concessionary terms.

Insurance

Insurance provides a specialized source of liquidity. Instead of reserving cash, savings, or credit to counter the effects of hail damage, a crop producer might buy an insurance contract that allows for a reserve of funds contingent upon the occurrence of the insured event. Commercial insurance indemnifies an asset or flow of income against the occurrence of specified events. Some examples are life, crop, loan, and household insurance against death or accident, yield loss, default, and fire, respectively. If certain statistical properties are met relative to the event and the indemnifiable property, the insurance company can maintain the reserve more cheaply than the individual farmer. In general, events with a very low and highly predictable probability of occurrence, but with a large potential loss, are well suited to insurance coverage.

An event's probability of occurrence to a producer is the same whether insurance or other liquid reserves are provided, as long as moral hazard is not a problem (i.e., the insured responding to insurance with actions that increase the likelihood of adversity). The major concern is the best means of providing liquidity for responding to an unfavorable event. Elements of an insurance decision for a producer include (1) the insurance premiums, (2) the liquid reserve needed without insurance, (3) the earnings rate of the liquid reserve, and (4) the earnings rate for investing the reserved funds in the business. These elements can be specified in the following model.

$$I = S(b - e) - P \tag{8.13}$$

where   I = the gain from insurance

               S = the size of needed reserve

               P = the annual insurance premium

               b = the opportunity cost of the reserve (i.e., its earning rate if funds were invested in the business)

and   e = the earnings rate of the liquid reserve

Decision rules are specified as:

If $I > 0$, insure.

If $I = 0$, be indifferent.

If $I < 0$, do not insure.

Consider a manager who is evaluating insurance for a $40,000 (S) building with a $600 annual premium (P). The manager can earn a 6 percent return (e) on a liquid reserve in a savings account. However, the funds will earn a 10 percent return (b) if invested in the business. Substituting in equation 8.13 yields:

$$I = 40,000(0.10 - 0.06) - 600$$

$$I = 1,600 - 600 = 1,000$$

Thus, the manager earns an additional $1,000 by insuring and investing the $40,000 directly in the business. In fact, the manager should insure if the difference in earnings rates $(b - e)$ of the reserves exceeds $P \div S$ — in this case, 0.015.

Most agricultural producers do not have liquid reserves available for every event that may be insurable, nor do they have such reserves available for new investments. An insurance company can generally provide the reserve more cheaply. While the premium for insurance is an obvious cash outflow, the protective returns from insurance are intangible and more difficult to measure. Yet, the consequences of many insurable events may jeopardize the firm's ability to survive. Generally, the dollar cost of insurance exceeds the dollar return in the long run; however, the utility or security generated by equity protection can make insurance quite favorable.

Unfortunately, insurance does not exist for many farm circumstances. Examples are prices of farm products and farm outputs, some weather events, failures in contractual relationships, and human performance. For circumstances such as these, the manager must depend on other liquid reserves to meet the consequences.

Summary

Business and financial risks when combined magnify potential losses in equity capital and net income and create inefficiencies in resource use by hampering business planning. Risk management involves the selection of methods for reducing, transferring, and/or bearing business and financial risks. Often, however, risk management involves a trade-off between risk and expected returns. That is, efforts to reduce risk may be accompanied by reductions in expected returns. The fundamentals of diversification strongly underlie risk analysis. Under proper conditions, engaging in multiple enterprises, holding more assets, spreading transactions across time, and other forms of diversification can yield significant reductions in risk without much loss in expected returns. Other highlights are as follows:

- Diversification as a risk-reducing strategy becomes more effective as the covariation (i.e., correlation) among investments is lower.

- Diversification becomes more effective as the number of assets, activities, or transactions increase, although risk reduction increases at a diminishing rate.

- Diversification potential may be limited by diseconomies of scope in which costs per unit of production increase with increases in the number of activities, assets, or transactions.

- Applications of portfolio analysis to financial servicing capacity under conditions of risk estimate how the probability of meeting financial obligations is affected by the extent of diversification, correlation values, and financial leverage.

- Marketing alternatives for responding to risk include forward contracts, hedging, options, inventory management, and participation in government community programs.

- Achieving greater flexibility in a business organization enables the manager to respond more quickly as new information becomes available.

- Major financial responses to risk include adjustments in the pace of investments, disinvestments, and withdrawals; management of credit reserves; and utilizing commercial or public insurance as a specialized source of liquidity.

Topics for Discussion

1. Explain the relationship between risk management and the financial control process.

2. Explain the concept of a risk management strategy. How is a strategy related to a specific method of risk management?

3. Explain how trade-offs may arise among various methods of managing risks. Why, for example, may hedging reduce the need for holding liquid reserves? Suggest and evaluate some other examples.

4. What conditions must exist for diversification to be an effective risk response? How is the correlation effect involved? The number of investments effect?

5. What factors limit farmers' use of enterprise diversification and flexibility in responding to risks?

6. How can the portfolio model be used to evaluate financial servicing capacity under risk?

7. What factors affect the use of hedging and forward contracting as methods of risk management?

8. Contrast asset sales and credit reserves as sources of liquidity. What are their advantages and disadvantages? Which, if either, is preferred?

9. A farmer is evaluating insurance for a $50,000 facility with a $400 annual premium. A 12 percent return can be earned on a liquid reserve and a 16 percent return can be earned if funds are invested in the business. Is insurance a profitable choice? Why or why not?

10. Given the following information:

	Investment A	Investment B
Expected rate-of-return	0.15	0.20
Standard deviation	0.12	0.15
Correlation coefficient	0.5	

   a. Find the expected value and standard deviation of a portfolio with equal proportions of investments A and B. How do these results compare with holding only A or B?

   b. Suppose the correlation coefficient declines to − 0.5. What are the impacts on portfolio risk? How do these results compare to those in a?

c. Suppose investment C is considered with expected return 0.25, standard deviation 0.20, and correlation of 0.5 with A and B. Using the data in a, evaluate a portfolio with equal proportions of the three investments. Contrast the results with those in a and b.

References

1. Alexander, G. J., and J. C. Francis, *Portfolio Analysis*, 3rd ed., Prentice-Hall, Inc., Englewood Cliffs, New Jersey, 1986.

2. Barry, P. J., "Capital Asset Pricing and Farm Real Estate," *American Journal of Agricultural Economics*, 62(1980):549 – 553.

3. Barry, P. J., ed., *Risk Management in Agriculture*, Iowa State University Press, Ames, 1984.

4. Barry, P. J., C. B. Baker, and L. R. Sanint, "Farmers' Credit Risk and Liquidity Management," *American Journal of Agricultural Economics*, 63(1981):216 – 227.

5. Barry, P. J., and D. R. Fraser, "Risk Management in Primary Agricultural Production: Methods, Distribution, Rewards and Structural Implications," *American Journal of Agricultural Economics*, 58(1976):286 – 295.

6. Boehlje, M. D., and L. D. Trede, "Risk Management in Agriculture," *Journal of the American Society of Farm Managers and Rural Appraisers*, 41(1977):20 – 29.

7. Crisostomo, M. F., and A. Featherstone, "A Portfolio Analysis of Returns to Farm Equity and Assets," *North Central Journal of Agricultural Economics*, 12(January 1990):9 – 22.

8. Fama, E. F., *Foundations of Finance: Portfolio Decisions and Securities Prices*, Basic Books, Inc., Publishers, New York, 1976.

9. Gabriel, S. C., and C. B. Baker, "Concepts of Business and Financial Risk," *American Journal of Agricultural Economics*, 62(1980):560 – 564.

10. Haley, C. W., and L. D. Schall, *The Theory of Financial Decisions*, 2nd ed., McGraw-Hill Book Company, New York, 1979.

11. Heady, E. O., *Agricultural Production Economics and Resource Use*, Prentice-Hall, Inc., Englewood Cliffs, New Jersey, 1952.

12. Knoeber, C. R., "A Real Game of Chicken: Contracts, Tournaments

and the Production of Broilers," *Journal of Law, Economics, and Organizations*, 5(1989):271 – 292.

13. Leuthold, R. M., and P. E. Peterson, "Using the Hog Futures Market Effectively While Hedging," *Journal of the American Society of Farm Managers and Rural Appraisers*, 44(1980):6 – 12.

14. Mapp, H. P., et al., "An Analysis of Risk Management Strategies for Agricultural Producers," *American Journal of Agricultural Economics*, 61(1979):1071 – 1077.

15. Markowitz, H. M., *Portfolio Selection*, John Wiley & Sons, Inc., New York, 1959.

16. Moss, C. B., A. M. Featherstone, and T. G. Baker, "Agricultural Assets in an Efficient Multiperiod Investment Portfolio," *Agricultural Finance Review*, 47(1987):82 – 94.

17. Musser, W. N., and K. G. Stamoulis, "Evaluating the Food and Agriculture Act of 1977 with Firm Quadratic Risk Programming," *American Journal of Agricultural Economics*, 63(1981):447 – 456.

18. Patrick, G. F., *Risk and Variability in Indiana Agriculture*, Bull. 234, Agricultural Experiment Station, Purdue University, West Lafayette, Indiana, July 1979.

19. Patrick, G. F., P. W. Wilson, P. J. Barry, W. G. Boggess, and D. L. Young, "Risk Perceptions and Management Responses: Producer-generated Hypotheses for Risk Modeling," *Southern Journal of Agricultural Economics*, 17:2(1985):207 – 214.

20. Paul, A. B., R. G. Heifner, and J. W. Helmuth, *Farmers' Use of Forward Contracts and Futures Markets*, Agricultural Economics Report 320, ERS – USDA, Washington, D.C., 1976.

21. Pflueger, B. W., and P. J. Barry, "Crop Insurance and Credit: A Farm-Level Simulation Analysis," *Agricultural Finance Review*, 46(1986):1 – 15.

22. Rhodes, V. J., *U.S. Contract Production of Hogs*, University of Missouri Agricultural Economics Report No. 1990 – 1, Columbia, Missouri, 1991.

23. Young, R. P., and P. J. Barry, "Holding Financial Assets as a Response to Risk: A Portfolio Analysis of Illinois Cash Grain Farms," *North Central Journal of Agricultural Economics*, 9:1(January 1987):77 – 84.

SECTION FOUR

Capital Budgeting and Long-Term Decision Making

9

The Time Value
of Money

The introductory discussion of profits and risk in Chapter 2 indicated that economic values of assets are largely determined by (1) the level, timing, and risk of their projected annual profits and (2) the time and risk attitudes of investors. We briefly showed how these factors provide a basis for long-term financial decision making.

This chapter develops the basic ideas underlying the time value of money and presents the necessary tools for determining the effects of time on financial decisions. Emphasis is given to time preferences, compound interest, and the use of discounting and compounding techniques in valuing flows of payments at a common point in time. Various applications of these concepts and tools to the financial management of agricultural firms are found in the following chapters.

CONCEPTS OF TIME VALUE

An interest rate serves as the pricing mechanism for the time value of money. It reflects the collective effects of all investors' time preferences for money and the productive uses of such money. The rate per period (i) is considered an exchange price between present and future dollars. Thus, $1 today exchanges for 1 + i dollars one period in the future. Alternatively, a $1 payment made one period in the future exchanges for 1 ÷ (1 + i) dollars now.

In financial markets, interest rates play the vital role of equating present and future claims for financial assets of different maturities. Interest rates may also account for risk and inflation, but for now the focus is on the time differences. These rates respond to changes in supply and demand for alternative financial assets, including money, just as other commodity prices respond to changes in their supply and demand.

Nearly all individuals display positive time preferences for money, wealth, and other desired objects. Hence, the time preference component of interest rates is always positive. The positive time preference means that the sooner the money is available, the greater its value. A dollar received today is preferred to a dollar received tomorrow or, alternatively, a dollar cost is better postponed from today to tomorrow.

These time preferences occur because there are always other valuable opportunities for using the money. Interest rates reflect the opportunity costs of not immediately putting the money into the best of these other uses. For example, if we lend money, the rate (i) is paid to us for our foregone consumption or foregone earnings on other investments. If we borrow, the rate (i) is paid by us to the lender to compensate for his or her foregone consumption or earnings. The level of interest rates reflects the value of these opportunities and thus the strength of time preferences.

COMPOUND INTEREST

The time value concept uses compound interest as a basis for determining present and future values. Compound interest differs from simple interest. **Simple interest** means that *only the original principal, or amount of money, earns interest over the life of the transaction.* Consider the purchase of a bond for $1,000 that pays 5 percent interest each year. The interest payment is $50 at the end of the first year. Since only the principal earns interest, the interest payments at the end of the second and third years are also $50. Hence, the total interest paid over the three years is $150.

Compound interest means that *each time interest is paid, it is added to or compounded into the principal and thereafter also earns interest.* As the principal increases through time from compounding, so do the interest payments that provide the source of the compounding. At the end of the transaction period, the total principal available is called the compound amount; the difference between the original principal and the compound amount is called compound interest. The rate of interest is called the compound interest rate. The magnitude of the compound amount is determined by the amount of the original principal, the number of compound or conversion periods, and the rate of interest per conversion period.

Suppose you have $1,000 to invest in a bank paying interest annually at a 5 percent compound rate (i = 0.05). After one year you will have:

1,000 + (1,000)(i) = 1,050

or:

1,000(1 + i) = 1,000(1.05) = 1,050

After two years, you will have:

1,000(1 + i)(1 + i) = 1,000(1.05)(1.05) = 1,102.50

or:

$1,000(1 + i)^2 = 1,000(1.05)^2 = 1,102.50$

The $1,102.50 is the compound amount, or future value, of the $1,000 principal, or present value, invested at a 5 percent compound interest rate for two years. The compound interest earned is $102.50. If the investment is left for a third year, its future value will be $1,158, yielding compound interest of $158 over the three-year period. The compound interest of $158 exceeds the simple interest of $150 earned when the compounding process was not used.

FUTURE VALUE OF
A PRESENT SUM

A general formula that uses compounding to determine the future value of a present value is:[1]

---

[1]Notation N is used to designate the number of compound periods, while n refers to the series of payments n = 0, 1, 2, . . . N. The conversion factors taken from the Appendix are rounded to the third decimal point in this chapter.

$$V_N = V_0(1 + i)^N \qquad \qquad (9.1)$$

or:

$$\frac{\text{Future}}{\text{value}} = \frac{\text{Present}}{\text{value}} \times \left(1 + \frac{\text{Interest rate}}{\text{per conversion}}\right)^{\text{number of conversion periods}}_{\text{period}}$$

In the preceding example, the future value of $1,000 compounded at 5 percent for 10 years is:

$$V_{10} = 1,000(1.05)^{10} = 1,629$$

The term $(1 + i)^N$ can always be computed. However, the procedure becomes tedious for high values of N. Fortunately, numerical results of such equations are tabulated for many values of i and N. Appendix Table II shows the conversion factors for $1 at present for selected values of i and N.

These values are substituted into equation 9.1 and multiplied by the number of present dollars for convenient solutions. Turn to Appendix Table II, and locate the appropriate conversion factors used above for i = 0.05 and N = 2 and 10. These values should correspond to the values used in the examples. Using the table, confirm that V_{20} in the preceding problem is $2,653.

The same procedures can also be quickly performed on an inexpensive pocket calculator. Moreover, a number of pocket calculators have built-in business programs that perform many types of compounding and discounting procedures at the touch of a few buttons. Microcomputers also may be used to carry out these calculations, as well as more sophisticated forms of compounding and discounting to be developed as follows.

PRESENT VALUE OF
A FUTURE PAYMENT

Suppose that we already know the future value, the interest rate, and the number of conversion periods. The goal then is to solve for the present value that is consistent with these known values. The present value is found by solving equation 9.1 for V_0. The result is a general formula for determining the present value of a future sum:

$$V_0 = \frac{V_N}{(1 + i)^N} \tag{9.2}$$

or:

$$\frac{\text{Present}}{\text{value}} = \frac{\text{Future value}}{\left(1 + \begin{array}{c} \text{Interest rate} \\ \text{per conversion} \\ \text{period} \end{array}\right)^{\text{number of conversion periods}}}$$

Equation 9.2 shows that the present value equals the future value divided by the conversion factor for interest rate (i) and conversion period (N). *The process of finding present values* is called **discounting** because the future value is discounted to a lower present value to account for the effects of the time value of money. The discount occurs because the investor must wait to receive the future payment and cannot invest it at present in the alternative investment opportunity yielding interest rate (i).

What is the present value of a $1,102.50 payment available two years from now, if the annual interest rate is 5 percent?

$$V_0 = \frac{1,102.50}{(1.05)^2} = 1,000$$

The present value is $1,000 and the discount for time is $102.50. Note that this level of discount equals the compound interest earned by investing $1,000 for two years at an interest rate of 5 percent. Discounting to compute present values is precisely the inverse of compounding, as was demonstrated by deriving equation 9.2 from equation 9.1. Hence, the present value (V_0) of a sum available in N years (V_N) is the amount which, if invested now and compounded for N years, yields V_N— provided, of course, that the interest rates used in the compounding and discounting are the same.

The appropriate discount factor could be taken from Appendix Table II and substituted directly into equation 9.2. However, multiplication is usually easier to visualize and perform than division. So Appendix Table III lists conversion factors for present values of $V_N = \$1$ at selected values of i and N, for:

$$V_0 = \frac{1}{(1 + i)^N} = (1 + i)^{-N}$$

These values are substituted into equation 9.2 and multiplied by V_N to give the solution. Using an annual interest rate of 5 percent, consider the present value of $1,629 available 10 years from now. Appendix Table III lists the conversion factor for $i = 0.05$ and $N = 10$ as 0.614. Therefore:

$V_0 = 1,629(0.614) = 1,000$

Confirm that the present value of $2,653 available 20 years from now is $1,000 — again, assuming an interest rate of 5 percent.

APPLICATIONS OF SINGLE-PAYMENT FORMULAS

Situation

You can purchase a note that will pay $10,000 five years from now. What is the note's present value with an annual interest rate of 10 percent?

Answer: Use Appendix Table III.

$V_0 = 10,000(1.10)^{-5}$

$\quad = 10,000(0.621)$

$\quad = 6,210$

Situation

Suppose you borrow $6,210 now and agree to repay the loan in five years plus interest charged at a 10 percent annual rate. Find the future value of the note at the end of year 5.

Answer: Use Appendix Table II.

$V_5 = 6,210(1.10)^5$

$\quad = 6,210(1.611)$

$\quad = 10,000$

Situation

Farm land in Champaign County, Illinois, has been selling for $2,000 per acre. If it is expected to increase in value at a compound rate of 3 percent per year, what will its value be in 25 years? *(Note:* In this situation, the interest rate is interpreted as a growth rate.)

Answer: Use Appendix Table II.

$V_{25} = 2,000(1.03)^{25}$

$\quad = 2,000(2.094)$

$\quad = 4,188$ per acre

Situation

If this same land has been increasing at an annual rate of 5 percent, what was its price six years ago?

Answer: Discount the $2,000 value to a present value for six years ago.

$V_0 = 2,000(1.05)^{-6}$

$\quad = 2,000(0.746)$

$\quad = 1,492$ per acre

Situation

Ms. Young Lender recently joined the Agricultural National Bank at a salary of $20,000. What will be her salary at retirement after a 40-year career and an average annual raise of 5 percent?

Answer:

$$V_{40} = 20{,}000(1.05)^{40}$$
$$= 20{,}000(7.040)$$
$$= 140{,}800$$

Situation

Mr. Retiring Lender observes that his starting salary 40 years ago was $2,800. How would this compare with today's $20,000 salary, assuming a 5 percent annual increase?

Answer: Discount the $20,000 salary to a present value for 40 years ago.

$$V_0 = 20{,}000(1.05)^{-40}$$
$$= 20{,}000(.142)$$
$$= 2{,}840$$

ANNUAL INTEREST
AND COMPOUNDING

It is always important to determine the interest rate per conversion period when you are compounding or discounting for time. Sometimes the interest rate is expressed as an annual rate, while compounding occurs more frequently, perhaps on a semi-annual, quarterly, or daily basis. Under these circumstances, the interest rate must be changed from an annual rate to a rate per conversion period. The result is:

$$V_N = V_0\left(1 + \frac{i}{m}\right)^{mN} \tag{9.3}$$

where m is the number of conversion periods per year and N is the number of years. Suppose, for example, that a bank offers a savings account paying a 6 percent annual rate of interest with interest compounded quarterly (every three months). The number of conversion periods each year is four, and the interest rate per period is 1.5 percent. By investing $1,000 for one year, you will have:

$$V_i = 1,000 \left(1 + \frac{0.06}{4} \right)^{(4)(1)} = 1,061.40$$

or an annual yield of 6.14 percent rather than the annual quoted rate of 6 percent. As will be seen later, this procedure is especially important in determining interest rates on many kinds of loans.

PRESENT VALUE OF A SERIES OF PAYMENTS

One of the most common problems of financial management is to determine the present value of an investment that generates a series of payments at the end of each of several years in the future. The procedure for determining the present value of a series of payments is an extension of the procedure for finding the present value of a single future sum. That is, discount each future payment in its respective year to a present value and add them all up.

$$V_0 = \frac{P_1}{(1 + i)} + \frac{P_2}{(1 + i)^2} + \ldots + \frac{P_N}{(1 + i)^N} \tag{9.4a}$$

or:

$$V_0 = \sum_{n=0}^{N} \frac{P_n}{(1 + i)^n} \tag{9.4b}$$

where:

V_0 = the present value of the payment series

P_n = the payment for each conversion period (n) and (n = 0, 1, 2, . . . N)

i = the interest rate

Σ = the summation for n = 0 to N

To illustrate, find the present value of an oil royalty expected to pay $3,000 at the end of year 1, $4,000 in year 2, $5,000 in year 3, $6,000 in year 4, and $7,000 in year 5. Use an interest rate of 8 percent. The present value model is set up as follows:

$$V_0 = \frac{3,000}{1.08} + \frac{4,000}{(1.08)^2} + \frac{5,000}{(1.08)^3} + \frac{6,000}{(1.08)^4} + \frac{7,000}{(1.08)^5}$$

Calculations involve multiplying the respective payments by the conversion factors in Appendix Table III for i = 0.08 and values of n ranging from 1 to 5:

$$V_0 = 3,000(0.926) + 4,000(0.857) + 5,000(0.794) + 6,000(0.735) + 7,000(0.681)$$

$$= 19,353$$

The present value of the oil-royalty payments is $19,353, based on an interest rate of 8 percent. The present value of the series, using a zero interest rate (i.e., undiscounted), is $25,000. Hence, the difference of $5,647 represents the size of discount for delayed receipt of the payments. It reflects the earnings lost by not investing the payments in an alternative investment yielding the 8 percent return. Anyone paying $19,353 for the right to receive this five-year series of payments would realize an annual compound rate-of-return (or yield) of 8 percent. Using a higher discount rate (10 percent) yields a lower present value, while using a lower discount rate (6 percent) raises the present value.

PRESENT VALUE OF
A UNIFORM SERIES

The discounting procedure is simplified if the payments and interest rates are equal in each conversion period. Then we can determine the present value of a *uniform* series of payments (an annuity), and equation 9.3 reduces to:

$$V_0 = A\left[\frac{1 - (1 + i)^{-N}}{i}\right] = A[USPV_{i, N}] \qquad (9.5)$$

where A equals the annuity payment in each period. $USPV_{i, N}$ is used as a simplified notation for a **uniform series present value** over N periods at interest rate (i).

Assume now that the oil-royalty payments in the preceding section are a uniform series of $5,000 paid at the end of each of the next five years. The present value is expressed as:

$$V_0 = 5,000 \left[\frac{1 - (1.08)^{-5}}{0.08} \right] = 19,965$$

We can use Appendix Table III for values of $(1 + i)^{-N}$. For i = 0.08 and N = 5, the conversion factor is 0.681. This value can be inserted into equation 9.5 to solve for V_0. The answer is $19,965. As a short cut, however, Appendix Table V lists directly the present values of an annuity of $1 for selected values of i and N. For i = 0.08 and N = 5, the conversion factor is 3.993. Multiplying this factor by the annuity (A) of $5,000 also gives the present value of $19,965.

Thus, the present value of a series of $5,000 payments received at the end of five successive years is $19,965, assuming an 8 percent interest rate. Alternatively, one can say that $19,965 invested currently at 8 percent interest will provide an annuity of $5,000 per year for five years.

Using Appendix Table V and equation 9.5 confirms that the present value of a $5,000 annuity payable over 20 years is $49,090 based on the 8 percent discount rate; a similar annuity for 50 years has a present value of $61,167.

PRESENT VALUE OF
AN INFINITE UNIFORM SERIES

As N increases for a given annuity, the present value of the annuity increases at a decreasing rate. Each additional conversion period adds another payment to the income stream. However, that payment is discounted over an additional time period. At some point in time, the present value of the last annuity payment approaches zero. In equation 9.5, this condition is signified when the term $(1 + i)^{-N} = 0$. When this condition occurs (i.e., as N approaches infinity), the present value of the annuity approaches its absolute limit, and equation 9.5 reduces to:

$$V_0 = \frac{A}{I} \qquad\qquad (9.6)$$

This is the standard capitalization equation used by real estate appraisers. It is applicable whenever a uniform series can be projected to occur perpetually into the future.

Actually, most situations involving the time value of money have measurable time periods. The projected economic life of a machine may be only 5 years, or the planning horizon for a farm family considering a real estate investment may be 10 to 20 years or less. Consequently, the capitalization equation (9.6) has limited application in financial analysis.

FUTURE VALUE OF
A SERIES OF PAYMENTS

Not all valuation occurs as present values. It is often appropriate to value in the future and to define procedures for determining the future value of payments. Earlier, we observed that present values (V_0) and future values (V_n) are related to each other through the interest rate i and the number of compounding periods. This same procedure applies to estimating the future value of a series of payments. Using equation 9.1, you can determine the future value of each payment and then sum the payment values as:

$$V_N = P_0(1 + 1)^N + P_1(1 + 1)^{N-1} + P_2(1 + 1)^{N-2} \ldots + P_N \qquad\qquad (9.7)$$

$$= \sum_{n=0}^{N} P_n(1 + i)^{N-n}$$

where:

V_N = the future value of the series of payments

P_n = the payment for each conversion period (n) and (n = 0, 1, 2, . . . N)

i = the interest rate

To illustrate, using an 8 percent interest rate, find the future value of the oil royalty expected to pay $3,000, $4,000, $5,000, $6,000, and $7,000 at the end of each of the next five years, respectively. The future value model is set up as follows:

$$V_5 = 3{,}000(1.08)^4 + 4{,}000(1.08)^3 + 5{,}000(1.08)^2 + 6{,}000(1.08) + 7{,}000$$

Note that there is no payment in period zero; the first payment is made at the end of year 1 and is compounded over the four remaining years. The fifth payment is received at the date of future valuation and is not compounded.

Calculations involve multiplying the respective payments by the conversion factors in Appendix Table II for i = 0.08 and values of n ranging from 1 to 4:

$$V_5 = 3{,}000(1.360) + 4{,}000(1.260) + 5{,}000(1.166) + 6{,}000(1.08) + 7{,}000$$

$$= 28{,}430$$

The future value of the series of oil-royalty payments is $28,430, based on an interest rate of 8 percent. The future value of the series based on a zero interest rate (i.e., uncompounded) is $25,000. The difference of $3,430 represents the compound interest earned by reinvesting the payments in the investment opportunity yielding 8 percent.

Note the linkage between the derivation of present and future values for this series of payments. The present value of this nonuniform series of oil-royalty payments is $19,353. The future value of the $19,353 figure compounded for five years at 8 percent interest is:

$$V_5 = 19{,}353(1.08)^5 = 19{,}353(1.469) = 28{,}430$$

which is also the future value of the series as calculated with equation 9.7.

The valuation of the payment series can occur at any point in time, with the choice largely resting on ease of calculation and on the intended use of the derived values. Clearly, however, to compare the values of two series of payments, they must be valued at the same point of time — either at the present, which is generally the case, or at the same future date.

FUTURE VALUE OF
A UNIFORM SERIES

Again, the procedure for finding the future value of a series of payments is simplified if the payments are equal. Then, we can determine the future value of a uniform series of payments, and equation 9.7 reduces to:

$$V_N = A\left[\frac{(1 + i)^N - 1}{i}\right] = A[USFV_{i,\ N}] \tag{9.8}$$

where A is the annuity payment in each period and $USFV_{i,\ N}$ stands for **uniform series future value** over N periods at interest rate (i).

Now the future value of the five-year uniform series of $5,000 oil-royalty payments is:

$$V_5 = 5{,}000\left[\frac{(1.08)^5 - 1}{0.08}\right] = 5{,}000\ [USFV_{0.08,\ 5}] = 29{,}335$$

Appendix Table IV lists conversion factors for the future value of an annuity of $1 for selected values of i and N. For i = 0.08 and N = 5, the conversion factor is 5.867. Multiplying 5.867 by the annual payments of $5,000 yields the future value of $29,335. Using Appendix Table IV, confirm that V_{10} for this annuity is $72,433.

APPLICATIONS OF
SERIES FORMULAS

The formulas for present and future values of a series of payments are used in numerous kinds of capital budgeting models that will be introduced in Chapter 10 and applied in subsequent chapters. However, making some applications now will help you to gain experience with their use. The formulas themselves have only mathematical content. They acquire financial relevance when they are applied to problems in investment and financing.

Situation

Ms. Young Lender who accepted the $20,000 salary from the Agricultural National Bank had passed up another banking alternative with a $17,000 starting salary. Assuming the salary differential persists over her 40-year employment, what is the annual difference of $3,000 worth now, using a 5 percent interest rate?

> *Answer:* Find the present value of a uniform series of $3,000 payments for $i = 0.05$ and $N = 40$.

$V_0 = 3,000[\text{USPV}_{0.05,40}]$

$\quad = 3,000[17.1591] = 51,477$

Situation

Cletus Hoopster is a professional basketball player who has just signed what the newspapers describe as a $50 million contract. The terms are 10 annual payments of $5,000,000 each. Determine the present value of the contract to Cletus, assuming his interest rate is 10 percent.

> *Answer:* Find the present value of a uniform series of $5,000,000 payments for $i = 10$ percent and $N = 10$.

$V_0 = 5,000,000[\text{USPV}_{0.10,10}]$

$\quad = 500,000[6.145] = 30,725,000$

Financial managers may need to determine the rate of savings necessary to accumulate a given sum or the number of annual payments required to repay an installment loan. Equations 9.5 and 9.8, which measure the present and future values of annuities, are used to answer these questions. In these cases, the manager knows the present or future values (V_0 or V_N), the interest rates (i), and the number of periods (N). Given this information, the manager must determine the level of annuity (A) per period.

Situation

The manager projects a need for $15,000 ($V_N$) in five years and can invest savings at a 6 percent annual interest rate, compounded semi-annually. How much must be saved each period to reach the objective?

> **Answer:** Find the annuity which, when invested for 10 periods at $i = 0.03$ per period, yields a future value of $15,000. (*Note:* the semi-annual compounding defines 10 conversion periods in the five years, with an interest rate of 3 percent per period.) Substitute this information into equation 9.8.

$$15,000 = A[USFV_{0.03,\,10}]$$

Solve for A by first finding the conversion factor for $i = 0.03$ and $N = 10$ from Appendix Table IV, then divide that factor into 15,000.

$$A = \frac{15,000}{[USFV_{0.03,\,10}]} = \frac{15,000}{11.464} = 1,308.44$$

The manager must invest $1,308.44 each six months at a 6 percent annual rate, compounded semi-annually, in order to accumulate $15,000 in five years. This type of fund is generally called a **sinking fund.**

Situation

The manager is borrowing $6,000 ($V_0$) to purchase a car and will repay the loan in 30 equal monthly payments at an interest rate of 1 percent per month. What is the level of monthly payment?

> **Answer:** Find the level of annuity (monthly payment), which has a present value of $6,000 when discounted over 30 conversion periods at an interest rate of 1 percent per period. This information can be substituted into equation 9.5 and solved for A.

$$6{,}000 = A[\text{USPV}_{0.01,\ 30}]$$

$$A = \frac{6{,}000}{[\text{USPV}_{0.01,\ 30}]} \quad (\text{See Appendix Table V})$$

$$A = \frac{6{,}000}{25{,}808} = 232.49$$

Thus, monthly payments of \$232.49 for 30 months will completely repay the original loan plus the interest. This is called a capital recovery problem. It represents the process of amortization, in which equal payments are made each conversion period to repay a debt within a specified time.

PRESENT VALUE OF A CONSTANT-GROWTH SERIES

Previously we assumed that the payment series was uniform over time. Now let's suppose that the payments experience a constant rate of growth (g). Let P_0 be the current payment, and assume that it was just paid so that it does not enter into the present value computations. The series of growing payments thus starts with the payment at the end of the first period: $P_1 = P_0(1 + g)$. The payment at the end of the second period is $P_2 = P_0(1 + g)^2$, and so on.

The present value is:

$$V_0 = \frac{P_1}{1 + i} + \frac{P_2}{(1 + i)^2} + \ldots + \frac{P_n}{(1 + i)^n}$$

which is equivalent to:

$$V_0 = \frac{P_0(1 + g)}{1 + i} + \frac{P_0(1 + g)^2}{(1 + i)^2} + \ldots + \frac{P_0(1 + g)^n}{(1 + i)^n}$$

If the series of constant-growth payments is for a specified number of periods, you can solve the present value model most easily by using a new discount rate, r, such that:

$$1 + r = \frac{1 + i}{1 + g} \qquad \text{(9.9)}$$

or

$$r = \frac{1 + i}{1 + g} - 1$$

This new discount rate then is used in equation 9.5 (for the present value of a uniform series), with P_0 as the uniform series payment, i.e.:

$$V_0 = P_0[USPV_{r,N}]^2$$

If the series is perpetual and the rate of growth is less than the discount rate, i, then the present value reduces to:

$$V_0 = \frac{P_1}{i - g}, \text{ or } V_0 = \frac{P_0(1 + g)}{i - g} \qquad \text{(9.10)}$$

To illustrate, suppose the series of $5,000 oil-royalty payments is growing at a 3 percent annual rate. For a five-year period and an 8 percent discount rate, the new discount rate is:

$$1 + r = \frac{1 + i}{1 + g} = \frac{1.08}{1.03} = 1.0485, \text{ or } r = 4.85\%$$

Then

$$V_0 = 5.000[USPV_{0.04855}] = 21,737$$

If the payments occur in perpetuity, the present value is:

---

[2]Alternatively, the present value of a series of constant-growth payments for a specified number of periods can be obtained from:

$$V_0 = \frac{P_0 (1 + g)\left[1 - \left(\frac{1 + g}{1 + i}\right)^N\right]}{1 - g}$$

$$V_0 = \frac{5,000}{0.08 - 0.03} = 100,000$$

THE EFFECTS OF TIME AND INTEREST RATES ON PRESENT AND FUTURE VALUES

The important variables determining present and future values of single payments or a series of payments are (1) the number of conversion periods and (2) the interest rate per compounding period. Both factors interact to determine the total effects of discounting or compounding on present or future values. The impacts are clearly evident in the levels of conversion factors in Appendix Tables II through V. At low rates of interest, the number of time periods have only a modest effect on either present or future values. For example, the future value of $1,000 invested now at 1 percent interest, compounded annually, is $1,105 in 10 years and only $1,220 in 20 years. Similarly, the present value of $1,000 available 10 years from now, with a 1 percent interest rate, is $905, compared to $820 for a similar sum available in 20 years.

At higher interest rates, however, time has a more significant effect on present and future values. Confirm, for example, from Appendix Tables II and III that with an interest rate of 20 percent:

1. The future value of $1,000 is:

    a. $6,192 at the end of 10 years

    b. $38,338 at the end of 20 years

2. The present value of $1,000 is:

    a. $162 if it is to be received in 10 years

    b. $26 if it is to be received in 20 years

Figures 9.1 and 9.2 provide an alternative approach to visualizing the relationship between time and interest rates in determining future and present values. Figure 9.1 plots $V_N = (1 + i)^N$ for interest rates of 1, 10, and 20 percent, respectively, over a range of N values. Notice the acceleration in the slope of the curvilinear relation as i increases. Similarly, Figure 9.2 plots $V_0 = V_N(1 + i)^{-N}$

for interest rates of 1, 10, and 20 percent over a range of N values. Notice again the acceleration in the negative slope of the curvilinear relation as i increases.

The same results occur in the calculation of present or future values of an annuity. In fact, the impact is magnified, since a new payment is received each conversion period. Again, the influence of time on the future value of an annuity increases as the interest rate increases. At 1 percent compounded annually, an annuity of $1,000 has a value of $5,101 after 5 years and $16,097 after 15 years. At a 20 percent rate, however, this annuity is worth $7,442 after 5 years and $72,035 after 15 years. The reverse is true, however, for present values, as indicated in Table 9.1, for varying rates and years.

Notice again that at high interest rates, the capitalization equation gives reasonably accurate estimates of present values for the series of payments when the number of time periods exceeds about 20. For low interest rates, however, the accuracy of the capitalization equation is unacceptable.

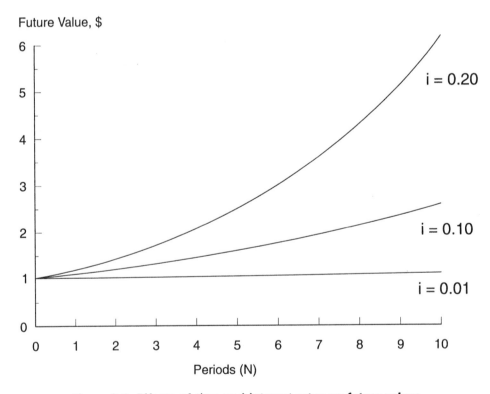

Figure 9.1. Effects of time and interest rates on future values.

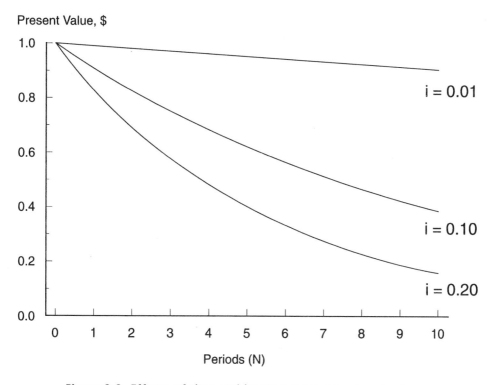

Figure 9.2. Effects of time and interest rates on present values.

Table 9.1. Present Values of an Annuity of $1,000 for Selected Times, Conversion Periods, and Discount Rates

Selected Number of Conversion Periods	Selected Discount Rates (i)			
	1%	10%	20%	25%
5	$ 4,853	$ 3,791	$2,991	$2,689
15	13,865	7,606	4,675	3,859
30	25,808	9,427	4,979	3,995
50	39,197	9,915	4,999	4,000
Capitalization equation $1,000 \div i$	$100,000	$10,000	$5,000	$4,000

Summary

Present value and future value concepts are fundamental to the valuation of payments received in the future or to a series of future payments. The pricing mechanism for determining present and future values is represented by interest rates that portray the time value of money and serve as discount or compounding rates in present and future value calculations, respectively. The time value concept uses compound interest to determine present and future values. As explained earlier, compound interest means that each time interest is paid, it is added to or compounded into the principal and thereafter also earns interest. The important variables determining present values are the magnitude and time patterns of payments, the number and frequency of time periods, and the size of the discount rate.

Other chapter highlights are as follows:

- The process of finding present values is called **discounting** because the future payments are discounted to a lower present value to account for the time value of money.

- Discounting to determine present values is precisely the inverse of compounding to determine future values.

- Simplified present value models can be utilized when the payments series is uniform or exhibits constant growth.

- It is important to determine the interest rate per conversion period when you are compounding or discounting for time.

- Uniform series models for present and future values can be used to solve for the levels of payment series needed to amortize a present value (a capital recovery problem) or to accumulate a future amount (a sinking fund problem).

- The length of the payment series and the level of interest rates interact to produce greater effects on present and future values as the payment series is longer and interest rates are greater.

Review of Formulas

The formulas used in compounding and discounting are presented in summary form to facilitate their use in following chapters.

Notation Summary

V_0 = present value of a future payment(s)

P_0 = payment at time period 0

P_n = payments in other conversion periods

V_N = future value of a present payment or series of payments

n = a conversion period (n = 0, 1, 2, . . . N)

i = interest rate per conversion period

A = uniform annuity payment made each conversion period

g = rate of payment growth per conversion period

1. The present value of a future value is:

$$V_0 = \frac{V_N}{(1 + i)^N} = V_N(1 + i)^{-N}$$

2. The present value of a series of payments is:

$$V_0 = \frac{P_1}{(1 + i)} + \frac{P_2}{(1 + i)^2} + \cdots + \frac{P_N}{(1 + i)^N} = \sum_{n=0}^{N} \frac{P_n}{(1 + i)^n}$$

3. The present value of a uniform series of payments is:

$$V_0 = A\left[\frac{1 - (1 + i)^{-N}}{i}\right] = A[USPV_{i, N}]$$

4. The present value of a perpetual uniform series of payments is:

$$V_0 = \frac{A}{i}$$

5. The future value of a present value is:

$$V_N = V_0(1 + i)^N$$

6. The future value of a series of payments is:

$$V_N = \sum_{n=0}^{N} P_n(1 + i)^{N-n}$$

7. The future value of a uniform series of payments is:

$$V_N = A\left[\frac{(1 + i)^N - 1}{i}\right] = A[\text{USFV}_{i, N}]$$

8. The present value of a constant-growth series is obtained by finding a new discount rate, which is:

$$r = \frac{1 + i}{1 + g} - 1$$

and substituting r for i in equation 3 above. Then:

$$V_0 = P_0\left[\frac{1 - (1 + r)^N}{r}\right] = P_0[\text{USPV}_{r, N}]$$

9. The present value of a perpetual constant-growth series is:

$$V_0 = \frac{A}{i - g}$$

Topics for Discussion

1. Explain how interest rates serve as the pricing mechanism for the time value of money.

2. Distinguish between compound interest and simple interest.

3. Distinguish between compounding and discounting.

4. You have two opportunities to invest $5,000 for 10 years. The first provides a yield of 8 percent annually, compounded quarterly. The second provides a yield of 8.5 percent annually, compounded annually. Which of these investments provides the higher returns? By how much?

5. Which would you prefer — $10,000 now, $20,000 10 years from now, or $30,000 20 years from now, assuming:

   a. a 6 percent annual interest rate?

   b. an 8 percent annual interest rate?

   c. a 10 percent annual interest rate?

6. You have the opportunity to purchase a note that still has four years to run. It pays $750 per year, with the first payment due a year from now. It can be purchased for $2,700. Should you buy the note, assuming you can earn 6 percent on an alternative investment?

7. A farm can finance a tractor purchase with a $40,000 loan at the local bank. The financing terms include a four-year loan, a 12 percent annual interest rate, and equal quarterly repayments of principal and interest. Given this information, determine the level of the quarterly payments.

8. Margaret Putter is a professional golfer who has just signed a $2 million contract. The terms are five annual payments of $400,000 each. Determine the present value of the contract to Margaret, assuming her interest rate is 10 percent.

9. Suppose that Margaret is able to reinvest her annual payments of $400,000 in an investment yielding 10 percent per year. How much will she have after five years?

References

1. Ayres, F., Jr., *Mathematics of Finance* (Schaum's Outline Series), McGraw-Hill Book Company, New York, 1963.

2. Bierman, H., Jr., and S. Smidt, *The Capital Budgeting Decision: Economic Analysis of Investment Projects*, 8th ed., Macmillan Publishing Co., Inc., New York, 1992.

3. Boudreaux, K. J., and H. W. Long, *The Basic Theory of Corporate Finance*, Prentice-Hall, Inc., Englewood Cliffs, New Jersey, 1977.

4. Van Horne, J. C., *Financial Management and Policy*, 8th ed., Prentice-Hall, Inc., Englewood Cliffs, New Jersey, 1989.

5. Weston, J. F., and E. F. Brigham, *Essentials of Managerial Finance*, 10th ed., Dryden Press, Hinsdale, Illinois, 1993.

10

Capital Budgeting Methods

A procedure for evaluating the effects of a farm manager's investment choices on a business's profitability, risk, and liquidity is developed in this chapter. This procedure is called **capital budgeting,** or **investment analysis.** *Investment* is defined as *the addition of durable assets to a business.* Alternatively, **disinvestment** is *the withdrawal of durable assets from the business.* **Financing** refers to *the means of acquiring control of assets: ownership by cash purchase or borrowing, or leasing.* An investment may be financed in several ways, and each may affect the firm's risk and liquidity positions.

Capital budgeting is *an orderly sequence of steps that produces information relevant to an investment choice.* These steps are (1) identification of investment alternatives, (2) selection of an appropriate method, (3) collection of relevant data, (4) analysis of the data, and (5) interpretation of results. These steps are developed in the following sections.

IDENTIFICATION OF ALTERNATIVES

Searching for profitable investments is an important managerial function that needs systematic and thorough attention. While profitable investments may contribute to firm growth, capital obsolescence and price declines may significantly diminish the value of a recent investment. Similarly, using high

269

financial leverage to acquire fixed assets can significantly modify the firm's risk and liquidity. Nonetheless, the manager must generate adequate investment and savings opportunities to efficiently utilize all retained income. Hence, the identification of investment opportunities is a crucial function of management.

Investment opportunities fall into several categories. They are:

1. Maintenance and replacement of depreciable capital items.

2. Adoption of cost-reducing investments to produce a given volume of output.

3. Adoption of income-increasing investments, including either durable business assets or financial assets.

4. A combination of the preceding.

As an example, consider a cattle feeder with a labor-intensive operation who is planning some changes. Some worn-out facilities need to be replaced. Due to high labor costs, the cattle feeder should consider the adoption of a mechanized feeding system. However, in order for the mechanized system to be fully utilized, the size of the operation must be increased. At the same time, relatively low returns from cattle feeding are influencing the cattle feeder to consider a hog enterprise in which some of the assets could be purchased or leased. This example involves many investment alternatives that need to be analyzed with appropriate capital budgeting methods.

CAPITAL BUDGETING METHODS

The next step is to choose a method of capital budgeting for ranking, accepting, or rejecting various investment alternatives. Four methods often used by, or suggested for, business planners are (1) simple rate-of-return, (2) payback period, (3) net present value, and (4) internal rate-of-return (and the new modified internal rate-of-return). Only the net present value and internal rate-of-return methods directly account for the time value of money. Hence, they are generally considered superior methods. However, the other methods are also developed because they are commonly used and are applicable in selected situations.

Different investments have unique time patterns of cash flow for several reasons. First, the inherent characteristics of agricultural enterprises cause differences in the timing of returns. Investments to establish orchards, improve pastures, and expand beef-cow herds generally have long payoff periods. On the other hand, investments in livestock-feeding facilities generally pay off more quickly. Second, new investments by inexperienced producers may exhibit a buildup in cash flow caused by the growing expertise of management and improved production of assets. Third, the terms of financing — downpayment, length of loan, and method of repayment — affect the timing of cash flows and thus affect the financial feasibility of an investment. Fourth, inflation and risk may influence the time pattern of returns. Lastly, alternatives in tax management also influence the timing of cash flows. All these factors indicate the need for capital budgeting methods that account for both the level and the timing of expected cash flows.

To illustrate these capital budgeting methods, consider the evaluation of the following three agricultural investment alternatives. Each alternative requires an initial outlay of $20,000 and will be evaluated over a five-year period. Initially, we assume no income tax obligations, no inflation, no risk, and complete depreciation over the five-year period. Projected cash flows for the three investments are reported in Table 10.1. Time period 0 reflects the current date, and the negative $20,000 is the initial investment.

Table 10.1. Cash Flows for Three Investments

Year	Investment A	Investment B	Investment C
0 – present	$-20,000	$-20,000	$-20,000
1	2,000	5,800	10,000
2	4,000	5,800	8,000
3	6,000	5,800	6,000
4	8,000	5,800	3,000
5	10,000	5,800	1,000

Notice that investment A exhibits increasing returns over the planning period. It might reflect an investment in beef-cow production requiring pasture establishment, range improvements, and a buildup of expected returns over time. Investment B exhibits uniform returns and might reflect an investment in grain production, where the returns are about the same each year. Investment C might reflect irrigated crop production from an exhaustible water supply and thereby exhibits declining returns over time. Each of these investments will be evaluated with the four methods of capital budgeting.

Simple Rate-of-Return

The simple rate-of-return expresses the average profits generated each year by an investment as a percent of either the original investment or the average investment over the investment's expected life. There are several ways to compute simple rates-of-return, and one is expressed as:

$$SRR = \frac{Y}{I} \qquad\qquad (10.1)$$

where Y is the average annual profits (depreciation has been subtracted) projected for the new investment, I is the initial investment or the average investment over the investment's life, and SRR is the simple rate-of-return.

Individual investments are ranked according to their simple rates-of-return and judged as to profitability by comparison with the required rates-of-return (RRR). The investment choice is then based on the following decision rules: If the simple rate-of-return exceeds the required rate-of-return, accept the investment; if the simple rate-of-return equals the required rate-of-return, be indifferent; if the simple rate-of-return is less than the required rate-of-return, reject the investment. Of course, before making the final decisions, consider information on risk, liquidity, and other factors.

The projected returns and average annual rates-of-return for investments A, B, and C are summarized in Table 10.2. In the case of investment A, for example, the average annual profits are $2,000.

$$2,000 = \frac{2,000 + 4,000 + 6,000 + 8,000 + 10,000 - 20,000}{5 \text{ years}}$$

Dividing the average annual profits by the original investment of $20,000 yields SRR = 10 percent. Dividing by the average investment of $10,000 yields SRR = 20 percent.

Based on the simple rate-of-return method, investment A is selected first, followed by investment B, with investment C ranked last. All are profitable as long as the required rate-of-return is less than 8 percent. Note also that if the average investment ($10,000) rather than the initial investment is used, the simple rates-of-return are much higher.

The simple rate-of-return method does not consider the timing of cash flows. This limitation is crucial when the differences in time patterns of cash

**Table 10.2. Simple Rate-of-Return
for Three Hypothetical Investments**

Project	Total Gross Returns	Total Depreciation	Average Annual Profits	Simple Rate-of-Return	
				Original Investment	Average Investment
A	$30,000	$20,000	$2,000	10%	20%
B	29,000	20,000	1,800	9%	18%
C	28,000	20,000	1,600	8%	16%

flow are large. The simple rate-of-return can lead to erroneous conclusions, as illustrated later.

The Payback Period

The payback period is also a widely used method of investment analysis. It estimates the length of time required for an investment to pay itself out or to recover the initial outlay of funds.

The payback period (P) is determined as:

$$P = \frac{I}{E} \tag{10.2}$$

if the projected cash flows (E) are uniform, or as the value of n, giving:

$$P = \frac{I}{\sum\limits_{n=0}^{N} E_n} = 1 \tag{10.3}$$

when the projected cash flows are nonuniform with I being the initial investment; E being the projected cash flow per period (n) from the investment; and P being the payback period, expressed in number of periods.

Individual investments are ranked according to their payback periods, with the shortest being the most favored; acceptability is determined by comparison with the investor's required payback period (RPP). The investment choice is then based on the following decision rules: If the payback period is less than the required payback period, accept the investment; if the payback period equals the required payback period, be indifferent; if the payback period exceeds the required payback period, reject the investment.

For the cash flows projected in Table 10.1, the payback period for investment A is four years, for investment B about four years, and for investment C about three years. Hence, with the payback period, C is preferred to B, which is preferred to A — the reverse of the rankings based on the simple rate-of-return.

The appeal of the payback criterion rests on its simplicity, on its relevance for firms with low liquidity that must concentrate on quick cash recovery, and on the contention that future returns beyond some required payback period are too uncertain to rely on. However, the payback method has been severely questioned for several reasons. First, it does not consider earnings after the payback date. This penalizes investments with returns that increase over time, as is the case with investment A. These large future returns are often quite important. Second, the payback method does not systematically account for differences in the timing of cash flows prior to the end of the payback period. Third, the payback period does not really measure an investment's total profitability; rather, it measures the speed of recovery of the initial investment, which may not be a true indicator of profitability.

Net Present Value

The net present value (NPV) method uses the discounting formulas for a nonuniform (9.4) or a uniform (9.5) series of payments to value the projected cash flows for each investment alternative at one point in time. In this fashion, the net present value criterion directly accounts for the timing and magnitude of the projected cash flows. An important step in implementing the net present value method is the identification and collection of appropriate data. Five types of data are needed. They are:

1. INV is the initial investment.

2. P_n is the net cash flows attributed to the investment that can be withdrawn each year.

3. V_N is any salvage or terminal investment value.

4. N is the length of planning horizon.

5. i is the interest rate or required rate-of-return, also called the cost of capital or discount rate.

The present value model is then set up as follows:

$$NPV = -INV + \frac{P_1}{1 + i} + \frac{P_2}{(1 + i)^2} + \cdots + \frac{P_N}{(1 + i)^N} + \frac{V_N}{(1 + i)^N} \qquad (10.4)$$

Alternatively, if the projected net cash flows are a uniform series of payments (A), the NPV model is:

$$NPV = -INV + A\left[USPV_{i, N}\right] + \frac{V_N}{(1 + i)^N} \qquad (10.5)$$

The NPV models indicate that each projected net cash flow is discounted to its present value, and then the resulting values are all added together to yield a net present value. Any terminal investment value is included as a cash flow in the last year of the planning horizon. The projected cash flows (P_n) can be either positive or negative. Hence, a projected operating loss in year (n) enters the cash flow with a negative sign and is discounted to a present value. The initial investment (INV) is negative because it reflects a cash outlay.

The sign and size of an investment's net present value determine its ranking and acceptability. If the investments under consideration are all income-generating, as is the case in Table 10.1, then the investment with the largest net present value is the most favored, with the next net present value being second favored, etc.

The acceptability of each investment depends on whether the net present value is positive or negative and is expressed in the following decision rules: If the net present value exceeds zero, accept the investment; if it equals zero, be indifferent; if it is less than zero, reject the investment. Note that acceptance of an investment implies that it is profitable, but only relative to the required rate-of-return implied by the discount rate. Rejection of an investment based on a negative net present value implies that the alternative investment with rate-of-return (i) is more profitable than the investment being evaluated.

If the investments under consideration are all cost-reducing, then the projected cash flows would reflect cash outlays (i.e., expenses). When cost-reducing investments are compared, the choice criterion is then based on the minimum net present value of cash outlays.

Suppose that the investor evaluating investments A, B, and C in Table 10.1 requires an 8 percent rate-of-return. The net present value of investment A is:

$$NPV = -20,000 + \frac{2,000}{1.08} + \frac{4,000}{(1.08)^2} + \frac{6,000}{(1.08)^3} + \frac{8,000}{(1.08)^4} + \frac{10,000}{(1.08)^5}$$

The conversion factors for $(1.08)^{-n}$ are taken directly from Appendix Table III for i = 0.08 and values of n ranging from 1 to 5.

$$NPV = -20,000 + 2,000(0.9259) + 4,000(0.8573) +$$
$$6,000(0.7938) + 8,000(0.7350) + 10,000(0.6806)$$

$$= 2,730$$

Similar calculations yield net present values of $3,158 for investment B and $3,766 for investment C. Hence, all three investments are acceptable in terms of the NPV decision rules, although C is preferred to B, which is preferred to A. Note that the payback criterion gave the same ordering, while the simple rate-of-return gave precisely the reverse ordering. These capital budgeting results are summarized in Table 10.3.

Now, suppose that the investor's required rate-of-return, or interest rate, is not 8 percent. In fact, let's start with an interest rate of zero and explore how the ranking and profitabilities change as the interest rate is increased. The net present values for several such interest rates are indicated in Table 10.3.

For a zero interest rate, the net present values of each investment are simply the sums of their undiscounted cash flows — $10,000 for A, $9,000 for B, and $8,000 for C. Now A is preferred to B, which is preferred to C. As the interest rate increases, the profitability declines, as reflected by lower net present values, although the rankings remain the same through an interest rate of 4 percent. However, for a 5 percent rate, the ranking changes in that C is preferred to A, which is preferred to B. Moreover, at a 6 percent rate, the ranking changes again in that C is preferred to B, which is preferred to A. For still higher rates, the rankings remain the same, although eventually, all the investments become unprofitable.

Why did the changes in rankings occur? Clearly, they are tied to the increasing interest rate, but how? The answer lies in the time patterns of the cash flows and in the implied rates-of-return (i) earned when the cash flows are reinvested as they are received. The net present value method assumes that the payments are reinvested in the alternative investment opportunity, whose rate-of-return is the interest rate (i). Raising the interest rate increases the rate of earnings on reinvesting the payments and favors those investments with larger payments coming in sooner. Clearly, investment B earns larger returns faster than investment A. Moreover, investment C earns larger returns faster than B. Raising the interest or reinvestment rate makes these larger early payments more valuable and accounts for the eventual change in rankings.

Table 10.3. Capital Budgeting Results for Three Investments

Criterion	Investment A	Investment B	Investment C
Simple rate-of-return	10%	9%	8%
Payback	4.0 years	3.4 years	2.3 years
Net present value			
i = 0.00	$10,000	$9,000	$8,000
i = 0.02	$ 7,907	$7,338	$6,824
i = 0.04	$ 6,013	$5,820	$5,732
i = 0.05	$ 5,132	$5,111	$5,215
i = 0.06	$ 4,294	$4,432	$4,715
i = 0.08	$ 2,730	$3,158	$3,766
Internal rate-of-return	12.01%	13.82%	17.57%

Internal Rate-of-Return

In determining the net present value of an investment, we used equation 10.4 or equation 10.5 to determine the net present value, given the projected cash flows and the rate of interest. These equations are also used to determine the internal rate-of-return (IRR) of the investment. This rate is called by various names — the *discounted rate-of-return*, the *marginal efficiency of capital*, the *yield of an investment*. It is that rate of interest that equates the net present value of the projected series of cash flow payments to zero.

To find the internal rate-of-return for an investment, simply set up the net present value model, equation 10.4, with the appropriate projected cash flows (INV, P_n, V_N), set the net present value equal to zero, and solve for i:

$$0 = -INV + \frac{P_1}{1 + i} + \frac{P_2}{(1 + i)^2} + \cdots + \frac{P_N}{(1 + i)^N} + \frac{V_N}{(1 + i)^N} \qquad (10.6)$$

The interest rate that satisfies the equation is called the internal rate-of-return (IRR). In effect, the internal rate-of-return equates the present value of the cash flow series to the initial investment (INV), since equation 10.6 can be rewritten as:

$$INV = \frac{P_1}{1 + i} + \frac{P_2}{(1 + i)^2} + \ldots + \frac{P_N}{(1 + i)^N} + \frac{V_N}{(1 + i)^N} \qquad (10.7)$$

To illustrate, find the internal rate-of-return for an investment requiring an initial outlay of $1,000 and yielding a $1,300 payment one year in the future. The IRR model is set up as:

$$0 = -1,000 + \frac{1,300}{1 + i}$$

Modifying the equation yields:

$$1 + i = \frac{1,300}{1,000} = 1.3$$

and

$$i = 0.3$$

Hence, IRR = 30%.

Suppose the $1,300 payment is received at the end of year 2 instead of year 1. The model is respecified as:

$$0 = -1,000 + \frac{1,300}{(1 + i)^2}$$

Modifying the equation yields:

$$(1 + i)^{-2} = \frac{1,000}{1,300} = 0.7692$$

and

$i = 0.14$

In this case, use Appendix Table III to find the value of i, which, for N = 2, gives a conversion factor of 0.7692. To use the table, first locate the appropriate N — year 2, in this case. Then, proceed to the right along row 2 until you reach the conversion factor nearest to 0.7692. In this case, the conversion factor for N = 2 and i = 0.14 is 0.7695. Therefore, the internal rate-of-return of this investment is approximately 14 percent. When the value of the conversion factor falls between two of the interest rates in the table, then linear interpolation is used to estimate the internal rate-of-return.

To illustrate the derivation of the internal rate-of- return for a series of payments, return to the investment choices in Table 10.1. Consider the internal rate-of-return for investment B. Since its payments are uniform, we can use equation 10.5 and solve for i as follows:

$$0 = -20,000 + 5,800[\text{USPV}_{i,5}]$$

or

$$[\text{USPV}_{i,5}] = 3.4483$$

In this case, use Appendix Table V to find the value of i, which, for N = 5, gives a conversion factor of 3.4483. For N = 5, the value listed for i = 0.14 is 3.4331, while for i = 0.13, it is 3.5172. Therefore, the internal rate-of-return for investment B lies between 13 and 14 percent. We can estimate more precisely by linear interpolation:

IRR	*Conversion Factor*		
13 percent	3.5172		
		689	
?	3.4483		841
14 percent	3.4331		

The linear estimate is $13^{689}\!/_{841}$, or 13.82 percent.

Finding the internal rate-of-return for investment A is more difficult because its payments are a nonuniform series requiring the use of equation 10.6:

$$0 = -20,000 + \frac{2,000}{1 + i} + \frac{4,000}{(1 + i)^2} + \frac{6,000}{(1 + i)^3} + \frac{8,000}{(1 + i)^4} + \frac{10,000}{(1 + i)^5}$$

The procedure is essentially a trial-and-error approach in search of the interest rate (i) yielding a zero net present value. A good way to start is to select a rate-of-return that you would like to have and discount the cash flows to a net present value at this interest rate. If the net present value is positive, your rate is too low. Discount again at a higher rate. If the net present value is negative, your rate is too high. In this fashion, you search for the interest rate that yields a zero net present value.

For investment A, an interest rate of 8 percent yields a net present value of $2,730. Try again with a higher rate. For i = 0.12, the net present value is $3.58, and for i = 0.13, the net present value is – $605.05. Linear interpolation yields an internal rate-of-return of 12.01 percent.

As with the other capital budgeting criteria, investments can be ranked and accepted or rejected on the basis of their internal rates-of-return. Ranking is based on the relative sizes of the internal rates-of-returns, with the largest being the most favored. The acceptability of each investment depends upon the comparison of its internal rate-of-return with the investor's required rate-of-return. Acceptability is based on the following decision rules: If the internal rate-of-return exceeds the required rate-of-return, accept the investment; if the internal rate-of-return equals the required rate-of-return, be indifferent; if the internal rate-of-return is less than the required rate-of-return, reject the investment. As with other capital budgeting criteria, these decision rules are still subject to consideration of risk and liquidity. For the investments in Table 10.1, C is preferred to B, which is preferred to A, according to the IRR criterion, and all are acceptable for a required rate-of-return of 8 percent.

Comparing Net Present Value and Internal Rate-of-Return

The net present value and internal rate-of-return methods are closely linked because each uses the same discounting procedure. However, the NPV method requires a specified discount rate, while the IRR method solves for the discount rate that yields a zero net present value. This linkage is portrayed in Figure 10.1, which graphs the net present values of investments B and C at alternative discount rates. At a zero discount rate, the net present value is highly positive. Increasing the discount rate lowers the net present value until it eventually becomes negative (i.e., below the horizontal axis). The internal rate-of-return is found at the point where the NPV line crosses the horizontal axis. It is the discount rate yielding zero net present value.

The IRR method gives the same ranking of investments as the NPV method under most circumstances. Both account for differences in the time pattern of cash flows. However, occasional differences in ranking can arise because of different assumptions about the rate-of-return on reinvested net cash flows. The IRR method implicitly assumes that net cash inflows from an investment are reinvested to earn the same rate as the internal rate-of-return of the investment under consideration. The NPV method, on the other hand, assumes that these funds can be reinvested to earn a rate-of-return that is the same as the firm's discount rate.

The possible differences in rankings are illustrated in Figure 10.1. Both the NPV and the IRR methods rank investment C over B as long as the discount

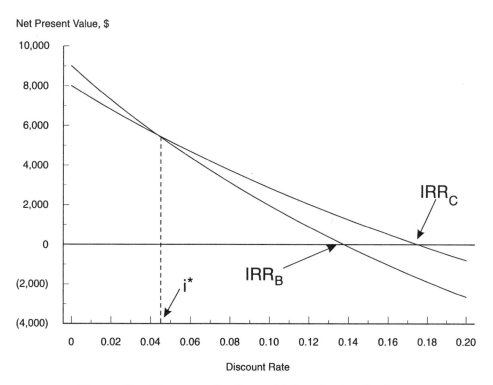

Figure 10.1. Net present value and internal rate-of-return.

rate exceeds i*. At i*, the present value lines intersect, indicating equal net present values. Below i*, investment B has the higher net present value, but a lower internal rate-of-return.

Which reinvestment rate is more realistic? The NPV rate has the advantage of being consistently applied to all investment proposals. In addition, the NPV rate may be more realistic if the discount rate is determined by the opportunity cost of capital. On the other hand, the internal rate-of-return from each investment alternative can be compared against a common required rate-of-return. The internal rate-of-return also represents profitability in percentage terms, which is often preferred by business managers, even though the increases in wealth measured by the net present values more directly reflect the objectives sought by the firm.

To solve equation 10.6 for the internal rate-of-return, you must have at least one sign reversal (e.g., one negative sign) within the cash flows. Generally, the initial cash flow is negative because of the cash outlay for the initial investment. Multiple internal rates-of-return result if more than one sign reversal occurs. In fact, an additional internal rate-of-return exists for every sign reversal; hence, when finding an internal rate-of-return under these circumstances, be cautious.

The IRR approach is also tedious to calculate when the payments are a nonuniform series. Fortunately, many pocket calculators and computers are programmed for discounted cash flow analysis according to the preceding procedures. These technological developments greatly simplify the calculations and enhance the usefulness of these capital budgeting methods.

Modified Internal Rate-of-Return

The discussion in the preceding section indicated that the IRR method may yield inconsistent rankings of mutually exclusive investment alternatives that have disparities in the time patterns of their cash flows, sizes of initial investments, and lifetimes. The inconsistent rankings reflect differences between the IRR and the NPV methods in the implied rates of earnings generated by reinvestment of the cash flows — reinvestment at the cost of capital for the NPV method and reinvestment at the internal rate-of-return for the IRR method. Between these two assumptions, the NPV method seems more valid and realistic — it avoids the implication that cash flows from an investment could be perpetually reinvested in the same investment to yield the same internal rate-of-return.

Despite this conceptual preference for the NPV method, numerous surveys of businesspersons, financial analysts, and other practitioners have indicated significant, sustained use of and preference for rate-of-return measures for evaluating and comparing investment alternatives. Consequently, analysts have sought to modify the calculation procedures for the IRR method in order to resolve its ranking and other technical problems [10]. The result is a modified internal rate-of-return (MIRR) method. It is modified in the sense that the cost of capital is used explicitly as the reinvestment rate for the project's cash flows.

The calculation procedures for finding the MIRR are straightforward. First, find the present value (PV_{CO}) of all negative cash outflows, in periods when they occur, using the cost of capital as the discount rate. Negative cash outflows usually represent capital expenditures. For example, the initial investment outlay (INV) is a negative cash outflow, and negative cash flows could occur in other periods as well.

$$PV_{CO} = -INV + \sum_{n=0}^{N} \frac{-P_n}{(1 + i)^n} \tag{10.8}$$

Second, using the cost of capital as the compounding rate, find the future value (FV_{CI}) of all positive or zero cash flows,

$$FV_{CI} = \sum_{n=0}^{N} P_n(1 + i)^{N - n} \tag{10.9}$$

Then, solve for the rate of discount (i_m) that equates the present value of the cash outflows to the future value of the cash inflows, or solve for i_m in:

$$PV_{CO} = \frac{FV_{CI}}{(1 + i_m)^N} \tag{10.10}$$

The resulting value of i_m is the MIRR.

To illustrate, using a cost of capital of 4 percent, find the MIRR for investment a in Table 10.1. The present value of the cash outflows is the initial investment of $20,000, or:

$PV_{CO} = 20,000$

The future value of the cash inflows is:

$FV_{CI} = 2,000(1.04)^4 + 4,000(1.04)^3 + 6,000(1.04)^2 + 8,000(1.04) + 10,000$

$= 31,649$

The MIRR then is:

$$20,000 = \frac{31,649}{(1 + i_m)^5}$$

$(1 + i_m)^5 = 1.5824$

$i_m = 0.0961$, or 9.61%

Table 10.4 indicates modified internal rates-of-return for investments A, B, and C in Table 10.1, using the same costs of capital (i = 0.00, 0.02, 0.04, 0.05, 0.06, and 0.08) found in Table 10.3 for the NPV results. The important point is that the rankings for the three investments under the modified internal rate-of-return are the same as the NPV rankings for the various costs of capital. Thus, the MIRR method can resolve ranking disparities between the IRR and the NPV methods that are caused by differences in the time patterns of cash flows.

The modified internal rate-of-return also resolves other technical problems associated with the internal rate-of-return. However, the one problem that the modified internal rate-of-return may not resolve is any ranking disparities associated with differences in the size of investment (note that investments A, B, and C each required the same initial investment of $20,000). When size differences occur, the rankings of the NPV and MIRR methods usually are the same, but not always. In these cases, the net present value continues to provide a better indicator of the effect of a new investment on the wealth position of a business.

**Table 10.4. Modified Internal Rates-of-Return
for Three Investments**

Cost of Capital	Investment A	Investment B	Investment C
i = 0.00	8.45%	7.71%	6.96%
i = 0.02	9.03%	8.58%	8.17%
i = 0.04	9.61%	9.45%	9.38%
i = 0.05	9.91%	9.89%	9.98%
i = 0.06	10.20%	10.33%	10.58%
i = 0.08	10.80%	11.21%	11.79%

DATA NEEDS

This section discusses the kinds of data needed for using the NPV and the IRR methods of capital budgeting. These data include the initial investment requirements, net cash flows, terminal value, discount rate, and planning horizon. First, however, a distinction must be established between returns to assets and returns to equity as the measurement objectives of capital budgeting. This distinction influences the types of financing data needed for the investment analysis.

Returns to Assets or Returns to Equity?

As indicated in Chapter 4, returns to assets and returns to equity are two important measures of profitability. Returns to assets measures profitability before interest is paid to a firm's lenders and before any withdrawals or retained earnings are claimed by the firm's owners. Returns to equity measures profitability after the costs of borrowed funds are accounted for. These same concepts apply when the NPV or IRR method is used to measure the profitability of individual investments.

First, consider that the capital budgeting objective is to measure the profitability of the assets committed to an investment project. Measure the returns to assets by projecting the expected payment flow for the investment without deducting any charges for interest or loan repayments and then discounting the payments to a present value, using the weighted average cost of the firm's debt and equity capital as the discount rate. This approach accounts for the effects of the firm's overall financial structure (leverage) in the discount rate. It is based on the assumptions that the firm's leverage determines the financing costs of an individual investment and that the financing for an individual investment does not influence the firm's overall cost of capital.

The returns-to-assets approach is used by large corporate firms that can establish and maintain rigorous leverage policies and that have efficient access to the national markets for debt and equity capital. These firms' leverage and weighted average costs of capital are relatively stable and predictable over time. Their capital budgeting procedures need not account for a specific investment's financing terms.

Now, consider that the capital budgeting objective is to measure the profitability of the equity capital committed to an investment project. Measure the return to equity by projecting the payment flow, net of the cash outflows for principal and interest on debt, and then discounting the payments to a present value, using the firm's cost of equity capital as the discount rate. This approach explicitly accounts for each investment's method of financing. It is based on the assumption that an investment's financing costs may strongly influence the firm's leverage and cost of capital. This approach is applicable for smaller, non-corporate firms whose leverage fluctuates over time and that lack access to the national markets for debt and equity. In smaller firms, the large size and financing terms of new large-scale investments often may significantly change the make-up of assets and liabilities, with the effects lasting several years.

The choice of measuring returns to assets or returns to equity in capital budgeting requires careful judgment, based on the characteristics of the business and the investments being analyzed. Either approach may be used. Here, and in the following chapters, we use the returns-to-equity approach because it is consistent with the smaller-scale, non-corporate structure of most farm businesses. Moreover, it enables the analyst to develop capital budgets that include all factors affecting cash flows, thus making the long-term discounted cash flow analysis more compatible with the cash flow budgeting outlined in Chapter 5. However, because the weighted average cost of capital approach to capital budgeting is widely used in corporate finance and is gaining applicability in agricultural finance, we illustrate its use in the appendix to this chapter.

Initial Investment Requirements

By initial investment, we refer to the initial equity the investor commits to the investment. The initial equity depends upon the value of assets needed to implement the investment and upon the method of financing. It is important to determine the acquisition cost of the durable assets because it represents the cost basis for annual depreciation charges. Charging depreciation influences annual cash flows through its effect on income tax obligations. It is also important to account for the type of financing. Use of borrowed funds may spread the cash outflow to pay for the investment over several years, due to the terms of financing. In the returns-to-equity approach to present value analyses, these cash outflows for the initial asset acquisition are combined with other cash flows arising from the asset's productive use. This combination is illustrated in the following section.

One must be sure that all costs necessary to make the investment operational are included in the initial investment requirement. These costs include freight installations, sales taxes, new components, modifications in buildings, roads, fences, as well as the cost of the asset. They may not all occur in the initial period. Later outlays must be reflected in the periods when they occur.

The trade-in or salvage value of a replaced asset should be subtracted from the purchase price of the new asset in determining the net investment cost. Only the net cash outflow associated with the new investment — the total purchase price plus installation charges less any trade-in — and, when applicable, the income taxes saved through capital losses on the old asset are relevant. For example, consider the replacement of an old machine by a new one that costs $15,000 at the factory. The old machine has a trade-in value of $5,000. Freight costs are $800 on the new machine. The net replacement cost is $15,000 + $800 – $5,000 = $10,800. If the machine is financed with a 25 percent downpayment of equity capital and 75 percent debt, then the initial equity outlay is $2,700.

Net Cash Flows

All capital budgeting methods require some measure of expected future returns for each investment alternative. In the NPV and IRR methods, net cash flow from the business rather than accounting profits is used as the measure of returns. Net cash flows include all cash inflows relating to an enterprise and all cash outflows for operating expenses, capital expenditures, income taxes, and financing. Thus, a net cash inflow is the stream of cash that the owner can withdraw for consumption or reinvestment elsewhere.

Cash flows are not identical to profits or income as derived from an income statement. In fact, changes in income can occur without any corresponding changes in cash flows. These changes can happen because the accrual accounting system (see Chapter 3) differs from the cash basis of accounting. Receipts and expenses are assigned to the fiscal year in which they are earned or incurred rather than when they are received or paid for in cash. Annual depreciation and inventory changes are examples of income statement items that do not directly represent cash flows. However, it is still important to determine the depreciation and other changes in accounting income, since they determine the income tax obligation — a cash flow item.

Under the returns-to-equity approach, cash flows also account for borrowing and debt repayments associated with the investment. These financing terms can influence investment profitability and modify, often substantially, the pattern of net cash flows. If the machine described in the preceding section were purchased for cash, the initial investment would be $10,800. Thus, $10,800 would be entered as a cash outflow at time zero (n = 0). Suppose, however, that the machine were purchased with a 25 percent downpayment or equity commitment, with the debt repaid in three equal annual installments. Only –$2,700 would enter the cash flow for n = 0. Additional entries of –$2,700 for each of the next three years (n = 1 , n = 2, n = 3) would cover annual repayments on principal. Interest on the loan would also be added as a cash outflow.

In addition to cash outlays to pay for the assets, most capital investments also affect annual operating costs and receipts. However, some farm investments are cost-reducing rather than output-increasing. A new milk-handling system might permit the reduction of one employee in a large dairy operation, thus saving $6,000 in wages. If the cash operating expenses for the machine are $1,000, the annual net reduction in cash outflow would be $5,000.

When several cost-reducing investments are compared, the choice criterion is then the *minimum* net present value of cash outlays. Similarly, a farm manager might be considering investments in buildings or machinery. Each investment has about the same effect on revenue but involves different investment requirements and operating costs. In this case, the analysis can ignore the returns, since they are assumed to be equal. Again, the object is to select the investment with the lowest present value of projected cash outflows.

Only future cash flows are relevant. Past records are relevant only insofar as they can guide projections of the future. However, the historical data must be supplemented by new information regarding expected prices, costs of production, changes in technology, etc. Accounting for the effects of inflation on revenues and expenses is also important. Most capital budgets involve payment series projected using "nominal" values. Thus, capital budgets include the effects of inflation.

It is also important to incorporate income taxes in projections of future cash flows. At high levels of income, before-tax income may differ substantially from after-tax income. After-tax income is the important measure. Not all cash outflows are tax-deductible. New asset purchases and principal payments on debt fall into this category. Similarly, not all cash inflows generate taxable income. The sale of an asset at its depreciated value or a payment received on a debt owed the business is a nontaxable inflow. Finally, some tax-deductible items are not part of the cash flow. The most notable example is depreciation.

In essence, the payment series involves the projection of nominal net cash flows that are specified after payment of taxes, interest, and loan principal.

Terminal Value

At the end of the manager's planning horizon, the assets involved in the investment may have a terminal or residual value. For depreciable assets such as machinery, this is often considered a salvage value. Some appreciation in value may also occur, resulting in capital gains at the end of the planning period. Historically, this has been the case with farm land.

If the manager foresees that part or all of the investment could be liquidated to recover a salvage value, then the terminal value should be considered as an expected cash inflow in the final year of the payment series. Only the equity portion should be included. Any debt outstanding against the assets should be repaid when they are sold. Moreover, the tax effects of capital gains or losses must also be included.

The Discount Rate

One of the dominant variables affecting present values is the discount rate. For these capital budgeting procedures, this rate is a firm's required rate-of-return on its equity capital. It is best understood as an opportunity cost. Thus, the relevant cost of capital for a particular investment is the rate such equity capital could earn in its most favorable alternative use.

The discount rate (i) used in capital budgeting is generally considered to contain three components: a real risk-free rate (i_t) for time preference, a risk premium (i_r) that reflects the riskiness of the expected net cash flows, and an inflation premium (i_f) that reflects an anticipated rate of inflation. Hence, the discount rate (i) is expressed in general by the following functional relationship:

$$i = f(i_t, i_r, i_f)$$

Chapter 11 will further develop the risk and inflation premiums and show how their effects are included in capital budgets.

The Planning Horizon

Analyzing an investment in durable assets requires a multi-year planning horizon because these assets influence cash flows for several years. One factor influencing the relevant length of planning horizon is the productive life of the assets. Good estimates of productive life for most durable assets are available from engineering data. Of course, land that is properly maintained has an infinite lifetime, unless a significant change in land use (e.g., urbanization) occurs.

The discounting process also influences the relevant length of planning horizon. The mathematics of discounting implies the declining present value of economic events far into the future. Moreover, the higher the discount rate, the lower the present value. Thus, an economically relevant planning horizon can be thought of as the time span needed to make an appropriate decision for the first period. A future event is irrelevant at present if it can assume any expected value, but yet be of no importance for determining actions during the first period.

Suppose, for example, that an investor expects the value of a tract of land to be $100,000 in a future year, although the potential variation of the land's value is from $90,000 to $110,000 — a $20,000 range. If the discount rate is 14 percent and the planning period is 10 years in length, the present value of this range is $5,395. For a 15-year period, the present value of the variation is $2,802, and for a 20-year period, the present value of the variation is $1,455. As the planning period lengthens, the present value of the expected variation of land value becomes smaller and of decreasing current importance.

An economically relevant planning horizon may also be influenced by the investor's risk attitude. Adding a risk premium to the discount rate tends to shorten the planning horizon. The greater the risk, the shorter the economically relevant horizon. Accurate and comprehensive forward planning should help to dispel the tendency of uncertainty to shorten planning horizons.

ANALYSIS AND INTERPRETATION: A CAPITAL BUDGETING APPLICATION

The logic and method of capital budgeting under alternative financing methods are illustrated in the following investment situation. Consider a cash-grain farmer who is investing in a hog farrow – finish enterprise to expand the size of the operation. Numerous systems of hog production are available and should be evaluated; however, here we will consider a confinement hog operation. It is a capital-intensive system that utilizes the latest technology in feed distribution, waste disposal, and animal care. The system has a capacity of 240 litters per year, with 120 sows farrowing every six months.

Using the NPV and IRR methods, evaluate the investment, under alternative combinations of debt and equity financing. The analysis begins with a cash purchase of the initial investment, with only the investor's equity capital used. Then, the analysis is rerun under the assumption that the investor uses equity capital to finance 20 percent of the initial investment and borrows the remaining 80 percent with an intermediate-term loan. Finally, combinations of debt and equity capital falling between these two cases are considered.

Tables 10.5 and 10.6 give a summary of the basic data, projected net cash flows, and the profitability measures for the full equity and 20 percent equity – 80 percent debt cases, respectively. This information is described in greater detail as follows.

The analysis requires the following data:

1. *Initial investment*

    The initial investment totals $140,000, with $90,000 required for buildings, $38,000 for equipment, and $12,000 for livestock. The breeding livestock fall in the three-year class for tax depreciation purposes, while the buildings and equipment are in the seven-year class. Consistent with the provisions of the Accelerated Cost Recovery System for tax treatment, depreciation is calculated according to the 200 percent declining balance method with a half-year convention for the first and last years and a switch to straight-line depreciation at a time to maximize depreciation deductions in each year.

2. *Planning horizon*

   The investor uses a 10-year planning horizon.

3. *Terminal value*

   The terminal value in year 10 is projected to be $21,000 and is fully taxable.

4. *Required rate-of-return*

   The investor stipulates a required rate-of-return of 11 percent without the use of debt financing. The required return increases as the use of debt financing increases in response to the greater financial risk from leveraging.

5. *Net cash flows*

   Net cash flows to the investor are determined by deducting projected operating expenses, income tax obligations, and, where appropriate, principal and interest payments from projected operating receipts in each year of the planning period. The initial investment is a negative cash outflow at present. The terminal value is considered part of the cash flow in the final period. In response to inflation, both the operating receipts and expenses are projected to increase at 2 percent annual rates. For simplicity, an income tax rate of 20 percent is assumed.

   Part B of Table 10.5 shows a worksheet approach to projecting the operating receipts, expenses, taxes, and net cash flow to the investor over the 10-year period for the full equity financing case. The net cash flow in line 12 is the series of payments to be discounted to a net present value. Annual net cash flow without debt financing is the sum of operating receipts and capital sales less operating expenses and income taxes paid. Income tax obligations are derived from taxable income, which is found by deducting operating expenses and depreciation from operating receipts.

   Equation 10.4 is used to calculate net present values.

$$NPV = -INV + \sum_{n=1}^{N} \frac{P_n}{(1 + i)^n} + \frac{V_N}{(1 + i)^N}$$

When a discount rate of i = 0.11 is applied to the net cash flows in Table 10.5, the result is:

$$\text{NPV} = -140{,}000 + \frac{22{,}889}{1.11} + \frac{26{,}137}{(1.11)^2} + \ldots + \frac{38{,}829}{(1.11)^{10}} = 2{,}698$$

Similarly, equation 10.6 is used to calculate the internal rate-of-return, which is 11.45 percent, and the modified internal rate-of-return, which is 11.21 percent. Hence, the hog enterprise is a marginally acceptable investment based on the positive net present value for an 11 percent discount rate and based on the internal rate-of-return and the modified internal rate-of-return, which exceed the required rate-of-return.

Part C of Table 10.5 shows the calculation of these profitability measures as well as the sensitivity of the net present value to a range in the discount rate from 5 to 17 percent. As expected, the net present value declines as the discount rate increases and becomes negative for discount rates that exceed the internal rate-of-return.

Table 10.6 shows the financial data for the 20 percent equity – 80 percent debt case. The basic difference from Table 10.5 is that net cash flows are calculated after the payments for interest and principal on the loan and the related tax effects are accounted for. The following financing terms are specified. The investor's initial equity requirement is 20 percent of the initial investment, or $28,000. The investor obtains an intermediate-term loan for $112,000, payable in 10 equal annual amortized installments of $19,822, based on an interest rate of 12 percent [*note:* with a 20 percent tax rate, the before-tax interest rate of 12 percent is equivalent to an after-tax rate of 9.60 percent = 12(1 – 0.2)]. The amortization schedule (annual interest and principal) is shown in lines 7 and 8 of Part B of Table 10.6.

Annual net cash flow, as projected in line 12, is now the sum of the operating receipts and terminal value less operating expenses, income taxes, principal, and interest paid in each year of the planning horizon. Interest (line 7) is calculated as 12 percent of the remaining loan balance, as adjusted by principal payments (line 8) made in previous years. Income tax obligations are derived from taxable income (line 9), which is found by deducting operating expenses, depreciation, and interest from operating receipts.

Due to the greater financial risk associated with leveraging, the investor now stipulates 19 percent as the required rate-of-return for the investment. As shown in Part C, applying a discount rate of i = 0.19 and equation 10.4 to the net cash flows in Table 10.5 yields:

**Table 10.5. Investment Data, Net Cash Flows,
for a Hog Enterprise**

PART A: Investment Data

Initial investment. $140,000

      3-year property 12,000

      7-year property 128,000

Terminal value. $21,000

PART B: Net Cash Flows

Projections of Net Cash Flows

Item		Year			
	0	1	2	3	4
1. Operating receipts		172,800	176,256	179,781	183,377
2. Terminal value					
3. Total cash inflow (1 + 2)	0	172,800	176,256	179,781	183,377
4. Initial equity	140,000				
5. Operating expenses		149,760	152,755	155,810	158,926
6. Depreciation		22,286	36,680	24,169	16,882
7. Loan interest		0	0	0	0
8. Loan principal		0	0	0	0
9. Taxable income (3 − 5 − 6 − 7)	0	754	(13,179)	(198)	7,569
10. Income taxes (9 × t)		151	(2,636)	(40)	1,514
11. Total cash outflow (4 + 5 + 7 + 8 + 10)	140,000	149,911	150,119	155,770	160,440
12. Net cash flow (3 − 11)	(140,000)	22,889	26,137	24,011	22,937

and Profitability Measures
Without Debt Financing

Growth rates in receipts and expenses 2%

Income tax rate (t) 20%

Number of litters 240

Gross income/litter 720

Direct costs/litter 624

% financed 0%

Interest rate on debt 12%

Discount rate 11%

for Hog Enterprise

Year					
5	6	7	8	9	10
187,045	190,786	194,602	198,494	202,464	206,513
					21,000
187,045	190,786	194,602	198,494	202,464	227,513
162,105	165,347	168,654	172,027	175,468	178,977
11,424	11,424	11,424	5,712	0	0
0	0	0	0	0	0
0	0	0	0	0	0
13,516	14,015	14,524	20,755	26,996	48,536
2,703	2,803	2,905	4,151	5,399	9,707
164,808	168,150	171,559	176,178	180,867	188,684
22,237	23,636	23,043	22,316	21,597	38,829

(Continued)

Table 10.5

PART C: Profitability Measures
      Internal Rate-of-Return 11.45%

			Net Present Value Sensitivity
Year	Cash Flow	Net Present Value @ 11%	5%
0	(140,000)	(140,000)	(140,000)
1	22,889	20,621	21,799
2	26,137	21,213	23,707
3	24,011	17,556	20,741
4	22,937	15,109	18,871
5	22,237	13,196	17,423
6	22,636	12,102	16,891
7	23,043	11,099	16,376
8	22,316	9,683	15,104
9	21,597	8,443	13,922
10	38,829	13,675	23,838
Net Present Value		2,698	48,672

$$\text{NPV} = -28{,}000 + \frac{5{,}755}{1.19} + \frac{8{,}850}{(1.19)^2} + \ldots + \frac{19{,}432}{(1.19)^{10}} = -728$$

Similarly, the internal rate-of-return is 18.26 percent and the modified internal rate-of-return is 18.69 percent. Hence, the hog enterprise now is judged marginally unprofitable under this financing arrangement because of the negative net present value and an internal rate-of-return that is lower than the revised required rate-of-return.

Notice, however, that the use of debt capital. or leveraging, raises the internal rate-of-return from 11.45 percent to 18.26 percent. Moreover, if the same discount rate is used to evaluate both the full equity case and the 20 percent equity – 80 percent debt case, the net present value increases for debt financing. For example, using the 11 percent rate, the net percent value is $2,698 for the full equity case and $9,333 for the 20 percent equity – 80 percent debt case.

Why does the change in type of financing alter the profitability? The answer involves the principles of leveraging and the cost relationships between debt and equity. For the leveraged case, the internal rate-of-return in-

(Continued)

---

to Discount Rates

7%	9%	13%	15%	17%
(140,000)	(140,000)	(140,000)	(140,000)	(140,000)
21,392	20,999	20,256	19,904	19,563
22,829	21,999	20,469	19,763	19,093
19,600	18,541	16,641	15,787	14,991
17,499	16,249	14,068	13,114	12,240
15,855	14,452	12,069	11,056	10,142
15,083	13,497	10,872	9,786	8,824
14,350	12,605	9,795	8,663	7,678
12,988	11,200	8,394	7,295	6,355
11,747	9,944	7,189	6,139	5,257
19,739	16,402	11,439	9,598	8,078
31,081	15,888	(8,808)	(18,895)	(27,777)

creased because lower-cost debt capital was, in effect, being substituted for higher-cost equity. Recall that the tax savings on the payment of loan interest makes the stated loan rate lower on an after-tax basis. In this situation, the before-tax interest rate on debt of 12 percent is equivalent to an after-tax rate of 9.60 percent, reflecting the 20 percent tax rate. As an exercise, the reader could confirm the following internal rate-of-return for the different financing plans:

Equity	*Debt*	*IRR*
100%	0%	11.45%
80%	20%	11.89%
60%	40%	12.63%
40%	60%	14.08%
20%	80%	18.26%

Of course, the principles of leverage involve more than the profitability effects alone. Financial risks increase and the firm's liquidity tends to decrease

Table 10.6. Investment Data, Net Cash Flows, for a Hog Enterprise

PART A: Investment Data

Initial investment.	$140,000
3-year property	12,000
7-year property	128,000
Terminal value.	$21,000

PART B: Net Cash Flows

Projections of Net Cash Flows

Item	Year				
	0	1	2	3	4
1. Operating receipts		172,800	176,256	179,781	183,377
2. Terminal value					
3. Total cash inflow (1 + 2)	0	172,800	176,256	179,781	183,377
4. Initial equity	28,000				
5. Operating expenses		149,760	152,755	155,810	158,926
6. Depreciation		22,286	36,680	24,169	16,882
7. Loan interest		13,440	12,674	11,816	10,856
8. Loan principal		6,382	7,148	8,006	8,966
9. Taxable income (3 − 5 − 6 − 7)	0	(12,686)	(25,853)	(12,014)	(3,287)
10. Income taxes (9 × t)		(2,537)	(5,171)	(2,403)	(657)
11. Total cash outflow (4 + 5 + 7 + 8 + 10)	28,000	167,045	167,406	173,229	178,091
12. Net cash flow (3 − 11)	(28,000)	5,755	8,850	6,552	5,286

as more debt is used relative to equity capital. Thus, the increase in profitability from leveraging is accompanied by greater risk. For the hog example, business risks already are high due to possible unanticipated variations in prices, production, disease, and so on. Leveraging magnifies these business risks because the fixed repayment obligations increase the variability of returns to the equity investor and raise the potential loss of equity capital. Moreover, borrowing

and Profitability Measures
With Debt Financing

Growth rates in receipts and expenses 2%

Income tax rate (t) 20%

Number of litters 240

Gross income/litter 720

Direct costs/litter 624

% financed 80%

Interest rate on debt 12%

Discount rate 19%

for Hog Enterprise

		Year			
5	**6**	**7**	**8**	**9**	**10**
187,045	190,786	194,602	198,494	202,464	206,513
					21,000
187,045	190,786	194,602	198,494	202,464	227,513
162,105	165,347	168,654	172,027	175,468	178,977
11,424	11,424	11,424	5,712	0	0
9,780	8,575	7,225	5,713	4,020	2,124
10,042	11,247	12,597	14,109	15,802	17,698
3,736	5,440	7,299	15,042	22,976	46,412
747	1,088	1,460	3,008	4,595	9,282
182,674	186,257	189,936	194,857	199,885	208,081
4,371	4,529	4,666	3,637	2,579	19,432

(Continued)

depletes the firm's credit reserve and can hamper repayment capacity on other investment projects.

Thus, these capital budgeting methods indeed generate useful information for decision making. *However, the numbers themselves do not make decisions — people do!* Clearly, factors such as risk and liquidity need to be considered along with the profitability measures.

Table 10.6

PART C: Profitability Measures
      Internal Rate-of-Return 18.26%

Net Present Value Sensitivity

Year	Cash Flow	Net Present Value @ 19%	13%
0	(28,000)	(28,000)	(28,000)
1	5,755	4,836	6,093
2	8,850	6,249	6,931
3	6,552	3,888	4,541
4	5,286	2,636	3,242
5	4,371	1,832	2,372
6	4,529	1,595	2,175
7	4,666	1,381	1,983
8	3,637	904	1,368
9	2,579	539	858
10	19,432	3,412	5,724
Net Present Value		(728)	6,288

Summary

Capital budgeting refers to a procedure for evaluating how one or more investment alternatives will affect a business's profitability, risk, and liquidity. The possible capital budgeting methods include the simple rate-of-return, payback, net present value (NPV), and internal rate-of-return (IRR), although only the net present value and the internal rate-of-return directly account for the level and timing of future payments and the time value of money. The data needed for present value methods of capital budgeting include initial investment requirements, projected net cash flows, terminal values, the discount rate, and the length of planning horizon. The procedures and data can be organized to measure either the profitability of the assets or the equity capital committed to an investment.

Other chapter highlights are as follows:

- Because many investments exhibit irregular patterns of projected cash

(Continued)

---

to Discount Rates

15%	17%	21%	23%	25%
(28,000)	(28,000)	(28,000)	(28,000)	(28,000)
5,004	4,919	4,756	4,679	4,604
6,692	6,465	6,044	5,849	5,664
4,308	4,091	3,698	3,521	3,354
3,023	2,821	2,466	2,310	2,165
2,173	1,994	1,685	1,553	1,432
1,958	1,766	1,443	1,308	1,187
1,754	1,555	1,229	1,096	979
1,189	1,036	791	694	610
733	628	464	400	346
4,803	4,042	2,886	2,452	2,086
3,637	1,315	(2,534)	(4,139)	(5,572)

flows, the capital budgeting methods based on present value concepts (NPV, IRR) are most appropriate to use.

- The NPV and IRR methods generally lead to similar acceptance – rejection decisions and to similar rankings of mutually exclusive investments, although inconsistent rankings may occur because of implied differences in reinvestment rates for the two methods.

- Under the NPV method, reinvested net cash flows are implicitly assumed to earn a rate-of-return equal to the firm's discount rate.

- The IRR method implicitly assumes that net cash inflows from an investment are reinvested to earn the same rate as the internal rate-of-return of the investment under evaluation.

- The modified internal rate-of-return can resolve potential inconsistencies in rankings for net present value and internal rate-of-return for investments of similar size by solving for the discount rate that equates the present value of all negative net cash flows with the future value of all positive net cash flows.

- Under the returns-to-asset approach to net present value, the payment series does not include the effects of financing, and the discount rate is a weighted average cost of debt and equity capital.

- Under the returns-to-equity approach to the net present value, the payment series is the net of cash outflows for principal and interest on debt, and the discount rate is the cost of equity capital. Thus, investments' specific financing arrangements may influence profitability under this approach.

Topics for Discussion

1. Contrast and critique the NPV and IRR methods of capital budgeting. What are their similarities and differences? Advantages and disadvantages? Which is generally preferred in capital budgeting? In what ways does the modified internal rate-of-return become helpful?

2. Under what conditions is the payback period a useful capital budgeting method?

3. Explain why changes in discount rates may influence the ranking and profitability of several investment alternatives.

4. Explain the alternative ways of accounting for financing transactions in capital budgets.

5. Discuss some of the problems in obtaining high-quality data for capital budgeting analysis. What kinds of information do managers need to make better investment and financing decisions in their businesses?

6. A farmer is considering replacing a labor-intensive machine system with a more capital-intensive one. Adopting the new system is estimated to increase machinery operating expenses by about $1,000 per year and to replace one hired laborer, whose annual salary is $6,000. The new machinery costs $30,000; however, the trade-in value of the old system is $10,000. Adopting the new machinery will increase annual depreciation by $4,000.

   Further data indicate an eight-year planning horizon, zero salvage value, a 20 percent tax rate, and a 10 percent after-tax cost of equity capital. Use the NPV method to evaluate the new machinery system's profitability.

7. A new hog investment requires an initial outlay of $50,000 and is expected to yield annual net cash flows of $8,850 over the investor's 10-year planning horizon. Assuming no salvage value, no taxes, and a 10 percent discount rate, and using the NPV, IRR, and MIRR methods, evaluate the investment's profitability.

8. In problem 7, determine how the net present value and internal rate-of-return are influenced by the following financing terms: 20 percent equity downpayment with repayment of the loan balance in eight equal annual principal payments plus interest charged at 8 percent annually on the outstanding loan balance.

9. Suppose the hog investor is in the 30 percent tax bracket. How will this tax obligation influence the capital budgeting results?

10. A large investment company is analyzing the profitability of a farming investment for one of its corporate clients. The investment requires a $10 million outlay to acquire the necessary assets and is projected to yield an annual payment flow of $2 million (before interest payments) over a 20-year horizon.

    The analysts figure a 12 percent cost of equity capital, a 10 percent cost of debt capital, a 20 percent salvage value (of the initial outlay), financing by a bond sale at the 10 percent rate with principal due at the end of the twentieth year, and constant financial leverage as indicated by the corporation's equity-to-asset ratio of 0.5.

    a. What capital budgeting method would you use to evaluate this investment? Why?

    b. Show clearly how you would set up the capital budgeting model, and show its solution.

    c. Explain your procedures for handling the investment's financing plan.

11. Sally Investor is offered the opportunity to invest $10,000 with a promise of receiving $11,400 one year later. Her response is: "I reject this investment. I have only $2,000 of my own money, and for an $8,000 loan, the bank wants 10 percent interest so that my net return would be only $10,600. Would you be satisfied with a 6 percent return these days?" Did Sally make the right decision? Critique the analytical approach.

12. Two mutually exclusive investments have projected cash flows as follows:

Period

Investment	0	1	2	3	4
A. Crop land	−10,000	5,000	5,000	5,000	5,000
B. Forest land	−10,000	0	0	0	30,000

a. Assuming a required rate-of-return of 10 percent, determine the net present value for each investment.

b. Determine the internal and modified rates-of-return for each investment.

c. Which investment would you select? Why? What assumptions influenced your decision?

References

1. Aplin, R. D., G. L. Casler, and C. P. Francis, *Capital Investment Analysis*, 2nd ed., Grid, Inc., Columbus, Ohio, 1977.

2. Bierman, H., Jr., and S. Smidt, *The Capital Budgeting Decision: Economic Analysis of Investment Projects*, 6th ed., Macmillan Publishing Co., Inc., New York, 1984.

3. Boudreaux, K. J., and H. W. Long, *The Basic Theory of Corporate Finance*, Prentice-Hall, Inc., Englewood Cliffs, New Jersey, 1977.

4. Brealey, R., and S. Myers, *Principles of Corporate Finance*, 4th ed., McGraw-Hill Book Company, New York, 1991.

5. Brigham, E. F., *Financial Management: Theory and Practice*, 6th ed., Dryden Press, Hinsdale, Illinois, 1988.

6. Copeland, T. E., and J. F. Weston, *Financial Theory and Corporate Policy*, 3rd ed., Addison-Wesley Publishing Co., Inc., Reading, Massachusetts, 1988.

7. Haley, C. W., and L. D. Schall, *Theory of Financial Decisions*, 2nd ed., McGraw-Hill Book Company, New York, 1979.

8. Hirshleifer, J., *Investment, Interest and Capital*, Prentice-Hall, Inc., Englewood Cliffs, New Jersey, 1970.

9. Levy, H., and M. Sarnat, *Capital Investment and Financial Decisions*, 3rd ed., Prentice-Hall, Inc., Englewood Cliffs, New Jersey, 1986.

10. McDaniel, W. R., D. E. McCarty, and K. A. Jessell, "Discounted Cash Flow with Explicit Reinvestment Rates: Tutorial and Extension," *The Financial Review*, 23(1988):369 – 385.

11. Van Horne, J. C., *Financial Management and Policy*, 8th ed., Prentice-Hall, Inc., Englewood Cliffs, New Jersey, 1989.

12. Weston, J. F., and E. F. Brigham, *Essentials of Managerial Finance*, 10th ed., Dryden Press, Hinsdale, Illinois, 1993.

Appendix to Chapter 10

Capital Budgeting Using the Weighted Average Cost of Capital

In this appendix, we illustrate and compare the net present value method under the returns-to-assets and returns-to-equity approaches. Both approaches yield a measure of investment profitability that represents the anticipated increase in the firm's wealth position from undertaking the investment. However, the two approaches are distinguished by differences in the treatment of financing transactions and costs in both the payment series and the discount rate.

The returns-to-equity approach is based on projecting the after-tax payment flow net of the cash outflows for repaying principal and interest on debt, using the firm's cost of equity capital as the discount rate. It is based on the assumption that an investment's financing costs may strongly influence the firm's leverage and costs of financial capital. The net present value for the returns-to-equity approach is expressed as:

$$NPV = -INV_{dp} + \sum_{n=1}^{N} \frac{(P_n - i_d D_n)(1 - t) - L_n}{(1 + i_e)^n}$$

where INV_{dp} is the downpayment of equity capital, P_n is the net cash flow in period n from the investment before income taxes and loan payments, i_d is the interest rate on debt, D_n is debt outstanding in period n, t is the income tax rate, L_n is the payment on loan principal in period n, and i_e is the cost of equity capital.

The returns-to-assets approach is based on projecting the after-tax payment flow without deducting any charges for interest or loan payments and then discounting the payments to a present value, using a weighted average cost of the firm's debt and equity capital as the discount rate.

This approach accounts for the effects of the firm's overall financial structure in the specification of the discount rate. It is based on the assumptions that the firm's leverage determines the financing costs of an individual investment and that the financing arrangements for an individual investment do not substantially influence the firm's overall cost of capital.

The net present value for the returns-to-assets approach is expressed as:

$$NPV = -INV + \sum_{n=1}^{N} \frac{P_n(1 - t)}{(1 + i_a)^n}$$

where $i_a = (i_d)(1 - t)(D \div A) + (i_e)(E \div A)$, and INV is the total investment, $D \div A$ and $E \div A$ are the ratios of debt and equity to assets, respectively, i_a, is the weighted average cost of capital, and the other variables are as defined previously.

To illustrate the returns-to-assets approach, consider a firm that is undertaking a single investment project that will last for five years.[1] The total initial investment is $9,000; the annual cash flow before the payments for interest, principal, and taxes is $5,000; the income tax rate is 30 percent; the interest rate on debt is 10 percent; and the cost of equity capital is 20 percent. The firm's financial structure contains equal amounts of debt ($4,500) and equity ($4,500), and these proportions of debt and equity are assumed to remain constant over time. Thus, the weighted average cost of capital is:

$$i_a = (0.10)(1 - 0.3)(0.5) + (0.20)(0.5)$$

$$i_a = 0.135$$

The project's net present value for the returns-to-assets approach then is:

$$NPV(A) = -9,000 + \sum_{n=1}^{5} \frac{5,000(1 - 0.3)}{(1.135)^n}$$

---

[1]This example and the following procedures are drawn from an article by J. R. Fiske, "Comparative Analysis of the Return to Equity and Weighted Average Cost of Capital Approaches to Capital Budgeting," *Agricultural Finance Review*, 46(1986): 48 – 57.

NPV(A) = 3,162

Now, suppose the returns-to-equity approach is used to evaluate the project. Moreover, assume that the initial debt of $4,500 is repaid in five equal annual principal payments of $900 plus interest on the remaining loan balance at the 10 percent interest rate. The cash flows for this situation are shown in Table 10A.1. The resulting net present value for the return to equity is:

$$NPV(E) = -4,500 + \frac{2,285}{1.20} + \frac{2,348}{(1.20)^2} + \frac{2,411}{(1.20)^3}$$

$$+ \frac{2,474}{(1.20)^4} + \frac{2,537}{(1.20)^5}$$

$$= 2,643$$

For both approaches, the net present values represent the anticipated increases in the firm's wealth position from undertaking the investment. Thus, the net present value emerging from either formulation indicates the change in the firm's equity capital. If the assumptions and numerical specifications are the same for both approaches, then the resulting net present values should be the same as well.

Table 10A.1. Projected Cash Flows for Investment Project

Item	Year				
	1	2	3	4	5
Cash inflow	$5,000	$5,000	$5,000	$5,000	$5,000
Interest expense.	450	360	270	180	90
Income taxes	1,365	1,392	1,419	1,446	1,473
Cash flow after taxes and interest	3,185	3,248	3,311	3,374	3,437
Principal payments	900	900	900	900	900
Net cash flow.	2,285	2,348	2,411	2,474	2,537

Source: Fiske, J. R., "Comparative Analysis of the Return to Equity and Weighted Average Cost of Capital Approaches to Capital Budgeting," *Agricultural Finance Review*, 46(1986):48 – 57.

The reason the net present values differ in the preceding example stems from the differences in the implied specification of the firm's financial structure over time.

In this case, the approach in which the weighted average cost of capital is used assumes that the ratios of debt and equity to total assets remain constant over the five-year period. In contrast, the returns-to-equity approach assumes a decline in leverage (i.e., a lower D ÷ A and a higher E ÷ A) over time as the initial indebtedness is repaid. Thus, the NPV(A) for the weighted average cost of capital is higher because lower-cost debt capital comprises a relatively higher proportion of the firm's financial structure over the five-year horizon, and no adjustments are made in the cost of equity capital to reflect the changing leverage position over time.

It is important to explore the conditions under which the two approaches are made comparable to one another. This is accomplished by adjusting the weights on debt and equity in the financial structure to reflect the changing mix of these sources of financial capital as loan repayments occur. The adjusted weights yield a revised weighted average cost of capital. In turn, the revised average cost of capital yields a net present value for the series of after-tax returns to assets that is the same as the net present value for the series of after-tax returns to equity.

To illustrate, consider in Table 10A.2 the projected beginning-of-year values of debt, equity, and total assets for the project whose cash flows are described in Table 10A.1. The debt levels in row 1 indicate the decline in debt due to the annual principal payments of $900. The equity values in row 2 indicate the sum of the initial equity of $4,500 plus the net present values of the after-tax series of returns to equity (row 6 in Table 10A.1), calculated at each time point over the remaining periods of the horizon. Thus, the $7,143 value in row 2 for year 1 is:

$$7,143 = 4,500 + \left[-4,500 + \frac{2,285}{1.20} + \frac{2,348}{(1.20)^2} + \frac{2,411}{(1.20)^3} + \frac{2,474}{(1.20)^4} + \frac{2,537}{(1.20)^5} \right]$$

Similarly, the equity value of $6,286 for year 2 is the sum of the initial equity plus the net present value, calculated over the remaining four years:

$$6,286 = 4,500 + \left[-4,500 + \frac{2,348}{1.20} + \frac{2,411}{(1.20)^2} + \frac{2,474}{(1.20)^3} + \frac{2,537}{(1.20)^4} \right]$$

Table 10A.2. Beginning of Period Balance Sheet Values and Weighted Average Cost of Capital for an Investment Project

Item	Year 1	2	3	4	5
1. Debt value	$ 4,500	$3,600	$2,700	$1,800	$ 900
2. Equity value	$ 7,143	$6,286	$5,195	$3,823	$2,114
3. Asset value	$11,643	$9,886	$7,895	$5,623	$3,014
4. Debt-to-asset ratio	0.3864	0.3642	0.3420	0.3201	0.2986
5. Equity-to-asset ratio	0.6136	0.6358	0.6580	0.6799	0.7014
6. Weighted average cost of capital (i_a')	0.1498	0.1527	0.1555	0.1584	0.1612

Source: Fiske, J. R., "Comparative Analysis of the Return to Equity and Weighted Average Cost of Capital Approaches to Capital Budgeting," *Agricultural Finance Review*, 46(1986):48 – 57.

The total asset values in row 3 are the sum of the debt and equity values in rows 1 and 2, respectively. The financial ratios in rows 4 and 5 represent debt and equity, respectively, divided by total assets. The discount rates in row 6 are the revised weighted average costs of debt and equity capital for each year, using the weights in rows 4 and 5. For year 1, the revised discount rate is:

$$i_a' = (i_d)(1 - t)(D \div A) + (i_e)(E \div A)$$

$$= (0.10)(1 - 0.3)(0.3864) + (0.20)(0.6136)$$

$$= 0.1498$$

and for year 2, the revised discount rate is:

$$i_a' = (0.10)(1 - 0.3)(0.3642) + (0.20)(0.6358)$$

$$= 0.1527$$

Notice that the weighted average cost of capital is increasing over time in this example, reflecting the project's declining leverage position.

Finally, the net present value of the returns to assets for the project is recalculated according to the revised discount rates in row 6 of Table 10A.2. The result is:

$$NPV = -9,000 + \frac{5,000(1 - 0.3)}{1.1498} + \frac{5,000(1 - 0.3)}{(1.1498)(1.1527)}$$

$$+ \cdots + \frac{5,000(1 - 0.3)}{(1.1498)(1.1527)(1.1555)(1.1584)(1.1612)}$$

$$= 2,643$$

This result is the same as the net present value when the returns-to-equity approach is used.

The choice of measuring returns to assets or returns to equity in capital budgeting requires careful judgment, based on the characteristics of the business and the investments being analyzed. Either approach can be used, although each has different implications about the financial structure of the firm over time.

To illustrate, care must be taken in applying the returns-to-equity approach in evaluating two investments that have similar returns to assets but differ substantially in their use of debt and equity capital. The resulting profitability measures might differ as well, solely because of the different financing practices, even though the financing effects would be combined and perhaps offset if both projects were undertaken.

By considering project-specific financing, the returns-to-equity approach runs the danger of attributing value to a project that is rightly attributable to the condition of the firm's existing balance sheet. This approach has no systematic way of considering the effects of highly leveraged investments on the firm's overall financial position, except through the specification of a higher cost of equity capital to reflect the greater financial risk from high leverage.

11

Capital Budgeting: Inflation, Risk, and Financial Planning

In this chapter we extend the capital budgeting concepts to account for the effects of inflation and risk and to evaluate an investment's financial feasibility. Procedures are also established to evaluate recurring investments with different economic lives. Moreover, the issue of optimal timing of replacement of depreciable assets is treated in the appendix to this chapter. These extensions make the capital budgeting procedures more complex. However, the effects of inflation, risk, leverage, and financing arrangements have become very important for financial management in agriculture. They must be carefully considered in long-term financial planning.

INFLATION AND CAPITAL BUDGETING

A basic principle of capital budgeting is to consistently account for the effects of inflation both on the flow of payments and on the discount rates used to determine present values. Both of these components of capital budgeting are influenced by inflation. As the following discussion will show, the effects of expected inflation on capital budgeting can occur in one of two ways. One way is to project cash flows in real terms and discount the projected cash flows to present values by using a real interest rate that is free of any effects of inflation. A second way is to project cash flows in nominal terms and discount the projected cash flows to present values by using a discount rate that includes a premium for inflation.

Inflation Concepts

Inflation refers to an increase in the general level of prices for all goods and services in an economy. In contrast, deflation is a decline in the general price level. A rate of inflation refers to the percentage rate of increase (e.g., 6 percent, 7 percent) in the general price level. Most countries develop one or more indexes for measuring inflation. In the United States, the consumer price index and the gross national product (GNP) deflator are commonly used measures.

A positive rate of inflation means (1) the purchasing power of a projected constant payment declines annually by the designated rate of inflation or (2) the nominal or money value of the projected constant payment increases annually by the designated rate of inflation. "Nominal," or "money," values are the actual amount of U.S. currency (or other types of currency) making up the cash flows. In contrast, "real" values reflect the purchasing power today of future cash flows. Real values are found by adjusting nominal values for the rate of inflation.

The process of converting from nominal to real values, or vice versa, uses the same discounting and compounding techniques developed earlier. The main difference is that the inflation rate is used as the compound rate. We will illustrate these effects in two ways. The first example shows how inflation affects nominal values of future cash flows. The second example shows how real present values are found by discounting for both inflation and time.

Suppose that the expected annual rate of inflation is 7 percent. For an investment with annual cash flows that respond to inflation, the nominal values of the cash flows increase annually by the factor $(1 + i_f)$. Hence, a projected payment of $100 is increased to $100(1.07) = $107.00 in one year, $100(1.07)^2 = $114.49 in two years, and so on, to account for inflation. This process uses the future value formula, although the compound rate (i_f) is the expected rate of inflation rather than the interest rate for time.

Now, suppose an investment has cash flows already projected in nominal terms. Let the investment occur in an annuity, a loan, or another fixed-payment financial asset. Under these conditions, an annual inflation rate of 7 percent implies that the purchasing power of the fixed payment declines annually by 7 percent. Thus, the real value, or purchasing power, of a $100 payment received in one year is $P_1 \div (1 + i_f) = \$100 \div 1.07 = \$93.46$.

Real values are not the same as present values. To determine present values, discount real values again at the time preference rate, or at the time preference rate plus a premium for risk. If the time preference rate is 5 percent, then the real present value of the $100 payment is $\$93.46 \div 1.05 = \$89.01$.

Modeling Inflation Effects

Finding real present values involves a double-discounting procedure that accounts for both inflation and time. It is expressed as follows for payment P_1 due one year in the future:

$$V_0 = \frac{\dfrac{P_1}{(1 + i_f)}}{(1 + i_t)} = \frac{P_1}{(1 + i_t)(1 + i_f)} = \frac{P_1}{1 + i_t + i_f + i_t i_f} \tag{11.1}$$

The multiplicative term $(i_t i_f)$ in the denominator increases the complexity of this present value model. Assuming that $i_t = 0.05$ and $i_f = 0.07$, the discount factor is:

$$\frac{100}{1 + 0.05 + 0.07 + (0.05)(0.07)} = \frac{100}{1.1235} = 89.01$$

To illustrate further, investment B in Table 10.1 indicates a uniform series of $5,800 payments over a five-year period and an initial investment of $20,000. Suppose that the projected payments are nominal values, the rate for time and risk ($i_t + i_r$) is 8 percent, and a 4 percent annual rate of inflation is expected. The procedures for finding the real net present value (NPV$_r$) are outlined in Table 11.1 and are indicated in model form as:

$$NPV_r = -20,000 + \frac{5,800}{(1.04)(1.08)} + \frac{5,800}{(1.04)^2(1.08)^2} + \frac{5,800}{(1.04)^3(1.08)^3} \qquad (11.2)$$

$$+ \frac{5,800}{(1.04)^4(1.08)^4} + \frac{5,800}{(1.04)^5(1.08)^5}$$

$$= 742$$

The real net present value is $742. Hence, it is still a profitable investment.

Often the discount factor for finding real net present values is approximated by simply adding together the rates for time and inflation. As is evident in equation 11.1, adding i_t and i_f to yield a discount factor $(1 + i_t + i_f)$ ignores the multiplicative term $(i_t i_f)$. As a result, this approximation procedure slightly underestimates the actual discount rate and therefore overestimates real present values.

To illustrate, suppose the projected nominal net cash flows in Table 11.1 are discounted by the factor $1 \div (1.12)^n$ rather than $1 \div (1.04)^n (1.08)^n$. The resulting real net present value is $907.70, which exceeds the $742 real net present value found in equation 11.2.

Table 11.1. Real Net Present Value

Year	Nominal Net Cash Flow	Real Net Cash Flow	Real Net Present Flow
0	$-20,000	$-20,000	$-20,000
1	5,800	$5,800 \div (1.04) = 5,577$	$5,577 \div (1.08) = 4,597$
2	5,800	$5,800 \div (1.04)^2 = 5,362$	$5,362 \div (1.08)^2 = 4,093$
3	5,800	$5,800 \div (1.04)^3 = 5,156$	$5,156 \div (1.08)^3 = 3,644$
4	5,800	$5,800 \div (1.04)^4 = 4,958$	$4,958 \div (1.08)^4 = 3,644$
5	5,800	$5,800 \div (1.04)^5 = 4,767$	$4,767 \div (1.08)^5 = 3,233$
			Real net present value = $742

The overestimation of real net present values increases as the rates of inflation and time preference increase. The overestimates could be significant for investments with very lengthy and large net cash flows. Nonetheless, for most investments, simply adding the rates for inflation and time for discounting purposes is satisfactory.

Capital Budgeting Procedures

We can now restate the general principle for capital budgeting under inflation conditions. Both the payment series and the discount rate must be specified either in nominal values or in real values, but not in both values concurrently. If the payments are projected in real terms, they must be discounted with a real rate. If the payments are projected in nominal terms, they must be discounted with a nominal rate that includes an inflation premium. It is inappropriate to discount a payment series specified in real (nominal) terms at a discount rate specified in nominal (real) terms.

In practice, is it better to use real or nominal values? Using nominal values is probably more common. Market interest rates already contain a premium for anticipated inflation that protects the lender's debt claim from loss in its purchasing power. The payment series may also have several cost and return items that respond differently to inflation, and debt servicing requirements are specified in nominal terms. Finally, income tax obligations are based on nominal values rather than real values. Thus, it is important to reflect anticipated inflation in the payment series, if the discount rate is based on market interest rates.

Inflation and Present Values

To show how projections in real or nominal values lead to the same present value, we specify an investment situation in which the payment series and the discount rate respond equally to a constant inflation rate (i_f). Let P_0 be the current payment and assume that it was just paid; thus, it does not enter the present value computations. The series of payments starts with the first-year payment $P_1 = P(1 + i_f)$, followed by the second-year payment $P_2 = P_0(1 + i_f)^2$, and so on. The discount factor is based on a nominal rate (i), which contains a real rate for time (i_t) and the inflation premium. Following the double discounting concept, the discount factor is expressed as:

$$1 + i = 1 + i_t + i_f + i_t i_f = (1 + i_t)(1 + i_f) \tag{11.3}$$

For a perpetual series of payments, the present value model is formulated in nominal terms as:

$$V_0 = \frac{P_0(1 + i_f)}{(1 + i_t)(1 + i_f)} + \frac{P_0(1 + i_f)^2}{(1 + i_t)^2(1 + i_f)^2} + \frac{P_0(1 + i_f)^3}{(1 + i_t)^3(1 + i_f)^3} + \dots \tag{11.4}$$

Cancelling the expressions for the inflation adjustment in the numerator and denominator reduces the model to real terms:

$$V_0 = \frac{P_0}{(1 + i_t)} + \frac{P_0}{(1 + i_t)^2} + \frac{P_0}{(1 + i_t)^3} + \cdots \tag{11.5}$$

Since this expression involves the present value of an infinite uniform series, it further reduces to:

$$V_0 = \frac{P_0}{i_t} \tag{11.6}$$

Hence, anticipated inflation has no effect on the present value (V_0) at the current time. This present value is simultaneously valued in both real and nominal terms. This makes intuitive sense because at any time the present value is the amount for which assets can be "cashed in" and represents their purchasing power at current prices.

Inflation does, however, cause the anticipated present value to increase from period to period at the inflation rate i_f, assuming, of course, that the asset markets are highly efficient and that all other factors stay constant. This change in value is shown by valuing the payment series one period later:

$$V_1 = \frac{P_0(1 + i_f)^2}{(1 + i_t)(1 + i_f)} + \frac{P_0(1 + i_f)^3}{(1 + i_t)^2(1 + i_f)^2} + \cdots \tag{11.7}$$

Cancelling terms and reducing yields:

$$V_1 = \frac{P_0(1 + i_f)}{i_t} \tag{11.8}$$

Recall from equation 11.6 that $V_0 = P_0 / i_t$. Substituting equation 11.6 into equation 11.8 yields $V_1 = V_0(1 + i_f)$, which means that the present value in period 1 equals the present value in period 0, adjusted for the rate of inflation.

These modeling concepts provide the basis for capital budgeting under inflation and for practical tests about the effects of inflation and other factors on investment and asset values. These concepts underlie the detailed examples of capital budgeting in other chapters.

RISK AND CAPITAL BUDGETING

To this point, we have assumed complete certainty in the projections of investment performance and cash flows. The net cash flows for each future period have been single-value projections arising from our best estimate of future outcomes. However, in earlier chapters we identified sources of business and financial risk in agriculture, evaluated their impacts on managerial goals, and considered management responses to risks. The effects of risks are even more pronounced in long-term investment decisions. Hence, accounting for risks in capital budgeting is an important part of investment analysis.

Two of the commonly accepted methods of accounting for risk in capital budgeting are adjusting the discount rate and converting the payment series to certainty equivalents. These methods are considered in this section.

Adjusting the Discount Rate

So far, the discount rate has solely reflected an investor's time preference and an inflation premium. It has been considered a risk-free rate. However, when investments are known to be risky and the investor is risk-averse, a simple method of accounting for risk in capital budgeting is to add a risk premium to the risk-free rate. Now the discount rate i has three components — a risk-free rate for time preference i_t, an inflation premium i_f, and a risk premium i_r. The risk premium is the investor's reward for risk-bearing; it reflects the difference in rates-of-return required on a risky versus safe investment.

Adding the risk premium increases the discount rate, thus reducing an investment's net present value. Thus, the risk premium represents a cost of risk-bearing and, like other costs, reduces an investment's profitability. Consider an investment proposal requiring $8,000 initially and returning $2,500 per year for four years. At a risk-free rate of 8 percent, the net present value of the cash flow stream is:

$$NPV = -8,000 + 2,500[USPV_{0.08,4}]$$
$$= -8,000 + 2,500[3.3121] = 280.25$$

The positive net present value makes this investment favorable. If, however, a 4 percent risk premium is added to the 8 percent risk-free rate in response to the manager's perception of the investment risk and his or her level of risk aversion, the net present value for a 12 percent discount rate declines to –$406.75 — an unacceptable level.

In small businesses, the choice of an appropriate risk premium is a subjective process. It must reflect both the amount of risk added to the firm by an investment and the investor's risk aversion. Investments may be subject to both business risk and financial risk. Business risk refers to the unanticipated variability of the investment's returns and how these returns are correlated with other business enterprises. Higher variability means greater risk, and thus a greater risk premium. The variability could be offset however by low or even negative correlation of an investment's returns with the returns of other assets. In this case, the investment is a good source of diversification and may warrant a lower risk premium.

Financial risk arises from using debt capital to finance an investment. It is expressed by greater variability of returns to equity capital and reductions in liquid reserves of credit and financial assets. Investments that increase the firm's leverage position may warrant a higher risk premium to reflect the greater financial risk.

In practice, managers may let their experience help to formulate risk premiums that reflect their risk attitudes and the level of investment risk. Historic data may be helpful, since risky enterprises are expected to yield higher rates-of-return. Some managers define several risk classes for investments, each with a designated risk premium or required rate-of-return. Using the appropriate risk-adjusted discount rate, these managers assign new investments to the "most suitable" risk class for capital budgeting analysis. In larger corporate firms with shares of stock traded in well-developed equity capital markets, it is possible to estimate risk premiums based on differences in rates-of-return observed in the marketplace. However, these procedures are not possible for smaller noncorporate businesses.

Certainty-Equivalent Approach

Some analysts opt for a risk-free discount rate in capital budgeting, adjusting for risk in the numerator or cash flows of the present value model. This adjustment occurs by multiplying the cash flows (P_n) by a coefficient (α_n), having a value that varies inversely between 0 and 1 with the degree of risk:

$$V_0 = \sum_{n=0}^{N} \frac{\alpha_n P_n}{(1 + i)^n} \qquad \text{(11.9)}$$

where: $0 \leq \alpha \leq 1.0$

The adjustment coefficient (α_n) is selected in order to make an investment's projected cash flows equivalent to a completely certain cash flow. Hence, this method is called the **certainty-equivalent approach.**

For example, consider an expected net cash flow of $5,000 for a specific investment. The manager must designate a completely certain cash flow that he or she regards as equivalent to this expected return. If $5,000 is designated, then $\alpha_n = 1.0$, and the investment is riskless. If $3,000 is designated, then $\alpha_n =$ $3,000 \div 5,000 = 0.6$. The smaller value of α_n indicates the larger risk associated with the projected returns. The same procedure determines the level of α_n in all future periods for each investment alternative. Consider an $8,000 investment yielding $2,500 per year for four years. Let's assume that risk increases with time so that $\alpha_0 = 1.0$; $\alpha_1 = 0.95$; $\alpha_2 = 0.90$; $\alpha_3 = 0.85$; $\alpha_4 = 0.80$. The risk-free rate is 8 percent. The investment's net present value is:

$$\text{NPV} = -8,000 + \frac{(0.95)(2,500)}{1.08} + \frac{(0.90)(2,500)}{(1.08)^2} + \frac{(0.85)(2,500)}{(1.08)^3} + \frac{(0.80)(2,500)}{(1.08)^4}$$

$$= -715$$

On a certainty-equivalent basis, the project is rejected, as it was with the risk-adjusted discount rate method.

The certainty-equivalent approach is conceptually similar to the risk premium approach. In fact, we can determine the risk-adjusted discount rate that is comparable to the values of α_n. This rate is the discount rate that equates the net present value of the payment series to – $715.00. This discount rate is 14 percent. Hence, the risk premium is 6 percent — the difference between the 8 percent risk-free rate and the 14 percent risk-adjusted rate.

Both the certainty-equivalent and the risk-premium approaches consider the adjustment in returns needed to make an investor feel indifferent between a risky and a safe investment. Hence, the designation of a certainty equivalent income is subjective, as is the choice of risk premium. In both methods, the investor must express the risk adjustment by a quantitative measure based on limited judgment. One important difference in the two methods is that the risk-adjusted discount rate implies that risks increase exponentially over time, even when the discount rate is constant. Despite their subjective bases, these two methods are for most purposes the most practical ways of including risk in capital budgeting for firms of relatively small size that do not have their equity capital valued by shares of stock traded in well-developed equity capital markets.

Probability Analysis

A more comprehensive approach to risk analysis in capital budgeting uses probability distributions to measure risks. Recall from the risk section of Chapter 2 that risk measures should reflect how an investor perceives risk. We assumed that risks are perceived as probability distributions and that investors express their expectations in probabilistic terms. These distributions are used to estimate three statistical measures — expected value, standard deviation, and coefficient of variation — which help describe an investment's risk. Equations 2.2, 2.3, and 2.4 and the related text describe these risk measures.

To illustrate probability analysis in capital budgeting, assume the manager uses a risk-free rate to derive net present values for two investments under three kinds of forecasts: optimistic, most likely, and pessimistic. Moreover, the manager has assigned probabilities to each forecast. The basis for the probabilities could be historic performance or personal judgments based on outlook information, advice from peers or other experts, or even "hunches." The result is the discrete probability distribution of net present values shown in Table 11.2.

Note that the expected value of investment 2 is greater than that of investment 1; however, the standard deviation of investment 2 is also larger:

$$\sigma_1 = [(0.30)(19,000 - 15,300)^2 + (0.40)(15,000 - 15,300)^2 + [(0.30)(12,000 - 15,300)]^{1/2}$$

$$= 2,722$$

$$\sigma_2 = [(0.10)(27,000 - 18,000)^2 + (0.60)(20,000 - 18,000)^2$$

$$+ [(0.30)(11,000 - 18,000)^2]^{1/2} = 5,020$$

The relative risks of the two investments are indicated by their coefficients of variation:

$$CV_1 = \frac{\sigma_1}{E(\overline{V})_1} = \frac{2,722}{15,300} = 0.178$$

$$CV_2 = \frac{\sigma_2}{E(\overline{V})_2} = \frac{5,020}{18,000} = 0.279$$

Table 11.2. Probability Distribution of Net Present Values (NPV) for Two Investments

Forecast	Investment 1		Investment 2	
	V_i	P_i	V_i	P_i
Optimistic	$19,000	0.30	$27,000	0.10
Most likely	15,000	0.40	20,000	0.60
Pessimistic	12,000	0.30	11,000	0.30

Since the coefficient of variation for investment 2 is larger, it is considered the riskier venture. The higher coefficient of variation indicates greater relative dispersion of outcomes around the expected value. Thus, the likelihood of loss may be greater. The choice between these two investments depends on the manager's risk – return preference. A highly risk-averse manager might prefer investment 1; a less risk-averse manager might prefer investment 2.

This method of analyzing risks could be extended by estimating probability distributions for returns in each year and for each relevant factor determining cash flows (i.e., variations in prices and production). Both expected values and standard deviations could then be discounted to present values. These methods are complex. However, managers who consider this risk framework — expected returns, probabilities, risk aversion — should have a sound basis for decision making.

FINANCIAL FEASIBILITY OF SINGLE INVESTMENTS

Financial feasibility analysis for single investments uses present value methods to test the effects of variations in loan length, interest rate, loan size, and level of payment on the investment's ability to meet these financing terms. If the terms are met, then the investment is judged financially feasible. If the terms cannot be met, then the plans are financially infeasible, and actions to achieve feasibility must be undertaken.

Present value methods are used because the cash flows associated with the investment occur over a number of years. Hence, the magnitude and timing of these cash flows are important factors influencing financial feasibility. To illustrate, recall the equation for the present value of a uniform series of annual payments:

$$V_0 = A[USPV_{i,N}]$$

where V_0 = present value

A = the annual cash flow per period

i = the interest rate period

N = the number of conversion periods

In financial feasibility analysis, the data in this equation reflect the financing terms associated with the investment. Variable V_0 represents the size of the loan, A is the level of debt payment, i is the loan interest rate, and N is the length of the loan. Changes in the values of each variable will influence financial feasibility.

One approach to evaluating financial feasibility is to assume that three of the four data requirements are known. The value of the fourth variable can then be derived. Hence, the total amount of the loan can be determined and evaluated if the annual loan payments, interest rate, and loan length are known. Similarly, the annual loan payment can be evaluated if the loan size, length, and interest are known. Or, the necessary loan length can be determined if the loan amount, interest rate, and annual payment are known. The feasibility of the interest rate can be similarly evaluated.

To illustrate, consider an agricultural producer who is improving pasture land and expanding a beef-cow herd by 100 cows. The budget indicates that $20,000 of net cash income is available each year for debt servicing, consumption, or other uses. This is part of the information needed to evaluate the financial feasibility of the investment. Additional information includes loan amount, interest rate, and loan length. Suppose that the investor formulates estimates for two of these variables. The evaluation is as follows:

Case A

The beef investor projects the application of $20,000 annually to debt servicing. For an annual interest rate of 12 percent and a seven-year loan, what is the maximum size loan that can be serviced?

Model

$$V_0 = \$20,000[USPV_{0.12,\ 7}]$$
$$= 20,000[4.5638]$$
$$= 91,276$$

Answer: $91,276

Evaluation

If a loan greater than $91,276 is needed, then the investment is financially infeasible.

Case B

What is the annual debt (principal and interest) payment on a loan of $91,276 with an interest rate of 12 percent and a seven-year maturity?

Model

$$91,276 = A[USPV_{0.12,\ 7}]$$
$$A = 20,000$$

Answer: $20,000 per year

Evaluation

If less than $20,000 is available for debt servicing, then the investment is financially infeasible.

Case C

What is the maximum interest rate for a loan of $91,276 for seven years when $20,000 is available annually for debt servicing?

Model

$91,276 = 20,000[\text{USPV}_{i,\,7}]$

$i = 0.12$

Answer: 12% per year

Evaluation

If the interest rate exceeds 12 percent, then the investment is financially infeasible.

Case D

What is the minimum length of loan needed for a loan of $91,276 when $20,000 is available annually for debt servicing and the interest rate is 12 percent?

Model

$91,276 = 20,000[\text{USPV}_{0.12,\,N}]$

$N = 7$

Answer: 7 years

Evaluation

If the available loan is less than seven years in length, then the investment is financially infeasible.

The financial feasibility analysis can be extended to evaluate how changes in the values of one or more variables influence financial feasibility. Interactions or trade-offs among these variables can also be evaluated. Examples of possible trade-offs include reduction in interest rates as loan maturity is reduced, reduction in maturity as loan payments are larger, and reduction in interest rates as loan size is increased. Thus, one can jointly analyze these effects on financial feasibility.

Table 11.3 demonstrates analysis for selected values of i, N, A, and V_0. Suppose, for example, that the beef-cow investor actually needs a loan in the amount of $110,000. The result of the feasibility analysis for i = 12 percent, N = 7 years, and A = $20,000 indicates a maximum loan size of $91,276.

What are the options for achieving financial feasibility? One option, of course, is to reduce the total borrowing to $91,276. However, lower borrowing may not be possible. Other options include (1) lengthening the loan maturity beyond seven years, (2) increasing annual debt servicing above $20,000, (3) lowering the interest rate below 12 percent, and (4) using a combination of all three of these options.

As Table 11.3 demonstrates, increasing the loan maturity to eight years for

Table 11.3. Maximum Loan Size for Different Interest Rates, Loan Maturities, and Payment Levels

	Interest Rates					
	10 Percent		12 Percent		14 Percent	
Loan Maturity (Years)	$20,000 Payment Level	$22,000 Payment Level	$20,000 Payment Level	$22,000 Payment Level	$20,000 Payment Level	$22,000 Payment Level
6	$ 87,105	$ 95,816	$ 82,228	$ 90,451	$ 77,773	$ 85,551
7	97,368	107,105	91,276	100,402	85,766	94,343
8	106,698	117,368	99,353	109,288	92,777	102,055
9	115,180	126,698	106,565	117,222	98,927	108,820
10	122,891	135,180	113,004	124,305	104,322	114,754

i = 12 percent and A = $20,000 increases the loan size to $99,353. Or, increasing debt-servicing capacity to $22,000 with i = 12 percent and N = 7 increases the loan size to $100,402. Or, lowering the interest rate to 10 percent for A = $20,000 and N = 7 increases the loan size to $97,368. None of these changes yield financial feasibility. However, feasibility is attained by increasing A to $22,000 and lengthening N to 9 with i = 12 percent. Alternatively, feasibility is attained by lowering i to 10 percent and lengthening N to 9 with A = $20,000. Moreover, if payments increase to $22,000 and borrowing occurs at i = 10 percent, a loan of $117,368 could be serviced with an eight-year maturity. Other combinations of variables that provide feasibility can be found in Table 11.3. Furthermore, many other values of i, N, A, and V_0 can be evaluated.

These results clearly indicate that managers need to carefully evaluate the financing terms available from their lenders and their debt-servicing capacity. More favorable financing terms increase the prospects for financial feasibility. In addition, managerial actions that increase yields, upgrade product quality, reduce costs, reduce other debt-servicing obligations, reduce family withdrawals, or increase profits in any other way will increase the serviceable size of loan.

Nonetheless, investments that appear financially infeasible at the outset are still undertaken. Clearly, other factors must offset the apparent infeasibility. These factors likely include one or more of the following conditions:

1. The investor uses personal equity capital instead of or in addition to debt capital to finance part or all of the investment.

2. The investor has other sources of income that can be used to service debt.

3. The investor obtains more favorable financing terms than those budgeted, perhaps calling for partial amortization with a balloon payment due at loan maturity or a graduated payment plan.

4. The investor's planning horizon exceeds the length of the loan maturity.

5. The investor expects increasing annual returns from the investment in the future.

6. The investor expects to realize capital gains from the investment in the future.

COMPARING INVESTMENTS WITH DIFFERENT ECONOMIC LIVES

We conclude this chapter by considering the procedures to be used when investments with different economic lives are compared. If these investments are recurring, they must be placed on a common time basis for proper evaluation. As shown below, two methods may be used to achieve this comparison.

To illustrate this situation, consider a manager who must choose between two machines of equal effectiveness. However, they differ as follows:

	Machine X	*Machine Y*
Original cost	$10,000	$12,000
Projected economic life	4 years	6 year
Projected after-tax annual cash outflows	$ 3,200	$ 3,500

Neither machine is expected to have a salvage value.

The differences in economic lives mean that the machines must be replaced at different intervals. The manager's cost of capital is 10 percent. The present values of cash flows are:

$$V_0(X) = -10,000 - 3,200[\text{USPV}_{0.10,\ 4}] = -20,144$$

$$V_0(Y) = -12,000 - 3,500[\text{USPV}_{0.10,\ 6}] = -27,243$$

Based on the present value of cash outflows over their respective economic lives, machine X appears preferable. However, because these values represent different numbers of years of service, they are not directly comparable. Even dividing the two present values by their years of service would not produce a satisfactory comparison, since it would not properly reflect the timing of cash flows.

The investments can be made comparable by placing the present value method on the least common denominator of time for each investment or by converting the preceding present values to annuity equivalents. Both methods lead to the same decision.

The least common denominator of time for investments with 4- and 6-year lives is 12 years. The present value of cash flows for 12 years with machine X including replacement outlays in years 4 and 8 is:

$$V_0(X) = -10{,}000 - 10{,}000(1.10)^{-4} - 10{,}000(1.10)^{-8}$$
$$- 3{,}200[USPV_{0.10, 12}] = -43{,}304$$

For machine Y, the present value for 12 years and one replacement in year 6 is:

$$V_0(Y) = -12{,}000 - 12{,}000(1.10)^{-6} - 3{,}500[USPV_{0.10, 12}]$$
$$= -42{,}617$$

On this basis, machine Y is slightly preferred to machine X.

The annuity-equivalent method determines the size of annual annuity for the economic life of the investment that could be provided by a sum equal to the present value of its projected cash flow stream, given the firm's cost of capital. It is given by:

For X: $20{,}144 = A_x[USPV_{0.10, 4}]$

$$A_x = \frac{20{,}144}{3.1699} = 6{,}355$$

For Y: $27{,}243 = A_y[USPV_{0.10, 6}]$

$$A_y = \frac{27{,}243}{4.3553} = 6{,}256$$

Again, machine Y is preferred since the annuity equivalent of its cost stream is lower than that of machine X.

Summary

Modifications to capital budgeting models in order to account for the effects of inflation, risk, financing terms, and unequal lives of investment alternatives under consideration have been considered in this chapter. In each case, specific adjustments to present value procedures have been identified and illustrated. The general effects are to enrich the capital budgeting framework to handle more realistic investment specifications and to add complexity to the process as well. Other specific highlights are as follows:

- The alternative methods of handling inflation in capital budgeting are to project cash flows in real terms and to discount to a present value using a real discount rate, or to project cash flows in nominal terms and to discount to a present value using a nominal discount rate.

- A nominal discount is made up of a real rate plus an inflation premium, plus an interaction term between the real rate and the inflation premium. In practice, the interaction term often is not considered.

- If a payment series and the discount rate are affected equally by inflation, then the two effects cancel out, leaving no effect of inflation on present values. However, an asset's value is expected to increase over time at the anticipated rate of inflation.

- Two of the commonly accepted methods of accounting for risk in capital budgeting are adjusting the discount rate and converting the payment series to certainty equivalents. Probability distributions of net present values and internal rates-of-return also may be employed.

- Financial feasibility analysis for single investments uses present value methods to test the effects of variations in loan length, interest rate, loan size, and level of payment on the investments' ability to meet the financing terms.

- Investments with different economic lives may be compared in present value terms by utilizing the least common denominator of time method or the annuity equivalent method.

Topics for Discussion

1. Explain the basic principles of capital budgeting under conditions of inflation.

2. Distinguish between real and nominal values of payments. Which is more commonly used in capital budgeting? Why?

3. For a time preference rate of 3 percent and an anticipated inflation rate of 6 percent, find the real value of a $10,000 payment due three years in the future. Find the real present value also.

4. Find the value of an asset yielding a perpetual series of payments

of $1,000 per year when the nominal interest rate is 10 percent and anticipated inflation is 5 percent.

5. Find the value of the asset in problem 4 when its series of payments starts at $1,000 and increases annually at the general inflation rate of 5 percent (use the 10 percent nominal interest rate).

6. At what annual rate will the values found in problems 4 and 5 increase over time? Why?

7. Explain the logic and method of using a risk-adjusted discount rate to account for investment risk in capital budgeting. Contrast the risk-adjusted rate approach with the certainty-equivalent approach.

8. Identify some ways to quantify the risk premium used in capital budgeting. How is the risk premium influenced by (a) the investment's risk, (b) the investor's risk aversion, (c) the degree of financial leverage, and (d) the investment's correlation with other business enterprises?

9. What sources of information are used in estimating probability distributions on outcomes of risky events?

10. In choosing between two risky investments, would choosing the investment with the higher coefficient of variation imply a relatively high or low degree of risk aversion?

11. An investor is appraising an investment under the following conditions:

Forecast	NPV	p_i
1	30,000	0.10
2	20,000	0.15
3	15,000	0.50
4	10,000	0.25

a. Find the investment's expected value, standard deviation, and coefficient of variation.

b. Explain the meaning of each of these measures.

c. Would the risk-averse investor prefer this investment to a risk-free investment with a net present value of $10,000? Explain.

d. Would the risk-averse investor prefer this investment to another investment with an expected net present value of $12,000 and a standard deviation of $12,000? Explain.

12. A new investment requires an initial outlay of $40,000 and is projected to yield annual net cash inflows of $8,500 over a 10-year period with no terminal value. Evaluate the profitability of the investment, assuming a 3 percent rate for time, a 4 percent risk premium, and a 7 percent expected rate of inflation.

13. The purchase of a new irrigation system is requiring a loan of $61,450 at a 10 percent annual interest rate. The investor figures on making annual payments of $10,000. Find the length of loan needed to make the investment financially feasible.

    Suppose the loan length is limited to 8 years. What is the effect on feasible loan size? What other actions will affect feasibility?

14. An investor is choosing between the following mutually exclusive investments:

Investment

	1	2	3	4
Initial outlay	$48,000	$60,000	$60,000	$36,000
Annual payments	$20,000	$12,000	$16,000	$10,000
Lifetime, years	5	15	10	15

    The discount rate is 16 percent. Rank the investments according to profitability, assuming they can be repeated in permanent replacement chains.

References

1. Anderson, J. R, J. L. Dillon, and J. B. Hardaker, *Agricultural Decision Analysis*, Iowa State University Press, Ames, 1977.

2. Aplin, R. D., G. L. Casler, and C. P. Francis, *Capital Investment Analysis*, 3rd ed., Grid, Inc., Columbus, Ohio, 1977.

3. Bierman, H., Jr., and S. Smidt, *The Capital Budgeting Decision, Economic Analysis of Investment Projects*, 8th ed., Macmillan Publishing Co., Inc., New York, 1992.

4. Boudreaux, K. J., and H. W. Long, *The Basic Theory of Corporate Finance*, Prentice-Hall, Inc., Englewood Cliffs, New Jersey, 1977.

5. Brigham, E. F., and L. C. Gapenski, *Intermediate Financial Management*, 4th ed., Dryden Press, Hinsdale, Illinois, 1993.

6. Cooley, P. L., R. L. Roenfeldt, and I. Chew, "Capital Budgeting Procedures Under Inflation," *Financial Management*, 4:4(Winter 1975):18 – 27.

7. Copeland, T. E., and J. F. Weston, *Financial Theory and Corporate Policy*, 3rd ed., Addison-Wesley Publishing Co., Inc., Reading, Massachusetts, 1988.

8. Hertz, D. B., "Risk Analysis of Capital Investment," *Harvard Business Review*, 42(January – February 1964):95 – 106.

9. Levy, H., and M. Sarnat, *Capital Investment and Financial Decisions*, 3rd ed., Prentice-Hall, Inc., Englewood Cliffs, New Jersey, 1986.

10. Perrin, R. K., "Asset Replacement Principles," *American Journal of Agricultural Economics*, 54(1972):60 – 67.

11. Van Horne, J. C., *Financial Management and Policy*, 8th ed., Prentice-Hall, Inc., Englewood Cliffs, New Jersey, 1989.

Appendix to Chapter 11

Asset Replacement

Resource control by ownership raises the issue of the optimal time for replacing assets. By definition, depreciable assets eventually wear out or become obsolete and must be replaced. Should they be replaced only when they are fully worn out, or is an earlier replacement warranted? Other types of enterprises, such as forests and orchards, which are eventually harvested and replaced, have similar problems of optimum harvest time determination. Even feedlots that turn over several lots of livestock within a year must determine the length of feeding period and timing of sale.

The processes of compounding and discounting are quite helpful in developing replacement rules. In this section, we will develop the capital budgeting criteria used in a replacement decision and indicate a few examples of their application. Once again, the discount rate plays an important role. (The material to follow draws heavily on principles and applications reported by R. K. Perrin in reference 10.)

Replacement Criteria

We still assume that the manager's objective is to maximize (minimize) the present value of the entire stream of returns (costs) arising from the use of an asset, although now the analysis is extended to evaluate the present value of a series of asset replacements in which the time of replacement becomes the decision variable. That is, a replacement period is sought that yields the optimal (maximum or minimum) present value.

This problem can be evaluated in two ways. One way consists of finding the present value of the series of returns (costs) for each replacement period over the full set of replacement periods and then choosing the period that yields the maximum present value of returns or the minimum present value of costs. This present value model is expressed as:

$$V_0 = \frac{1}{1 - (1 + i)^{-s}} \left[\sum_{n=1}^{s} \frac{R_n}{(1 + i)^n} + \frac{M_s}{(1 + i)^n} - M_0 \right] \tag{11A.1}$$

where R_n is the return (cost) in year n, M_s is the asset value in year S, M_0 is the original asset value, and S is the year of replacement. The expression in brackets is the present value of the returns (costs) of the asset over one life interval (that is S years long). Thus, it is received every S years. The discount factor has been adjusted to represent the present value of a series of bracketed expressions, which, in effect, is equivalent to a perpetual annuity received every S years. Thus, V_0 is the present value of a perpetual annuity received every S years, while the bracketed expression is the amount of each annuity payment.

The second method of evaluating the problem, which is closely related to the first, is based on marginal analysis. The gains from keeping an asset for another time period are compared with the opportunity gains (i.e., marginal costs) that could be realized from a replacement asset during the same period.

Defining opportunity gains in this case is complex because it involves the annualization of the value of a stream of returns (or costs) that depends upon the replacement policy for the asset. As above, the effect of the entire stream of replacements over time must be considered, rather than just the current asset and its first replacement.

The marginal analysis is implemented by mathematically solving the present value model (11A.1) for its marginal revenue and marginal cost at each period of time and searching for the replacement period at which these values are equal to each other. The marginal equilibrium condition is portrayed as:

$$R_{s+1} + \Delta M_{s+1} = \frac{1}{\text{USPV}_{i,s}} \left[\sum_{n=1}^{s} \frac{R_n}{(1 + i)^n} + M_s - M_0 \right] \tag{11A.2}$$

where ΔM is the annual change in asset value and other variables are defined previously.

The term on the left of the equality represents the marginal returns of keeping the asset an additional year. The terms on the right of the equality represent the marginal opportunity cost. The bracketed portion represents the present value of an upcoming replacement cycle at the moment of replacement where M_s minus M_0 is realized, and the summation yields the present value of earnings from the new asset until the time it is replaced. Thus, once the series of replacement cycles begins, the amount in brackets is received as a perpetual annuity every S years.

To convert this periodic annuity into a continuous marginal cost reflecting an opportunity flow of earnings, multiply it by the capital recovery factor $[1 \div USPV_{i,s}]$ where:

$$USPV_{i,s} = \left[\frac{1 - (1 + i)^{-n}}{i} \right]$$

In a case in which the discount rate is zero, equation 11A.2 reduces to:

$$R_{s+1} + \Delta M_{s+1} = \frac{1}{s} \left[\sum_{n=1}^{s} R_s + M_s - M_0 \right] \tag{11A.3}$$

which indicates that optimal replacement age occurs at age s, where marginal returns equal the average undiscounted returns over time. This is also the age that maximizes average undiscounted revenue over time.

Examples of Replacement Decisions

Let's consider two examples of replacement decisions. In the first example, the harvesting and replacement of a forest tract, a measurable revenue is produced. In the second, the replacement of machinery, replacement policies are based on projected cost streams.

In the forestry example, the expected harvest values (M_s) per acre of trees are indicated in column 2, Table 11A.1, for alternative harvest years. All trees in the stand are assumed to be cut and sold in the same year. There are no other annual costs, revenues, or restarting costs. Thus, R_n and M_0 are zero. The harvest value increases each year, and, in the absence of replacement considerations, it would appear to be profitable to hold the trees beyond year 30. However, opportunity gains and the pattern of growth rate of replacement stands require that we evaluate the optimal replacement period in terms of the two methods indicated previously.

First, consider the present value method. Using equation 11A.1 and assuming the discount rate of 5 percent (i = 0.05), find the present value for the 28-year replacement period, which is:

$$V_0 = \frac{1}{1 - (1.05)^{-28}} \left[\frac{409.99}{(1.05)^{28}} \right] = 140.40$$

as shown by the entry in column 3 of Table 11A.1 in the row for year 28. Following the same procedure, the present value for a 27-year replacement period is $140.07, and it is $140.26 for a 29-year replacement period.

Moreover, the present values for other replacement periods are still lower. Thus, the optimal replacement period for the 5 percent discount rate is 28 years. A higher discount rate will shorten the optimal replacement period, since the marginal cost of not replacing is higher. This is confirmed for a 10 percent discount rate (i = 0.10) by

Table 11A.1. Information Required for a Forest Harvest and Replacement Decision

Age of Forest	Harvest Value (M_s)	Present Value		Growth Rate $(\Delta M_{s+1} \div M_s)$	Capital Recovery Factor	
		i = 0.05	i = 0.10		i = 0.05	i = 0.10
(1)	(2)	(3)	(4)	(5)	(6)	(7)
14	0	0	0	—	—	—
15	$ 58.91	$ 54.60	$18.54	—	0.101	0.136
16	83.03	70.19	23.10	0.410	0.096	0.131
17	108.34	83.85	26.72	0.305	0.092	0.128
18	134.55	95.65	29.51	0.242	0.089	0.125
19	161.36	105.67	31.54	0.199	0.086	0.122
20	188.63	114.09	32.93	0.169	0.083	0.120
21	216.19	121.05	33.78	0.146	0.080	0.117
22	243.94	126.70	34.16	0.128	0.078	0.116
23	271.80	131.21	34.17	0.114	0.076	0.114
24	300.00	134.82	33.90	0.103	0.074	0.113
25	327.49	137.23	33.30	0.092	0.072	0.111
26	355.16	138.97	32.53	0.085	0.071	0.110
27	382.69	140.07	31.60	0.078	0.070	0.109
28	409.99	140.40	30.55	0.071	0.068	0.108
29	437.03	140.26	29.40	0.066	0.067	0.107
30	463.80	139.26	28.20	0.061	0.066	0.107

Source: Perrin, R. K., "Asset Replacement Principles," *American Journal of Agricultural Economics*, 54(1972): 60 – 67; data are adjusted for price changes.

observing that the maximum present value in column 4 of Table 11A.1 is now $34.17 for a 23-year period.

Next, consider the marginal analysis method, in which marginal returns are compared with marginal opportunity costs as reflected by the annualized present value of future earnings from an S-year replacement policy. In applying equation 11A.2 and recalling that R_n and M_0 are zero in this case, find the replacement, which occurs when:

$$\Delta M_{s+1} = \left[\frac{1}{USPV_{i,\,s}} \right] M_s \qquad\qquad \text{(11A.4)}$$

$$\frac{\Delta M_{s+1}}{M_s} = \frac{1}{USPV_{i,\,s}} \qquad\qquad \text{(11A.5)}$$

Thus, the optimal age for harvest and replacement occurs when the rate of growth of harvest value $\Delta M_{s+1} + M_s$, column 5, Table 11A.1, most nearly equals the capital recovery factor $[1 + USPV_{i,\,s}]$. Capital recovery factors are listed in columns 6 and 7 for the discount rates of 0.05 and 0.10, respectively. As indicated, optimal harvest and replacement are in years 28 and 23 for the 0.05 and 0.10 discount rates, respectively.

Thus, the manager knows in year 27 that the marginal return from holding an acre of trees for one more year would be $27.30 (= $409.99 − $382.69). The marginal opportunity cost of not replacing the acre, as reflected by the average expected flow of net return from a new 27-year stand, would be $26.02 (= $382.69 × 0.068). Because marginal returns exceed marginal costs, the trees should be held until year 28. In year 28, the marginal returns from holding the stand until year 29 would be $27.04 (= $437.03 − $409.99), while the marginal opportunity cost of not replacing would be $27.47 (= $409.99 × 0.067). Since the marginal costs of not replacing in year 28 exceed the marginal returns, the manager should plan to harvest and sell the trees in year 28.

The second example deals with the replacement of durable equipment. The flow of services provided by the equipment is assumed to be constant over time and thereby not relevant to the replacement decision. The objective is to minimize the present value of the cost stream of the original equipment and its replacements. Again, the two methods of evaluating optimal replacement may be used.

Table 11A.2 contains cost information estimated from engineering data for a truck valued at $6,930. Column 1 indicates the various ages (n) of the truck, column 2 gives the truck's depreciated values in the respective years, column 3 lists the expected repair costs for each year, and column 4 specifies the total marginal costs in each year minus annual depreciation plus repair cost. Because R and ΔM are costs, their values have negative signs.

The present values over the series of replacement at the various replacement ages are shown in column 5. To illustrate, using equation 11A.1 and a discount rate of i = 0.10, find the present value for a seven-year replacement policy.

Table 11A.2. Information Required for a Truck Replacement Decision

Age	Depreciated Value (M_s)	Annual Repairs (R_n)	Annual Cost ($R_n + \Delta M_s$)	Present Value (V_0)	Bracketed Term (V_n)	$[USPV_{i, s}]^{-1}$	$V_n[USPV_{i, s}]^{-1}$
(1)	*(2)*	*(3)*	*(4)*	*(5)*	*(6)*	*(7)*	*(8)*
0	$6,930	–	–	–	–	–	–
1	5,329	$ –231	$–1,832	$–25,250	$ –1,811	1.099	$–1,990
2	4,096	–462	–1,695	–23,835	–3,426	0.576	–1,973
3	3,146	–739	–1,689	–22,974	–4,931	0.402	–1,982
4	2,419	–970	–1,697	–22,358	–6,321	0.315	–1,991
5	1,857	–1,201	–1,763	–21,980	–7,628	0.264	–2,014
6	1,428	–1,478	–1,907	–21,844	–8,891	0.230	–2,045
7	1,095	–1,663	–1,996	–21,796	–10,077	0.205	–2,066
8	846	–1,894	–2,143	–21,859	–11,210	0.187	–2,096

Source: Perrin, R. K, "Asset Replacement Principles," *American Journal of Agricultural Economics*, 54(1972):60 – 67; data are adjusted for price changes.

$$V_0 = \frac{1}{1 - (1.10)^{-7}} [-4,242.96 + 561.91 - 6,930]$$

$$= -21,795.68$$

The – $4,242.96 entry in brackets is the present value of the series of annual repair costs for years 1 through 7 in column 3 of Table 11A.2, while the $561.91 entry is the present value of the truck's depreciated value in year 7. As shown in column 5, the present value of – $21,795.68 is the minimum cost value for the various replacement periods, indicating that a seven-year replacement period is optimal.

The same result occurs when marginal analysis is used, based on the conditions shown in equation 11A.2. The entries in column 6 are the information needed for the left side of equation 11A.2: $R_{s+1} + \Delta M_{s+1}$. Column 7 indicates values for the bracketed term on the right side of equation 11A.2: the present value of an upcoming replacement cycle at the moment of replacement. The – $1,811 entry for year 1 is calculated as:

$$V_0 = \sum_{n=1}^{s} \frac{R_n}{(1 + i)^n} + M_s - M_0$$

$$= -1,811 = \frac{-231}{1.10} + 5,329 - 6,930$$

Similarly, the value for year 2 is:

$$-3,426 = \frac{-231}{1.10} + \frac{-462}{(1.10)^2} + 4,096 - 6,930$$

The values in column 6 must be converted to a continuous flow of expected opportunity costs of nonreplacement for comparison with actual marginal costs in column 4. This conversion is accomplished by multiplying the data in column 6 (V_n) by their respective capital-recovery factors [1 ÷ USPV$_{i, s}$] as listed in column 7 for i = 0.10. The resulting product is listed in column 8. The replacement decision is made by comparing the total marginal cost for year (s + 1) in column 7 with the annualized cost in column 8, which would be incurred by replacing the truck and all future trucks when they reach age S. From the data in Table 11A.2, it is apparent that in year 7, the marginal cost of keeping the truck until year 8 (– $2,143) exceeds the annualized cost of – $2,096 for a seven-year replacement policy. Thus, the replacement plan should be to trade in year 7.

Additional Considerations

These examples are based on replacement by similar items under constant prices, rates of productivity, and discount rates. Changes in any of these items could alter the manager's replacement policy. Typically, replacement occurs with technologically improved assets (more efficient machinery, higher-producing cows, etc.). The improved replacements produce a greater stream of benefits in the future and may call for earlier replacement.

Higher discount rates will generally be associated with a preference for earlier replacement, although this depends on the levels of M_s and M_0 and the time pattern of R_n. In any case, managers may not agree about the preferred replacement policy for the same asset simply because they use different discount rates.

Finally, the perceptive manager will consider the risk of involuntary replacement due to random phenomena. Animals may die, machines may disappear or accidentally break, and buildings may burn. Such occurrences become a real part of replacement decisions; however, they are not easily handled in capital budgeting. Generally, one must utilize some of the methods discussed earlier in this chapter — adjustment of the discount rate and estimation of probabilities of life — to introduce the consideration of uncertainty into the replacement decision.

12

Controlling Farm Land

Farm land is unique among farm assets; thus, it warrants special attention in long-term decision making. In this chapter we consider its unusual characteristics and the factors affecting farm land values. To evaluate land values, we apply asset valuation procedures, and we consider the profitability and financial feasibility of farm land investments under various conditions. We also deal with characteristics of the leasing market for farm land.

CHARACTERISTICS OF FARM LAND

Land is a durable, immobile resource. Its basic properties do not change over time, or if they do change, they change so slowly that land is considered to have an infinite life. In contrast to other farm assets, land is not used up in the production of farm commodities. Because of its durability and immobility, land has special treatment in institutional arrangements such as taxation, leasing, and government programs. It has a specific legal description, with ownership recorded by local governments for taxation and other controls. Specific realty law regulates its use. The availability of land-related resources, such as water, minerals, oil, and buildings, as well as conservation practices and recreational uses, may influence land values and the land investment decision. Land also has specific financing instruments and lenders who are specialized in real estate lending.

Land markets typically exhibit a low rate of market transactions. Only about 2 to 3 percent of the total farm land in the United States is sold each year. Land-leasing markets also have low activity. The terms of farm leases change slowly, and leases change hands infrequently. When tracts of land are sold or leased, they generally occur in large units that involve substantial amounts of capital. Thus, borrowed capital is usually employed to finance such purchases. Moreover, the lower capital investment involved in leasing explains its importance in a producer's financial strategies.

Land values are also influenced by special factors that may differ among potential buyers. To illustrate, an agricultural producer with excess machinery capacity may place greater value on a new tract of land than will a neighbor who must buy more machinery to operate the added land. Some non-monetary factors are pride of ownership, family tradition, hobby farming, and rural living. Some of these factors can be included in the present value analysis. Others cannot.

ALTERNATIVES IN
LAND CONTROL

Expansion in farm size and income-generating capacity for most agricultural units requires an increase in the land acreage under the manager's control. The basic choices for controlling additional land are ownership and leasing. Both choices may substantially influence the business's profitability, risk, and liquidity. Thus, analysis of land control decisions must consider these effects.

Numerous options are also available within the purchase or lease alternatives. A cash purchase of land may be the simplest arrangement. However, cash purchases are rare because most land transactions involve large sums of money. Data published by the U.S. Department of Agriculture show, for example, that from 1980 to 1991, about 78 percent of farm real estate transfers were credit-financed, with buyers borrowing about 75 percent of the funds used to purchase land. A credit purchase is characterized by a downpayment or pledge of equity by the buyer and a note exchanged for the land title — the note being secured by a mortgage or deed of trust on the real estate (see Chapter 19). Alternatively, the buyer could use a land-purchase contract (see "Seller Financing" in Chapter 17), making a small downpayment and paying the remainder to the seller over several years. Title to the land is typically held by the seller or in escrow until the contract terms are fulfilled.

Farm real estate leases may be short- or long-term, and they may differ substantially in their rental arrangements. Cash rents, share rents, and combinations thereof are common. Moreover, the swings in inflation, business risk, and interest rates that characterized the 1970s and 1980s brought financial innovations and creative financing in both purchase and lease arrangements for farm land. Thus, a careful, complete analysis is needed for any land control situation, in which investors consider (1) the market price of land, (2) the profitability of the land investment *to the investor,* and (3) the financial feasibility of the investment.

LAND VALUES

Determining the market price of land is complicated by the limited availability of timely information about land values and by the effects of many factors that influence these values. In this section we consider the sources of price information and the approaches to estimating land values.

Sources of Price Information

Most potential buyers and sellers are not able to keep current on precise price levels in the farm land market. The relatively low activity in land transactions limits the availability of pricing information for specific properties. In addition, the size, location, soil quality, water supply, and other factors affecting land values may vary from one tract to another. Local businesspersons, such as lenders, real estate agents, extension personnel, lawyers, and farmers, may have good, timely impressions of the land market, but their information is largely based on opinions, observations, and judgments. Thus, a prospective buyer might want to hire a professional real estate appraiser to make a "market appraisal."

The professional appraiser estimates the price at which land would be exchanged by a willing buyer and a willing seller when each is equally knowledgeable of the situation. Appraisers may use several methods of estimation. First, they might observe recent land values in sales of comparable tracts of land in the same neighborhood. This "comparable sales" method is based on the premise that the values of neighboring tracts of land are similar, assuming the sales do not occur under extraordinary circumstances or between family members.

In addition, appraisers use the capitalization approach, in which the value of property to the "average" or "typical" manager is estimated under average production and price conditions. The perpetual series of returns to land, net of other costs, is capitalized into a land value, using equation 9.6: $V_0 = A \div i$.

With any appraisal method, the estimated value is adjusted to account for unique features of the land or neighborhood. These include field sizes, dwellings on the property, nearness to roads, markets, schools, and so on. Mineral rights, water supplies, and recreational uses also may be considered.

The appraised value provides useful information to both the buyer and the seller in determining price objectives and negotiating strategies. However, the appraisal does not constitute the actual price, which is determined only when the land in question is sold.

Capitalization Models

Economists, appraisers, policy makers, and other analysts have extensively researched the factors that affect farm land values and the efficiency of farm land markets. Indeed, the asset structure of farming is uniquely characterized by the dominance of farm land and the tendency for farm income to be capitalized into land values that appear high compared to current earnings. Profitability in farming has long been characterized by relatively low rates of current (cash) returns to farm assets and a tendency for land values and net worth to increase over time.

This situation has made farm land values seem high when compared to the current earnings. It also makes highly leveraged investments in land appear financially infeasible when expected cash returns in early years fall short of the debt-servicing requirements. These points are illustrated by levels of cash rent for farm land in the U.S. Corn Belt falling in the $80 per acre range, compared to land values in the $1,500 to $2,000 per acre range, providing current rates-of-return of 4 to 6 percent.

However, the proper application of valuation concepts under inflation and empirical evidence suggest that farm land prices can be explained in terms of long-term patterns of earnings growth. These concepts and evidence indicate that unrealized capital gains or losses in land and other assets should logically be treated as part of the land owner's total returns. Moreover, accounting rates-of-return on farm land are not directly comparable to rates-of-return on other types of farm and nonfarm assets because of the nondepreciability characteristic of farm land [3].

The concepts involved here are similar to the "growth stock" phenomenon in the markets for equity capital. These important concepts are illustrated in the following valuation model that uses the equations and inflation guidelines established in Chapters 10 and 11.

The basic valuation approach is to estimate an asset's current market value by capitalizing its flow of expected future earnings at an appropriate interest (or capitalization) rate. Recall from equation 11.6, and the discussion in the text, that the present value of a perpetual series of payments (P_n), which responds to inflation the same as does the nominal discount rate, is:

$$V_0 = \frac{P_0}{i_t} \tag{12.1}$$

This expression means that the asset (land) value is found by dividing the asset's real earnings (P_0) by the real capitalization rate (i_t). Suppose, for example, that an acre of farm land is expected to yield a real rent of $80 in perpetuity and that the real capitalization rate, net of general inflation, is 0.04, or 4 percent. Then the land value per acre is $2,000 (= 80 ÷ 0.04).

Recall that the real interest rate is approximated by subtracting the anticipated inflation rate from the nominal market interest rate ($i_t = i - i_f$). Thus, the land's present value is essentially found by dividing the current rent by the difference between a market interest rate and an inflation premium.

Moreover, the asset (land) is expected to increase in value at the general inflation rate. Thus, in a market equilibrium setting, the total annual returns (i) of investors can be approximated as a real rate-of-return ($i_t = P_0 ÷ V$), plus a nominal rate of capital gain (i_f) or:

$$i = i_t + i_f = \frac{P_0}{V} + i_f \tag{12.2}$$

The real rate-of-return consists of a current cash return (adjusted for inflation). Adding the real return to the capital gain from inflation gives the total nominal return.

For the numerical example, suppose that the anticipated inflation rate is 6 percent. The investor's nominal annual return is 10 percent; it consists of the real return of 4 percent (80 ÷ 2,000) plus the inflation-induced capital gain of 6 percent.

Next, suppose that the asset's real earnings grow over time, at rate g, in addition to the growth in response to general inflation. In this case, the expected rental payment to land might grow at a 7 percent nominal rate, while the expected inflation rate is 6 percent. The real rate of growth is approximated as the difference between the nominal rate and the inflation rate, which is 1 percent in this case.

Real growth in the asset's payment series partitions the real rate-of-return into two parts: (1) a current real return and (2) a real capital gain. To show this effect analytically, we combine the time value equation for a growing payment series — equation 9.10 — with the inflation-adjusted valuation model in equation 11.4. The result is:

$$V_0 = \frac{P_0(1 + i_f)(1 + g)}{(1 + i_t)(1 + i_f)} + \frac{P_0(1 + i_f)^2(1 + g)^2}{(1 + i_t)^2(1 + i_f)^2} \quad\quad (12.3)$$

$$+ \frac{P_0(1 + i_f)^3(1 + g)^3}{(1 + i_t)^3(1 + i_f)^3} + \cdots$$

First, cancel the terms for $1 + i_f$ in the numerator and denominator of each term. Then, follow the adjustment indicated in Chapter 9 for a constant growth model to yield:

$$V_0 = \frac{P_0(1 + g)}{i_t - g} \quad\quad (12.4)$$

Now, divide the asset's real earnings in period 1 by the difference between the real capitalization rate and the real growth rate to find the asset value. To illustrate, suppose as before that the current land rent is $80 per acre, the real capitalization rate is 4 percent, and the rent is expected to experience real growth at 1 percent per year. Then, the land value per acre is:

$$V_0 = \frac{80(1.01)}{0.03} = \$2,693.33$$

The partitioning of the real return into a current real return plus a real capital gain (g) is shown by solving equation 12.4 for i_t.

$$i_t = \frac{P_0(1 + g)}{V_0} + g \qu\quad (12.5)$$

For the numerical example, the real return of 4 percent contains a current gain of 3 percent ($0.03 = 80.8 \div 2{,}693.33$) plus the 1 percent capital gain.

Higher rates of growth of real payments mean that a higher proportion of real return (i_t) consists of capital gain g and a lower proportion as current return. Finally, if the valuation process in equation 12.3 is repeated a year later, the land value V_1 is found to increase at the compounded rates of inflation and real growth.

$$V_1 = V_0(1 + i_f)(1 + g) \qquad (12.6)$$

The combined market effects on investors' returns from inflation and real growth are interpreted as follows. The total annual rate-of-return i contains a real return i_t, plus a capital gain due to inflation:

$$i = i_t + i_f \qquad (12.7)$$

However, when the payments experience real growth, the real return also consists of a current return plus a real capital gain:

$$i_t = \frac{P_0(1 + g)}{V_0} + g \qquad (12.8)$$

Substituting equation 12.8 into equation 12.7 shows that the total return under real growth contains a real current return plus a real capital gain plus a capital gain due to inflation:

$$i = \frac{P_0(1 + g)}{V_0} + g + i_f \qquad (12.9)$$

$$\begin{array}{ccccc}
\text{Total} & = & \text{Real} & \text{Real} & \text{Inflationary} \\
\text{return} & & \text{current} + & \text{capital} + & \text{capital} \\
& & \text{return} & \text{gain} & \text{gain}
\end{array}$$

For the numerical example, the total return of 10 percent consists of a current return of 3 percent, plus a real capital gain of 1 percent, plus an inflation-induced capital gain of 6 percent.[1]

---

[1]These calculations ignore the interaction terms between i, g, and i_f. Thus, the calculations are close approximations to the true values. Moreover, the term "real growth" could be a real loss if a g were negative.

Implications for Land Values

The conceptual models described in the preceding section show how some of the important variables may interact to influence land values and the resulting mix of current returns and capital gains or losses experienced by investors. These types of models serve as a guide in evaluating the response to changes in these variables.

Evidence [3, 12], for example, has indicated that farm earnings, and thus returns to land, experienced real growth during the 1960s and 1970s. Moreover, the real earnings growth appears to have been capitalized into higher land values that have increased at rates roughly comparable to the growth in farm earnings. To the extent that these growth patterns are comparable, then farm land is not necessarily "overpriced" relative to the levels and growth of its expected earnings. Instead, land values were adjusted during these times so that investors earned more of their return as capital gains.

Similar, but opposite, effects may occur in response to a decline in the level of expected earnings or in the growth rate of expected earnings. In the early 1980s, for example, levels of farm income dropped sharply in the United States, while real interest rates reached record highs with much uncertainty about future prospects for farm income and financial market conditions. These factors suggest that the level $(P_1, P_2, \ldots, P_n)$ of projected returns to land likely declined, while capital costs (i) were increasing. Expectations on earnings growth (g) may have changed as well. The combined effects of these conditions suggest that land values should have declined rather quickly, as indeed was the case for U.S. farm land nationwide and especially midwestern farm land in the 1981 – 1985 period.

Further study and evidence are clearly needed about factors affecting land values. The effects of tax policies, government programs, financing arrangements, risk, and expectations are all involved [6, 8]. So are nonfarm demands for agricultural land and the efficiency of land markets.

Market efficiency refers to the speed and strength with which changes in factors affecting land values actually are reflected in the land market. The effects of market efficiency are illustrated as follows. Suppose a sharp change in international trade conditions increases the expected earnings growth for farm land. In an efficient land market, land values would go up immediately to reflect the higher earnings growth and then should continue to grow over time at the new growth rate. However, market imperfections may cause the initial change in land values to be distributed over time, thus blending it with the longer-term growth in land values. Again, similar, but opposite, effects may occur if the land earnings decline and are expected to experience negative real growth ($g < 0$) over time.

ANALYZING LAND INVESTMENTS

In this section we analyze land investments from the individual investor's standpoint, under the assumption that the individual's actions do not influence the aggregate level of land values or other market factors. Individual investors are concerned with the profitability of land investments under their own set of circumstances, such as each investor's expectations of market prices for resources and products, financing arrangements, inflation rates, tax obligations, length of planning horizon, costs of equity capital, and other relevant factors. Combining information about these variables in a capital budgeting model will show the profitability (NPV) of a land investment.

Any analysis of land investments, simple or complex, is still expressed by the basic capital budgeting model shown in equations 10.4 and 10.5. With the nonuniform series version being used and with no debt financing assumed, a typical land investment situation is modeled as:

$$\text{NPV} = -\text{INV} + \sum_{n=1}^{N} \frac{(1 - t)P_n}{(1 + i)^n} + \frac{V_N - T_N}{(1 + i)^N} \tag{12.10}$$

The variables are identified as follows:

NPV = the net present value of the land investment

INV = the asking price of the land, perhaps resulting from a land appraisal

i = the after-tax cost of the investor's financial capital

N = the length of planning horizon

P_n = the annual net cash flow projected for the land investment. [If subject to growth, then $P_n = P_0(1 + g_p)^n$, where g_p = the nominal growth in the payment series including the effects of both anticipated inflation and real growth.[2]]

---

[2]Thus, the nominal growth rate g_p is related to inflation and real growth through the following expression: $g_p = (1 + g)(1 + i_f) - 1$.

V_N = the terminal land value. [If land values are growing, then $V_N = INV(1 + g_1)^N$, where g_1 = the nominal growth in land values.] *(Note:* depending on the investor's expectations, the expected growth in the payment series, g_p, may differ from the expected growth in land value, g_1.)

t = the ordinary income tax rate

T_N = the tax obligation on the change in land values, where $T_N = (V_N - INV)(t)$

If the investment involves a purchase of land, then V_N represents the land value (net of any outstanding debt for a credit purchase) at the end of the planning period, less any tax obligation on capital gains. If land is controlled under a long-term lease, then V_N represents any terminal value of the lease. If land is leased on an annual basis, V_N would likely be zero.

The net present value could conceivably be modified to include intangible values in a land investment, such as pride of ownership and prestige in the community. Although these values may not be quantifiable directly, it is often possible to estimate how large they must be to justify acquiring high-cost land. In the following analysis, we will disregard these intangible values and focus on the capital budgeting procedures.

We will begin with a simple investment situation that ignores the effects of taxes, inflation, and external financing; then, we will introduce the effects of these variables to show the increasing realism and complexities of the analysis.

Case 1 Land Purchase:
No Taxes, No Inflation,
No External Financing

Consider a farmer who is evaluating the purchase of additional land. The farmer operates 800 acres but with present labor and machinery can handle another 200 acres. A nearby 200-acre farm is being sold from an estate for an asking price of $1,500 per acre. The farmer projects the gross income per acre to be $180 per year. The payment of operating costs and property taxes leaves an annual net cash flow of $90 per acre. Other data at this stage of the analysis include a 15-year planning horizon and a 6 percent cost of capital. Taxes, inflation, and financing are not considered.

The net present value model is specified as:

$$\text{NPV} = -1{,}500 + 90[\text{USPV}_{0.06,\ 15}] + \frac{1{,}500}{(1.06)^{15}} \tag{12.11}$$

$$= -1{,}500 + 874.10 + 625.90$$

$$= -1{,}500 + 1{,}500$$

$$= 0$$

The present value of the uniform series of A = $90 per year for 15 years is $874.10 ($90 × 9.7122). The conversion factor is found in Appendix Table V for n = 15 and i = 0.06. The present value of the terminal value of the land ($1,500 in year 15) is $625.90, which results from $1,500(1.06)$^{-15}$ = $1,500(0.4173). The sum of these two present values is $1,500, which, when added to the initial investment of $1,500, gives a net present value of 0.

The zero net present value means that the land investment has breakeven profitability to the investor. If an investor indeed pays $1,500 per acre for the land, the investor will experience a 6 percent yield over the 15-year period, given these data projections. That is, the investment's internal rate-of-return is 6 percent.

If the asking price were lower than $1,500, then the investment would be a profitable one, yielding more than the required rate-of-return. A higher asking price would yield a lower rate-of-return. Alternatively, if the investor used a lower cost of capital, the net present value would exceed zero (assuming a $1,500 asking price), indicating a profitable investment. A higher cost of capital would yield an unprofitable investment. Confirm, for example, that an 8 percent cost of capital would yield a net present value of – $256.79, if $1,500 were bid for the land.

For this situation, changing the length of planning horizon does not change the capital budgeting results. Suppose the planning horizon is lengthened to 20 years. The net present value is:

$$\text{NPV} = -1{,}500 + 90[\text{USPV}_{0.06,\ 20}] + \frac{1{,}500}{(1.06)^{20}} \tag{12.12}$$

$$= -1{,}500 + 1{,}032.29 + 467.71$$

$$= 0$$

The net present value is the same as for the 15-year horizon, although a higher proportion of the returns is attributed to annual earnings than to the terminal value. In any case, the terminal value plays an important role in the investment's profitability. This phenomenon is typical in land investments.

Case 2 Land Purchase: with Inflation, No Taxes, No External Financing

Introducing the effects of inflation and growth will influence the investment's payment series, terminal value, and cost of capital. Investors might expect inflation to affect each of these values differently, as indicated in the following example. Let's assume that the anticipated rate of general inflation is 4.72 percent. This inflation premium increases the cost of capital to a nominal rate of 11 percent, i.e., $0.11 = (1.06)(1.0472) - 1$. Moreover, the case farmer believes that the net cash flow will grow at a 6 percent annual rate and that land values will grow at a 5 percent annual rate.

Under these conditions the net present value model is specified in nominal terms as:

$$NPV = -1,500 + \sum_{n=1}^{15} \frac{90(1.06)^n}{(1.11)^n} + \frac{1,500(1.05)^{15}}{(1.11)^{15}}$$ (12.13)

$$= -1,500 + 952.30 + 651.76$$

$$= 104.06$$

The calculation procedures using nominal values are more tedious because the payment series is nonuniform. The payments are calculated each year as P for the preceding year ($P_0 = 90$) multiplied by $(1 + g_p)$. Discounting occurs separately for each year, with the discounted values then summed for all years in the planning period. In this case, the present value of the payment series is $952.30. The present value of the terminal value of the land is found by compounding the asking price to a future value at the 5 percent growth rate and then discounting to a present value at the 11 percent cost of capital. This present value is $651.76, which, when added to the discounted earnings value, gives a present value of $1,604.06. The net present value is $104.06, which indicates that the land investment is profitable relative to the 11 percent required return.

An easier calculation procedure is to adjust the discount rate for each component of the valuation model by the compound growth rate found in the respective numerators. This approach was shown in equation 9.9 for a constant-growth earnings series. For the earnings flow, divide 1.11 (1 + discount rate) by 1.06 (1 + growth rate) and then subtract 1 to yield a new discount rate of 4.717 percent. Similarly, for the terminal value, divide 1.11 (1 + discount rate) by 1.05 (1 + growth rate) and then subtract 1 to yield a new discount rate of 5.714 percent. Then, use the new discount rates to resolve the net present value model, which yields the same value as in equation 12.13.

$$\text{NPV} = -1{,}500 + 90[\text{USPV}_{0.04717,\,15}] + \frac{1{,}500}{(1.05714)^{15}} \tag{12.14}$$

$$= -1{,}500 + 952.30 + 651.76$$

$$= 104.06$$

Table 12.1 reports several net present values for land to illustrate the effects of different assumptions about growth in net cash flows and land values, length of planning horizon, and costs of capital. Arranging the data in this way allows an analyst to consider the sensitivity of bid prices and net present values to changes in individual variables or combinations of variables. This approach could account for some of the risk characteristics of the investment.

The results in Table 12.1 show that a higher cost of capital, with all other variables held constant, will always lower the net present values. But changing

Table 12.1. Net Present Values of Land Investment for Different Levels of Key Variables, No Taxes or Debt Financing

Growth in Net Cash Flow	Horizon	Growth in Land Value					
		4%			6%		
		8% Cost of Capital	10% Cost of Capital	12% Cost of Capital	8% Cost of Capital	10% Cost of Capital	12% Cost of Capital
4 percent	10 years	$264	$ 25	$–173	$480	$205	$–23
	15 years	363	34	–221	645	248	–58
6 Percent	10 years	342	94	–112	558	274	38
	15 years	518	163	–113	800	377	51

the planning horizon may have contrasting effects on the present values, depending on the values of other variables. For high costs of capital and lower growth rates for net cash flows and land values, net present values may decline as the planning horizon lengthens. Otherwise, net present values increase with longer planning horizons. Trade-offs between growth rates for cash flow and land values, given the cost of capital, may also be considered. The analyst should remember that inflation tends to affect growth rates and costs of capital in similar ways. Thus, sharp differences between changes in the values of these variables need careful justification.

Case 3 Land Purchase: with Taxes and Inflation, No External Financing

Introducing income tax obligations means that the appropriate variables in the capital budget must be expressed on an after-tax basis. This is shown in equation 12.10. Annual earnings and the investor's cost of capital are measured after taxes. The terminal value is also net of any tax obligations on capital gains or ordinary income, as illustrated by the general model in equation 12.10.

Assuming an ordinary income tax rate of 20 percent (t = 0.20), the after-tax cost of capital in the example is expressed as i(1 − t), or 0.088 = 0.11(1 − 0.2). Other data are the same as in case 2. Table 12.2 summarizes the capital budgeting calculations. The net present value at the bottom of the table is $170.91. It is composed of the sum of present values for the after-tax cash flows, plus the terminal value, minus the initial investment and the capital gains tax.

Comparing the net present value ($170.91) for this situation to the net present value ($104.06) in case 2 suggests the interesting result that the presence of tax obligations increases the investment's profitability. This is not strictly the case, however. The increase in net present value that takes place here is attributed to the sheltering of tax obligations on a major portion of the investment's terminal value, compared to the implicit assumption that the earnings from an alternative investment, expressed through the discount rate, are fully taxable.

Tax obligations on the land's terminal value occur only if land experiences long-term capital gains, as is the case here because of the positive growth rate specified for land values. (Prior to 1987, U.S. tax laws permitted sheltering part of the capital gains from taxation, thus yielding an even larger tax savings.)

Since the terminal value in land investments represents a major portion

**Table 12.2. Net Present Value of Land Investment
with Inflation, Taxes, and No Debt Financing**

Item	Present Value
1. Initial investment (INV)	$-1,500.00
2. Net cash flow series	
$$\sum_{n=1}^{N} \frac{(1 - t)(P_0)(1 + g_p)^n}{(1 + i)^n}$$	
or	
$(1 - t)(P_0)[USPV_{r,N}]$	882.22
where $r = (1 + i)/(1 + g_p) - 1$	
$(1 - 0.2)(90)[USPV_{0.02642,15}] = 882.22$	
3. Terminal land value (V_N)	
$INV(1 + g_1)(1 + i)^{-N}$	880.03
$1,500(1.05)^{15}(1.088)^{-15} = 880.03$	
4. Capital gains tax (T_N)	
$$\frac{(V_N - INV)(t)}{(1 + i)^N}$$	-91.34
$$\frac{(3,118.39 - 1,500)(0.20)}{(1.088)^{15}}$$	
5. Net present value (NPV)	$ 170.91

of the investor's returns, the relationships between changes in land values, taxation, and investment profitability are especially important to consider.

Case 4 Land Purchase: with Debt Financing

Thus far, we have assumed that the investor is financing the land investment only with equity capital. In practice, this is seldom the case. Instead, a downpayment of from 10 to 50 percent of the purchase price usually is required, with the remaining amount borrowed and repaid over a series of years. Using the returns-to-equity approach developed in Chapter 10, we will now consider how the use of borrowed capital affects the capital budgeting analysis.

The net present value model in equation 12.10 must be modified to include the effects of financing terms on the initial investment, on annual net cash flows, and on the terminal equity in the investment. The model is reformulated as:

$$NPV = -INV_{dp} + (1 - t)A[USPV_{i,N}] - \sum_{n=1}^{N} \frac{[P_n + (1 - t)I_n]}{(1 + i)^n} \qquad (12.15)$$

$$+ \frac{V_N - T_N - D_N}{(1 + i)^N}$$

where INV_{dp} is the downpayment of equity capital, P_n is the repayment of loan principal in each period, $(1 - t)I_n$ is the after-tax interest payment, and D_N is the debt outstanding at the end of year N.

To evaluate debt financing, consider again the land purchase in cases 2 and 3. The farmer could purchase a tract of land for $1,500 per acre, which would increase the net cash flow by $90 per year, with the payment series and land value growing at annual rates of 6 percent and 5 percent, respectively. Assuming an 11 percent cost of capital, a 15-year planning horizon, a zero tax rate, and no external financing, the investment's net present value was $104.06, as shown in equation 12.14.

Now, let's assume that the financing terms include a cash downpayment of 20 percent of the asking price, or $300, with the remaining $1,200 repaid in equal annual installments over 15 years at 11 percent interest. The size of payment required to amortize $1,200 over 15 years is found by solving for A in:

$$1,200 = A[USPV_{0.11,\ 15}] \qquad (12.16)$$

$$A = \frac{1,200}{USPV_{0.11,\ 15}} = 166.88$$

as calculated with the conversion factor for $i = 0.11$ and $N = 15$ in Appendix Table V.

Since the tax rate is assumed to be zero, the total after-tax loan payment (principal plus interest) is $166.88 each year, so that the debt-servicing requirement is a uniform series. Including the debt-servicing requirement of $166.88 in equation 12.16 and noting that the debt outstanding at the end of the planning horizon is zero results in the following net present value:

$$NPV = -300 + \sum_{n=1}^{15} \frac{90(1.06)^n}{(1.11)^n} - 166.88[USPV_{0.11,\ 15}] + \frac{1{,}500(1.05)^{15}}{(1.11)^{15}}$$

$$= -300 + 952.30 - 1{,}200 + 651.76$$

$$= 104.06$$

Not surprisingly, the net present value is $104.06, which is the same as the net present value when no external financing was used. Thus, in the limiting case, in which (1) the interest rate on the land debt equals the investor's cost of equity capital, (2) the planning horizon equals the length of the loan, and (3) income taxes are disregarded, the terms of financing do not affect the profitability of the land investment.

However, the interest paid on debt is tax-deductible, and the cost of equity capital should be more than the after-tax interest rate on debt to reflect the higher risk from leveraging. Moreover, the investor's planning horizon may be shorter or longer than the loan maturity. Thus, the size and pattern of debt payments do affect cash flows and investment profitability.

Consider the following financing terms for the land investment to illustrate these effects. A 20 percent downpayment again is required, with the loan balance of $1,200 per acre repaid in equal annual payments of principal and interest over 30 years at a 10 percent interest rate. The investor's expectations on cash flows and growth remain the same, the cost of capital is 11 percent, the planning horizon is still 15 years, and the tax rate is zero. Thus, the initial investment of equity capital (INV_{dp}) is $300, and the annual debt servicing is $A = \$127.30 = (1{,}200 \div USPV_{0.10,\ 30})$. The outstanding loan balance (D_N) in year 15 is found by solving for the present value (in year 15) of the remaining 15 loan payments, with the loan rate used as the discount rate. Thus:

$$D_{15} = 127.30[USPV_{0.10,\ 15}] = 968.25$$

The net present value model is formulated as:

$$\text{NPV} = -300 + \sum_{n=1}^{15} \frac{90(1.06)^n}{(1.11)^n} - 127.30[\text{USPV}_{0.11, \, 15}] + \frac{1,500(1.05)^{15} - 968.25}{(1.11)^{15}}$$

$$= -300 + 952.30 - 915.40 + 449.39$$

$$= 186.29$$

The positive net present value indicates that the investment is profitable under these financing terms. Moreover, the net present value has increased as a result of partial debt financing, compared to complete equity financing. The increase in profitability reflects the effects of leveraging, in which lower-cost debt capital (10 percent) is substituted for higher-cost equity capital (11 percent). Moreover, as shown below, the tax deductibility of interest payments further favors the use of debt capital. Of course, leveraging involves greater financial risk too; if a risk premium were added to the cost of equity capital, the payoff from leveraging would be less.

When the tax rate is non-zero, the computations are more tedious, since the amount of the annual loan payment allocated to interest, which is tax-deductible, will vary each year. As a result, we can no longer use the convenient uniform series factors of Appendix Table V. Instead, the present values of the after-tax debt-servicing requirements must be computed separately for each year and then summed to a total present value.

To illustrate, consider the introduction of income taxes at a 20 percent rate ($t = 0.20$) as in case 3. Table 12.2 showed the calculation procedures and the resulting net present value of $170.91 for the land investment under equity financing. If the investment is financed in part with debt capital according to the terms used in this section, then the effects occur on the initial investment, the present value of the loan payments, and the terminal debt. The after-tax present value of the loan payments is represented by the third term of equation 12.15:

$$\sum_{n=1}^{15} \frac{[P_n + (1 - t)I_n]}{(1 + i)^n}$$

In Table 12.3, we project a 15-year portion of the amortization schedule of principal and interest payments for these financing terms (the construction of

Table 12.3. Present Value of Principal and After-Tax Interest Payments

Year	Principal	Interest	After-Tax Interest $2 \times (1 - 0.2)$	Discount Factor $(1/(1.088)^n)$	Present Value $[(1 + 3)(4)]$
	(1)	*(2)*	*(3)*	*(4)*	*(5)*
1	$ 7.30	$120.00	$96.00	0.919	$ 94.93
2	8.03	119.27	95.42	0.845	87.41
3	8.83	118.47	94.78	0.776	80.40
4	9.72	117.58	94.06	0.714	74.10
5	10.69	116.61	93.29	0.656	68.21
6	11.76	115.54	92.43	0.603	62.83
7	12.93	114.37	91.50	0.554	57.85
8	14.23	113.07	90.46	0.509	53.29
9	15.65	111.65	89.32	0.468	49.13
10	17.21	110.09	88.07	0.430	45.27
11	18.93	108.37	86.70	0.395	41.72
12	20.83	106.47	85.18	0.363	38.48
13	22.91	104.39	83.51	0.334	35.54
14	25.20	102.10	81.68	0.307	32.81
15	27.72	99.58	79.58	0.282	30.28
Total					$852.26

an amortization schedule is shown in Chapter 14) and show the present value of the after-tax payments to be $852.26.

The summary of the NPV calculations is adapted from Table 12.2 and shown in Table 12.4. The net present value at the bottom of the table is $245.40. It is composed of the sum of present values for the after-tax cash flows plus the terminal value minus the initial investment, loan payments, capital-gains tax, and terminal loan balance. The positive net present value indicates a profitable investment when borrowed capital is used and when the effects of inflation and taxation are considered. Moreover, the net present value for debt financing is higher than the comparable net present value for full equity financing in Table 12.2, indicating the effects of leveraging. The after-tax net present value also exceeds the before-tax net present value, indicating the combined effects on profitability of tax shelters associated with the terminal value and interest payments. Of course, as stated earlier, the financial risks associated with the use of debt capital should be considered as well.

Table 12.4. Net Present Value of Land Investment
with Inflation, Taxes, and External Financing

Item	Present Value
1. Initial investment (INVdp)	$-300.00
2. Net cash flow series (see Table 12.2)	882.22
3. Loan payment series (see Table 12.3)	-852.26
$\sum_{n=1}^{15} \dfrac{[P_n + (1 - t)I_n]}{(1 + i)^n} = 852.26$	
4. Terminal land values (see Table 12.2)	880.03
5. Capital gains taxes (see Table 12.2)	-91.34
6. Terminal loan balance	-273.25
$\dfrac{D_N}{(1 + i)^N} = \dfrac{(958.25)}{(1.088)^{15}}$	
6. Net present value (NPV)	$ 245.40

Case 5: A Bid Price Model

In cases 1 through 4, the focus was on determining the profitability (i.e., net present value) of a land investment, given data about the asking price, annual returns, tax rates, the cost of capital, and financing arrangements. Thus, the net present value was the unknown variable that was solved for. Often, however, an investor may be interested in determining another piece of information about the investment situation — that is, the maximum price he or she could bid for the land in order to earn a stipulated rate-of-return, given data about all of the other factors.

In a bid price model, the net present value is no longer the unknown variable. Rather, it is set equal to zero, and the unknown variable is the bid (or asking) price. A couple of approaches could be followed to find the maximum bid price. A trial-and-error approach would involve varying the asking price in the present value model until the net present value converged to zero. At this point (i.e., net present value equals zero), the level of the asking price would be the maximum bid price. The second approach would be to explicitly solve the present value model for the level of the asking price that equates net present value to zero. This second approach is developed as follows.

To show the set up and solution to the bid price model, we will return to equation 12.10 and to case 1 in which no taxes, no inflation, no external financing were considered. The bid price model then is formulated as:

$$NPV = 0 = -INV + A\left[USPV_{i,N}\right] + V_N \tag{12.17}$$

or, because $V_N = INV(1 + g_l)^N$

$$NPV = 0 = -INV + A\left[USPV_{i,N}\right] + \frac{INV(1 + g_l)^N}{(1 + i)^N} \tag{12.18}$$

The object now is to solve for the asking price, given the other variables in the model. Because "INV" appears in both the first and third terms to the right of the equals sign, a short series of algebraic steps is used to yield the following as the expression for the bid price model.

$$INV = \frac{A\left[USPV_{i,N}\right]}{\left[1 - \frac{(1 + g_l)^N}{(1 + i)^N}\right]} \tag{12.19}$$

Using the numerical values in case 1, where $A = 90$, $i = 0.06$, $N = 15$, and $g_l = 0.00$, will yield the maximum bid price, which is:

$$INV = \frac{90\left[USPV_{0.06, 15}\right]}{\left[1 - \frac{1}{(1.06)^{15}}\right]} = \frac{874.10}{0.5827} = 1,500$$

Thus, an investor who bids $1,500 per acre for the land investment will earn a 6 percent internal rate-of-return over the 15-year period. If the annual returns increased to $100, the bid price would be $1,666.67. Alternatively, a decline of annual returns to $80 per acre would lower the bid price to $1,333.33. In effect, the bid price represents a break-even value for the investment because the net present value is set equal to zero, and the internal rate-of-return will equal the required rate i.

As in cases 1 through 4, the form of the bid price model becomes more complex as inflation, taxes, and external financing are considered. Under case 2 conditions with the alternative growth rates for annual returns and land values, the bid price model is:

$$\text{INV} = \frac{\displaystyle\sum_{n=1}^{15} \frac{90(1.06)^n}{(1.11)^n}}{1 - \frac{(1.05)^{15}}{(1.11)^{15}}} = \frac{952.30}{0.5655} = 1,684.01$$

In this case a bid price of $1,684.01 per acre will yield an 11 percent internal rate-of-return (i = 0.11), given the other data characterizing the investment, including a zero asking price.

When debt financing is utilized, it is still possible to derive a maximum bid price model that will yield a net present value of zero, although the model formulation is cumbersome and exceeds the space dimension of this text. Thus, the derivation of a maximum bid price model under conditions of debt and equity financing is left as an exercise for the reader (see reference 16 by Robison, et al., for further details).

FINANCIAL FEASIBILITY
OF LAND INVESTMENTS

Financial feasibility refers to *the ability of an investment to satisfy the financing terms and performance criteria that are agreed upon by both the borrower and the lender.* A profitable investment may not always be financially feasible, if the financing plan does not accommodate the magnitude and timing of the investment's returns. As we shall see, land investments often present special challenges in financial feasibility and liquidity management because of the role of capital gains.

Several types of liquidity problems arise for leveraged investors who purchase farm land, subject to expected earnings growth. One liquidity problem is attributed to the non – self-liquidating characteristics of land investments. Because land is not used up in the farm production process, no inherent set-aside of cash occurs to meet debt-servicing requirements, as happens when depreciation is charged on durable farm assets and operating expenses on other inputs. Instead, the funds for servicing land debt come from net business income, thus reducing the funds available for family living or other investments.

A second liquidity problem arises because current earnings from land investments often are a relatively small portion of the investor's total return compared to capital gains. The capital gains provide no cash flows unless the land is sold or used as a basis for refinancing. The lesser role of cash earnings reduces the investment's liquidity.

Another liquidity problem arises from differences in the kind of compensation required by lenders and experienced by equity investors in the land. Lenders require all of their scheduled compensation (principal and interest) in a cash payment, of which part is a real return and part is an inflation premium, that is needed to compensate for the anticipated loss of purchasing power in their debt claim. Borrowers, however, experience only part of their return as a current payment; the rest is capital gain. Thus, if leverage and inflation are high enough, borrowers will experience a financing gap because there will be insufficient cash from the land's early earnings to meet the debt payments. The borrowers' equity is growing favorably, but their cash position is not. Only after several years of earnings growth will the financing gap disappear.

This situation arises in part because the traditional repayment schemes based on either constant principal payments or equal installment payments make no direct allowances for the pattern of cash flows arising from the land investment. As shown as follows, these repayment practices contribute to financing gaps and trigger various liquidity responses by borrowers.

Financial feasibility can be analyzed in several ways. One method, introduced in Chapter 11, is based on the use of present value methods to determine the maximum feasible loan size that can be serviced given the annual cash flow from the investment, the length of loan, and the interest rate. To illustrate, consider the numerical examples in the previous section, and suppose that a constant annual cash flow series of $90 per acre is available to repay a 30-year loan at 10 percent interest. The maximum feasible loan size then is $848.42 = $90[USPV_{0.10, 30}]$. If a larger loan is needed, the investment is financially infeasible without other sources of cash to meet the loan payments. If the cash flow were growing at a 5 percent annual rate, then the maximum feasible loan size would increase to $1,421.86, assuming the loan contract allowed graduated payments.

Another approach to analyzing financial feasibility is to compare the cash flow from the investment with the loan repayment requirement in the first year of the planning horizon. If the repayment requirement exceeds the net cash flow, then the investment is financially infeasible. Again, consider the data for the land investment in the preceding section.

The financing terms on the $1,500 per acre purchase included a $300 downpayment with annual payments of $127.30, fully amortizing the $1,200 loan over 30 years at 10 percent interest. However, the nominal value of the land earnings in year 1 is $95.40 = $90(1.06). Thus, ignoring the effects of income taxes, the investor experiences a cash deficit in the first year of – $31.90 = $95.40 – $127.30. The cash deficit, or financing gap, continues to diminish until year 6, when the nominal value of the payment series has grown to $127.67 = $90(1.06)^6.

This deficit means that the investment is financially infeasible. It does not generate sufficient cash from its earnings alone to meet the debt payment. Recall, however, that the investment is still profitable (positive net present value), although the profitability is largely because of the growth in earnings and the terminal land value.

How can a leveraged investor in farm land, especially a young, first-time farm borrower, make the investment financially feasible when only part of the nominal returns occur as cash flow and the rest as capital gains? One approach is to postpone the investment until additional equity is accumulated, but this may severely curtail the growth potential.

Another approach is for the investor to generate cash from other earnings, or perhaps from liquidating other assets, to pay the lender. These sources of cash might be earnings on rented land, other owned land, livestock enterprises, nonfarm wages, foregone consumption, or sales of capital assets. Short-term borrowing or carryover loans from non – real estate lenders also can be used to meet the long-term debt payments, thus creating additional financial risk. Older, well-established operators or nonfarm investors are more likely to use these approaches than are younger farmers. However, for all types of investors, these approaches, when measured in real terms, may substantially alter their leverage and reinvestment policies.

An alternative approach under inflation is for the lender to modify payment policies on debt so that the growth in a borrower's repayment obligations more closely corresponds with the expected growth in debt-servicing capacity. These modifications directly involve the lender's participation in the design of the financing plans.

At one extreme, repayments could be waived, with loan principal and interest indexed to the expected growth in the borrower's returns. This would result in permanent indebtedness, with the amount of debt growing at about the same rate as the investor's equity. As a result, the liquidity problems would be resolved, and the real financial positions of both the lender and the borrower would remain the same, without the need to divert other earnings or obtain short-term loans to meet long-term payments.

Another alternative to inflation-indexing is to specify full amortization of long-term debt according to an inflation-adjusted repayment scheme. Most such schemes are designed to alleviate cash deficits and financing gaps early in the repayment period. They have also been used to finance residential housing, and they warrant consideration in farm lending too. Numerous schemes — graduated payments, purchasing power mortgages, partial amortization, interest only, and combinations thereof — are possible.

To illustrate, a graduated payment plan might start with the borrower's repayment obligations set at the level of expected land earnings in year 1 ($90 in the example). The payment obligation would then increase at the expected growth rate (6 percent) for several years (5, 7, or 10), after which the remaining loan balance would be refinanced on a conventional fixed-payment loan.

These types of plans may be more administratively feasible to lenders, since they involve full amortization of the loan rather than permanent indebtedness. They also may be tailored to selected types of borrowers and tied to loan insurance or other kinds of risk protection for the parties involved. Even equity participation loans have been used by some long-term lenders. Called a "shared appreciation mortgage," this type of loan compensates the lender in part by sharing the appreciated value of the property being financed.

All of these financing programs involve liquidity responses for land investments that depart sharply from traditional amortization methods based on fixed-payment or declining-payment loans. They also shift the emphasis in credit evaluations more toward evidence of managerial skills and financial progress and away from repayment history. These innovative financial programs respond directly to the financial feasibility problems of land investments. They warrant careful consideration, although providing risk protection from swings in farm earnings and possible declines in land values is important too.

LEASING FARM LAND

Leasing farm land is a widespread method of financing, especially for farm operators in the growth stage of their life cycle. Nearly 40 percent of the farms in the United States and nearly 65 percent of the farm land are operated by farmers who lease part or all of the land they operate. Most of the leasing occurs by part owners, with a typical pattern of land control characterized by heavy reliance on leasing by younger operators, followed by combined ownership and leasing for farmers ages 30 through 60, with greater reliance on ownership thereafter.

Leasing also varies according to the type of farm. Full owners are dominant for fruit and nut, vegetable, and dairy farms. Ownership is also more important in most types of livestock operations, in which land contributes less to total asset value. Leasing is most common for crop farms where grain, oilseeds. cotton, etc., are the major products.

Lease Contracts

The legal document for renting land is the **land lease,** *a contract conveying control over use rights in real property from one party to another without transferring title.* A contract usually stipulates the property's intended use, with extent and conditions of payment for this use. Thus, leasing is a means of financing that enables a manager to control the use of assets belonging to others without making a downpayment or incurring other ownership obligations.

There are important reasons for using written leases in land rental arrangements. Specifying the details of the lease in writing improves communication between the parties involved and helps to prevent misunderstandings. On occasions, new participants become involved in leasing transactions. Specifying the terms of the lease in writing insures that these terms will be binding on all parties and reduces the chances of misunderstanding.

Types of Land Leases

Among the more important types of land leases are the **cash lease,** the **crop-share lease,** and the **livestock-share lease.** Cash leases predominate in those states where much of the rented land is publicly owned grazing land. In the Corn Belt, crop-share leasing prevails, along with livestock-share leasing and combinations of cash and share leases.

Under the cash lease, a predetermined cash fee — either per acre or for the total farm unit — becomes a fixed fee to the tenant, irrespective of the resulting yields or product prices. This fee may be payable in several payments during the year, some occurring prior to harvest. Thus, part of the cash rent may enter a farm's operating line of credit. However, the cash lease is sometimes modified whereby the cash rent is indexed to an average yield index or to an average product price index. Such a modification shifts some of the production and price risks from the tenant to the land owner, without creating the need for the land owner to share in the farm management responsibilities.

A fixed-fee cash lease may create substantial financial risk for the tenant when business risks from yield and price variations are high. The fixed financial claim associated with the lease is another form of financial leveraging that magnifies the possible loss of the tenant's equity as business risks increase. In poor years, cash renters may have difficulty meeting lease payments, compared to share-rent obligations; in contrast, in good years, they may receive surplus earnings.

A logical outgrowth of the high financial risk of cash leases is the product-share lease, which enables the land owner to share with the tenant the risk of variability in yields and prices and to furnish some of the management input. Share leases introduce perfect positive correlation between a farmer's crop returns and the rental obligations, including sharing the financing of annual operating costs. Thus, share leases are highly risk-efficient financing plans. However, on the average, share rents are likely higher than cash rents, reflecting the premiums needed to compensate the land owner for accepting greater risk and contributing part of the management.

With a crop-share lease, a specified percentage of the crop is paid to the land owner for use of the land. The share is set so that both parties are properly compensated for the resources that they contribute. Ordinarily, the tenant provides all labor and machinery, the land owner provides the real estate, and both parties share most of the variable costs of production.

Under a livestock-share lease, the land owner usually furnishes the land and buildings, while either party or both parties jointly own the livestock. Returns from livestock sales, net of any shared cost, are divided between the two parties according to the share arrangements. Generally, the land owner owns half of the livestock and equipment, pays half of the operating expenses (including all feed inputs), shares in management decisions, and receives half of the farm receipts.

Share-Leasing Terms

Share leases promote maximum economic efficiency when both parties' shares in farm production are proportionate to their contributions to the business and when all truly variable expenses are shared in the same proportion as crops are shared. What, then, determines satisfactory rent shares? In the case of land-intensive crops, such as corn, wheat, soybeans, grain sorghum, and rice, why do we find typical rental payments of one-half of the crops in central Illinois, two-fifths of the crops in southwestern Ontario, or one-third or one-fourth of the crops in Texas? Or, why does some land rent for $120 per acre and other land for $60 per acre?

Differences in the quality and value of resources contributed by the tenant and land owner are especially important. Differences in land quality probably have the greatest impact on rent levels. Many of the differences in land quality are attributable to soil productivity. However, lease terms may also reflect differences in water supply, drainage, slope, proportion of tillable land, buildings, and other features that affect the land's earning capacity. For some specialty crops, acreage allotments and market accessibility may be important factors.

The differences in shares arise because the tenant's costs for planting, cultivating, and harvesting an acre of most types of crops should be about the same regardless of location or land quality. To illustrate, suppose we compare the share-lease arrangements on two tracts of land located in geographic regions with significant differences in land productivity. Tract A averages a yield of $400 gross return per acre; tract B averages a $300 gross return per acre. In both cases, the tenants' "costs" average $200 per acre. For tract A, the tenant should receive one-half (200 ÷ 400) of the crop to cover costs. For tract B, the tenant needs two-thirds (200 ÷ 300) of the crop to cover the same costs. The land owner receives the residual income as rent. Differences over time in this residual income are capitalized into land values that, in turn, reflect the differences in land productivity.

Failure to share variable expenses in the same proportion as products are shared will reduce the economic efficiency of the lease arrangement. The tenant and land owner may then have different incentives for using the variable inputs. To illustrate, a land owner who does not share the cost of fertilizer would prefer that the tenant apply a yield-maximizing level of fertilizer. However, this would not be a rational level for the tenant, who must bear the full marginal cost of the fertilizer.

Some inputs are difficult to classify. If herbicide applications, for example, only substitute for the tenant's usual cultivation practices, then herbicide costs should not be shared. But if yields increase from the herbicide, then sharing its costs would be consistent with sharing its returns.

Leasing terms and their impact on managerial decisions are largely a result of custom or tradition. Share-rent levels, for example, tend to persist over time, even though changes occur in farm practices and economic conditions. The persistence of 50–50 crop-share leases in the Midwest is a good example. These leasing terms may constitute the only expression of market price for the services of unpaid inputs such as land, operating labor, and management. In share leasing, the economic rent for land is disguised behind a number of associated nonland inputs. Thus, the adoption of new methods of crop harvesting, storage, feeding systems, or herbicides becomes involved in the land-rental problem.

The general use of cash leases fluctuates over time with changing financial conditions in agriculture. However, the conditions for more widespread use are improving as better technology reduces yield variability and as tenants need greater freedom in managing leased land. Cash rents also allow more flexibility in bidding for leased land. Levels of cash rents have likely adjusted to changing conditions more readily than have share rents. This adjustment results in part from increased bargaining between land owner and tenant, a factor that may raise tenure risks if land control shifts frequently to the highest bidder.

Analyzing Land Leases and Financing Arrangements

The framework for analyzing leasing as a means of financing the control of farm land is similar to ownership. The type and length of lease may affect cash inflows and outflows over the manager's planning horizon. The net cash flows may be discounted to a present value for comparison with other land financing methods. Non-economic factors may be considered as well.

With share leases, the major impacts on a farmer's cash flow are the percentage reductions in output and the variable costs that are allocated to the land owner. Cash leases have no effect on the farmers' returns, only on the projected costs from the land investment. In either case, estimates of all cash flow items influenced by the lease should be considered in the capital budgeting analysis on an after-tax basis.

The rental arrangements may also affect the short-term financing of the farm operation. In cash leases, many land owners require more than one payment during the year. If part of the cash rent is due prior to harvest, then the tenant must finance it through the operating line of credit or with cash from other sources. Thus, the timing of the rental payments will influence short-term financial planning.

The timing of rental payments should also influence the level of rent, since earlier payments of rent shift the financing burden from the land owner to the tenant. Suppose, for example, that the typical cash rent in a community is $100 per acre, based on payment at harvest in November. A land owner who requires a full payment six months in advance should be prepared to accept a lower rent to reflect the time value of money. If the interest rate is 8 percent annually, then the discount should be $4, based on the six-month time period. The negotiated rent would then be $96 per acre.

Share rents also influence short-term financing through the practices followed in sharing variable expenses. In principle, the land owner and the tenant should cover, through cash payment or credit, their respective shares of operating expenses as they are incurred. Hence, two lines of credit — one for the tenant and one for the land owner — might be needed. In practice, however, the arrangements for sharing the financing of operating inputs vary widely. In many cases, the tenant and the land owner receive separate billings as expenses are incurred. In other cases, the tenant may act as an agent for the land owner in acquiring and financing operating inputs. A final settlement for expenses (including interest) occurs at harvest time. This latter arrangement is especially useful for farmers having large operations, with land leased from several land owners. In any case, leasing arrangements clearly bring added demands for financial management.

Other indirect costs of leasing are associated with tenure risk and the effects of leasing on credit. When short-run leases are used, with no option for renewal, tenure risk is a special concern to both land owner and tenant. Most farm leases, whether written or unwritten, are limited to one year. Few investment programs become feasible in a one-year horizon. Consequently, tenants may be reluctant to undertake investments that would be profitable over longer periods. Lenders may be reluctant to finance such investments for similar reasons.

Tenants and land owners can usually gain from increasing the tenure security of the lease contract, especially after a favorable experience with a one-year contract. An example is a three- to five-year contract that is renewable each year. This arrangement extends the contract into the future at least two years. The terms of the lease also should enable either party to terminate the contract if the other party fails to perform satisfactorily. In some states, annual leases are extended automatically if notice of discontinuance is not given prior to the termination date. Arrangements whereby one party compensates the other for residual values of investments when the lease terminates may also be developed.

Numerous examples arise of land owner – tenant arrangements that have continued for many years without long-term contracts. The tenant who regularly attains above-average yields and product quality usually has little risk of losing a lease. In fact, such a tenant usually has a waiting list of interested land owners. Similarly, land owners who have a reputation for fairness, integrity, and management skills tend to attract better candidates as tenants.

Leasing adds to the control of resources for a firm with a minimum of financial disturbance. The manager thus avoids the lower liquidity that tends to be associated with ownership through cash purchase or borrowing. No liabilities are created beyond the rent obligation. However, a lender may be reluctant to finance the addition of durable assets on land that may not be available to the tenant in the future. The reclaimability of assets ranks high on lender preferences for loan purposes. However, the degree of reclaimability for the same asset may vary with the tenure of the operator: owner, part-owner, or full tenant. Thus, the credit available for financing durable assets may be less favorable for the tenant than for the owner-operator.

Summary

Farm real estate is a dominant and unique asset in the agricultural sector. Its durable and immobile attributes make it well suited as security in financing transactions, and it is subject to specific laws and tax requirements as well. Because land investments usually involve significant amounts of capital and the use of borrowed funds, leasing of farm land has become a widespread means of gaining control over the use of farm land without incurring the costs and risks of ownership. In addition, the low rates of market transactions for the sale and leasing of farm land mean that appraisals of farm land are often used to estimate land values. Such appraisals ultimately are based on present value procedures applied to anticipated flows of returns from investing in farm land.

Other highlights are as follows:

- The durability of farm land means that investors receive both current returns and capital gains (or losses), with current returns being relatively low compared to capital gains, especially when inflation rates are relatively high.

- Also important in the pricing of farm land and profitability of farm land investments are the effects of tax policies, government programs, financing arrangements, risk, and the degree of efficiency in farm land markets.

- Capital budgeting analyses of farm land investments should systematically account for the effects of the level of returns to land, growth rates, financing costs and terms, tax rates, terminal values, and the length of the investors' planning horizons.

- A maximum bid price for land is the purchase price that would yield a zero net present value for a land investment. The maximum bid price would allow the investor to earn the rate-of-return stipulated by the discount rate, given data about all of the other factors.

- Because of relatively low current rates-of-return, a leveraged investor in farm land often may experience financing gaps early in the investment period when repayment obligations exceed the returns from the land. While nominal growth in earnings eliminates the financing gaps over time, other sources of liquidity are needed to meet the repayment obligations early in the investment period.

- Leasing of farm land is another form of financial leveraging because of the fixed rental obligation on the lease.

- Major types of land leases are the cash lease, the crop-share lease, and the livestock-share lease.

- Share leases allow a sharing of business risk between the land owner and the tenant, with the relative shares set to properly compensate both parties for the resources that they contribute.

- Leasing also may affect other financing arrangements through the timing and level of rental payments, sharing of operating costs, and tenure risks associated with the possible loss of the use of leased acreage in the future.

Topics for Discussion

1. Identify the characteristics of farm land that make it unique from other resources. What are the implications for financing land investments?

2. How active is the market for farm land? How does this activity affect land values and their responses to changes in farm income, interest rates, and inflation?

3. Estimate the appraisal value of a tract of land with a current cash rent of $80 per acre, a nominal capitalization rate of 10 percent, and an inflation rate of 5 percent. How would you validate this value?

What is the anticipated land value one year in the future? Two years?

4. How would the appraised value in problem 3 change if the current cash rent is expected to experience real growth of 1 percent annually? What would be a land investor's current rate-of-return and rate of capital gain under this condition?

5. How would the appraised value in problem 3 change if the current cash rent were expected to experience a real decline of 1 percent annually?

6. An individual investor is evaluating the profitability of an anticipated land investment that has an asking price of $2,000 per acre and a current net cash flow of $120 per acre. The investor will pay cash, plans on holding the investment for 10 years, and has a 5 percent cost of capital. No inflation or tax obligations are involved. Evaluate the profitability of the investment.

7. Evaluate the investment in problem 6 under an assumption that the anticipated inflation rate is 6 percent, which equally affects net cash flows, land values, and the cost of capital.

8. Evaluate the investment in problem 7 under the same conditions, except that land values are expected to grow at a 5 percent annual rate.

9. Evaluate the investment in problem 7, assuming that the investor's tax rate on ordinary income is 20 percent.

10. Evaluate the investment in problem 9 with the use of debt financing, assuming a 20 percent downpayment and a loan at 11 percent interest with equal payments of principal and interest amortized over 30 years. Discuss the possible risks associated with this investment.

11. Evaluate the financial feasibility of the land investment in problem 10. If the land investment is found to be infeasible, how might feasibility be attained?

12. Explain how an investment can be profitable, yet financially infeasible.

13. Using the data in question 6, find the maximum bid price the investor can afford to pay in order to earn the 5 percent cost of capital rate. Also find the maximum bid price for the conditions of question 7.

14. Discuss some of the reasons why leasing is an important method of financing the control of farm land.

15. Contrast leases using cash rents with those using share rents. Which is more risk-efficient from the tenant's view? Why?

16. Explain some of the financing implications of cash leases and share leases. How do the rental arrangements affect operating lines of credit?

17. Explain why the percentage shares of products between land owner and tenant may differ according to geographic regions.

18. Which is the proper observation: "Land rents determine land values" or "land values determine land rents"? Why?

19. Explain how land values should respond to (a) an increase or a decrease in the level of returns to land, (b) an increase or a decrease in the expected earnings growth for land, and (c) an increase or a decrease in the cost of financial capital.

References

1. Alston, J. M., "An Analysis of Growth of U.S. Farmland Prices, 1963 – 82," *American Journal of Agricultural Economics*, 68(1986):1 – 9.

2. Baron, D., "Agricultural Leasing in the Cornbelt: Some Recent Trends," *Journal of The American Society of Farm Managers and Rural Appraisers*, 46:1(1982).

3. Barry, P. J., and L. J. Robison, "Economic Versus Accounting Rates of Return on Farmland," *Land Economics*, 62:4(November 1986): 388 – 401.

4. Burt, O. R., "Econometric Modeling of the Capitalization Formula for Farmland Prices," *American Journal of Agricultural Economics*, 68(1986):10 – 26.

5. Castle, E. N., and I. Hoch, "Farm Real Estate Price Components, 1920 – 1978," *American Journal of Agricultural Economics*, 64(1982): 8 – 18.

6. Economic Research Service, *Farm Real Estate Market Developments* (annual series), USDA, Washington, D.C.

7. Falk, B., "Formally Testing the Present Value Model of Farmland Prices," *American Journal of Agricultural Economics*, 73(1991):1 – 10.

8. Feldstein, M., "Inflation, Portfolio Choice, and the Prices of Land and Corporate Stock," *American Journal of Agricultural Economics,* 62(1980):910 – 916.

9. Klemme, R. M., and R. A. Schoney, "Economic Analysis of Land Bid Prices Using Profitability and Cash Flow Considerations," *North Central Journal of Agricultural Economics,* 6:2(1984):117 – 127.

10. Lee, W. F., and N. Rask, "Inflation and Crop Profitability: How Much Can Farmers Pay for Land?," *American Journal of Agricultural Economics,* 58(1976):984 – 990.

11. Lins, D. A., N. E. Harl, and T. L. Frey, *Farmland,* Century Communications, Inc., Skokie, Illinois, 1982.

12. Melichar, E. O., "Capital Gains vs. Current Income in the Farming Sector," *American Journal of Agricultural Economics,* 61(1979): 1082 – 1092.

13. Reiss, F. J., *Farm Leases for Illinois,* Rev., Agricultural Extension Service Circular 960, University of Illinois at Urbana – Champaign, September 1972.

14. Reiss, F. J., *What Is a Fair Crop-Share Lease?,* Agricultural Extension Service Circular 918, University of Illinois at Urbana – Champaign, October 1965.

15. Robison, L. J., and J. R. Brake, "Inflation, Cash Flows, and Growth: Some Implications for the Farm Firm," *Southern Journal of Agricultural Economics,* 12(1980):131 – 138.

16. Robison, L. J., S. A. Koenig, and M. P. Kelsey, "Farm Real Estate Prices and the Tax Reform Act of 1986," *Agricultural Finance Review,* 47(1987):21 – 30.

17. Tweeten, L. G., "Macro-Economics in Crisis: Agriculture in an Underachieving Economy," *American Journal of Agricultural Economics,* 62(1980):853 – 865.

13

Leasing Non – Real Estate Assets

A **lease** is *a contract by which control over the right to use an asset is transferred from one party (the lessor) to another party (the lessee) for a specified time in return for a rental payment to cover the lessor's costs of ownership.* Thus, **leasing** is *a method of financing the control of an asset that separates its use from its ownership.* Several types of leases are used specifically in agriculture. Especially important is the leasing of farm land, which was addressed in Chapter 12. Also important are operating leases, custom hiring, capital leases, and leveraged leases, which are discussed here.

In the following sections, we describe the types of non – real estate leases used in agriculture, evaluate their advantages and disadvantages, and apply present value techniques to analyze the economics of leasing for both the lessee and the lessor. Most of the discussion focuses on capital leases for farm machinery and equipment, because of their growing importance. However, the basic principles and tools of lease analysis apply to many other types of agricultural assets.

TYPES OF LEASES

Leasing of non – real estate assets has grown in agriculture, but it is still less extensive than in other industries such as the trucking, manufacturing,

375

computer, and airline industries. Generally, leasing in agriculture involves machinery, equipment, some buildings, and breeding livestock. Most other types of non – real estate assets are less suited for leasing because they are not easily reclaimed when the leases expire.

Most manufacturers of farm machinery and equipment offer leasing programs to farmers through dealerships, although these programs receive relatively moderate use. In addition, financial institutions, such as commercial banks and the Farm Credit System, and independent leasing companies and individual firms or persons offering specialized services provide leasing services. In large agribusinesses, more complex formal lease arrangements, called **leveraged leasing,** may develop involving the lenders with the lessor and lessee. In the following sections, we consider the characteristics of these various types of leases.

The Operating Lease

The **operating lease** is *usually a short-term rental arrangement (hourly, daily, weekly, monthly) in which the rental charge is calculated on a time basis.* Rental cars, trailers, and retail rental stores are common nonfarm examples. The lessor owns the asset and performs nearly all the functions of ownership, including maintenance. The lessee pays the direct costs, such as fuel and labor. These terms vary with the type of machine and the length of rental period. When a trench digger is leased for a day or two, the lessor likely pays for all maintenance. In contrast, the lessee might carry most of the maintenance costs on a tractor leased for an entire cropping season. These terms are often negotiated between the two parties.

Companies offering operating leases are usually those that perform maintenance tasks, such as manufacturers or their subsidiaries, dealers, or other specialized businesses. The greatest success for operating leases is with general-purpose items, which have user demands spread over various time periods, or with specific-purpose items, whose demand varies by region or by type of service. Innovation in operating leases may help increase the dealer's merchandising activities and provide a supply of relatively new items for the popular "used" market, especially for machinery.

The operating lease may also be combined with a purchase program. Lease-purchase arrangements in a contract essentially give the lessee the option to apply part or all of the lease payment to an asset's purchase price. Thus, the operating lease may be changed to a cash or credit purchase.

Custom Hiring

Custom hiring is *a form of leasing that combines the hiring of labor services with the use of the tangible asset.* Common examples include custom harvesting for many crops, custom applications of chemicals and fertilizer, and custom or contract feeding of livestock in commercial feedlots. These specific functions are performed by individuals or firms that control the needed machines or facilities and will supply the services of both the asset and their own labor.

For some locations and farming types, machinery and equipment dealers may offer custom crop services that utilize their labor and equipment. With good quality of service and established maintenance departments, they can offer custom services on favorable terms. Chemical suppliers may also provide custom applications, thus assuring expert quality of work. In some cases, environmental regulations limit the application of chemicals to licensed operators.

In many areas, land owners can hire the custom operation of their entire farm — soil tilling, planting, cultivating, harvesting, etc. Where cash crop farming is prevalent, "custom farming" thus offers an alternative to leasing arrangements on farm real estate, although the latter is much more prominent. In livestock areas, especially in the Plains states, many cattle and some hogs are custom-fed in commercial feedlots that own neither the animals nor the feed. The cattle owner pays a fee to the feedlot for the production and marketing services.

Custom hiring resembles other types of leasing in that it avoids the financial drain of capital investments and the obsolescence risk of owned capital. It may also improve the accuracy of cost estimates for the various services, thus facilitating projections of cash flows and short-term financing needs. Custom operations may also relieve limitations on the availability of family labor or skilled hired labor for many agricultural enterprises. Custom labor that is specialized to a specific task may perform more efficiently than other types of labor. The timeliness of operations may also improve, although there may be uncertainties about the availability of custom operators. Finally, custom hiring may be cheaper than either owning or leasing an asset for a specialized task.

Custom services have a long history in agriculture. In the U.S. Corn Belt, for example, the services evolved from the practice of neighborhood work-sharing for machines that were too large for single farms to own. Threshing and corn shelling are historical examples. Much custom work is still done by farmers with excess machine capacity. However, in the cotton and wheat areas, custom services tend to come from specialized operators.

The Capital Lease

The **capital lease** is *a long-term contractual arrangement in which the lessee acquires control of an asset in return for rental payments to the lessor.* The contract usually runs for several years and cannot be cancelled. Except for price variations of the asset, the lessee acquires all of the benefits, risks, and costs of ownership without having to make the usual investment of equity capital. In many ways, the capital lease is comparable to a credit purchase financed by an intermediate-term loan that provides 100 percent financing, although prepayments of rent have the same cash flow effects as downpayments in credit purchases. The rental charge is established to cover the lessor's expected costs of ownership plus profit.

Capital leases have become very important in agriculture and may have more use in the future. Thus, the remainder of this chapter focuses on their characteristics and methods of analysis.

Leveraged Leasing

Leveraged leasing is an extension of capital leasing that has developed in large agribusinesses and other corporate firms to finance projects with large capital expenditures and long economic lives. The main feature of **leveraged leasing** is *the formal involvement of a lender in providing debt capital to finance the lessor's purchase of the leased asset.* As with a regular capital lease, the lessee selects the asset to be leased and negotiates the leasing terms with the lessor. The difference in a leveraged lease, however, is that the lessor purchases and becomes owner of the asset by providing only part (20 to 40 percent) of the capital needed. The rest of the purchase price is borrowed from one or more institutional lenders, with the loan secured by a first lien on the assets, by an assignment of the lease contract, and by an assignment of the rental payments. Arrangements are carefully developed to assure that the various transactions between the lessee, lessor, lender, equipment manufacturer, and any other parties are soundly conceived and effectively carried out.

In agriculture, an example of leveraged leasing is the participation of the Banks for Cooperatives as lenders in leasing arrangements between large agricultural cooperatives, lessors or their trustee representatives, and manufacturers. Examples of capital projects that have been financed through leveraged leasing by cooperatives are sugar beet refineries, fertilizer plants, rural electric systems, vessels, production equipment, and commodity handling systems. Farmers and other agricultural producers who patronize and own these cooperatives benefit indirectly from the use of leveraged leasing by the cooperatives to finance their operations.

ISSUES IN CAPITAL LEASING

The popular view of capital leasing highlights several basic advantages: the conservation of working capital, nearly 100 percent financing, the use of modern equipment and facilities, and possible tax benefits for the lessee and lessor. These advantages and many other issues are important in evaluating the payoffs of using capital leases as a method of financing non – real estate assets. They are discussed under the following subheadings of this section.

Profitability of Leasing

Leasing may be a lower-cost method of financing than controlling assets through ownership and borrowing. Profitability depends on the comparison of the present values of after-tax cash flows for the various financing methods. The effects of income tax obligations on these cash flows are also very important. The analytical procedures involved are discussed later in the chapter.

Leasing and Borrowing Capacity

Leasing generally does not restrict a firm's borrowing capacity. As indicated in the following explanation, leasing activities seldom are included in a farmer's balance sheet and thus do not directly influence many financial ratios (see "Leasing and Financial Statements"). Indeed, leasing may even increase a farmer's credit over time if working capital is conserved and income expectations improve. However, the effects of leasing on financial and credit analyses should be integrated with the effects of other assets and liabilities in order for a business' overall financial feasibility to be evaluated properly. Acquiring a leased asset may on occasion be worse than not acquiring the asset at all, if the financial demands on cash flows are too strong. Under these conditions, a farmer's credit would be adversely affected too.

The relationships between leasing and credit also involve a lender's regulatory and liquidity positions. Some farmers who borrow from agricultural banks, for example, have loan requests that exceed the banks' legal lending limit. These banks then need overline loan participation from other lenders to fully fund the loans. An alternative is for the farmers to lease some of their assets, perhaps from the agricultural banks' leasing subsidiaries, thus reducing the need for overline participation from higher-cost sources.

Similarly, if a bank has a tight liquidity position, as indicated by a high loan-to-deposit ratio, it may encourage its farm customers to lease some assets in order to relieve loan demands. The lender can then include the rental obligation in the farmer's operating line of credit, rather than providing a term loan. Of course, lenders who provide a leasing service must still finance their own leasing activities.

Leasing and Total Financing

Capital leasing is often considered to provide the lessee with total financing that does not require the commitment of equity capital through cash purchases or downpayment on a loan. This is not always true, however. The terms of leases often result in cash payments at the beginning of the lease contract, which have cash flow consequences similar to downpayments on a credit purchase. Prepayments on rent and security deposits are common examples. They must be met by the lessee's own funds. Nonetheless, these initial cash payments usually are lower than the downpayment requirements on a credit purchase.

Leasing and Financial Statements

The concepts and practices involved in accounting for leasing activities on a firm's financial statements provide an interesting contrast. In concept, a firm's financial statements should report the full range of activities affecting its assets and liabilities. With this approach, assets and liabilities are broadly defined. Included are the values of everything the business controls, whether owned outright or leased. Liabilities are all the claims on assets and income, including those of lenders, lessors, and equity capital. The value of leased assets is thus offset by the value of rental claims.

In practice, however, the effects of leasing only appear as operating expenses for rental payments on farm income statements and flows of funds' accounts. No accounting of leasing is shown on the firm's balance sheets, so that leasing is often characterized as "off – balance sheet" financing. Sometimes minimal disclosure of leasing activities might occur in footnotes to the financial statements.

Under these arrangements, financial analysis of firms that lease assets has yielded ratios and other measures showing stronger liquidity and solvency than actually is the case. The same situation applies to farmers who lease farm land. Most real estate leases are short-term written or oral contracts that resemble operating leases more than capital leases. Consequently, real estate leasing is also considered off – balance sheet financing.

Generally accepted accounting principles now call for capitalization on the balance sheet of certain types of leases. A leased asset should be capitalized on the balance sheet if any one of the following conditions is met:

1. The lease transfers ownership of the asset to the lessee by the end of the lease contract.

2. The lease contains an option for the lessee to purchase the asset at a bargain price.

3. The lease period equals or exceeds 75 percent of the asset's estimated economic life.

4. At the start of the lease, the present value of the rental payments exceeds 90 percent of the value of the leased assets.

In general, if the lessee acquires essentially all of the economic benefits and risks of owning the leased asset, then the value of the leased asset is shown as an asset and the rental obligation is shown as a debt. These values roughly offset each other on the balance sheet and are reflected in the various measures of balance sheet liquidity. Moreover, information about the leasing terms and the methods of capitalization should be given in footnotes to the financial statements.

The capitalization process involves both the balance sheet and the income statement. The procedures are summarized as follows. First, determine the present value of the lease by discounting its flow of rental payments to a present value, using an appropriate discount rate (the lessee's borrowing rate or the lessor's implicit interest rate in the lease contract). Enter this value as both an asset and a liability on the firm's balance sheet, distinguishing between the current and noncurrent portions, as appropriate. Then, establish procedures for amortizing these lease values over the contract so that the asset and liability values reach zero by the end of the contract period. The lease's asset value is usually amortized according to the firm's depreciation practices (straight-line or accelerated). The lease's liability value is amortized like a long-term debt obligation into interest and principal portions each year. The leasing charge in the firm's income statement for any year is then the sum of the annual asset amortization plus the calculated interest charge.

These accounting procedures are illustrated in Table 13.1 for a farmer who is leasing a $5,000 microcomputer system on either an operating lease or a capital lease for a three-year period. In either case, the lessor figures the annual rental charge at $2,010 based on a 10 percent implicit interest rate. If the farmer acquires the asset on an operating lease, the $2,010 rent is charged as an expense against each year's income, and no entries occur on the balance sheet. For a capital lease, the asset and liability values for the lease are $5,000, which is the present value of a uniform series of $2,010 payments for three years at 10 percent interest. The asset value is amortized on a straight-line basis, yielding annual amortization charges of $1,667 (column d). The liability value is reduced according to the amortization schedule shown in columns f and g for imputed interest and principal. The annual expense (column h) is the sum of annual amortization (column d) and imputed interest (column f).

Table 13.1. Financial Accounting for an Operating Lease Versus a Capital Lease[1]

End of Year	Operating Lease		Capital Lease					
	Rental Payment	Annual Expense	Asset Value	Annual Amortization	Liability Value	Imputed Interest	Principal	Annual Expense (d + f)
	(a)	(b)	(c)	(d)	(e)	(f)	(g)	(h)
0			$5,000		$5,000			
1	$2,010	$2,010	3,333	$1,667	3,489	$500	$1,511	$2,167
2	2,010	2,010	1,666	1,667	1,827	348	1,662	2,015
3	2,010	2,010	0	1,666		182	1,827	1,848
Total	$6,030	$6,030						$6,030

[1]Note: Some rounding of numbers occurs in the figures reported in the table.

As Table 13.1 shows, the total expenses over the lease contract are the same ($6,030) for the operating and capital leases. However, the time pattern of the expenses differs, with larger charges occurring earlier for the capital lease. Clearly, the accounting procedures for capital leases are more complex than for operating leases and make leasing an "on – balance sheet" method of financing. But, these procedures also account more completely for leasing's effects on the firm's total financial position.

Leasing and Tax Considerations

The lessee's rental payments are a deductible expense for income tax purposes on operating or capital leases, as long as the transaction clearly qualifies as a lease. The conditions that qualify a transaction as a capital lease, in contrast to a credit purchase, have evolved over time in response to changes in federal laws and guidelines and rulings issued by the U.S. Internal Revenue Service. A major focus of these laws and rulings has been that lease transactions not be used just for the purpose of transferring tax benefits.

The following guidelines have been in effect to qualify a transaction as a lease for tax purposes:

1. The lessor must have a minimum "at risk" investment of at least 20 percent in the property.

2. The lessee may neither provide investment capital for acquiring the property nor lend to or guarantee the loans of the lessor.

3. The lessee may not have an option to purchase the property at the end of the lease except at the existing fair market value. In no case can the lessee be required to purchase the property.

4. The lessor must expect to receive a profit and positive cash flow from the transaction, independent of the tax benefits.

5. The property should not be limited in use to only the lessee.

In addition, 1982 legislation created a new category of tax lease, called a **finance lease,** which applies to limited-use property and which *allows the lessee a fixed-price purchase option at 10 percent or more of the property's original cost.*[1] Thus, a finance lease is considered to occur in lease agreements that contain a 10 percent fixed-price purchase option or that involve limited-use property, where otherwise the lessor is treated as the property owner for tax purposes. The major feature of the finance lease is the provision for the fixed-price purchase option. In the past, the requirement that purchases at lease termination occur only at fair market value created much uncertainty for both lessee

---

[1]Legislation passed in 1986 repealed the general use of the finance lease, although its availability for use with farm property was continued.

and lessor. The fixed-price option reduces the uncertainty and allows residual values to be considered explicitly in setting lease payments and other terms.

Leasing and Financing Terms

The variability and length of payment obligations, security requirements, and credit worthiness factors also warrant consideration in the evaluation of lease versus credit-purchase transactions. In most capital leases, the rental obligation is fixed over the term of the lease contract, in contrast to the use of variable (or floating) interest rates on intermediate-term loans. The risks to the lessor associated with a fixed-rate lease tend to be passed on to the lessee in the form of higher rental payments. Since variable-rate leasing contracts are possible, they may receive greater use, as lessors respond to changes in their own financing costs.

The comparability of maturities for capital leases and term loans likely varies with the source of leasing and the source of credit. In general, these maturities should be similar, as lessors and lenders increasingly establish contract lengths that approach the economic life of the assets involved. Security requirements and other credit factors should also be similar for capital leases and term loans.

With most capital leases, the lessee must demonstrate lease worthiness the same as a borrower demonstrates credit worthiness. Showing repayment capacity, pledging collateral, meeting net worth requirements, and other credit factors may be necessary.

Other Leasing Issues

Many other factors unique to the specific situation need consideration in leasing transactions. Leasing may reduce the risk of obsolescence by allowing more rapid replacement of equipment at lower transaction costs, although this is less true for capital leases than for operating leases. Leasing may provide financing for the noninterest costs of acquiring an asset. Sales taxes, delivery and installation costs, and other costs may be included in the rental payments on a lease.

Leasing may also introduce "tenure risks" if the lease is not renewed when the contract expires. Loss of a leased asset could jeopardize other parts of a business operation, although this risk is greater for leasing farm land than for other types of non – real estate assets.

ECONOMICS OF LEASING

In this section we demonstrate procedures for evaluating the economics of leasing for both the lessee and the lessor, using hypothetical numerical examples. For the lessee, leasing is contrasted to ownership, based on intermediate-term financing. For the lessor, leasing is contrasted to other forms of investment, with evaluation based on capital budgeting procedures. In both analyses, present value methods are used in order to account for leasing's effects on the timing and magnitude of cash flows.

Lease Versus Credit Purchase: Lessee's Position

Consider a farmer who has made the decision to acquire the services of a new tractor. The farmer's main concern at this point is choosing the best financing method. The financing alternatives are numerous. The tractor services could be hired on a custom basis, rented each year on an operating lease, rented on a capital lease with subsequent purchase, or purchased now. If the tractor is purchased, numerous loan sources are also available.

We will assume, however, that the farmer has narrowed the financing choices to two alternatives: (1) a capital lease from a local machinery dealer followed by a purchase when the lease expires and (2) a purchase financed by an intermediate-term loan from a local financial institution. The preferred choice is based on the lower-cost method of financing. Thus, the appropriate "costs" must be identified and measured over the planning period. This necessitates projecting the cash flows associated with the two financing methods.

The computational procedures are simplified by eliminating all the cash flow elements that are similar for the two alternatives. In this case, machinery operating costs, property taxes, maintenance, and labor are the same for the farmer whether the tractor is owned or leased. So, they can be ignored in the comparison. Only the cash flows that differ between the two financing methods are considered.

These cash flows are projected over the life of the transactions and discounted to their present values. The financing method with the lower present value of net cash outflows is the preferred choice. As the following example will show, the income tax consequences of the lease versus purchase decision are very important.

The relevant data associated with the tractor acquisition are as follows: The tractor is valued at $50,000, and the farmer intends to keep it for seven years. The farmer is in the 30 percent tax bracket and has a 10 percent after-tax cost of capital. Variations in the tax rate and discount rate will be considered later.

The terms for purchasing the tractor with credit financing are a 25 percent downpayment with equally amortized annual payments over six years at an interest rate of 11 percent on the remaining loan balance. Thus, the downpayment is $12,500, leaving a loan balance of $37,500 and annual amortization payments of $8,864. Depreciation is based on the seven-year class of the Modified Accelerated Cost Recovery System (MACRS), with a shift from double-declining balance to straight-line in the fourth period to maximize the allowable depreciation deductions (see "Accounting for Income Tax Obligations" in Chapter 3). The residual market value projected for the tractor is $2,000.

The terms of the capital lease include a seven-year contract, with annual rental payments of $9,600 due at the beginning of each year. No purchase of the tractor is specified at the end of the lease contract, implying that the farmer would renew the lease to maintain the services of a new tractor. Net cash flows associated with the lease and purchase are projected on an after-tax basis in Table 13.2.

Net cash outflows for leasing in column c consist of the seven annual payments less the tax credit (or savings) at 30 percent of the rental payment, lagged one year to reflect the farmer's practice of accounting for tax obligations in the year following the transactions. The tax credit is determined by multiplying the tax-deductible expenses for each alternative by the relevant tax rate. The tax-deductible expenses reduce taxable income, thereby reducing the cash outflow for taxes. Since the table expresses net cash outflows, the entries in parentheses in columns c and i represent cash inflows.

The net cash flow for the credit purchase is based on entries in columns d through h in Table 13.2. Entries in column d indicate the 25 percent downpayment in year 0 and the principal portions of the amortized loan payments for years 1 through 6. The interest portion of the amortized payments is shown in column e. Annual depreciation is in column f, and the residual value is in column g. The series of tax credits in column h represents the sum of interest (column e) plus depreciation (column f), multiplied by the appropriate tax rate (30 percent), and lagged one year. In addition, the tax effect of the residual

Table 13.2. Net Cash Flows for Lease Versus Purchase of a $50,000 Tractor

Year	Capital Lease			Purchase					
	Rent	Tax Credit	Net Cash Outflow (a – b)	Down-payment and Principal	Interest	Depreci-ation	Residual Value	Tax Credit	Net Cash Outflow (d + e – g – h)
	(a)	(b)	(c)	(d)	(e)	(f)	(g)	(h)	(i)
0	$9,600		$9,600	$12,500		$5,355			$12,500
1	9,600	$2,880	6,720	4,739	$4,125	9,565		$1,607	7,258
2	9,600	2,880	6,720	5,260	3,604	7,515		4,107	4,757
3	9,600	2,880	6,720	6,839	3,025	6,125		3,336	5,529
4	9,600	2,880	6,720	6,480	2,383	6,125		2,745	6,119
5	9,600	2,880	6,720	7,194	1,670	6,125		2,552	6,312
6	9,600	2,880	6,720	7,986	878	6,125		2,338	6,626
7		2,880	(2,880)			3,065	$2,000	2,101	(4,101)
8								320	(320)

value is reflected in year 8, along with the final half-year of depreciation. Thus, the tax credit for the end of year 1 of $1,607 is 30 percent of the first-year depreciation of $5,355, which is shown as a year 0 entry in the table. The tax credit of $4,107 for year 2 is 30 percent of the sum of interest ($4,125) plus depreciation ($9,565) for year 1. These specifications place the timing of the tax effects for leasing and purchase on the same basis.

The series of net cash outflows in column i is the sum of the entries in columns d plus e minus g minus h. The irregular pattern of the net cash flows reflects the combined effects of the financing terms and the tax implications of interest, depreciation, and the residual value. Clearly, these nonuniform patterns in the timing and magnitude of the cash flows for both the purchase and the lease require present value methods for accurate evaluation.

Table 13.3 shows the present values of the net cash flows for the lease and purchase, based on an after-tax discount rate of 10 percent. The net present values are $37,388 for the capital lease and $36,712 for the credit purchase. Thus, the credit purchase is preferred, because it provides a lower present value of the net cash outflows.

Table 13.4 shows the present values of net cash outflows and the resulting net advantage to the purchase for a range of values of the discount rates and the tax rates. For these data, the purchase alternative is preferred for lower

Table 13.3. Present Value of Net Cash Flows
(10 Percent Discount Rate)

	Capital Lease			Purchase		
Year	Net Cash Outflow	Present Value Factor $[1 \div (1 + i)^n]$	Present Value	Net Cash Outflow	Present Value Factor $[1 \div (1 + i)^n]$	Present Value
0	$9,600	1.0000	$ 9,600	$12,500	1.0000	$12,500
1	6,720	0.9091	6,109	7,258	0.9091	6,598
2	6,720	0.8264	5,553	4,757	0.8264	3,932
3	6,720	0.7513	5,049	5,529	0.7513	4,154
4	6,720	0.6830	4,590	6,119	0.6830	4,179
5	6,720	0.6209	4,173	6,312	0.6209	3,919
6	6,720	0.5645	3,793	6,526	0.5645	3,684
7	(2,880)	0.5132	(1,478)	(4,101)	0.5132	(2,104)
8				(320)	0.4665	(149)
Total			$37,388			$36,712

Table 13.4. Present Values of Lease Versus Purchase

Discount Rate	0 Percent		5 Percent		10 Percent	
	Lease	Purchase	Lease	Purchase	Lease	Purchase
0 percent	67,200	63,685	63,840	60,500	60,480	57,316
5 percent	58,327	56,070	55,549	53,429	52,772	50,787
10 percent	51,411	50,079	49,074	47,851	46,737	45,623
15 percent	45,931	45,294	43,934	43,388	41,937	41,481
20 percent	41,525	41,420	39,795	39,767	38,064	38,115
25 percent	37,934	38,242	36,416	36,794	34,899	35,346
30 percent	34,970	35,607	33,625	34,325	32,280	33,043

Net Advantage

Discount Rate			
0 percent	3,515	3,340	3,164
5 percent	2,257	2,120	1,985
10 percent	1,332	1,223	1,114
15 percent	637	546	456
20 percent	105	28	(51)
25 percent	(308)	(378)	(447)
30 percent	(637)	(700)	(763)

discount rates and lower tax rates, although the patterns of change as the discount rates and tax rates increase are not unambiguous.

First, consider increases in the discount rate for a given tax rate. In all cases, higher discount rates yield a lower advantage to the purchase alternative, with leasing eventually becoming the lower-cost financing method at a relatively high discount rate. To illustrate, for the 30 percent tax rate, the purchase alternative yields the lower present value of net cash flows up to the 25 percent discount rate, where the preference shifts to leasing.

Next, consider increases in the tax rate. For all discount rates (e.g., 0 percent, 5 percent, 10 percent), increases in the tax rates reduce the advantage of purchasing, even when leasing is the lower-cost method of financing.

These results clearly indicate the close and sensitive relationships of the financing methods to changes in discount rates and tax rates, given the irregular patterns of the after-tax net cash flows. In general, higher discount rates should favor those financing methods with larger net cash outflows occurring later in the payment period. But, the irregularities caused by the time patterns of the cash flows make it difficult to draw general results from this example.

We must stress that the emphasis here is on the methods of evaluating lease versus purchase decisions. These numerical results are specific to the

for Different Tax and Discount Rates (in Dollars)

15 Percent		20 Percent		30 Percent	
Lease	**Purchase**	**Lease**	**Purchase**	**Lease**	**Purchase**
57,120	54,132	53,760	50,948	47,040	44,579
49,994	48,146	47,217	45,504	41,662	40,221
44,400	43,395	42,063	41,167	37,389	36,712
39,940	39,575	37,943	37,668	33,949	33,855
36,334	36,462	34,604	34,810	31,144	31,505
33,382	33,898	31,864	32,449	28,830	29,553
30,935	31,761	29,950	30,479	26,900	27,916

to Purchase

2,988	2,812	2,461
1,848	1,713	1,441
1,005	896	677
365	275	94
(128)	(206)	(361)
(516)	(585)	(723)
(826)	(889)	(1,016)

terms and arbitrary specifications of the tractor example. Other terms and investment situations would yield different results. Moreover, changes in tax laws may shift the economic balance of leasing versus purchase over time; a recent example is the 1986 tax legislation in the United States that repealed the investment tax credit and changed depreciation procedures, thus affecting the tax position of both lessors and lessees. In any case, once the necessary data are known, lease and purchase alternatives can be evaluated based on the present values of the net cash flows.

Leasing as an Investment: The Lessor's Position

In this section we consider the profitability of capital leases from the lessor's standpoint. Leasing is an income-generating activity for the lessor; thus, the analytical approach uses capital budgeting procedures to estimate a net present value (NPV) or an internal rate-of-return (IRR) for leasing. The results can be used to accept or reject leasing as an investment activity or to rank its profitability relative to other investments. The budgeting procedures are also helpful in evaluating changes in terms (e.g., rental payments, residual values) on leases that will enhance the lessor's competitive position while also improving the lessee's cash flow.

To illustrate the analytical methods, consider the case of a large commercial bank that is contemplating the adoption of a leasing program for its agricultural customers in addition to the lending program that it presently offers. The bank's task is to establish a basis for comparing the profitability of leasing to lending. The comparison is based on the following terms and assumptions for capital leasing and intermediate-term lending.

- *Lending:* A $1 million term lending program with repayments based on equal annual amortized payments over seven years at an interest rate of 10 percent. Assuming the bank's costs of funds and operations total 7 percent per dollar of loan, the bank's debt-to-equity ratio is 10.0, and its corporate income tax rate is 35 percent, the after-tax rate-of-return on equity from lending is 21.45 percent.[2]

---

[2]The net yield on lending is the 10 percent loan rate minus the 7 percent lending costs, or 3 percent. The after-tax net yield is 3 percent times 1 minus the tax rate (1 – 0.35), or 1.95 percent. Multiplying the after-tax net yield by the ratio of assets to equity (1 plus the debt-to-equity ratio) gives the after-tax rate-of-return on equity capital. In this case, the after-tax return on equity is 1.95 percent times 11, or 21.45 percent.

- *Leasing:* A $1 million leasing program with equal rental payments due at the end of each of the next seven years. Rental payments are set at $190,000. The asset is fully depreciated over seven years. Depreciation allowances are figured according to the MACRS, beginning with 150 percent declining balance and then shifting to straight-line according to the allowable tax provisions. The estimated residual value is 15 percent of the initial asset cost, or $150,000, which is fully taxable at the 35 percent tax rate. Using the 7 percent cost of funds and operations as the cost of debt capital and the 21.45 percent rate-of-return to equity from lending as the cost of equity, the weighted average after-tax cost of debt and equity capital to use in evaluating the leasing program is:

$$i_a = (i_e)(E \div A) + (1 - t)(i_d)(D \div A)$$

$$= (0.2145)(1 \div 11) + (1 - 0.35)(0.07)(10 \div 11)$$

$$= 0.0609, \text{ or } 6.09\%$$

where the weights on equity and debt reflect the bank's debt-to-equity ratio of 10.0.

The net present value procedures could be illustrated with a worksheet, as in the preceding section, that identifies the various sources of cash flow. In this example, however, the capital budgeting procedures are expressed in a present value model to show an alternative approach to the analysis. The present value model for the capital lease is:

$$NPV = -INV + (1 - t)A[USPV_{i, n}] + \sum_{n=1}^{N} \frac{tD}{(1 + i)^n} + \frac{(1 - t)V_N}{(1 + i)^N} \qquad (13.1)$$

The variables are:

i   = The discount rate, measured here as the 6.09 percent after-tax weighted average cost of debt and equity capital.

N   = The length of the lease contract, or 7 years.

A   = The annual rental payments, or $190,000.

V_N = The residual value of the leased assets, or $150,000.

t = The income tax rate, or 35 percent.

INV = The asset's initial cost, or $1 million.

D = The annual depreciation, projected as:

Year	
1	$107,100
2	191,300
3	150,300
4	122,500
5	122,500
6	122,500
7	122,500
8	61,300

The values of these variables are included in equation 13.1 in order to determine the net present value. Table 13.5 summarizes the calculation procedures. The net present value at the bottom of the table is $28,097. It is composed of the sum of the present values for the after-tax rental payments, the depreciation tax credit, and the residual value minus the initial investment.

The positive net present value indicates that the capital lease program is marginally profitable, relative to the alternative of intermediate-term lending, given these particular data specifications. A similar result occurs when the internal rate-of-return is found — the discount rate that yields a zero net present value in equation 13.1. The internal rate-of-return on the capital lease is 6.83 percent. Thus, compared to a 6.09 percent weighted average cost of capital for intermediate-term lending, the capital lease again is of slightly higher profitability according to the internal rate-of-return criterion.

Table 13.6 indicates the sensitivity of the net present value for capital leasing to a range of values for the discount rate and the income tax rate. The wide range of profitability measures in the table shows the importance of conducting this type of analysis.

The net present value model can also be used to evaluate how alternative leasing terms affect the lessor's profit position. Suppose, for example, that the commercial bank is willing to set the seven rental payments at a level that

Table 13.5. Net Present Value of Capital Leasing
for the Lessor

Item		Present Value
1. Initial investment	INV =	$-1,000,000
2. Rental payments	$(1 - t)A[USPV_{i, N}]$	
	$(1 - 0.35)(190,000)[USPV_{0.0609, 7}] =$	687,224
3. Depreciation tax credit	$\sum_{n = 1}^{N} \dfrac{tD}{(1 + i)^n}$	
	Year 1: $(0.35)(107,100)(1.0609)^{-1} = 35,333$	
	Year 2: $(0.35)(191,300)(1.0609)^{-2} = 59,489$	
	Year 3: $(0.35)(150,300)(1.0609)^{-3} = 44,056$	
	Year 4: $(0.35)(122,500)(1.0609)^{-4} = 33,846$	
	Year 5: $(0.35)(122,500)(1.0609)^{-5} = 31,903$	
	Year 6: $(0.35)(122,500)(1.0609)^{-6} = 30,072$	
	Year 7: $(0.35)(122,500)(1.0609)^{-7} = 28,345$	
	Year 8: $(0.35)(61,300)(1.0609)^{-8} = 13,370$	
	TOTAL $= \$276,414$	276,414
4. Residual value	$\dfrac{(1 - t) V_N}{(1 + i)^N}$	
	$\dfrac{(1 - 0.35)(150,000)}{(1.0609)^7} =$	64,459
5. Net present value		$ 28,097

Table 13.6. Net Present Value of Capital Leasing
Under Different Discount and Tax Rates

Discount Rate	Tax Rates				
	0 Percent	10 Percent	20 Percent	30 Percent	35 Percent
0 percent	$480,000	$432,000	$384,000	$336,000	$312,000
5 percent	206,013	167,619	129,225	90,831	71,634
7.5 percent	96,767	62,170	27,572	(7,025)	(24,324)
10 percent	1,973	(29,352)	(60,677)	(92,001)	(107,664)
15 percent	(153,130)	(179,157)	(205,184)	(231,211)	(244,224)
20 percent	(273,265)	(295,254)	(317,242)	(339,230)	(350,224)

yields the same after-tax rate-of-return as intermediate-term lending; that is, 6.09 percent. This is accomplished by finding the level of rental payments (A) in equation 13.1 that yields NPV = 0, assuming that the values of all other variables remain the same. For the data in Table 13.5, the present value of rental payments (entry 2) would have to be $659,127 to yield an NPV of 0. The after-tax value of A is then:

$$A = \frac{659,127}{USPV_{0.0609, \, 7}} = 118,450$$

Dividing A by (1 – t) yields a before-tax rental payment of $182,232.

Similar procedures may be followed to determine the rental payments needed to yield a 6.09 percent return when a 10 percent residual value is projected for the leasing assets. In this case, the before-tax rental payments are $188,172. Finally, assume in the original specification of the problem that the lessor requires the rental payments at the beginning of the year instead of at year end. Confirm in Table 13.5 that these advance payments would increase the present value of the rental payments to $729,076 and the lessor's net present value to $69,949.

Thus, the effects of numerous changes in leasing terms may be evaluated in order to find mutually acceptable arrangements for both lessor and lessee, based on their cash flow and tax positions.

Summary

Leasing of non – real estate assets in agriculture, while less common than in other industries, offers an alternative means of financing asset control. Leasing programs primarily are offered by manufacturers and independent dealers and by major financial institutions. The basic method of evaluating lease versus purchase decisions is based on present value methods in which the preferred choice has the lowest net present value of cash outflows. Leasing also has important effects on a firm's borrowing capacity, financial statements, liquidity, and tax position.

Other specific highlights are as follows:

- An **operating lease** is *usually a short-term rental arrangement in which the rental charge is calculated on a time basis.*

- **Custom hiring** is *a form of leasing that combines the hiring of labor with the use of a tangible asset.*

- The **capital lease** is *a long-term leasing contract.*

- **Leveraged leasing** is characterized by the *formal involvement of a lender in providing debt capital to finance the lessor's purchase of the leased asset.*

- Among the potential advantages of capital leasing are the conservation of working capital, nearly 100 percent financing, the use of modern equipment and facilities, and possible tax benefits for both contracting parties.

- In principle, the effects of leasing transactions should be reflected on each of the major financial statements of a firm. In practice, however, leasing often is considered an off – balance sheet form of financing for farm businesses, with only rental payments appearing as an expense on the income statement.

- Lease – purchase analysis often is strongly sensitive to changes in discount rates and tax rates, given irregularities in the time patterns of after-tax net cash flow for these financing methods.

Topics for Discussion

1. Distinguish between operating leases and capital leases. How do they represent methods of financing? How is custom hiring distinguished? Explain the concept of leveraged leasing.

2. Critically evaluate the proposition that "leasing provides 100 percent financing."

3. Identify the effects of leasing on a farm's borrowing capacity.

4. Critically evaluate the proposition that "leasing is an off – balance sheet form of financing."

5. A farmer leases a tractor for five years with annual rental payments of $11,225 due at the end of each year. Using a 15 percent interest rate, show the impacts for year 1 on the farmer's balance sheet and income statement.

6. Identify the guidelines to qualify a transaction as a lease for tax purposes. What additional features characterize a finance lease?

7. How do lessors evaluate the lease worthiness of lessees? Contrast this process with the lenders' evaluation of the credit worthiness of borrowers.

8. Compare a capital lease with a credit purchase for a $150,000 asset, depreciable over five years, with a zero residual value. Lease payments are $37,000, due at the beginning of each of the next five years. Credit terms are 25 percent downpayment and a loan payable in equal payments of principal and interest over five years at

14 percent interest. The investor's tax rate is 30 percent, and the after-tax cost of capital is 10 percent. Investment tax credit and the MACRS apply.

9. A Farm Credit System association is developing a capital leasing program for its farm customers, along with its current lending program. The lending terms on farm tractors are a five-year loan at 13 percent interest, with equally amortized payments. Assuming the tax rate is 34 percent, find the annual rent the Farm Credit System would charge on the lease to give the same yield as the loan rate. Assume a 10 percent residual value on tractors, a five-year term, lease payments at year end, full use of ACRS, an FCS equity-to-asset ratio of 0.15, and a 7 percent cost of funds. Figure on a per dollar of original tractor value.

How does this rent compare with the farmer's cash payments for a credit purchase? What are the implications for capital leasing by the Farm Credit System and other lenders?

References

1. Barnes. G. D., "An Ag Lender Looks at Leasing," *AgriFinance*, 23(1980):40 – 48.

2. Brealey, R., and S. C. Myers, *Principles of Corporate Finance*, 4th ed., McGraw-Hill, Inc., New York, 1991.

3. Brigham. E. F., and L. Gapenski, *Financial Management: Theory and Practice*, 6th ed., Dryden Press, Hinsdale, Illinois, 1991.

4. Crawford, P. J., C. P. Harper, and J. McConnell, "Further Evidence on the Terms of Financial Leases," *Financial Management*, 10:3(1981):7 – 14.

5. LaDue, E. L., "Financial Leasing of Dairy Cattle," *North Central Journal of Agricultural Economics*, 5:2(July 1983):55 – 64.

6. Levy, H., and M. Sarnat, *Capital Investment and Financial Decisions*, 3rd ed., Prentice-Hall, Inc., Englewood Cliffs, New Jersey, 1986.

7. Pritchard, R. E., and T. Hindelang, *The Lease – Buy Decision*, AMACOM, American Management Associations, New York, 1980.

8. Sorenson, J. W., and R. E. Johnson, "Equipment Financial Leasing Practices and Costs: An Empirical Study," *Financial Management*, 6:1(1977):33 – 40.

9. Van Horne, J. C., *Financial Management and Policy*, 8th ed., Prentice-Hall, Inc., Englewood Cliffs, New Jersey, 1989.

14

Costs of
Financial Capital

In this chapter, procedures for measuring a firm's costs of debt and equity capital are developed. In previous chapters, we primarily took these costs as given, in order to evaluate their effects on a firm's profitability, leverage, financial growth, capital budgeting, and financial feasibility. However, in financial planning and investment analysis, these costs must be determined from available data and sound judgment about future conditions.

In the following sections, the costs of debt capital are estimated under different financing terms and from different lenders, based on present value analysis. Costs of equity capital are distinguished on the basis of retained earnings and new outside equity brought into a farm operation. Effects of taxes, inflation, and nonmoney costs are also considered in evaluating a firm's average cost of capital.

MONEY COSTS OF
DEBT CAPITAL

Costs of debt capital include money costs to meet interest payments and other noninterest obligations and nonmoney costs associated with financial risks. These costs are generally measured on an after-tax basis with appropriate allowance for inflation. We begin with the money cost of borrowing.

The **money cost of borrowing** is *the difference between the total money that the borrower receives from the loan* (sometimes different from the total loan) *and the total money paid to retire the loan.* It is composed, primarily, of interest money costs, but it often includes noninterest money costs as well. Both interest and noninterest money costs are included when the percentage, or "rate," cost of borrowed capital is computed.

A discussion of procedures for determining interest rates is complicated by the different methods used to compute interest costs and by the differences in terminology used to identify different kinds of interest rates — the *simple* rate, the *compound* rate, the *real* rate, the *nominal* rate, the *actuarial* rate, the *annual percentage* rate, the *effective* rate, the *contractual* rate, and others. Hence, it is generally not wise to base a financing decision solely on an interest rate that is stated on a note.

The Actuarial Interest Rate

Our approach is to develop a basic model for determining money costs of debt capital that may be used for any kind of financing transaction — simple or complex. In our approach, we use the notion of compound interest developed in Chapter 9, together with the present value models, to determine an interest rate per conversion period on a loan transaction. This interest rate per conversion period is called the **actuarial interest rate.** The actuarial rate is *the interest rate, or discount rate, that equates to zero the sum of the present values of all cash flows associated with the loan transaction.* Or, expressed in another way, it is the interest or discount rate that equates the present value of all cash inflows associated with the loan transaction to the present value of all cash outflows.

The concept of the actuarial rate is similar to the definition of the internal rate-of-return discussed in Chapter 10. Only the application is different. The internal rate-of-return is used to evaluate investment profitability. The actuarial interest rate is used to evaluate money costs of debt capital. Each reflects the interest rate, yielding a zero net present value for a series of payments.

To demonstrate the procedures for calculating an actuarial interest rate, let's identify the appropriate present value model, keeping in mind that we seek a net present value of zero. For a loan with a single payment of principal and interest due at the end of N periods, the model is:

$$0 = +V_0 - \frac{P_N}{(1 + i)^N}$$

(14.1)

where V_0 is the cash proceeds of the loan received by the borrower, P_N is the repayment of principal and interest, and i is the actuarial interest rate. Notice that the signs of the variables are different from those used in Chapter 10 for analyzing investment profitability. Now V_0 has a positive sign, reflecting the cash inflow of the loan proceeds to the borrower, while the later payment (P_N) of principal and interest has a negative sign, signifying the cash outflow.

To illustrate, let's find the actuarial rate of interest on $1,000 borrowed on January 1 and repaid a year later with a payment of $1,120. It is:

$$0 = 1,000 - \frac{1,120}{1 + i} \tag{14.2}$$

$$i = \frac{1,120}{1,000} - 1 = 0.12, \text{ or } 12\%$$

Had the loan been repaid after two years with a payment of $1,254.40, the actuarial rate would be:

$$0 = 1,000 - \frac{1,254.40}{(1 + i)^2} \tag{14.3}$$

$$(1 + i)^2 = 1.2544$$

$$i = 0.12, \text{ or } 12\%$$

From Appendix Table II, an interest rate of 12 percent is found to be associated with N = 2 and a conversion factor of 1.2544. Hence, the actuarial interest rate is 12 percent per conversion period.

For an installment loan in which a series of payments of principal, interest, or both are made to fully repay the loan, the appropriate present value models are the nonuniform series if the payments differ between the periods:

$$0 = V_0 - \sum_{n-01}^{N} \frac{P_N}{(1 + i)^n} \tag{14.4}$$

or the uniform series if the payments are the same in each period:

$$0 = V_0 - A[\text{USPV}_{i, N}] \tag{14.5}$$

In each case, the object is to find the value of i that equates the sum of the present value of all the cash flows to zero.

To illustrate, let's find the actuarial rate of interest on a $1,000 loan that is repaid in four annual payments of $329.23 each. The total payments over the four years are ($329.23)(4) = $1,316.92. Hence, interest paid over the four years is $316.92. Since the annual payments are constant and the conversion periods are equal in length, we can use the uniform series model, which is specified as:

$$0 = 1,000 - 329.23 \, [\text{USVP}_{i,\,4}] \tag{14.6}$$

We must now determine the value of i, which, when substituted into equation 14.6, will equate the sum of the cash flows to zero. First, solve for the value of the conversion factor for $[\text{USPV}_{i,\,4}]$:

$$[\text{USPV}_{i,\,4}] = \frac{1,000}{329.23} = 3.0373$$

Then, look in Appendix Table V for the rate of interest that is associated with N = 4 and a conversion factor of 3.0373. In this case, the interest rate is found to be 0.12, or 12 percent. Hence, the actuarial interest rate on the loan is 12 percent. Of course, the problem could be solved with a financial calculator, but using the tables is beneficial as an instructional tool.

This kind of loan is called a fully amortized or **constant-payment loan,** because *equal payments composed of principal and interest are used to fully repay the loan and interest during its period to maturity.* It is commonly used on farm real estate loans and on consumer installment loans. Another repayment pattern is the **constant – payment-on-principal method.** This method requires *equal principal payments in each period, with interest paid on the remaining loan balance.* Because the interest payments are smaller in each successive period, the total payments of principal and interest decline over the period of the loan.

Suppose, for example, that the $1,000 loan is to be repaid in four annual principal payments of $250 each, plus interest in each year. Total payments are $370 at the end of year 1, $340 in year 2, $310 in year 3, and $280 in year 4. The $370 payment in year 1 is the $250 principal plus 12 percent interest times the $1,000 loan balance. To find the actuarial interest rate, use equation 14.4 to formulate the present value model as:

$$0 = 1{,}000 - \frac{370}{1 + i} - \frac{340}{(1 + i)^2} - \frac{310}{(1 + i)^3} - \frac{280}{(1 + i)^4} \qquad (14.7)$$

Again determine the value of i that, when substituted into equation 14.7, will equate the sum of the cash flows to zero. Because the cash flows are nonuniform, you must use Appendix Table III and follow the trial-and-error method.

For i = 0.11, or 11 percent, the net present value of equation 14.7 is –20.40. Hence, the interest rate must be increased to yield NPV = 0. In this case, an interest rate of 12 percent yields NPV = 0. Hence, the actuarial interest rate for this loan is 12 percent.

The Annual Percentage Rate (APR)

The annual percentage rate (APR) is found by expressing the actuarial interest rate on an *annual* basis. That is, the **annual percentage rate** is *the interest rate per year.* Clearly, the annual percentage rate and the actuarial rate are identical when the conversion periods are each one year long. This was the case in the previous examples. Hence, for the $1,000 loan to be repaid in four annual installments of $329.23 each, the actuarial rate of 12 percent is also a 12 percent annual percentage rate.

Differences between the actuarial rate and the annual percentage rate occur when each conversion period is not one year in length. As an example, monthly payments on a loan create 12 conversion periods within a year. Similarly, quarterly payments on a loan create four conversion periods within a year. Under these circumstances, compounding of loan interest happens more frequently than once a year, and the actuarial interest rate must be converted to an annual percentage rate.

To illustrate, consider the case of a loan for $1,000 to be repaid in four quarterly payments of $269.03 each. Again, present value equation 14.4 or 14.5 can be used. However, since compounding of loan interest occurs more frequently than once a year, we utilize equation 9.3, $V_N = V_0(1 + i \div m)^{mN}$, to reflect i as an annual percentage rate. In this case, the actuarial rate is expressed as i ÷ m, where m = 4 due to the quarterly payments. Multiplying the actuarial rate by m = 4 will yield the annual percentage rate.

The problem is set up as follows:

$$0 = 1,000 - \frac{269.03}{1 + i \div 4} - \frac{269.03}{(1 + i \div 4)^2} - \frac{269.03}{(1 + i \div 4)^3} - \frac{269.03}{(1 + 1 \div 4)^4} \quad \text{(14.8)}$$

or, since the payments are a uniform series:

$$0 = 1,000 - 269.03[\text{USPV}_{i \div 4, 4}]$$

Solve for the conversion factor:

$$[\text{USPV}_{i + 4, 4}] = \frac{1,000}{269.03} = 3.7171$$

Then, find the interest rate (i ÷ 4) in Appendix Table V that is associated with a conversion factor of 3.7171 and N = 4. The conversion factor for 3 percent is 3.7171.

Hence, the actuarial interest rate (i ÷ 4) for this loan is 3 percent per conversion period. Since each cash conversion period is one-fourth of a year, the actuarial rate is multiplied by four to yield the annual percentage rate. The result is (3)(4) = 12 percent.

Truth in Lending and Annual Percentage Rate

The procedures for calculating and comparing interest costs and rates on financing transactions have always been perplexing to many consumers and businesspersons. Often, the lender's quote of a contract interest rate differs considerably from the annual percentage rate, due to different methods of charging interest or to noninterest costs. As a result, federal legislation has been passed in the United States requiring lenders to inform borrowers precisely and explicitly of the total amount of the finance charge that they must pay and the annual percentage rate of interest to the nearest 0.1 percent. The purpose is to help borrowers have a better understanding of market rates and to reduce the likelihood of fraudulent practices.

Truth-in-lending compliance is required of creditors who (1) regularly make loans in the ordinary course of business; (2) levy a finance charge for more than four installment payments; or (3) make loans to natural persons, who use the proceeds primarily for personal, family, household, or agricultural purposes. Excluded transactions in the early legislation included loans for commercial or business purposes; loans to partnerships, corporations, cooperatives, and organizations; and loans greater than $25,000, except for real estate loans, where compliance is required irrespective of the amount.

The inclusion of interest disclosure on agricultural loans and not on commercial or business loans concerned many lenders who felt that these loans should receive comparable treatment. As a result, all agricultural loans were exempt from truth-in-lending requirements. Hence, for most agricultural operations that use large amounts of short- and intermediate-term debt, disclosure of annual percentage rates and finance charges is no longer required.

Lenders who comply with truth in lending must report an annual percentage rate, which is based on the logic and method of the actuarial rate and compound interest as previously outlined. The simplifying differences in truth in lending are that the lender need only report the rate to the nearest 0.1 percent and that the lender may use tables or computerized techniques in its estimation.

The Effective Interest Rate

As indicated in the preceding sections, the actuarial rate is the interest rate per conversion period, while the annual percentage rate is the actuarial rate expressed on an annual basis. An alternative annual rate is the **annual effective rate,** i_e, *which directly accounts for the compounding effects over the number of conversion periods within a year.* It is possible, for example, for two loans to have the same annual percentage rates, yet represent different annual costs of borrowing because of differences in the number of compounding periods. Under these conditions, the annual borrowing costs for the two loans can be compared in terms of their annual effective interest rates.

The annual effective interest rate is calculated by compounding the actuarial rate over the number of conversion periods within a year. The process is represented as follows:

$$i_e = [(1 + i \div m)^m - 1]$$

(14.9)

where i_e is the annual effective interest rate, i is the annual percentage rate, i ÷ m is the actuarial rate, and m is the number of compounding periods (see equation 9.3). Thus, the three interest rate measures — actuarial, annual, and effective — are mathematically related to one another.

To illustrate, consider two loans each having an annual percentage rate of i = 0.12, or 12 percent, but loan A is repaid quarterly (m = 4), and loan B is repaid monthly (m = 12). The actuarial rate for loan A is i ÷ m = 0.12 ÷ 4 = 0.03, or 3 percent, and for loan B it is i ÷ m = 0.12 ÷ 12 = 0.01, or 1 percent.

The annual effective interest rates are:

$$\text{Loan A: } i_e = \left(1 + \frac{0.12}{4}\right)^4 - 1 = 0.1255, \text{ or } 12.55\%$$

$$\text{Loan B: } i_e = \left(1 + \frac{0.12}{12}\right)^{12} - 1 = 0.1268, \text{ or } 12.68\%$$

Thus, loan A has a lower annual effective interest rate than loan B, even though their annual percentage rates are the same and B has the lower actuarial rate.

In general, as the frequency of compounding periods increases, the difference between the annual percentage rate and the annual effective rate also increases, although at a diminishing rate. In addition, as the actuarial rate increases, for a given number of conversion periods, so does the difference between the annual percentage rate and the effective rate. Finally, when the conversion or compound period is one year in length, the actuarial rate, the annual percentage rate, and the annual effective rate are the same. Table 14.1 shows how effective interest rates vary across different conversion periods and annual percentage rates.

The Contractual Interest Rate

At this point you should be familiar with the concepts of the actuarial rate, the annual percentage rate, and the annual effective rate. Now we introduce the contractual rate, which may or may not be identical to the actuarial rate, the annual percentage rate, or the effective rate. In simplest terms, the **contractual interest rate** is *the interest rate stated by the lender on the note* — 8 percent, 10 percent, 12 percent, etc. In the first example given earlier, in which a $1,000 loan was repaid in a single payment of $1,120 one year later, the actuarial rate, the effective rate, and the annual percentage rate were 12 percent. The contractual rate is also 12 percent per year. It would appear on the note, signifying

Table 14.1. Effective Interest Rates Across Different Conversion Periods and Annual Percentage Rates

Annual Percentage Rates (i ÷ m)	Conversion Periods (m)					
	1	**2**	**4**	**6**	**12**	**365**
	- - - - - - - - - - - - - - - - - *(Effective interest rates [i_e])* - - - - - - - - - - - - - - - - -					
APR = 6%	6.00%	6.09%	6.14%	6.15%	6.17%	6.18%
APR = 8%	8.00%	8.16%	8.24%	8.27%	8.30%	8.33%
APR = 10%	10.00%	10.25%	10.38%	10.43%	10.47%	10.52%
APR = 12%	12.00%	12.36%	12.55%	12.62%	12.68%	12.75%
APR = 14%	14.00%	14.49%	14.75%	14.84%	14.93%	15.02%

that the borrower promises to repay the loan in one year with interest calculated at a 12 percent annual rate.

The actuarial rate, the annual percentage rate, and the effective rate may differ from the contractual rate when noninterest money costs are part of the loan transaction or when different methods of calculating interest are used. Recall that the actuarial rate is defined as the rate of interest equating the sum of the present value of all cash flows associated with the loan transaction to zero. Cash flows for the loan principal, for interest, and for noninterest costs included in the loan payments are part of the actuarial rate. Hence, the actuarial rate has a broader meaning than the contractual rate.

NONINTEREST MONEY COSTS

A variety of noninterest money costs are often associated with loans. Although there are no fixed rules on which costs to include in deriving the actuarial rate, some guidelines are offered in paragraph 226.4 of the Rules and Regulations of the Federal Reserve System pertaining to the Truth in Lending Act. With few exceptions, all charges that are payable directly or indirectly by the borrower and that are imposed directly or indirectly by the lender as incident to or as a condition of the extension of the loan should be included. In addition to interest, these charges consist of (1) service, transaction, activity, or carrying charges; (2) loan-fee points or finder's fees; (3) fees for investigations and credit reports (except those related to real property); and (4) premiums for special insurance required as a condition of the loan for either the lender or the borrower.

When itemized and disclosed to the customer, the following types of charges need not be included in the finance charge: (1) fees prescribed by law paid to public officials for perfecting or releasing any security (or insurance in lieu of perfecting any security interest if the premium does not exceed the latter cost); (2) license, certificate of title, and registration fees imposed by law; (3) taxes; and (4) delinquency, default, and reinstatement charges. In addition, the rules specifically exclude the following charges relating to real property transactions: (1) fees for title examination, title insurance, and similar purposes, and for property surveys; (2) fees for preparation and/or notarization of deeds and other documents; (3) required escrow payments for future payment of taxes, insurance, etc.; (4) appraisal fees; and (5) credit reports.

As a general rule, one should include all costs arising from borrowing. Moreover, as later examples will indicate, special requirements by different types of lenders, for example, compensating deposit balances often required by commercial banks as a loan condition and stock requirements associated with borrowing from the Farm Credit System, should be accounted for.

METHODS OF COMPUTING INTEREST

Computing interest rates is further complicated because lenders may use three methods of computing interest, and the interest cost may vary considerably according to the method being used. These methods are called the **remaining-balance, add-on,** and **discount** methods. The remaining-balance method is most commonly used, although the others, especially the add-on method, have considerable use as well. Each method yields a different time pattern of cash flows associated with the loan transaction, which in turn causes differences in the annual percentage rate.

Remaining-Balance Method

With the **remaining-balance method,** *interest is calculated by multiplying the principal outstanding by the contractual rate for the period in question.* The procedure is illustrated in the following formula, where P = loan principal, i_c = the contractual rate, and T = the length of period expressed as a fraction of a year. The interest cost (I) is:

$$I = (P)(i_c)(T) \tag{14.10}$$

For example, a $1,000 loan at a 12 percent annual rate for six months would carry an interest cost (I) of ($1,000)(0.12)(1 ÷ 2) = $60.

To illustrate further, let's return to the constant – payment-on-principal method of repayment indicated in equation 14.7. A $1,000 loan is being repaid in four annual principal payments of $250 each, plus interest at a 12 percent annual rate on the remaining balance. The interest payments in each of the four years are found as follows:

$$I_1 = (1,000)(0.12)(1) = 120$$

$$I_2 = (750)(0.12)(1) = 90$$

$$I_3 = (500)(0.12)(1) = 60$$

$$I_4 = (250)(0.12)(1) = 30$$

These interest payments are added to the $250 principal payments in each year to give total payments of $370, $340, $310, and $280 in each respective year.

The remaining-balance method may also be used in the equally amortized or constant-payment installment loan, although the procedures for calculating the interest paid in each installment are more detailed. To illustrate, consider the earlier example of a $1,000 loan to be repaid in four equal annual payments of principal and interest at a 12 percent annual rate calculated on the remaining-balance method. To find the annual payment, solve for the level of annuity (A) that will amortize a $1,000 loan over four periods at a 12 percent rate per period. The uniform series model is formulated and solved as follows:

$$1,000 = A[USPV_{0.12, 4}]$$

$$A = 329.23$$

The next step is to determine an amortization schedule that will show the breakdown of each payment into principal and interest. As indicated in Table 14.2, the simplest approach is to subtract the interest due in the respective year from the payment to determine the principal portion of the payment. The principal portion then is subtracted from the beginning of period loan balance to determine the end of period loan balance. The same procedure is repeated in subsequent years. Thus, the $329.33 payment in period 1 contains $120 of interest and $209.23 of principal.

Columns 4 and 5 in Table 14.2 show the interest and principal payments, respectively, while columns 2 and 6 show the beginning and remaining loan balances. Notice the pattern of interest and principal payments. As the remaining balance of the loan is reduced, the interest payment decreases. Consequently, a higher portion of each payment is principal, which reduces the loan balance. When the last payment is reached, very little is charged as interest.

These procedures for deriving an amortization schedule characterize any constant-payment installment loan with interest paid on the remaining balance. Procedures are the same whether the payments are made for annual, quarterly, monthly, or other periods. Amortization tables for different rates of interest and different time periods are readily available for the convenience of lenders or others who need this detailed information.

Table 14.2. An Amortization Schedule for a Fully Amortized Constant-Payment Loan

Period (T)	Balance Beginning of Period (P)	Payment (A)	Interest Portion [(2) × 12%]	Principal Portion [(3) – (4)]	Remaining Balance [(2) – (5)]
(1)	(2)	(3)	(4)	(5)	(6)
1	$1,000.00	$329.23	$120.00	$209.23	$790.77
2	790.77	329.23	94.89	234.34	556.43
3	556.43	329.23	66.77	262.46	293.97
4	293.97	329.25	35.28	293.97	0.00

Add-on Method

When lenders say that interest is charged at a 10 percent add-on rate, they mean that interest is calculated on the original loan for the entire period of the loan. The formula for calculating the total interest obligation (I_a) is:

$$I_a = (P)(I_c)(N) \tag{14.11}$$

where P is the loan principal, i_c is the contractual interest rate per period, and N is the number of periods. The sum of total interest and principal is divided by the number of payments to obtain the amount of each installment.

To illustrate, consider a $1,000 loan for four years with four annual payments, add-on interest, a 10 percent contractual rate, and no noninterest costs. The total interest obligation is:

$$I_a = (1,000)(0.10)(4) = 400$$

The total obligation of principal and interest is then $1,400, and each annual payment is $1,400 ÷ 4 = $350.

Using the present value model, find the actuarial rate for add-on interest as follows:

$$0 = \$1,000 - 350[USPV_{i,\ 4}] \tag{14.12}$$

and, using Appendix Table V, find that the actuarial interest rate is 14.96 percent. Since each conversion period is one year long, the annual percentage rate is also 14.96 percent. Hence, the 10 percent add-on rate with four annual installments is equivalent to a charge of 14.96 percent per year on the remaining loan balance.

The difference in rates occurs because the add-on method fails to account for the principal payments made on the loan prior to its maturity — payments that reduced the outstanding loan balance. The actuarial rate and the annual percentage rate for add-on interest can always be calculated by following the procedures described previously.

Discount Method

When money is borrowed as a "personal" loan, the discount method is sometimes used in computing interest. Here, the meaning of *discount* differs from its earlier use in net present value analysis. For discount loans, interest is calculated on the original amount of the loan for its full period, and this amount, plus any other loan costs, is subtracted from the amount of the loan at the beginning. The borrower receives the difference. Dividing the amount of the loan by the number of installment payments will give the size of each installment.

For the same contractual rate and length of time, the actuarial rate will be higher for the discount method than for the add-on method. This increase in the actuarial rate is because of the payment or discount of interest at the very beginning of the loan transaction. To determine the actuarial rate, as we have indicated before, first identify the timing and magnitude of all cash flows associated with the loan, and then determine the interest rate that makes the present value of the sum of all these cash flows equal to zero. In the preceding example, suppose the $400 interest obligation for the $1,000 loan was deducted immediately, leaving the borrower with $600 as loan proceeds to use. The loan is then repaid in four annual installments of $250 each. Again, using the present value model, find the actuarial rate as follows:

$$0 = 600 - 250[USPV_{i,4}]$$

The conversion factor is:

$$[USPV_{i,4}] = \frac{600}{250} = 2.400$$

and is associated with an actuarial interest rate of 24.10 percent, as determined in Appendix Table V. The annual percentage rate is also 24.10 percent. Clearly, for a given contractual rate, the annual percentage rate for the discount loan is greater than it is for the add-on loan, and the annual percentage rates for both of these methods are greater than the remaining-balance rate.

EFFECTS OF
BALANCE REQUIREMENTS

Commercial banks may require or recommend that borrowers hold part of their loans on deposit with the banks. This "compensating balance" is common in corporate financing by regional or money center banks. Some rural banks use the practice too. Moreover, even if deposit balances are not formally required, the balances provided by borrowers during the year are usually considered in loan pricing. A different, yet analogous, practice is the requirement by the Farm Credit System that its borrowers purchase stock in the system from the loan proceeds.

From the borrower's standpoint, these balance requirements increase the loan's annual percentage rate relative to the contractual rate. This increase occurs because the borrower forgoes use of part of the loan. To show the rate increase, consider a single-period loan for V_0 dollars at contractual rate i_c, with b percent of the loan held as a compensating balance. To solve for i (the APR), formulate the present value model as:

$$0 = (1 - b)V_0 - \frac{V_{0(1 + i_c)}}{1 + i} + \frac{bV_0}{1 + i} \tag{14.13}$$

The first term is the borrower's receipt of the loan, net of the balance requirement; the second term is the present value of the repayment of principal and interest one period later; and the third term is the later return of the balance to the borrower, discounted to a present value.

Solving equation 14.13 for i yields the following expression for the annual percentage rate:

$$i = \frac{i_c}{1 - b} \tag{14.14}$$

Thus, a balance requirement increases the borrower's annual percentage rate by the factor $1 \div (1 - b)$, relative to the contractual rate i_c. A loan with a contractual rate of 12 percent, subject to a 10 percent balance requirement, would have an annual percentage rate of 13.3 percent.

AN APPLICATION: DETERMINING THE MONEY COSTS OF DEBT CAPITAL

To illustrate the application of these procedures for determining borrowing costs, consider the case of Joseph and Julie Farmer who are purchasing new harvesting equipment valued at $50,000 and must determine the best source of financing. They have ruled out the manufacturer of the equipment as a financing source because of the high interest rates. The choice lies between a local office of the Farm Credit System (FCS) (Case 1) and a local bank (Case 2). The following financing terms are offered by each lender.

Terms	FCS	Bank
Downpayment (equity)	$10,000	$10,000
Equipment loan	$40,000	$40,000
Loan length	4 years	4 years
Contractual interest	9%	9%
Payment	annual and equally amortized	annual and equally amortized
Other	10% stock, $300 loan fee	15% compensating balance

While many of the terms are comparable, the differences in stock and compensating-balance requirements make it difficult to compare these two lenders solely in terms of their contractual interest rates. To account for the effects of noninterest costs of borrowing, we must derive the actuarial rate and the annual percentage rate. The first step is to project all cash flows associated with the loan transactions and to formulate the present value models.

Case 1

FCS loans are generally formulated to include the stock ownership requirement and any loan fees as part of the total loans to the farm borrowers. In effect, a borrower uses part of the total loan to pay any loan fees and to pay the stock requirement. The stock is owned by the borrower and can be recovered when the loan is repaid. However, the borrower loses the use of these funds during the period of the loan.

In this case, the Farmers must borrow enough to cover the $40,000 for the equipment, the $300 for the loan fee, and the stock requirement. The stock requirement is 10 percent of the total loan, which includes the stock purchase. To find the stock requirement, we must recognize that the rest of the loan — $40,300 — comprises 90 percent of the total loan. Hence, the total loan must be $40,300 ÷ 0.9 = $44,778, leaving a 10 percent stock requirement of $44,778 − $40,300 = $4,478.

The total loan of $44,778 is to be repaid in four equally amortized annual payments. The annual payments are found by solving for A in the uniform series present value model:

$$0 = 44{,}778 - A[\text{USPV}_{0.09,\,4}] \tag{14.15}$$

$$A = \frac{44{,}778}{[\text{USVP}_{0.09,\,4}]}$$

$$= 13{,}821.57$$

All the data are now available to formulate the present value model of the loan cash flows and to solve for the actuarial interest rate (i). The model is set up as follows:

$$0 = 40{,}000 - 13{,}821.57\,[\text{USPV}_{i,\,4}] + \frac{4{,}478}{(1 + i)^4} \tag{14.16}$$

The $40,000 entry reflects the cash made available to the Farmers to purchase the equipment; the – $13,821.57 entry reflects the uniform series of payments needed to fully amortize the total loan of $44,778; and the $4,478 entry reflects the value of the stock made available to the Farmers at the end of the four-year period.[1] Because all the cash flows are not a uniform series, we must use both Appendix Tables III and V in solving for i by trial and error. We anticipate that the rate will be higher than the contractual rate of 9 percent, so let's try 10 percent as the first estimate:

$$\text{NVP} = 40{,}000 - 13{,}821.77\,[\text{USPV}_{0.10,\,4}] + \frac{4.478}{(1.10)^4}$$

$$= 40{,}000 - 43{,}813.15 + 3{,}058.53$$

$$= -754.62$$

Because the net present value is negative, we know that 10 percent is too low. Using this same procedure, confirm that the net present value of the cash flows is $68.51 when i = 0.11. Thus, 11 percent is too high. By linear interpolation, the actuarial rate is found to be 10.92 percent, which is also the annual percentage rate.

---

[1]Technically, the borrower's stock investment could be reduced as the loan balance is reduced. This reduction would alter the annual cash flows. However, we assume that the stock investment remains constant until the loan is fully repaid.

Case 2

The compensating balance required by the local bank means that a specified percentage of the total loan must be held on account at the bank as a deposit through the length of the loan. When the loan is fully repaid, the deposit is available to the borrower. However, the borrower loses the use of the compensating-balance funds during the loan period.[2]

The Farmers must borrow enough to cover the $40,000 for the equipment and to meet the compensating balance. The compensating balance is 15 percent of the total loan, which includes the compensating balance. To find the compensating balance, we must recognize that the $40,000 for the equipment makes up 85 percent of the total loan. Hence, the total loan must be $40,000 ÷ 0.85 = $47,058.28, leaving the 15 percent compensating balance of $47,058.82 − $40,000 = $7,058.82.

The total loan of $47,058.82 is to be repaid in four equally amortized annual payments, which are found by solving for A in the uniform series present value model:

$$0 = 47{,}058.82 - A[\text{USPV}_{0.09,\,4}] \tag{14.17}$$

$$A = \frac{47{,}058.82}{[\text{USVP}_{0.09,\,4}]}$$

$$= 14{,}525.58$$

The cash flow model can now be formulated as:

$$0 = 40{,}000 - 14{,}525.58\,[\text{USPV}_{i,\,4}] + \frac{7{,}058.82}{(1 + i)^4} \tag{14.18}$$

The $40,000 entry reflects the cash made available to the Farmers to purchase the equipment; the −$14,525.58 entry is the uniform series of payments needed to amortize the $47,058.82 loan; and the $7,058.82 entry reflects the value of the compensating balance returned to the Farmers at the end of the four-year period.

---

[2]If the borrower's deposit were interest-bearing, then the interest payments would enter the present value model as positive cash flows to the borrower. As a result, the annual percentage rate for the loan transaction would be reduced.

As in Case 1, Appendix Tables III and V must be used to solve for i. Using i = 0.11 yields the following:

$$NPV = 40,000 - 14,525.58 \, [USPV_{0.11, \, 4}] + \frac{7,058.82}{(1.11)^4}$$

$$= 40,000 - 45,064.82 + 4,649.86$$

$$= -414.96$$

Because the net present value is negative, we know that 11 percent is too low. Using this same procedure for i = 0.12 yields NPV = 366.75. By linear interpolation, the actuarial rate is found to be 11.53 percent, which is also the annual percentage rate.

Hence, for these financing terms, the local bank provides the more costly source of financing, with costs reflecting all the cash flows associated with the loan transaction. This result was probably clear from the initial data. That is, we would expect the cost of the 15 percent compensating balance to outweigh the cost of the 10 percent FCS stock requirement and the relatively small service charge.

The actuarial rates confirm this expectation. Suppose, however, that all the financing terms (equity, length, contract rate, balances) differed between the two lenders. Then, the lower-cost source would be less clear, with even greater reliance placed on finding the actuarial rates.

OTHER COSTS

Interest Rates After Taxes

The preceding discussion has treated interest rates on a before-tax basis. However, both interest and noninterest costs of borrowed capital are deducted from taxable income, thereby reducing the after-tax net cash outflow from borrowing. Hence, it is important to express costs of borrowing on an after-tax basis. If the relevant tax rate is t, then the after-tax interest rate (i_t) is:

$$i_t = i(1 - t) \tag{14.19}$$

where i is the annual percentage rate. Suppose, for example, that the annual percentage rate is 12 percent and the tax rate is 20 percent. The after-tax interest rate is:

$$i_t = 0.12(1 - 0.2) = 0.096, \text{ or } 9.6\% \qquad\qquad (14.20)$$

Note that the after-tax cost of borrowing will differ among borrowers, depending on *their* particular tax positions.

The Real Interest Rate

The term **real interest rate** reflects *interest rates measured in real terms* (see Chapter 11). For example, assume that the annual percentage rate increased from 12 percent to 14 percent over a particular period. During the same period, suppose the rate of inflation also increased by 2 percent. We could then say that no change occurred in the real rate of interest.

The concept of the real interest rate is meaningful in financial management. Unanticipated inflation reduces the real cost of borrowing because fixed-debt obligations incurred in the past are being repaid with inflated dollars at present. In the preceding example, if interest rates had remained at 12 percent while the rate of inflation increased by 2 percent, then the real cost of borrowing would decline by 2 percent. As a result, the payoff from debt increases during such periods. Conversely, unanticipated reductions in the rate of inflation increase the cost of debt.

Nonmoney Costs of Debt Capital

Nonmoney costs of debt capital primarily reflect the premium associated with greater financial risk as leverage increases. In Chapter 7, we indicated that credit reserves (unused borrowing capacity) are an important source of liquidity for most firms. However, this source is consumed rapidly as leverage increases, causing borrowers to place an increasing liquidity value on the diminishing amount of credit reserve. (It may be helpful to review the discussion associated with Figure 7.1 to insure a clear understanding of the liquidity value of the credit reserve, identified as r in the figure.)

This premium for financial risk is important in determining the total cost of borrowed capital. Clearly, it is a "nonmoney" cost. Nonetheless, the difficulty of quantifying this cost does not diminish either its reality or its importance in financial management.

The liquidity premium may be determined in a couple of ways. One way is to consider it as a required rate-of-return, over and above the interest costs of borrowing, that is needed to justify further borrowing. Another way is to measure the cost of utilizing a source of liquidity other than credit. Consider, for example, the dairy manager who raises feed grains in an area subject to crop losses from hail damage. The manager has always held liquid reserves to buy feed in case the grain crop is lost because of hail. This year a credit reserve of $20,000 is maintained for that purpose. Alternatively, $20,000 worth of hail insurance could be purchased to set up a contingent reserve in the event of hail damage. If the annual premium on the $20,000 hail insurance is $420, then the credit reserve has at least a *minimum* value of $420 ÷ $20,000 = 2.1 percent. Because the credit reserve is more flexible in its availability and use than are most insurance reserves, we treat the insurance cost as the *minimum* value of the credit reserve.

Total Costs of Debt Capital

An investor's total costs of debt capital are represented by the real after-tax money costs of borrowing plus the liquidity premium. One component of this total cost is the actuarial interest rate expressed as an annual percentage rate. Other components include the borrower's tax rate, the rate of inflation, and the liquidity premium. Two of these components — the actuarial rate and the inflation rate — are determined in financial markets. The other two components — the tax rate and the liquidity premium — are unique to each investor and will cause differences among investors in their total costs of debt capital.

To illustrate the derivation of the total costs of debt capital, consider again Joseph and Julie Farmer who are purchasing the $50,000 worth of harvesting equipment. Their least-cost source of financing occurred from the Case 1 lender with an annual percentage rate of 10.92 percent. Suppose further that the tax rate is 20 percent, the inflation rate is 4 percent, and the liquidity premium on the credit reserve is 6 percent. The Farmers' after-tax cost of debt capital is $10.92(1-0.2) = 8.74$ percent. Their real after-tax cost of debt is estimated at 8.74 percent minus the 4 percent inflation rate, or 4.74 percent. The total cost of debt is then 4.74 percent plus the 6 percent liquidity premium, or 10.74 percent.

COSTS OF EQUITY CAPITAL

Our present concern is with the logic and method of estimating the costs of equity capital. New equity must come either from present owners or from new investors. The present owners may provide equity through reinvestment of earnings from the operation or by investing funds earned from other sources. Equity may be raised from outside investors by selling common and preferred stock or by adding partners.

Costs of Retained Earnings

Historically, most of the new equity capital in agriculture came from retained farm earnings and unrealized capital gains, although debt capital has played a more significant role since the mid-1940s. Because retained earnings do not carry an immediate and specific cash outflow, they may appear costless. However, this is not the case. There are important opportunity costs of retained earnings that reflect either earnings on alternative investments or, perhaps more importantly, family consumption that must be foregone to increase retained earnings.

These costs of retained earnings can be thought of as **required rates-of-return** needed to compensate for foregone consumption. As an example, agricultural producers might ask what rate-of-return they must have on the last increment (e.g., $100 or $1,000) of disposable income to divert it from consumption to savings. If the response is "a 15 percent rate-of-return," then 15 percent is the cost of retaining that increment of earnings. Retaining additional increments of earnings will likely have increasing opportunity costs. Moreover, the higher the business and financial risks associated with the farm operation, the higher the cost of equity capital, reflecting the premium needed to compensate for bearing the added risk.

Larger corporate firms experience similar costs of retained earnings. Conceptually, their costs of retained earnings are the owners' returns, if the retained earnings were paid as dividends. In practice, management tends to depend on stockholders to become vocal when they feel too much of the firm's earnings are being retained in the business. Income tax considerations also influence dividend versus retained-earnings decisions. If earnings are retained in the business, the stockholder avoids the immediate tax on dividends. Delaying present taxes on dividends by the stockholders makes retained earnings a more economical source of equity capital than selling common stock.

Costs of Outside Equity Capital

The costs of bringing outside equity capital into a farm proprietorship are generally high. A farmer might ask, "How high an expected return on capital would I need to make me invest in a farm like my own if it were operated by someone else?" The answer would help to indicate the costs of obtaining outside equity. The traditional farm proprietorship is seldom attractive to outside investors. Other forms of business organization are much better adapted to this purpose.

Larger corporate firms may consider sales of preferred or common stock as a means of raising new equity capital. The cost of preferred stock is closely tied to the fixed-dividend commitment that must be paid from the profits of the firm. Because of the fixed dividend, preferred stock can often be sold for a lower expected yield than can common stock. The cost of capital for preferred stock is given by:

$$k_p = \frac{D}{P_1 - C_1} \tag{14.21}$$

where D is the annual dividend, P_1 is the sale price of preferred stock, and C_1 is the cost of selling. For example, if preferred stock carrying an annual dividend yield of \$2.00 is sold for \$20.00 per share with sale costs of \$2.50 per share, then the seller will realize a net addition to capital of \$17.50 per share. Hence:

$$k_p = \frac{2.00}{20.00 - 2.50} = 11.43\% \tag{14.22}$$

Since dividends are not tax-deductible, this figure is already on an after-tax basis.

The cost of equity capital raised by selling common stock to investors is more challenging to estimate because of the greater importance of risk. Several approaches could be followed. For our purposes, the most straightforward approach is to express this cost of equity capital as the discount rate that equates the present value of the firm's series of expected future dividends to the current market price of the stock (net of any sales costs). The resulting discount rate is considered the rate-of-return required by investors who purchase the stock. The discount rate is found by solving the following present value model for rate k_c:

$$V_0 = \frac{D_1}{1 + k_c} + \frac{D_2}{(1 + k_c)^2} + \cdots \tag{14.23}$$

or

$$V_0 = \sum_{n=1}^{\infty} \frac{D_n}{(1 + k_c)^n}$$

where V_0 is the current market price of the stock and D_n is the expected dividend in period n.

If a portion of the firm's earnings is retained in the firm for reinvestment purposes, then it is logical to expect the series of dividends to grow over time. Under these circumstances, the present value model is modified to represent the constant-growth model, shown in Chapter 9, so that:

$$V_0 = \frac{D_1}{k_c - g} \tag{14.24}$$

where g is the expected growth rate. Solving for the cost equity yields:

$$k_c = \frac{D_1}{V_0} + g \tag{14.25}$$

Suppose, for example, that the firm's expected dividend on common stock at the end of period 1 is $3 per share, the expected growth rate is 10 percent, and the current market price is $60 per share. The cost of equity then is:

$$k_c = \frac{3}{60} + 0.10 = 0.15, \text{ or } 15\%$$

The 15 percent figure is considered to include a risk premium, since it is assumed that the current market price of the stock is already "discounted" to reflect the general consensus of investors about the riskiness of future earnings and growth rates. Other more sophisticated approaches to estimating costs of equity capital focus on breaking the cost into its risky and risk-free components, estimating these separate components, and summing them to a total cost. In practice, financial analysts with large corporate firms may use several approaches in estimating the costs of equity for common stock and utilize their judgment in making the final determination.

USING COSTS OF FINANCIAL CAPITAL

Although some of the discussion of the costs of capital has used terms and instruments that are especially relevant for large corporate businesses, the logic and methods also apply to farm proprietorships and other small businesses. Equity capital is more costly than debt capital on an after-tax basis, even for a farm proprietorship or partnership. Moreover, the cost of retained earnings to an agricultural producer is very real. Hence, the concept of costs of capital is an important determinant of a firm's optimal leverage. Moreover, the costs of financial capital serve as the basis for defining the discount rate or required rate-of-return to be used in capital budgeting.

Deriving the costs of debt capital considers a number of factors — the financing terms and potential changes in rates of inflation, tax obligations, and liquidity values. The derivation procedures are clearly established, although some factors are easier to quantify than others. Nonetheless, estimating the total costs of debt capital together with equity capital will help us to identify the firm's average costs of capital, to understand better the role of financial leverage, and to appropriately conduct capital budgeting analyses.

Summary

Costs of debt and equity capital make up the overall costs of financial capital for an agricultural business. These costs are estimated using present value procedures in which cash flows attributed to two types of capital — debt and equity — are measured. Debt costs, in particular, can vary considerably based on the methods of charging interest, time specifications, and on the use of noninterest money costs along with interest payments as part of the lender's compensation. The resulting estimates of costs of debt and equity play important roles in determining the firm's leverage position, as described in Chapter 6.

Other highlights are as follows:

- The **actuarial interest rate** is *the interest or discount rate that equates to zero* *the sum of the present values of all cash flows associated with the loan transaction.* This rate is expressed per conversion period, which may be specified on an annual, quarterly, monthly, or some other basis.

- The **annual percentage rate (APR)** is found by *expressing the actuarial rate on an annual basis.* Lenders are required to report the annual percentage rate on many types of loans, but not on agricultural, commercial, and various other types of loans.

- The **effective interest rate** is calculated by *compounding the actuarial rate over the number of conversion periods within a year.* As the number of conversion periods increases, the difference between the annual percentage rate and the annual effective rate increases, although at a diminishing rate.

- The **contractual interest rate** is *the interest rate stated on the note.*

- The major methods of computing interest are the **remaining-balance method,** the **add-on method,** and the **discount method.**

- A lender's use of fees, compensating deposit balances, and stock requirements increase the annual effective interest rate paid by the borrower.

- Estimated nominal costs of debt financing may be converted to real after-tax costs and further adjusted to account for risk or liquidity premiums as average increases.

- Costs of equity capital apply to retained earnings and outside corporate or partnership equity capital. Present value models based on projected dividend payments are used to quantify equity capital costs for common and preferred stock.

Topics for Discussion

1. Under what circumstances, if any, might the following statement be true? "There are no costs to using your own capital." Explain.

2. Explain the differences between actuarial rate, annual percentage rate, and annual effective interest rate.

3. Find the actuarial interest rate on a loan of $30,000 to be repaid in equal annual installments of $6,687.47 each over a six-year period.

4. Find the annual percentage interest rate (APR) on a $5,000 loan to be repaid in equal monthly payments of $180.76 each over a three-year period. What is the annual effective interest rate on this loan?

5. Discuss the importance of nonmoney costs of debt capital. What gives rise to those costs?

6. In general, what are the effects of the add-on and discount methods of computing interest on the annual percentage rate?

7. What are the impacts of income taxes and inflation on the costs of debt and equity capital?

8. A farmer is buying a new tractor valued at $25,000. The dealer will allow $5,000 on the trade-in of the old tractor, leaving a balance to be financed by one of the following three methods. Compute the annual percentage rate for each method. Which one is the most inexpensive?

   a. The local office of the Farm Credit System will make a four-year loan with equal annual payments of principal and interest at a 12 percent contractual interest rate with a $300 loan fee and a 10 percent stock requirement.

   b. The local bank will also make a four-year loan with equal annual payments of principal and interest at a 13 percent contractual interest rate with a 5 percent compensating balance requirement.

   c. The manufacturer will make a four-year loan with equal payments of principal and interest, with interest computed at 10 percent add-on.

References

1. Cissell, R., et al., *Mathematics of Finance,* 8th ed., Prentice-Hall, Inc., Englewood Cliffs, New Jersey, 1990.

2. Fisher, L., *The Rate of Interest,* MacMillan Publishing Co., Inc., New York, 1907.

3. Jones, B. L., and P. J. Barry, "Impacts of PCA Capitalization Policies on Borrowers' Costs," *Agricultural Finance Review,* 46(1986):15 – 26.

4. LaDue, E. L., "Influence of the Farm Credit System Stock Requirement on Actual Interest Rates," *Agricultural Finance Review,* 43(1983):50 – 60.

5. Solomon, E., *The Theory of Financial Management,* Columbia University Press, New York, 1963, pp. 93 – 98.

Financial Markets
for Agriculture

15

Financial Intermediation
in Agriculture

The financing needs of the agricultural sector are met by a variety of financial institutions. Understanding the costs, terms, and sources of financial capital for agriculture necessitates considering the organization of the financial markets, the process of financial intermediation, and the characteristics of the institutions participating in this process. In this chapter we review the development of financial markets in general, as well as the financial institutions serving agriculture. Following this review, we present a conceptual discussion of the intermediation process, the functions of intermediation, the regulatory environment, and the role of market competition. This chapter serves as an introduction to the discussion of management issues of financial institutions in Chapter 16 and to the agricultural financial institutions described in Chapter 17.

FINANCIAL MARKETS

In the United States, some institutions are local or regional in orientation, while others are directed toward the national financial market, which is centered in the financial district of New York City. There one finds offices of many of the nation's largest banks, insurance companies, and many other large financial institutions, as well as the leading underwriters and dealers in debt and equity securities and many large industrial corporations. Most of these large financial institutions also participate in international markets. Moreover, offices of large foreign-owned financial institutions also are located in the United States. Through a far-flung and flexible network of international, national, regional, and local offices and institutions, the financial market provides continuous linkages between suppliers and users of funds.

AN HISTORICAL PERSPECTIVE

Through the late 18th century and the 19th century, the commercial banking system in the United States developed. There was a strong orientation away from large national banks towards smaller localized banks, even though large "city" banks existed in the major trading centers. This orientation reflected a preference for avoiding large concentrations of power within the banking system. Thus, the chartering and regulation of banks largely was left to the individual states, and except for several "experiments" at the federal level, no central bank existed to provide monetary stability and regulatory control over the financial system. Moreover, the notes issued by the various commercial banks often served as currency in local markets, thus creating a chaotic currency situation in inter-regional trade.

Beginning in the 1860s, partly in response to Civil War conditions, Congress passed legislation that created a class of nationally chartered banks to be supervised by the comptroller of the currency. This period marked the beginning of the dual system of banking in the United States in which banks could have either a state or a national charter. Through the rest of the 19th century and into the early 20th century, some banks indeed grew stronger and acquired considerable market power in a variety of business endeavors. Such large and extensive growth eventually was constrained by antitrust legislation in the early 1900s.

The emphasis on a decentralized banking system came at the cost of greater instability of the banking system and difficulties in achieving monetary control at the national level. Accordingly, after lengthy study, the Federal Reserve Act of 1913 recognized the need for a more central institution that would exert monetary control and supervise the financial system. A system of 12 Federal Reserve Banks was created for these purposes, although the regional approach continued to reflect the country's concern about concentration of financial power and the preservation of states' rights.

Finally, the United States had a central bank, but its regional structure made it considerably different from the central banks in other countries. Nationally chartered banks were required to be part of the Federal Reserve System, but state-chartered banks were not. Reserve requirements differed between state and national banks and were not made similar until the 1980s. Thus, high levels of instability, failures, and monetary inflexibilities continued to occur in banking until the depression era of the 1930s brought greater, although still not complete, centralization to the Federal Reserve System.

Two other major pieces of banking legislation were the McFadden Act of 1927 and the Glass-Steagall Act of 1933. The McFadden Act focused on the geographic structure of banking. It prevented national banks from becoming interstate institutions by requiring them to observe local regulations in their home states. The Glass-Steagall Act segmented the financial industry by separating banking and securities businesses so that potential conflicts of interest between lending and underwriting of securities could be avoided. This separation has continued into the 1990s, although many of the constraints on products and services many banks offer have been relaxed.

The Glass-Steagall Act also strengthened the power of the Federal Reserve System and created the Federal Deposit Insurance Corporation to, partially at least, protect depositors against bank losses. Finally, other important legislation in the 1930s set maximum limits on the interest rates banks could offer on various types of deposit and savings accounts.

Taken together, the combination of geographic restrictions on banking, confinement of banking activities, interest rate limits, and deposit insurance represented a strong regulatory environment in banking that contributed to several decades of financial stability. At the same time, this stability was aided by the generally prosperous economic times of the 1950s and 1960s in the United States. In particular, interest rates were low and stable during these times. The rewards for financial innovation were minimal, and services and non-price factors played a more important role than interest rates in establishing and maintaining durable customer relationships between lenders and borrowers. Bank management focused mostly on the spread between the lending rate and the cost of funds. Rigorous management of assets and liabilities had little payoff. At the same time, however, regulated product differentiation in banking was leading to the formation of other types of financial institutions — savings and loan associations, mutual savings banks, mutual funds, securities brokers and dealers, pension funds, and life insurance companies were operating in different markets.

Conditions began to change in the 1970s as inflation pressures raised nominal interest rates, financial markets became more internationalized, new financial markets such as the Euro dollar market appeared beyond the control of U.S. regulators, currency exchange rates became flexible, negotiability of large certificates of deposits was introduced, securitization became widespread, and other new financial instruments and intermediation channels were developed to circumvent regulations. **Disintermediation** from banking *(deposit withdrawals in response to higher, unregulated rates-of-return in money market mutual funds or for direct placement in other money market instruments)* accelerated and added instability to the banking system. New types of financial services firms began to emerge, often part of or affiliated with other retail firms having effective access to U.S. consumers. Attitudes toward regulation began to relax, first with selective freedoms of some deposit rates and then in the early 1980s, with complete deregulation of interest rates (see Chapter 18 for further discussion).

In 1979, the Federal Reserve System shifted its monetary policy target emphasis away from interest rate levels toward growth rates of the nation's money supply. Interest rates were allowed to reach market levels. As recent history shows, this approach to dealing with inflation eventually was successful, although nominal and real interest rates reached record highs in the early 1980s, as a part of the adjustment process, with serious consequences for capital-intensive agriculture, increasingly dependent on foreign commodity demand.

In general, the stable non-price customer relationship – driven system in banking and other types of financial institutions shifted toward a price-driven system with greater unfettered interest rate movements. By the 1990s, price deregulation was complete in the financial markets, and both the product and the geographic restrictions were changing considerably.

IMPORTANCE OF
FINANCIAL INTERMEDIATION

Although commercial banks, thrift institutions, life insurance companies, finance companies, and government-sponsored financial programs are commonplace in advanced society, these institutions are of relatively recent origin. The growth of financial intermediation in the United States is evident from the substantial increase in assets of commercial banks, mutual savings banks, life insurance companies, and savings and loan associations during the 20th century. Nearly two-thirds of this increase occurred after 1975.

Changes in the ratio of assets held by financial institutions to the value of all tangible national assets are particularly revealing. At the beginning of the 19th century, this ratio was below 1 to 10 and had increased to only about 1 to 7 a century later. By 1920, however, it had advanced to nearly 1 to 4. Presently, the value of assets of all financial intermediaries represents about 60 percent of the total assets in the United States. This does not mean that financial intermediaries actually own 60 percent of the assets in the United States; rather, the intermediaries may have significant debt claims on various types of assets and the income they generate. Moreover, the liabilities of these financial intermediaries indicate that other economic units (businesses, individuals, government) both in and outside the United States have claims on the assets of financial intermediaries. In the case of depository institutions, these claims occur as transactions, savings, and time deposits.

Tables 15.1 and 15.2 provide further evidence on the growth in financial assets and the changing relative importance of various institutions and types of financial instruments. Outstanding credit debt increased from $427 billion in 1950 to $13,560 billion in 1990 — a 31.8 fold increase (see Table 15.1). The shares of total debt indicate a significant decline in the relative importance of U.S. government securities, substantial growth early in the time period of mortgage debt, modest growth in open market paper, and irregular patterns of change in other categories. The financial assets of selected financial institu-

Table 15.1. Outstanding Credit Market Debt
(Percent of Total)

Category	1950	1960	1970	1980	1990
	- *(%)* -				
U.S. government securities	51.1	31.2	21.5	21.8	28.8
Tax-exempt securities	5.6	9.1	9.0	7.5	7.7
Corporate & foreign bonds	9.1	11.6	12.7	10.6	11.9
Mortgages	17.1	26.9	29.5	31.1	28.9
Consumer credit	4.9	7.2	8.4	7.6	6.0
Bank loans, n.e.c.[1]	6.6	8.1	9.5	9.8	6.0
Open market paper[2]	0.2	0.8	2.5	3.5	4.5
Other loans	4.9	6.0	6.9	8.1	6.2
Total	100.0	100.0	100.0	100.0	100.0
Total debt (billion $)	427	778	1,596	4,666	13,560

[1]Bank loans, n.e.c. (not elsewhere classified), are nearly all loans to businesses.

[2]Includes commercial paper and bankers acceptances.

Source: Board of Governors of the Federal Reserve System, *Flow of Funds Accounts*, various editions.

Table 15.2. Financial Assets of Selected Financial Institutions
(Percent of Total)

Institution	1950	1960	1970	1980	1990
	- (%) -				
Commercial banks	51.2	38.3	38.6	36.7	31.6
Savings and loan associations	5.8	11.9	12.9	15.4	10.4
Mutual savings banks	7.6	6.9	5.9	4.2	2.5
Credit unions	0.3	1.0	1.3	1.7	2.1
Life insurance companies	21.3	19.3	15.0	11.5	12.9
Private pension funds	2.4	6.4	8.3	11.6	11.0
State and local pension funds	1.7	3.3	4.5	4.9	7.1
Other insurance companies	4.0	4.4	3.7	4.3	4.8
Finance companies	3.2	4.6	4.8	5.0	7.4
Real estate investment trusts	0.0	0.0	0.3	0.1	0.1
Mutual funds	1.1	2.8	3.5	1.5	5.5
Money market mutual funds	0.0	0.0	0.0	1.9	4.7
Securities brokers and dealers	1.4	1.1	1.2	1.1	2.5
Securitized credit issuers	0.0	0.0	0.0	0.0	2.1
Total	100.0	100.0	100.0	100.0	100.0
Total (billion $)	294	600	1,342	4,040	10,572

Source: Board of Governors of the Federal Reserve System, *Flow of Funds Accounts*, various editions.

tions increased from $294 billion in 1950 to $10,572 billion in 1990 — a 36-fold increase (see Table 15.2) — although a significant part of this growth was caused by inflation. The shares of institutional holdings indicate a substantial decline in the relative importance of commercial banks, a smaller decline for life insurance companies and mutual savings banks, and increasing shares for pension funds, finance companies, mutual funds, and, more recently, issuers of securitized credit instruments. Other institutions exhibited irregular patterns of change over this period.

Growth in intermediation also characterizes the development of financial markets in other countries and is represented as well by the globalization of international financial markets. The flows of funds and securities among the countries of the world have grown substantially in the last quarter of the 20th century, with expanded international trade, increasing economic integration, and advances in technologies for information transfer.

EVOLUTION OF
FINANCIAL MARKETS
FOR AGRICULTURE

The evolution of financial markets for agriculture largely parallels the general developments in finance. Prior to World War I, the capital needs of farmers were not especially great. New land was available through homesteading, and prices on established farm land were low. Machinery use was not widespread, and farmers used few other purchased inputs.

Farmers' short-term financing needs largely were met by merchants and dealers and by local banks, except during periods of tight money. However, the local banks and merchants generally were subject to the same kinds of risks as their farm customers. If farmers prospered and paid off their loans, the banks and merchants survived. If farmers and local merchants did not prosper, the banks risked failure and difficulty in meeting their own financial obligations.

Typical mortgages for farm land were short term (three to five years) and high cost, and they were not amortized. Hence, large lump sum repayments were required when the loans matured. Renewals of matured loans were common; however, dependence by farmers on renewals was risky because of changing conditions in financial markets, especially without a central banking system.

Concern over the farm credit situation led to the establishment of several commissions to study farm credit needs. Particular attention was given to cooperative farm credit systems in Europe. These studies ultimately led to the establishment of the Farm Credit System (FCS) and the Farmers Home Administration (FmHA) as they are known today. However, these institutions were not created overnight — their initial development covered more than four decades (from 1910 through the 1940s) — and they periodically are reviewed and modified.

Because the Farm Credit System was designed to be self-supporting, it was not well suited to financing high-risk borrowers with inadequate resources or for farmers who experienced drought or other disasters. As a result, various programs of direct government assistance have provided the needed emergency or supervised financing. The current means of providing such financing is through the Farmers Home Administration, established in 1946 to provide financing programs for farmers who could not be served on reasonable terms from other lenders. Later, the purposes of the Farmers Home Administration were broadened to include loans for rural housing, recreation, and rural development projects. More recently, greater emphasis has again been given to the Farmers Home Administration serving as a lender of last resort to farmers.

The earlier discussion about the historic evolution of commercial banking in the United States also applies to bank involvement in agricultural finance. This role by banks has been consistently strong, especially in non – real estate lending. A notable feature is the extensive involvement in agricultural lending of relatively small community banks located in rural areas of the Midwest, the Southwest, and other parts of the country as well. These small rural banks traditionally have been major suppliers of financial capital to agricultural producers, and their own financial performance is strongly influenced by the economic health of agriculture in the local market.

CREDIT SOURCES
FOR AGRICULTURE

The levels of farm debt since 1955 and the relative importance of different lenders nationally are indicated in Tables 15.3 and 15.4. These tables represent farm real estate debt and non – real estate farm debt, respectively — consistent with the data classifications reported by the U.S. Department of Agriculture.

Farm Real Estate Debt

The note for a farm real estate loan is nearly always secured by a pledge of farm real estate as collateral through either a real estate mortgage or a deed of trust. An alternative is the land purchase contract. Farm real estate debt normally has maturity in excess of 10 years and less than 40 years, usually ranging between 15 and 30 years. A recent exception involves adjustable rate loans, especially from commercial banks. Adjustable rate loans generally have maturities of less than 10 years, longer amortization periods, balloon payments due at maturity, and the intention of loan renewal at the current market interest rate.

Total farm real estate debt outstanding in the United States declined between 1935 and 1945, as farmers emerged from the depths of a prolonged depression into the improved financial conditions associated with World War II. As shown in Table 15.3, farm real estate debt increased steadily between

1955 and the early 1980s. Growth was especially strong in the 1970s, reached a peak in 1984, and subsequently declined.

The Farm Credit System (Federal Land Banks) emerged during the 1970s as the most important supplier of farm real estate debt, showing an annual rate of growth in outstanding debt of about 14 percent in the 1970 to 1980 period. In 1986, the Farm Credit System held about 42 percent of the outstanding farm real estate debt, although this figure varied considerably according to the region of the country — highest in the Southeast and lowest in the Northeast. The Farm Credit System share then declined to about 34 percent in 1990, reflecting the system's financial difficulties resulting from the financial crisis in agriculture.

Following World War II, life insurance companies were the largest supplier of farm real estate debt to agriculture. In 1955, they held about one-quarter of all outstanding farm mortgage debt. However, their relative position declined until it leveled out at the 11 to 13 percent range after 1980. The share of farm mortgage debt held by non-institutional lenders, primarily individuals who sell their own land either on contract or with a mortgage, also was historically strong — remaining in the 30 to 40 percent range through the 1970s. However, this share has declined substantially to about 20 percent in 1990.

Farm real estate lending by commercial banks is smaller than their non – real estate lending, but still significant, especially in the 1980s. The market shares of banks declined steadily over the 1950 to 1985 period, before recovering strongly between 1985 and 1990. Commercial banks' market shares of farm real estate debt also vary by regions — the highest shares are in the Northeast and Southeast, primarily because banks in these regions tend to use farm real estate as security when they are financing farmers' non – real estate loan requirements.

Non – Real Estate Farm Debt

As shown in Table 15.4, commercial banks have the largest volume of non – real estate debt, although their market share has fluctuated in recent years, primarily because of large swings in loan volume by government sources of credit. The latter include the Farmers Home Administration and the Commodity Credit Corporation (CCC). The relative importance of the Farmers Home Administration in supplying non – real estate loans to farmers declined following the 1940s but increased sharply during the 1980s, reflecting various

Table 15.3. Real Estate Farm Debt (Including Operator Households),

Year	Farm Credit System[1]		Farmers Home Administration		Life Insurance Companies	
	(mil. $)	(%)	(mil. $)	(%)	(mil. $)	(%)
1945	1,322	27.8	184	3.9	891	18.7
1950	991	16.2	257	4.2	1,353	22.1
1955	1,480	16.4	413	4.6	2,272	25.1
1960	2,539	19.7	723	5.6	2,975	23.1
1965	4,240	20.0	1,497	7.1	4,802	22.6
1970	7,145	23.4	2,440	8.0	5,610	18.4
1975	16,029	32.2	3,369	6.8	6,726	13.5
1980	36,196	37.1	8,163	8.4	12,928	13.3
1985	44,584	42.2	10,427	9.9	11,836	11.2
1990	26,885	34.3	8,093	10.3	10,186	13.0

Annual Rate

1950 – 1960	9.86	10.90	8.20
1960 – 1970	10.91	12.93	6.55
1970 – 1980	17.62	12.84	8.71
1980 – 1990	(2.93)	(0.09)	(2.36)

[1]FLB debt prior to 1988.

Source: *Economic Indicators of the Farm Sector, National Financial Summary, 1990*, USDA – ERS, ECIFS 10 – 1, November 1991.

types of emergency loan programs and the serious economic problems of agriculture during these times. Similarly, the price support loans of the Commodity Credit Corporation are low when market prices of farm commodities are well above their support levels; when market prices decline to approach the support level, CCC loan volume increases substantially.

The proportion of non – real estate farm debt held by the Farm Credit System increased steadily from the mid-1930s (when Production Credit Associations were established) until the mid-1970s. The market share remained in the 23 to 27 percent range during the 1970s and then declined to about 15 percent in 1990.

The "Individuals and Other" category in Table 15.4 is a smaller, yet still important group. This group mostly includes merchants, dealers, and other agribusinesses making short- and intermediate-term loans to farmers. Individuals and other miscellaneous lenders also were included in this category. Finally, regional differences also arise in the distribution of non – real estate

December 31, Selected Years, 1945 – 1990

Commercial Banks		Other Lenders		Total Real Estate Debt
(mil. $)	*(%)*	*(mil. $)*	*(%)*	*(mil. $)*
507	10.7	1,856	39.0	4,760
986	16.1	2,544	41.5	6,131
1,275	14.1	3,609	39.9	9,049
1,592	12.4	5,040	39.2	12,869
2,607	12.3	8,074	38.0	21,220
3,772	12.4	11,524	37.8	30,491
6,296	12.6	17,432	35.0	49,852
8,571	8.8	31,636	32.4	97,494
11,385	10.8	27,507	26.0	105,739
17,227	22.0	16,007	20.4	78,398

of Change (%)

Commercial Banks	Other Lenders	Total Real Estate Debt
4.91	7.08	7.70
9.01	8.62	9.01
8.55	10.63	12.33
7.23	(6.59)	(2.16)

farm debt. Noteworthy are the low market shares held by commercial banks in the Southeast and Delta states and the high shares held by the Farmers Home Administration in these regions.

EQUITY CAPITAL

Despite the relatively large use of farm debt, equity capital dominates the aggregate balance sheet for the agricultural sector. It also is the dominant source of capital for most individual farms, although a significant part of equity growth over the long term is attributed to unrealized capital gains on farm land.

Historically, the primary source of new equity capital for agriculture has been retained earnings. However, "outside" equity capital has become increasingly important, especially in poultry raising, cattle feeding, swine production,

Table 15.4. Non – Real Estate Farm Debt (Including Operator Households),

Year	Commercial Banks		Farm Credit System[1]		Farmers Home Administration	
	(mil. $)	(%)	(mil. $)	(%)	(mil. $)	(%)
1945	1,034	32.9	221	7.0	413	13.1
1950	2,524	36.5	513	7.4	329	4.8
1955	3,308	34.0	706	7.2	406	4.2
1960	4,991	37.5	1,568	11.8	420	3.2
1965	7,677	39.5	2,718	14.0	717	3.7
1970	11,102	46.3	5,515	23.0	795	3.3
1975	20,160	48.2	11,120	26.6	1,772	4.2
1980	31,564	37.1	20,539	24.1	11,397	13.4
1985	35,513	36.0	14,562	14.7	16,720	16.9
1990	32,913	46.3	10,103	14.9	10,652	15.0

Annual Rate

Year	Commercial Banks	Farm Credit System	Farmers Home Administration
1950 – 1960	7.06	11.82	2.47
1960 – 1970	8.32	13.40	6.59
1970 – 1980	11.01	14.05	30.51
1980 – 1990	0.42	(6.85)	(0.67)

[1]Prior to 1988, FCS loans were reported separately for PCA and FICR loans through other financial institutions.

[2]Estimates of loans outstanding from merchants, dealers, individuals, and other miscellaneous lenders.

and fruit and vegetable production. Such capital might be in the form of equity invested in farm operations by local nonfarmers as a joint venture or a general partnership or it might be invested under a more formal legal arrangement such as a limited partnership or a large industrial corporation that also is engaged in agriculture. Although the financial market for outside equity capital is not well developed for agriculture compared to many other industries, the trend is toward greater use of outside equity capital.

CONCEPTUALIZING FINANCIAL INTERMEDIATION

Financial intermediation is *the channeling of funds and securities between savers and investors (borrowers).* It is a two-way process. The flow of funds is

December 31, Selected Years, 1945 – 1990

Individuals and Others[2]		Commodity Credit Corporation		Total
(mil. $)	*(%)*	*(mil. $)*	*(%)*	*(mil. $)*
1,200	38.2	277	8.8	3,145
2,760	39.9	794	11.5	6,920
3,490	35.8	1,833	18.8	9,743
4,990	37.5	1,342	10.1	13,311
6,950	35.8	1,374	7.1	19,436
4,850	20.2	1,730	7.2	23,992
8,553	20.4	232	0.6	41,837
17,721	20.8	3,836	4.5	85,057
15,378	15.6	17,598	17.8	98,771
13,000	18.3	4,377	6.2	71,045

of Change (%)

6.10	5.39	6.76
(0.28)	2.57	6.07
13.83	8.57	13.49
(3.05)	1.33	(1.78)

from savers to investors (borrowers). The flow of securities is from investors (borrowers) to savers. The flow of funds arises because investors have needs for funds that exceed their present supply of funds. The flow of securities assures savers (i.e., provides security) that their funds will eventually be returned along with compensation payment for their use. Both of these flows are activated by transactions in financial markets. The aggregate effects of these transactions reflect a reconciliation of differences in the risk, return, and liquidity preferences of savers and investors.

The flow process is indicated in Figure 15.1. Funds flow along the top of the diagram, and securities flow in the reverse direction along the bottom. The funds are divided into two classes called debt and equity, both of which create financial claims on the assets of a business or other economic unit. Debt funds are distinguished by the investor's (borrower's) promise to repay the funds at a designated time along with payment of interest to compensate for using the

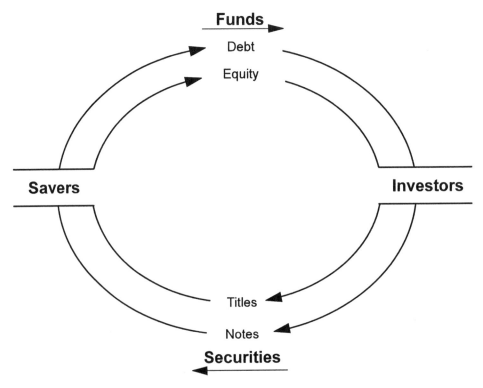

Figure 15.1. Flows of funds and securities in financial intermediation.

funds. Equity funds do not have the same kind of repayment promise; thus, they are considered more risky than debt funds. The payment for providing equity funds is a share of the business profit as represented by dividends and changes in owners' equity.

Securities also can be divided into the same two classes, debt and equity. While there are many kinds of securities, we will use the term *note* to refer to debt securities and the term *title* to refer broadly to equity securities. These securities also are called financial assets because they represent financial claims by their holders on the assets of those who issued the securities.

Financial assets originate in primary transactions between savers and investors or between savers or investors and a financial intermediary. Such participants constitute a primary market. In contrast, a secondary market is one in which existing financial assets are sold from one holder to another.

In an earlier, more primitive time, financial transactions occurred directly between savers and investors. Such a cumbersome process was inefficient and did not lend itself well to economic development. As a result, specialized financial intermediaries that could perform the intermediation process more efficiently and safely than could individual savers and investors, while still earning an acceptable rate-of-return on their own equity capital, began to develop. Along with the intermediaries came the development of laws and standard financial instruments that facilitated the important role of financial intermediation in economic development, commerce, and government and consumer finance.

Most intermediaries resulted from a demand for services sufficient to generate a profit for them. In other cases in which gaps in financial markets existed, government-owned or sponsored intermediaries were developed to fill the void. In either case, the success of the intermediaries depends in part on their efficiency in performing the intermediation process.

IDENTIFYING FINANCIAL INTERMEDIARIES

Financial intermediaries are distinguished from other types of businesses by two important characteristics: (1) claims (loans, securities, etc.) on other economic units dominate the assets in their balance sheets, and claims by others (deposits, Farm Credit bonds, etc.) dominate their liabilities and (2) their economic activities center on the purchase, transformation, and sale of such claims. Thus, transactions involving financial assets predominate in financial intermediaries, and virtually all of their planning, accounting, and control activities are directed toward the management of financial assets.

Financial Intermediation Services

Financial intermediaries emerge because they perform the joint flows of funds and securities more efficiently than can individual savers and investors who, in the absence of intermediaries, would have to locate each other, agree to contractual terms, and monitor the performance of the other parties relative to those terms. Greater efficiency in turn means greater profits for the owners of financial intermediaries, although competition among intermediaries helps to assure high efficiency and avoidance of monopolistic conditions.

The services performed by financial intermediaries include the following: (1) origination, (2) flows of funds, (3) risk-bearing, (4) liquidity, and (5) servicing.

Origination

Origination refers to *the primary transactions in which financial assets are created.* A major example is the note signed by a borrower when a loan transaction occurs. The loan, signified by the note, originates in such a transaction. Other cases of origination involve the issuances of bonds or other debt instruments and new shares of stock or other types of equity securities.

Flows of Funds

Flows of funds arise from the provision of funds as a part of the financial transaction — the transmission of funds from the saver to the user. In this process, the intermediary essentially must match up the specific characteristics of the funds supplied by the saver with the characteristics of funds needed by the user. In farm lending, for example, funds must be aggregated into larger units than are provided by most savers. In the absence of financial intermediaries, some agricultural producers would have to borrow directly from 50 to 100 individual savers to acquire the needed funds.

Funds also must have unique time dimensions to be of value to agriculture. Farmers must finance highly seasonal operating expenses. Their loans can be repaid only when their farm products are sold. Additionally, farmers need intermediate-term financing for acquisition of depreciable assets and long-term financing for real estate. The need for large blocks of funds with a unified time dimension to fit the special financing requirements gives added emphasis to the role of intermediation.

Finally, geographic differences between savers and farm investors must be reconciled by intermediaries who supplement investable funds in rural areas with funds collected from urban areas. Urban savings move predominantly into urban and suburban banks, thrift institutions, retirement and pension funds, and life insurance companies with accounts in large city banks. These funds are not available to agricultural borrowers until they are acquired by intermediaries serving agricultural areas.

Risk-Bearing

Risk-bearing refers to *the ability of financial intermediaries to efficiently manage and absorb the risks inherent in the intermediation process.* The risks in most agricultural loans must be modified if savers are to invest in agricultural securities. Most savers probably would not loan directly to farmers because such loans would be "too risky" (see, however, "Individuals and Seller Financing" in Chapter 17), especially compared to the federal deposit insurance available on accounts placed with most types of depository institutions.

A financial intermediary modifies risk in several ways. Most important, an intermediary substitutes its own financial strength for the financial strength of the borrower. Consequently, rural bank depositors do not look to local agricultural producers for deposit security, even though the local bank loans funds from those deposits to local farmers.

Intermediaries also may specialize in the collection and evaluation of information from borrowers. Included are the assessment of credit worthiness of individual borrowers and the monitoring of loan performance and financial progress during the term of the loan contract. Moreover, specialized ratings firms (e.g., Moody's, Standard and Poor's) have emerged to evaluate the credit worthiness and financial performance of larger corporate firms and major financial institutions themselves in order to convey such information efficiently to financial market investors.

Financial intermediaries also may rely on guarantees or insurance from private or public (e.g., Federal Deposit Insurance Corporation) sources to reduce the saver's risk. Lenders may also utilize government guarantees to secure loan repayment, as happens with the Farmers Home Administration, the Small Business Administration, and agencies involved in financing residential housing.

Finally, intermediaries can spread loan risks over wide geographic areas and over many borrowers. The banks of the Farm Credit System, for example, jointly sell bonds and notes in the financial market, based on portfolios of loans made to thousands of agricultural borrowers scattered throughout the United States. Similarly, statewide branch banking systems and interstate bank holding companies can spread weather risks and other risks of agricultural lending over widely dispersed geographic regions. In contrast, the loan portfolios of small rural banks are much less diverse than those of larger banks.

Liquidity

Liquidity services provided by financial intermediaries have several dimensions. First, the liquidity provided in most agricultural securities, reflecting a farmer's own liquidity position, must be modified to satisfy most savers. Loans that must be repaid prior to the achievement of the loan purpose would have little value to agricultural borrowers. Yet, few individual and corporate savers are willing to commit funds directly to such an inflexible schedule of recall. Thus, matching the liquidity needed by savers with the liquidity of agricultural securities is an important function of financial intermediation. This function is accomplished through the benefits of multilateral trade involving large numbers. It enables the intermediary to make longer-term loans (e.g., six months, one year, or more) funded by short-term deposits (e.g., demand depositions, savings accounts, short-term certificates of deposit).

The second liquidity dimension involves the possible sale of securities in secondary markets. A secondary market is a market in which existing financial assets are bought and sold by investors. It provides liquidity through the timely generation of cash from security sales. In the process, the presence of the secondary market mobilizes flows of funds and encourages the holdings of financial assets.

Well-developed secondary markets exist for many types of equity and debt securities (e.g., stock and bond markets, government securities markets, commercial paper markets). Secondary markets are much less developed for loans to individual borrowers. Exceptions include government guaranteed loans, residential housing loans, and the recently developed secondary market for farm real estate loans (see the discussion about the Federal Agricultural Mortgage Corporation in Chapter 17).

Servicing

Servicing refers to *the actions performed by a financial intermediary to carry out the terms of the loan transaction and assure timely loan performance.* Included are completing the loan application, evaluating credit worthiness, taking security in and appraising collateral, meeting legal requirements, monitoring borrower performance, accepting loan repayments, adjusting interest rates where applicable, and dealing with loan contingencies.

Separation of Services

A major development in recent decades has involved a separation in the carrying out of some of these intermediation services by different financial institutions. For example, the institutions originating and servicing a loan generally are one and the same, even if part or all of the loan is sold to another holder in a secondary market transaction. In the latter case, the flows of funds and liquidity functions are separated from origination and servicing. Sometimes, however, origination and servicing may be separated as well. If, as is common, only part of a loan is sold, then both funding and risk-bearing are separated, perhaps disproportionally if the loan is sold with recourse. Selling with recourse means the originating institution carries the initial risk up to its investment in the loan, with risk then shifting to the secondary holder.

THE REGULATORY ENVIRONMENT

A major feature of the intermediation process for channeling savings into investments is the intangible nature of financial assets and services. As a result, considerable confidence, trust, and stability are required among market participants in order for financial markets to develop and function effectively. Accordingly, these markets experience high degrees of government regulation in order to safeguard savers and investors, modify competition, respond to market imperfections, and facilitate effective monetary policy.

Regulations take many forms:

1. Restraints on geographic expansion of financial institutions, as in branching and holding company regulations.

2. Mandatory specialization in some services (housing or farm lending, transactions accounts).

3. Portfolio diversification through reserve and capital requirements, loan limits, and asset allocations.

4. Special borrowing privileges.

5. Fair trade practices.

6. Public programs for credit and insurance.

7. Laws affecting the design, trade, and negotiability of financial instruments.

Regulations yield many positive effects, as intended. However, they also may impose substantial costs on participants in financial markets as changes occur in economic, social, technological, and institutional conditions. Inequities that discourage savings, impede flows of funds, hamper the performance of some intermediaries, and reduce the efficiency of intermediation can arise. Regulations also have induced financial institutions to circumvent government policies, thus further distorting and destabilizing financial markets. When the costs of regulations become excessive, the stage is set for regulatory change.

In Chapters 17 and 18, we will consider the relationships between regulations and specific financial institutions in greater detail and identify some of the major regulatory issues affecting the cost and availability of financial capital to agricultural producers. And, in Chapter 19, we will address the legal aspects of agricultural finance.

COMPETITION IN FINANCIAL MARKETS

The degree of competition in a financial market is an important indicator of performance in carrying out the intermediation process. Generally, from society's point of view, the greater the competition, the better the market performance. In this section, we will consider competition in the financial markets for agriculture.

In evaluating competition, it is important to first consider the set of attributes that agricultural borrowers associate with a well-functioning credit market. These credit attributes include the following:

- *Low and stable interest rates* — so that interest rates and other borrowing costs for agricultural borrowers are comparable to those in other economic sectors and do not fluctuate excessively.

- *Easy accessibility* — so that transactions costs of having access to credit are not excessive.

- *Reliability* — so that credit and related services will be predictable and available through good and bad economic times, whether these times affect the agricultural sector, the financial sector, or both.

- *Versatility* — so that credit is available for a variety of uses and loan programs.

- *Tailored terms* — so that loan sizes, maturities, repayment patterns, security requirements, and so on, are consistent with the payoff periods, cash flow patterns, and other characteristics of the activities being financed.

- *Auxiliary services* — so that other financially related services (e.g., accounting, insurance, management counsel, leasing) are available as needed at reasonable costs.

Economics provides a set of criteria with which to evaluate the competitive performance of financial markets and the institutions that make up these markets. The basic approach is to array the possible market structures on a continuum ranging from perfectly competitive to pure monopoly — from the financial institution's standpoint — and to establish the perfectly competitive market as the ideal target from society's viewpoint.

As a concept, perfect competition assures a socially optimal mix of loans, securities, and other financial services. In perfect competition, the market is characterized by numerous, well-informed, profit-maximizing financial institutions that produce the same types of products and services, that can freely enter and leave the market, and that acquire their resources (e.g., loanable funds) in highly efficient markets. Over the long run, prices, (i.e., interest rates) will just cover the total cost of producing loans; the interest rates and costs will be at their minimum average levels; and a fair rate-of-return will be earned on all resources, including the lender's equity capital. Throughout the market, resources and products will flow so as to equalize returns at the margin, and institutional sizes will be optimal.

Many factors complicate the perfectly competitive market model and make it difficult to achieve. If risk is introduced, then the marginal returns may differ among loans according to their risk premiums. Loan funds then will be allocated to borrowers having the greatest risk-adjusted rates-of-return because these borrowers have the highest ability to repay. The result is equal risk-adjusted returns — but not equal interest rates among regions, loan types, and borrowers. Measuring and pricing the amounts of risk are challenging tasks.

Regulation is another factor complicating the market models. Regulations presumably respond to specific market imperfections and help to assure the safe and sound workings of credit markets. However, regulations may produce unintended adverse consequences as well if, for example, they turn out to stifle innovation, curtail competitive forces, restrain interest rate movements, or block entry and exit of market participants.

The two extremes of competitive conditions are perfect competition and monopoly in which one seller is the only market alternative. A monopolistic market is less effective and more costly to society. However, the extreme case of perfect competition is virtually impossible to achieve. Thus, movements toward greater competition generally are preferred.

An interesting irony in this process is the realization by individual financial institutions that they themselves may not all benefit from greater competition. To a businessperson, the economic concept of perfect competition really means no competition at all! In daily operations, businesspersons strive to outcompete their competitors in order to improve profitability, market share, growth, and so on, without unduly jeopardizing survival. Thus, strong efforts are observed in a "more competitive" environment to actually create less competition — by segmenting the market perhaps along customer lines, differentiating products, emphasizing marketing, and developing innovative pricing strategies in these segmented markets. These efforts are further complicated by changes in the attributes for an effective credit market cited previously.

Observing trade-offs between concentration and market power on the one hand and economies of size and efficiency on the other has been common in evaluating the structure and performance of markets and institutions. High levels of market concentration and power could lead to excessive profits and even collusion among institutions in terms of pricing, barriers to entry, and suppression of market information, all to the detriment of the public interest. At the same time, however, greater market concentration often is associated with larger sizes of firms and more efficient operations, which may contribute to the public interest. Establishing laws and public policy about market competition necessitates considering both criteria.

In general, greater competition is preferred in order to move a financial market closer to the perfect market model. However, it is important to remember that competition can be a double-edged sword — that is, both the benefits and the disadvantages of competition should be considered.

Benefits of Increased Competition

- Borrowers have greater choices — in selecting lenders, loan programs, and credit terms.

- Credit attributes are improved — lower interest rates, lower transactions costs, greater reliability, more versatile uses, additional services.

- Lender quality is heightened — as competition increases, lenders develop keener management, more market awareness, greater innovativeness, and more effective strategic planning.

- Better data about market conditions and firm performance are widely available at lower costs.

Disadvantages of Increased Competition

- Excessive competition could cause losses and detrimental actions for both borrowers and lenders.

- Excessive lending risks could result from overly aggressive or ill-conceived competitive actions of lenders.

- Smaller and less efficient sizes of some lending operations could occur.

- Instability in credit availability could result from lenders exiting the industry, redirecting their lending activities to other types of borrowers, or failing to identify all qualified borrowers.

- Unintended structural changes in borrowing units could take place.

Summary

Financial intermediation facilitates the channeling of financial capital from savers to investors in agricultural areas. The development of financial markets in general and of financial markets serving agriculture has been reviewed in this chapter. The conceptualization of financial intermediation included descriptions of the process and functions of intermediation, the regulatory environment, and the role of market competition.

Specific highlights include the following:

- The historic development of commercial banking in the United States has reflected the avoidance of large concentrations of power within the banking system.

- Nonbank financial institutions have played an increasing role in both domestic and international financial markets.

- Financial markets for agriculture have long had significant participation by commercial banks, especially smaller banks in rural areas, but other specialized financial institutions for agriculture have developed as well.

- The Farm Credit System is a specialized, privately owned government-sponsored institution that provides a full complement of short-, intermediate-, and long-term credit to agricultural producers.

- The Farmers Home Administration is a government lending agency of last resort for farmers who cannot qualify for credit from commercial sources.

- The major functions of financial intermediaries include loan origination, flows of funds, risk-bearing, liquidity, and loan servicing.

- The intangible nature of financial assets and services requires a strong regulatory environment in order for financial markets to develop and function effectively.

- A strong regulatory environment in banking is characterized by geographic restrictions on banks, confinement of banking activities, deposit insurance, and until the 1980s, limitations of interest rates on deposits and loans.

- Greater competition in financial markets benefits society by better meeting the needs of savers and investors.

Topics for Discussion

1. Describe the historic development of commercial banking in the United States. Why is the banking system decentralized?

2. What factors caused asset – liability management in banking to become more important beginning in the 1970s?

3. What are the key factors in the development of financial markets for agriculture?

4. Discuss the major changes in the levels and market shares of farm real estate debt and non – real estate debt since 1945.

5. What is financial intermediation and how are financial intermediaries developed?

6. Describe the major services provided by financial intermediaries and indicate trends in the separation of these services.

7. Why do financial markets experience high degrees of regulation? What forms do regulations take?

8. What attributes do agricultural borrowers associate with a well-functioning credit market?

9. What are the characteristics of a perfectly competitive financial market? What factors complicate the attainment of a perfectly competitive market?

10. Identify and discuss the benefits and disadvantages of increased commercial banking in the United States. Why is the banking system decentralized?

References

1. Henning, C. N., W. Pigott, and R. H. Scott, *Financial Markets and the Economy*, 5th ed., Prentice-Hall, Inc., Englewood Cliffs, New Jersey, 1988.

2. Hoag, W. G., *The Farm Credit System*, Interstate Publishers, Inc., Danville, Illinois, 1976 (out of print).

3. Hughes, D. W., S. C. Gabriel, P. J. Barry, and M. D. Boehlje, *Financing the Agricultural Sector: Future Challenges and Policy Alternatives*, Westview Press, Boulder, Colorado, 1986.

4. Johnson, H. J. *Financial Institutions and Markets: A Global Perspective*, McGraw-Hill Book Company, New York, 1993.

5. Kidwell, D. S., R. Peterson, and D. Blackwell, *Financial Institutions, Markets and Money*, 5th ed. Dryden Press, Fort Worth, 1993.

6. Lee, W. F., et al., *Agricultural Finance*, 8th ed., Iowa State University Press, Ames, 1988.

7. Madura, J., *Financial Markets and Institutions*, 2nd ed., West Publishing Co., New York, 1992.

8. Meerschwam, D. M., *Breaking Financial Boundaries*, Harvard Business School, Boston, 1991.

9. Rose, P., and J. Kolari, *Financial Institutions*, 4th ed., Richard D. Irwin, Inc., Homewood, Ilinois, 1993.

10. Van Horne, J. C., *Financial Market Rates and Flows*, 3rd ed., Prentice-Hall, Inc., Englewood Cliffs, New Jersey, 1990.

16

The Management Environment for Financial Intermediaries

Financial intermediaries generally are *corporate firms whose performance is motivated by profitability and wealth considerations for their owners.* These motivations are similar to the goals of other non-financial types of firms. Specific characteristics of financial intermediaries are the dominance of transactions involving financial assets, relatively high levels of financial leverage, and strong regulatory environments. These factors give rise to a unique perspective on the financial management of intermediaries, discussed in this chapter. Included are discussions of asset – liability management, risk management, and the basic concepts of loan pricing.

ASSET – LIABILITY MANAGEMENT

Financial management of intermediaries generally is characterized by the term *asset – liability management.* **Asset – liability (A – L) management** refers to *the simultaneous consideration of all interest-bearing assets and liabilities that make up the institution's portfolio in order to achieve desired levels of expected profitability, risk, and liquidity.* Included in asset – liability management are decisions about the volume, mix, maturity, cost, liquidity, and risk attributes of various types of financial assets.

Because the major assets and liabilities of a financial institution are interest-bearing, we direct our attention to *the margin of difference (or spread) between the interest earnings on assets and the interest costs of liabilities.* This margin of difference is called the **net interest margin.** The earnings from the net interest margin are used to pay institutional operating costs, cover loan losses, meet tax obligations, and provide a return to the institution's equity capital. An intermediary's objectives focus on the magnitude and stability of the net interest margin. Strategic plans center on making the net interest margin as large and as stable as possible.

The magnitude of the net interest margin largely reflects the degree of competition in the intermediary's market place. That is, pricing reflects what the market will bear in terms of interest rates on loans and other assets versus the cost of funding these assets. The greater the competition, the higher will be the intermediary's cost of funds because the intermediary must bid higher interest rates to attract an investor's funds. Similarly, the greater the competition, the lower will be interest rates on assets because other intermediaries are offering competitive rates to obtain the borrower's business. The net effect of greater competition is a smaller net interest margin.

Secondary markets for many types of bonds are well developed and highly competitive. The institution largely is a price taker in these markets. However, markets for customer loans and for local sources of funds (e.g., deposits) may be characterized by downward sloping demand curves and upward sloping supply curves for individual institutions, thus indicating less than perfectly competitive markets. That is, prices (i.e., interest rates) on these types of financial assets may be a function of the volumes of assets traded, with the intermediary, in part, becoming a price maker in the markets for these assets. Thus, financial institutions operate in markets with significantly different competitive conditions.

Sources of Risk

The stability of the net interest margin depends on the intermediary's major sources of risk and use of various risk management techniques. In broad terms, financial institutions may experience seven sources of risk:

1. *Credit or default risks* from potential delinquency or default on repayment by borrowers. Included are credit risks on loans to individuals and businesses and potential default on securities and other types of investments.

2. *Investment risks* from potential capital losses on securities sold prior to maturity.

3. *Liquidity risks* from possible loss of access to specific sources of funds (e.g., deposits, bond markets).

4. *Cost of funds risks* from unanticipated variation in the intermediary's cost of major sources of funds.

5. *Financial risks* from high financial leverage that characterizes most financial institutions.

6. *Regulatory risks* from unanticipated changes in the institution's regulatory environment that require costly adaptations in procedures and policies.

7. *Fraud and fiduciary risks* associated with the performance of the intermediary's personnel.

Several of these sources of risk may be interrelated. For example, the intermediary may experience a higher cost of funds or a loss in access to funds resulting from higher credit risks on loans. Higher credit risks could also lead to a tighter regulatory environment. Changes in costs of funds, reflecting variations in interest rates, also will cause changes in values of financial assets, thus linking together investment risk and cost of funds risk. This linkage is often called **interest rate risk.** In addition, the institution may adjust its leverage position or be required to do so by regulation in response to changes in other sources of risk.

GAP MANAGEMENT AND INTEREST RATE RISK

One approach to handling instability of net interest margins for a financial intermediary is called **gap management,** which primarily *stresses the effects of interest rate risk on the net interest margin.* The concepts and measures of gap management for financial institutions are similar to those of liquidity management for other types of firms. A short-term planning horizon is established, and the balance between assets and liabilities that can be repriced within the planning horizon is emphasized.

The gap measures employ the concepts of **rate-sensitive assets (RSA)** and **rate-sensitive liabilities (RSL).** Rate-sensitive assets and liabilities are *those* *that either mature or can be repriced within the planning horizon used in asset – liability management.* Assuming a planning horizon of six months, let's consider examples of rate-sensitive assets that include securities with remaining maturities of less than six months, fixed-rate loans with remaining maturities of less than six months, and floating-rate loans that can be repriced within six months, regardless of the remaining maturity. Examples of rate-sensitive liabilities include transaction accounts in depository institutions, time and savings deposits with maturities less than six months, other sources of fixed-rate borrowing with remaining maturities less than six months, and variable-rate liabilities that allow rate adjustment within the planning horizon.

The gap measures may be expressed as absolute dollar values or as ratios, similar to measures of working capital and current ratio, respectively, that are used in liquidity analysis for other types of firms. An absolute measure of gap is the difference between rate-sensitive assets and rate-sensitive liabilities, as follows:

$$\text{Gap \$} = \text{RSA} - \text{RSL}$$

The gap ratio is:

$$\text{Gap}_{\text{ratio}} = \frac{\text{RSA}}{\text{RSL}}$$

The gap ratio approach is the more general measure because it allows comparisons of gap positions among institutions of different sizes.

Another commonly used ratio in asset – liability management is the ratio of the difference between rate-sensitive assets and liabilities to earning assets (EA) employed by the institution.

$$\text{Gap to EA ratio} = \frac{\text{RSA} - \text{RSL}}{\text{EA}}$$

The gap to earnings assets ratio may be a more valid indicator of vulnerability to interest rate risk when the proportions of assets and liabilities that are rate-sensitive are small.

To illustrate, a commercial bank with $80 million of rate-sensitive assets, $60 million of rate-sensitive liabilities, and $100 million of earning assets would have a gap of $20 million (= $80 million – $60 million), a gap ratio of 1.33 (=

$80 million ÷ $60 million), and a gap to earnings assets ratio of 0.20 (= $20 million ÷ $100 million).

In principle, a financial institution's net interest margin is insulated from interest rate risk when the gap ratio equals 1.0:

$$\frac{\text{RSA}}{\text{RSL}} = 1.0$$

Or, when the dollar value of gap is 0:

$$\text{RSA} - \text{RSL} = 0$$

Under zero gap conditions, changes in interest rates should have offsetting effects on the revenues and costs of assets and liabilities so that the net interest margin will be stabilized. The effect is to achieve a perfectly (or nearly so) positive correlation between returns on assets and the costs of liabilities over the planning horizon in question. This strategy would only be modified if trends in interest rates were anticipated. For example, an upswing in interest rates suggests that the gap ratio should exceed 1.0; more assets then would be repriced than liabilities, thus increasing the net interest margin. An anticipated downward trend in interest rates would suggest a gap ratio of less than 1.0. Of course, in an efficient financial market, trends in interest rates are difficult to anticipate because all of the information about future rates is already reflected in the current rates.

The concepts and measures of gap are relatively simple and straightforward to use; however, like other widely used financial ratios, the gap approach has several significant shortcomings. Neither rate-sensitive assets nor rate-sensitive liabilities are homogeneous within their respective categories. As a result, interest rates on different financial assets and liabilities do not move perfectly together. Repricing opportunities within the planning horizon may differ as well, especially as the length of the planning horizon is extended. Thus, the choice of planning horizon (e.g., one month, three months, six months) is important. In addition, changes in net interest margins resulting from adjustments in the volume and mix of earnings assets and interest-bearing liabilities are not accounted for by the gap measures.

Competitive and structural characteristics of banks may influence gap strategies as well. For example, competition may constrain an institution's repricing of assets, or an institution may have to pay more for funds than desired. A manager of a small bank might choose to maintain a gap ratio greater than 1.0 because the bank has less control over the cost of funds but can adjust the return on loans. On the loan side, the smaller agricultural bank may be facing imperfect markets stemming from strong customer relationships, seasonality factors, and fewer loan sources available, while the market for sources of funds could be more competitive.

Finally, customers themselves could be resistant to the effects of some gap strategies. If interest rates are expected to increase, customers would prefer to lock fixed rates on loans, while the institution would favor variable rates that could be adjusted upward. Moreover, the use of variable-rate loans passes interest rate risks along to borrowers, who may view this action as an increase in credit risk on their loans.

DURATION GAP
AND INTEREST RATE RISK

While gap management focuses on stabilizing a financial institution's net interest margin, a more sophisticated tool called **duration gap management** focuses on stabilizing the market value of the financial institution's net worth from interest rate changes. The fundamental idea, based on present value concepts, is that *the market (present) value of a fixed-rate financial asset is inversely related to the level of current market interest rates.* As interest rates increase, and the asset's cash flow remains constant, the asset's market value decreases — and vice versa.

Viewed from an asset – liability perspective, an increase in market interest rates would decrease the values of both a financial asset and a financial liability. If the decreases in present values are of the same size, then the difference between the value of the asset and the value of the liability — i.e., net worth — would remain the same. The institution's net worth then would be insulated or immunized against interest rate risk.

The technique for immunizing net worth against interest rate risk involves structuring a portfolio so that the present values of the fixed-rate assets and liabilities will respond equally to changes in market interest rates. Matching maturities of assets and liabilities will not accomplish the goal because differences in time patterns and levels of cash flows within the maturity period will yield different present value responses to changes in interest rates. However, matching durations will work.

An asset's (or a liability's) **duration** is defined as *the weighted average* *maturity of its cash flows, with the present value of each cash flow used as a weight.* Duration is measured by the following equation:

$$D = \frac{\sum_{n=0}^{N} (n) \left(\frac{P_n}{(1 + i)^n} \right)}{\sum_{n=0}^{N} \frac{P_n}{(1 + i)^n}} \tag{16.1}$$

where n is the time period, P_n is the cash flow, and i is the market interest rate. For example, the duration of a series of \$1,000 payments (P_n) received for three years at an interest rate (i) of 8 percent is:

$$D = \frac{(1) \left(\frac{1,000}{1.08} \right) + (2) \left(\frac{1,000}{(1.08)^2} \right) + (3) \left(\frac{1,000}{(1.08)^3} \right)}{\sum_{n=0}^{N} \frac{1,000}{(1.08)^n}}$$

D = 2.71 years

The duration measure represents the asset's price elasticity — that is, it approximates the average percentage change in an asset's present value in response to a very small unit change in interest rate. Thus, two financial assets having the same duration should have the same relative increase or decrease in market value in response to a given decrease or increase in market interest rates. The strategic implication for net worth stabilization for a financial institution then is to match durations on assets to those on liabilities. If the durations are the same, then a percentage point increase in market interest rates should reduce the present value of assets and liabilities in the same proportion, leaving net worth essentially the same.

Duration measures are complex to derive and to use. Conceptual difficulties arise because an asset's duration is sensitive to the levels of interest rates. When interest rates change, so do durations. Frequent changes in portfolio holdings then may be needed as a part of the net worth immunization strategy. Moreover, detailed data and calculation procedures are necessary to implement the approach. Thus, duration strategies are mostly employed by larger financial institutions that have the necessary skills and resources, that have substantial holdings of fixed-rate financial assets, and that are concerned about short-term fluctuations in institutional net worth. In agricultural lending, for example, the smaller size and greater concentration of ownership of smaller agricultural banks suggests a somewhat greater focus by bank management on the stability of net operating margins relative to net worth.

Despite the limitations of gap and duration gap measures, these concepts play an important role in asset – liability management by financial institutions. They contribute significantly to structuring assets and liabilities in order to stabilize net interest margins and/or net worth and thus respond to interest rate risks. The measures serve as important signals to potential problems if interest rates change unexpectedly, and they contribute to strategic financial planning. They also complement other responses to risk employed by financial institutions.

OTHER RISK MANAGEMENT TECHNIQUES

Other risk management techniques used by financial institutions are similar in general to those of other types of firms. Diversification between loans and securities, among types of loans, and across customers within given types of loans helps to reduce potential income variations. The correlations among the returns on various types of financial assets are an important element in a diversification strategy. Generally, larger financial institutions or broader-based institutional systems (e.g., bank holding companies) are considered to have greater potential for diversification. Such institutions usually can carry more financial leverage. Some specialized lenders (e.g., the Farm Credit System; thrift institutions) may have broader geographic markets to help achieve diversification.

Liquidity management is an important institutional response to risk. Included are holdings of primary and secondary reserves. Primary reserves consist of cash, deposits held at other institutions, and other highly liquid financial assets. They are held for transactions purposes and to meet day-to-day changes in cash demands. Secondary reserves consist of holdings of government securities, agency bonds, corporate bonds, and other financial assets with well-developed secondary markets. These holdings respond to longer-term liquidity needs over the course of a year. Because liquid assets generally are less risky, they offer lower rates-of-return than other assets, especially loans. Thus, it is important for financial institutions to maintain enough liquidity to avoid a crisis, while not unduly sacrificing earnings. More specialized financial institutions usually hold larger liquidity in order to counter their greater risks.

Making adjustments in an institution's capital (i.e., leverage) position is another important risk management technique. In riskier times, financial institutions are expected to reduce their leverage positions, thus building a relatively stronger equity capital base to deal with downswings in income and asset values. Leverage positions generally increase in more favorable economic times. For depository and other publicly chartered financial institutions, regulatory requirements play important roles in both capital and liquidity management. As indicated earlier in this chapter, more specialized lenders generally have lower leverage positions.

Larger financial institutions may utilize financial futures contracts and options (and related methods called derivatives because they are derived from other securities) in responding to interest rate risks. However, just as futures contracts are not available for all agricultural commodities, futures instruments are not available for all types of financial assets, especially most types of loans. Usually a closely related financial asset that does have a traded futures contract is used as a proxy for a loan asset. Most involvement in financial futures contracts by financial institutions is concentrated in the securities portion of the contracts' portfolios.

When various lenders are involved in loan participation arrangements, the credit risks of these lenders are usually reduced and less concentrated. Smaller banks may seek loan participations with their correspondent banks to finance large-scale customers. Similarly, a consortium of banks may participate in financing a large-scale corporate project or in international lending. Loan insurance and government guarantees on some types of loans may be considered, thus shifting risk-bearing to the insurer or guarantor. At the individual borrower level, requirements for collateral, loan documentation, and customer monitoring can be tailored to the anticipated credit risk. Finally, some financial institutions may simply choose to avoid certain types of loans whose risk position has already increased.

LOAN PRICING

Loan pricing is an important element of asset – liability management for a financial institution because it is a key managerial control variable. Pricing is based on factors both external and internal to the institution. External factors include the strength of loan demand, lending risks, and the degree of competition in the institution's lending market. Internal factors involve costs of institutional resources, especially sources of funds, and efficiency of the lending program.

An easy way to consider these factors is to translate them into costs experienced by the lender in delivering loans and other financial services to customers. Viewed in this way, decisions about loan pricing basically entail establishing prices to cover the full set of lending costs.

Lending Costs

The major components of an institution's lending costs are as follows:

1. *Administrative costs,* including outlays for personnel salaries, documents, equipment, rent, advertising, legal services, computers, supplies, and other costs of running the loan program.

2. *Funding costs,* including costs of both debt funds and equity funds assigned to support the loan. Debt costs for the institution are the interest payments on funds purchased in the financial market. Examples are interest paid on deposits by depository institutions and interest paid on bonds, commercial paper, or other financial instruments sold in the financial market.

   A unique feature of lending by depository institutions is that borrowers may fund part of their own loans through deposits maintained at the institutions during the term of the loans. The institutions' cost of funding the borrowers' deposits then may be considered in loan pricing.

   Equity costs are the desired return on the institution's own equity capital that underlies the loan. The desired return on equity represents the "profit margin" in the loan.

3. *Risk-bearing costs,* including possible delinquency and default by borrowers (credit risks), unanticipated variations in the borrower's need for funds (liquidity risks), and the combined effects of other institutional risks and methods of responding to risks. Generally, credit risks are the most important and straightforward to estimate; they are reported in aggregate by financial institutions as a "provision for loan loss" to cover anticipated losses and to build a contingent reserve against possible losses in the future.

4. *Competitive costs,* reflected by the level of competition in the institution's loan market. Stronger (weaker) competition from other lenders generally results in lower (higher) prices and profits on loans.

5. *Non-loan costs,* covering services that are packaged or "bundled" into the loan. That is, the loan price may reflect a "bundling" of costs and revenues from various services and activities (e.g., financial counseling, market information) that are closely related to the lending function. Alternatively, the institution may charge separate fees for some of these services so that its compensation does not come from the loan interest rate.

Loan Pricing Model

Given these five cost components, an institution's loan pricing model may be specified as:

$$i = \frac{TC}{L_b} = \frac{A + I_{de} + R + N - F}{L_b} \tag{16.2}$$

where i is the interest rate that must be charged on the loan balance (L_b) to cover total projected lending costs (TC). Thus, the loan rate is found by dividing the total projected costs by the outstanding loan balance. The total lending costs include the sum of administrative costs (A), funding costs (I_{de}) for institutional debt and equity, risk-bearing costs (R), and non-loan costs (N) less any fees (F) paid by the borrower. At this point, no explicit provision is made in the model for the "costs of competition" for the loan.

Clearly, some of the costs of lending programs are more easily measured and quantified than others. Moreover, lenders also may take a long-term view of their relationships with key borrowers. That is, pricing concessions might be offered in the short run to financially distressed borrowers or to young borrowers with good business potential, recognizing that loan growth over the longer term will restore loan profitability.

The loan pricing model also may accommodate variable- versus fixed-rate lending and average cost pricing versus marginal cost pricing. A variable-rate loan is based on the premise that one or more of the key cost components (especially funding costs) may change during the term of the loan, thus affecting loan profitability. The variable-rate concept allows periodic recalculation of the loan rate (upward or downward) to cover the change in costs and to maintain the desired rate-of-return on the institution's equity capital.

Average cost pricing versus marginal cost pricing basically reflects whether the loan is funded by a pool of funds already in the institution or by new funds (a single source or a pool) to be purchased in the market. The average cost of a pool of in-house funds tends to change less frequently than the marginal cost of new purchased funds tailored to the maturity and repayment pattern of new loans. Moreover, average cost does not reflect the current costs of funds, and new funds seldom are available at average cost.

Because the basic credit evaluation functions are similar for various sizes of loans, administrative costs per dollar of loan generally are much smaller on larger loans. Thus, larger lines of credit may warrant lower interest rates, while rates and/or fees may be higher on smaller loans. Similarly, risk premiums will be lower on lower-risk loans than on higher-risk loans. Finally, borrowers who are funded by lower-cost funding sources and who are sought competitively by other lenders may experience lower loan rates as well. Viewed in this fashion, unequal loan rates across borrowers from a given institution are logical to expect and result in more equitable loan pricing.

Pricing Example

To illustrate these concepts, consider an agricultural lending institution that is formulating its pricing policies for the coming year. One of its lower-risk farm borrowers is projected to have an average loan balance of $100,000 over this time period. Administrative costs are estimated to be 2.5 percent of the loan balance (or $2,500). Funding costs are 6.0 percent ($6,000) of the loan balance for purchased funds and 1.5 percent ($1,500) of the loan balance for equity funds supporting the loan, yielding total funding costs of 7.5 percent ($7,500) of the loan balance. Risk-bearing costs are 0.5 percent ($500) of the loan balance. Non-loan costs and fees are each $250.

The data are summarized as follows:

$L_b = 100,000$

$A = 2,500 \ (2.5\%)$

$I_{de} = 7,500 \ (7.5\%)$

$R = 500 \ (0.5\%)$

N = 250

F = (250)

TC = 10,000

Using the loan pricing model, find the interest rate.

$$i = \frac{2{,}500 + 7{,}500 + 500 + 250 - 250}{100{,}000}$$

$$i = \frac{10{,}500}{100{,}000}$$

i = 0.105, or 10.5%

Thus, a 10.5 percent interest rate on the loan will yield sufficient revenue to cover all of the lending costs, for this borrower, including a $1,500 return to the lender's equity capital underlying the loan. If the $100,000 loan were financed with $90,000 of purchased funds and $10,000 of equity capital, then the $1,500 return to the $10,000 of equity would be a 15 percent rate-of-return on equity for the lender.

If a higher-risk borrower were being financed, the risk-bearing cost might increase from $500 to $1,500, yielding a loan rate of 11.5 percent. Thus, risk-adjusted interest rates may differ among borrowers from a given lending institution.

If the loan were a variable-rate loan, then it would be repriced periodically in response to changes in the lender's cost of purchased funds. Suppose, a month later, the cost of purchased funds increased from 6.0 percent to 7.0 percent. The loan rate then would increase from 10.5 percent to 11.5 percent for the lower-risk borrower and from 11.5 percent to 12.5 percent for the higher-risk borrower. Similarly, loan rates would decline if the cost of funds declined.

If the loan were for $1 million instead of $100,000, the administrative costs probably would not increase in the same proportion. Suppose, for example, that the administrative costs increase from $2,500 for the $10,000 loan to $10,000 for the $1 million loan. In percentage terms, these administrative costs per dollar of loan have actually decreased from 2.5 percent to 1.0 percent. Assuming the other lending costs remain proportional to the loan balance, the interest rate on the larger loan is now 9.0 percent versus 10.5 percent on the smaller loan. Thus, economies of size in lending costs are reflected in the interest rates of loans of different size.

Finally, if another lender were trying to attract the borrower's business away from the present lender by offering a lower interest rate, the present lender might have to reduce the loan rate, say from 9.0 percent to 8.5 percent, in order to meet the competition. This loss of 0.5 percent profit would represent a cost of competition for the loan.

Institutional Pricing Level and Profitability Analysis

These loan pricing concepts may be applied at any level in a financial institution — that is, to the aggregate loan portfolio, to major types of loans (e.g., commercial loans, consumer loans) perhaps organized as profit centers or departments and at the individual borrower level. In recent years, there has been a clear trend toward individualized loan pricing at the customer level, especially in commercial and agricultural lending. This trend is attributed to several factors: (1) greater competition in financial markets; (2) computerized accounting and information management by financial institutions; (3) improved financial accounting and reporting by borrowers; (4) greater lending risks; and (5) more sophisticated risk assessment and credit evaluations by lenders (see the credit-scoring discussion in Chapter 7). These developments allow financial institutions to tailor loan rates to the unique administrative, funding, risk-bearing, and competitive costs of individual borrowers.

Finally, the pricing process for individual loans is subject to ex post evaluation after the loans have been repaid to ascertain whether the original profit targets for the loans actually were achieved. This evaluation, called **customer profitability analysis,** determines the rate-of-return realized on the institution's equity capital allocated to the loan, based on actual revenues and expenses associated with the loan. Significant departures from ex ante profit targets can be identified, explained, and accounted for in future pricing practices.

Summary

The dominance of transactions involving financial assets, relatively high levels of financial leverage, and strong regulatory environments are specific characteristics of financial intermediaries that provide a unique perspective on the financial management of lending institutions. Included in this perspective is a comprehensive discussion on asset – liability management.

Specific highlights are as follows:

- Asset – liability management is centered on the magnitude and stability of the **net interest margin,** *the difference between the interest earnings on assets and the interest cost of liabilities.*

- The magnitude of the net interest margin reflects the degree of competition in the market.

- Stability of the net interest margin reflects the institution's vulnerability to risks, including credit risk, investment risk, liquidity risk, cost of funds risk, financial risk, and regulatory risk.

- The stability of the net interest margin is the focus of **gap analysis,** which considers *the relationship between rate-sensitive assets and rate-sensitive liabilities.*

- **Duration gap management** focuses on *insulating the market value of a financial institution's net worth from the interest rate risk.*

- Other risk management techniques for financial institutions include various forms of diversification, liquidity, leverage adjustments, financial futures, loan participation, insurance, guarantees, customer monitoring, and avoidance.

- Effective loan pricing basically involves setting interest rates to cover the major sources of lending costs: administrative costs, funding costs, risk-bearing costs, competitive costs, and non-loan costs.

- Differences in these costs among borrowers mean that differences in interest rates are logical to expect.

- Loan pricing and post-loan customer profitability analysis are important components of management by financial institutions.

Topics for Discussion

1. What are the major objectives of asset – liability (A – L) management?

2. Identify and explain the major sources of risk faced by financial institutions.

3. Define *gap,* explain the various measures of gap, and indicate the role of gap in responding to interest rate risk.

4. How is duration gap management distinguished from gap man-

agement? In what way does duration represent an asset's price elasticity?

5. What are the limitations of gap and duration gap management?

6. What other methods of risk management are employed by financial institutions?

7. Explain the relationship between loan pricing and costs of lending.

8. Distinguish between fixed- versus variable-rate lending and average versus marginal cost pricing.

9. Given the following data, find the loan price (interest rate): administrative costs, 2 percent; cost of funds, 5 percent; risk-bearing costs, 1 percent; non-loan costs, 0.5 percent; fees, 0.0 percent.

References

1. Barry, P. J., "Financial Stress for the Farm Credit Banks: Impacts on Future Loan Rates for Borrowers," *Agricultural Finance Review*, 46(1986):27 – 36.

2. Barry, P. J., and W. F. Lee, "Financial Stress in Agriculture: Implications for Agricultural Lenders," *American Journal of Agricultural Economics*, 65(1983):945 – 952.

3. Bierwag, G. O., and G. C. Kaufman, "Duration Gap for Financial Institutions," *Financial Analysts Journal*, 41(March – April 1985):68 – 71.

4. Ellinger, P. N., and P. J. Barry, "Interest Rate Risk Exposure of Agricultural Banks: A Gap Analysis," *Agricultural Finance Review*, 49(1989):9 – 21.

5. Gardner, M. J., and D. L. Mills, *Managing Financial Institutions: An Asset / Liability Approach*, 2nd ed., Dryden Press, New York, 1991.

6. Graddy, D., and A. Karna, "Net Interest Margin Sensitivity Among Banks of Different Sizes," *Journal of Bank Research*, 14(Winter 1984):283 – 290.

7. Kaufman, G. C., "Measuring and Managing Interest Rate Risk: A Primer," *Economic Perspectives*, Federal Reserve Bank of Chicago, 8(January – February 1984):16 – 29.

8. Reilly, F. K., and R. S. Sidhu, "The Many Uses of Bond Duration," *Financial Analysts Journal*, 36(July – August 1980):58 – 72.

9. Sinkey, J. F., *Commercial Bank Financial Management in the Financial Service Industry*, 4th ed., Macmillan Publishing Co., New York, 1991.

17

Financial Intermediaries in Agriculture

Most countries have several types of financial intermediaries that provide loans and other financial services to agriculture. These credit sources include:

1. Commercial banks that primarily rely on deposits for funds.

2. Specialized farm lending institutions having corporate or cooperative organizations that depend on money market sources of funds.

3. Government programs at the federal, provincial, and/or state level that primarily rely on tax sources of funds.

4. Credit unions composed of members with a common bond.

5. Farm-related trade or agribusiness firms.

6. Intermediaries that perform important fiduciary or trust functions, such as insurance companies, pension funds, and trust companies.

7. Individuals, such as family members, sellers of farm land, or "money lenders" in developing economies.

These credit sources differ in their organizational structures, operational characteristics, degrees of specialization, sources of funds, and relative importance. But, they all perform the basic intermediation services identified in Chapter 15. Moreover, they also appraise each borrower's credit worthiness to achieve a desired loan portfolio. In addition, government programs may be tailored to the credit needs of particular borrowers, often with concessionary terms.

In this chapter we review the role and characteristics of major agricultural lenders in the United States. In general, these lenders have achieved high performance in financing farmers and in making timely innovations of institutions, instruments, and practices for meeting farmers' capital and credit needs. The agricultural credit market evolved from strong reliance on local merchants and country banks a century ago to today's dependence on the cooperative Farm Credit System, U.S. government lending agencies and credit programs, local – regional – national credit programs of many trade firms and agribusinesses, and an extensive commercial banking system. Still, however, considerable financing continues to occur by individuals, especially sellers of farm land.

Major evolutionary features of the agricultural credit market are the relatively large size and the regional or national orientation of many intermediaries. The characteristics of the cooperative Farm Credit System are those of a large national branching organization. Life insurance companies have regional or national orientations in agricultural lending, as do many agribusiness firms involved in agricultural lending. Even local offices of the federal government are branches of a national organization. Money center banks, regional banks, many branch banks, and Federal Reserve Banks also are large and are part of the national financial markets. Even so, smaller community banks located in rural areas remain an important, although declining, source of agricultural lending.

COMMERCIAL BANKS

As discussed in Chapter 15, commercial banks play a major role in providing credit and related services to U.S. agriculture. Their involvement varies by bank size, location, specialization, and type of organization. The major money center banks (in New York, Chicago, San Francisco, Dallas, etc.) generally finance larger operations involved mostly in livestock and poultry production, although an increasing volume of specialized crop production is also being financed by the large money center banks. Money center banks in states with liberal branching laws may serve both large and small farming operations. These banks may also finance agribusinesses and international trade and participate in loans with regional and community banks. Moreover, they are important buyers of securities sold by the cooperative Farm Credit System.

A detailed analysis of balance sheet data for all insured commercial banks taken from the Federal Deposit Insurance Corporation (FDIC) reveals that important shifts are occurring in bank financing of agriculture. In 1964, for example, banks with less than $25 million in assets held about 70 percent of all bank farm debt outstanding. By 1991, the number of these small banks had declined significantly, and their share of bank farm loans had declined to about 13 percent. Over the past decade, many banks with less than 10 percent of their assets invested in farm loans have become important agricultural lenders.

Although the traditional agricultural banks (those with farm loans making up 15 percent or more of their assets) have lost ground to the nonagricultural banks, the agricultural banks are still an important and stable source of farm loans. They tend to be rural banks of sufficient size to achieve economies of scale. Moreover, those banks with over 40 percent of their assets in farm loans tend to be small, rural banks. These banks rely heavily on the economic viability of the local economy for their loans and for their sources of funds. In rural areas, that economy is directly linked to agriculture.

Structure of the Banking System

Federal laws about banking structure in the United States reflect an historic concern about undue concentration of economic power, competitive equality between state and national banks, and state sovereignty in setting restrictions on geographic limits in banking. These concerns have been expressed in the McFadden Act of 1927, which prohibits interstate branching and reserves intrastate branching policies to each state, and in the Douglas Amendment to the Bank Holding Company Act of 1956, which prevents bank holding companies from buying or establishing out-of-state subsidiaries unless authorized by the states. The result is a diverse set of state limitations on branching and holding companies. Thus, states may be designated as unit banking states, limited branching states, or statewide branching states. Moreover, states differ in their authorizations for single bank holding companies or multi – bank holding companies, although these differences have decreased significantly over time.

In a unit banking system, an individual bank traditionally maintains only one office or place of business. Rural banks in unit banking states generally are small, with their size being limited by the amount of business in their communities. Even in traditional unit banking states, however, drive-in windows and additional offices within specified distances are now being allowed. Moreover, as discussed later, these unit banks have correspondent relationships with larger banks to obtain various banking services.

Branching refers to the multiple offices that one bank firm has. In the

United States, branching cannot extend beyond state boundaries, and in some states, it is limited to specific areas, such as counties or groups of counties. In contrast, national branching is found in many other countries.

A bank holding company brings one or more banks under the control of a firm that owns a major portion of the bank's stock. A one bank holding company generally permits the bank to provide financially related services through a wholly owned subsidiary, thereby expanding the range of bank services. A multi – bank holding company is generally organized to provide some of the benefits of branch banking, such as economies of scale in furnishing some services, spreading risks over broader geographic areas with greater economic diversity, and enhancing flows of funds between the affiliate banks.

Restrictions on branching and holding companies are easing in response to technological and economic pressures for banks to expand toward interstate banking. Despite federal restrictions, banks are finding ways to expand interstate through loan production offices, nonbank subsidiaries, reciprocal bank expansion laws among states, and nonbank banks that do not simultaneously take deposits and make commercial loans.

Numerous states have reduced the constraints on branching and multi-bank holding companies. Bills to permit some form of interstate banking involving holding company affiliations have been passed by over 30 states in recent years. Moreover, a number of regional banking markets have developed. Zones in New England, the Midwest, and the Southeast have been formed, and others are near implementation. The legalizing of regional interstate banking was questioned initially, but in 1985, the U.S. Supreme Court ruled that state legislation permitting regional interstate banking is not unconstitutional.

Multi-office banking has significantly increased its role in farm lending during the past decade. Multi-office banking provides greater access to money markets, reduces legal lending limit problems, allows banks to more easily diversify into new markets and services, and promotes economies of size in functions such as electronic data processing and telecommunications. The trend toward increased branching and holding company affiliation, including interstate banking, will continue. In late 1994, the U.S. Congress passed legislation that will soon allow nationwide interstate branching by commercial lenders, unless individual states choose not to allow it.

Other Regulatory Issues

As discussed in Chapter 15, financial institutions in general and commercial banks in particular experience high degrees of regulation to assure desired levels of safety, stability, competitiveness, and responsiveness to monetary policy. Besides the structural issues considered previously, other major bank regulations include requirements for legal reserves, capital adequacy, legal lending limits, and deposit insurance. Banks are also subject to periodic audits and examinations to check regulatory compliance and to appraise their portfolio quality and management practices.

Reserve requirements established by federal and state legislation assure minimum levels of bank liquidity. The major emphasis in reserve requirements, however, has shifted to the federal level. The 1980 Depository Institutions Deregulation and Monetary Control Act (DIDMCA) established uniform reserve requirements at all depository institutions in the United States; depository institutions include commercial banks, savings and loan associations, mutual savings banks, and credit unions. The act requires depository institutions to hold reserves on all transactions accounts and on all nonpersonal time deposits. **Transactions accounts** are *those that are subject to withdrawals for transaction purposes.* They include demand deposits, negotiable orders of withdrawal (NOW), and automatic transfers from savings.

Effective in 1993, required reserves are specified as 3 percent of a bank's first $51.9 million of transactions balances. The reserve requirement on larger transactions balances is 10 percent. The Federal Reserve Board may vary these requirements within specified ranges and may impose supplemental reserves for monetary control purposes. Member banks may hold reserves as cash or as balances at Federal Reserve Banks. Banks that are not members of the Federal Reserve System may also place their reserves as pass-through balances in other depository institutions, which, in turn, maintain such funds as balances in Federal Reserve Banks. Thus, the required reserves must be held in a non-earning form.

Legal lending limits to any bank borrower are based on the bank's equity capital. They are designed to limit the concentration of lending risks. For nationally chartered banks, loans to an individual borrower cannot exceed 15 percent of the bank's unimpaired capital and unimpaired surplus fund (a measure of net worth). However, the limit increases to 25 percent for loans to purchase livestock or for loans fully secured by readily marketable collateral (other exceptions are also provided in the law). Loan limits for state-chartered banks vary among the states but, in general, are comparable to those of national banks. Agricultural banks in unit banking states often experience problems in meeting larger farm loan requests that exceed the banks' legal lending limits. These banks must develop loan participations with other lenders for these customers or risk losing their business.

Deposit insurance through the Federal Deposit Insurance Corporation (FDIC) is required of all banks that are members of the Federal Reserve System and is used by nearly all nonmember banks as well. As of 1994, the insurance covers deposits up to $100,000 per account.

In 1988, regulators from 12 countries signed a risk-based capital adequacy plan for commercial banks. This agreement, called the Basle Agreement, divided bank capital into two categories and classified assets into separate risk groups. A total value for risk-adjusted assets is determined, and a minimum level of capital is required, based on the value of risk-adjusted assets.

The two capital categories are termed tier 1, or "core capital," and tier 2, or "supplemental capital." Tier 1 is primarily composed of tangible equity capital, while tier 2 is predominantly made up of subordinated debt, loan loss reserves, and intermediate-term preferred stock.

Assets are classified into different risk groups and assigned weights to determine the total risk-based assets. Assets such as cash and U.S. government securities have no default risk and thus have zero weight. Items such as cash items in the process of collection and mortgage-backed U.S. government and U.S. government agency securities are given a 20 percent weight. Examples of assets in the 50 percent weight group are state and local revenue bonds and first mortgages on home loan. All other assets including loans and higher-risk securities are part of the 100 percent weight class. Off-balance sheet items are also converted into "credit equivalents" and are placed into one of the four asset weight categories.

The minimum capital standards required by the Basle Agreement by year-end 1992 are: (1) Tier 1 capital must be 4 percent of risk-adjusted assets and 3 percent of total assets and (2) total capital (tier 1 plus tier 2 capital) must be 8 percent of risk-adjusted assets.

Portfolio Management of Banks

Banks are commercial businesses, organized as corporations. They operate for profit purposes on behalf of their stockholders, with due consideration given to risk and liquidity. Since farm lending is a major income-generating activity, especially for smaller rural banks, we should consider its contribution to bank profitability, risk, and liquidity.

Figure 17.1 presents the various alternatives in managing a bank's assets (uses of funds) and liabilities (sources of funds) in order to generate profits, build equity capital, and achieve other bank goals. The major uses of funds are categorized as (1) **reserves,** (2) **loans,** (3) **investments,** and (4) **services.** In addition, a relatively small proportion of the bank's total assets are allocated to bank facilities and equipment. Funds also are withdrawn from the bank to pay dividends on bank stock and to meet income tax obligations. There are two types of reserves — primary reserves that satisfy legal requirements and secondary reserves that are held to meet longer-term liquidity needs brought about by seasonal or cyclical factors in the bank's market area. Secondary reserves are interest-bearing and generally consist of U.S. Treasury bills; bonds issued by "agencies" of the federal government (agency bonds), such as FCS bonds; or short-term loans (called federal funds) to other banks.

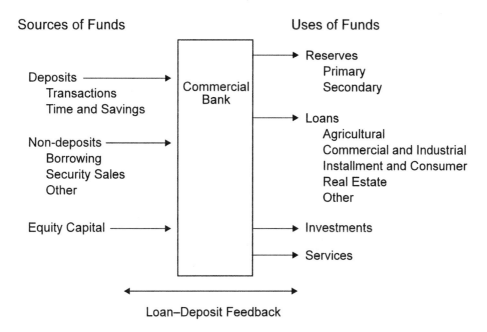

Figure 17.1. Sources and uses of bank funds.

Loans are usually categorized as commercial and industrial, agricultural, installment and consumer, real estate, and other. They usually constitute the major portion of a bank's assets. *Although lending tends to yield the highest profits, it also carries the highest risk and lowest liquidity of all bank investments.* Along with some deposits, loans provide the greatest opportunity for differentiating between the bank's "products" and competing for a greater market share. The bank's loan portfolio is generally determined by the size, the business orientation and other characteristics of the market area, as well as by bank policy. Larger banks operating in regional, national, or international markets usually have more diverse loan portfolios and more specialized loan officers.

Banks face two sets of policy choices in formulating their loan portfolios. One choice involves the allocation of loanable funds to the various loan categories. This major policy decision is determined by the executive officers and bank directors. It is a longer-run type of decision that is re-evaluated from time to time. The other policy choice involves the allocation of loan funds to the various borrower-applicants who make up the respective loan categories. These day-to-day types of decisions represent the operational component of lending activities. Both choices are based on profitability and risk considerations; however, credit decisions at the customer level emphasize the credit worthiness factors considered in earlier chapters.

Investments also include holdings of various types of securities: U.S. government bonds, U.S. agency bonds, municipal bonds, and others. Investment holdings usually represent a residual use of funds after reserve requirements and loan demands have been satisfied. Investments have longer maturities than reserves do. The expected returns are higher than those of reserves, but they usually are lower than those of loans and involve both interest returns and capital gains or losses if securities are sold prior to their maturity. Secondary markets for sales or purchases make investments highly liquid assets. Investment risks primarily consist of fluctuations in security prices as market interest rates change. A bank's management of income tax obligations centers on the investment portfolio because municipal bonds are exempt from federal income taxation and U.S. government and agency securities are exempt from state and local taxation.

Various services offered to banking customers are checks, safety deposit boxes, trust and farm management, credit cards, auto licenses, insurance, and many others, including stock brokerage in some situations. These services are generally self-supporting through the payment of fees or service charges. Although federal deregulation of commercial banks has not yet included the range of products and services banks may offer, many states now permit state-chartered banks to offer a number of financially related services.

A bank's assets will vary according to the size and type of bank. The typical agricultural bank located in a rural area with less than $100 million in total assets may have 40 to 60 percent of its total assets in loans, 30 to 50 percent in investments, and 5 to 15 percent in reserves.

The sources of funds that make up a bank's liabilities are mostly deposits, although non-deposit sources have become important, especially for larger banks. Deposits include transactions accounts such as demand deposits and negotiable order of withdrawals (NOW) accounts, as well as time and savings deposits. Nearly all deposits are interest-bearing (except for demand deposits), with rates that differ by size and length of deposit and that vary with market rates of interest.

Non-deposit sources of funds for banks include loans from other banks on a daily or a longer-term basis, loans from Federal Reserve Banks, loan participations with correspondent banks, and for larger banks, commercial paper or bankers' acceptances sold in the national financial markets. In addition, eligible banks may obtain loan funds for their farm customers from Farm Credit Banks by discounting loans, by direct borrowing, or by forming a subsidiary agricultural credit corporation for the same purposes.

Finally, banks must maintain an equity base that, along with deposit and non-deposit funds, forms the capital base for asset holdings. A bank's equity capital must grow over time along with the bank's increase in loans and other income-generating activities. For most smaller banks, the main source of equity capital is earnings retained from the profits of lending and investments. Banks must plan carefully to assure that equity grows along with other activities. Banks are highly leveraged firms with equity capital ranging from 5 to 10 percent of total assets and even lower for very large banks.

Banks manage their assets and liabilities to achieve high profitability subject to the risk, liquidity, and competitive conditions described in Chapters 15 and 16. Interest rate risk, in particular, has grown more crucial in recent years. In the past, most banks (and other financial institutions) obtained funds with short maturities at costs that became increasingly rate-sensitive and then committed these funds to loans or investments with longer maturities and fixed rates-of-return. This practice magnified the variability of the banks' profits and reduced bank profitability during periods of rising interest rates. Banks responded to this costs-of-funds risk with improved methods of asset – liability (A – L) management. Included in A – L management are greater use of the floating (or variable) interest rates on loans, more precise matching of maturities and durations on assets and liabilities, and use of financial futures contracts to hedge interest rate risks.

Floating interest rates have gained wide use on farm loans, commercial loans, and to a lesser degree, consumer loans. A floating-rate loan is periodically repriced prior to loan maturity as changes occur in the bank's costs of funds. This practice enables a bank to reduce its interest rate risk by achieving a higher correlation between its costs of funds and its returns on loan assets. In the process, interest rate risks are shifted to the borrower.

Farm Lending and Bank Performance

In general, farm lending has contributed favorably to bank goals. Although subject to the variations in economic performance of farms, agribusinesses, and other farm-related enterprises in their local communities, agricultural banks have exhibited strong profitability on the average over time. Most agricultural banks structure their portfolios to accommodate seasonal patterns of loans and deposits in their market area and possible swings in farm and farm-related income. Their ratios of loans to deposits and loans to total assets are generally less than the ratios of larger banks.

During the 1970s, losses and service costs associated with agricultural loans tended to be lower per dollar of loan than for installment and commercial loans. However, increased delinquencies and loss rates on farm loans during the agricultural depression of the 1980s increased costs and lowered returns from farm loans. As a result, the returns on assets to banks with heavy investments in farm loans dropped below those of nonagricultural banks for the first time in several decades and then came back to more normal levels at the beginning of the 1990s.

Banks are developing improved access to sources of funds for farm loans that should reduce the liquidity problems experienced in the past. Farm loan liquidity is improving too, as evidenced by the growing use of secondary markets for farm and other rural loans guaranteed by the Farmers Home Administration or the Small Business Administration. Through broker services, banks can sell the guaranteed portion of these loans while still retaining loan servicing and other contacts with the borrower. Moreover, the Agricultural Credit Act of 1987 authorized the creation of a secondary market for long-term farm mortgage loans that would improve the ability of smaller rural banks to originate and service such loans [see "Federal Agricultural Mortgage Corporation (Farmer Mac)" for a description of the secondary market].

Loan – Deposit Relationship

The loan – deposit relationship is a unique feature of banking that helps to explain the policies of banks in formulating their loan portfolios. This relationship is based on the importance of deposits as a primary source of a bank's capacity to lend and invest and the resulting importance of loan customers who hold deposits. Loans and deposits have an interactive feedback over time that affects both bank performance and business activity in the local community. This feedback relationship for an individual bank is a micro application of the deposit-creating process that characterizes an entire banking system.

The feedback is expressed when part of the loans to bank customers are returned to the bank as deposits. The feedback of deposits is the result of the combined effects of (1) the borrower's increased business growth arising from the financing, (2) the borrower's increased level of deposits prior to the spending of the loan's proceeds or the repayment of the loan, and (3) the increased level of business activity in the community. The feedback deposits provide the basis for additional loans whose demand may be strengthened by current loans. In contrast, there is no such feedback from most investments or from loans to nonlocal borrowers. Instead, funds allocated for these purposes leave the bank's market area.

The strength of this loan–deposit relationship is difficult to measure. It differs according to the type of loan, the borrower, and the bank structure and competition. It appears especially important in rural financial markets involving relatively few banks. The qualitative aspects of the feedback relationship are evidenced by (1) the practice of most banks whereby they lend only to their depositors, (2) the use of deposit balances in loan pricing, and (3) the value bankers place on attracting new, nonlocal sources of funds that contribute to the bank's deposit base and to community business activity. In recent years, the deregulation of interest rates and greater rate volatility have curtailed the traditional importance of customer relationships and made pricing issues more important. Nonetheless, customer relationships remain a significant factor in banking.

Correspondent Banking

Correspondent banking takes place when a large bank (the *correspondent)* gives various banking services to a smaller bank in return for compensation provided by the smaller bank's payment of fees or placement of a demand deposit with the correspondent. Correspondent services include check clearing, federal funds transactions, security safekeeping and transactions, international services, leasing services, electronic data processing, investment counseling, loans to bank directors and officers, and loan participation.

Rural banks have long relied on loan participation with correspondent banks to meet farm loan requests that exceed their legal limits, to meet total loan demands that are high relative to available loan funds, and as a means of sharing lending risks. Sometimes correspondent banks may originate large loans in which one or more smaller banks are invited to participate. Generally, each of the participating banks carries portions of the loan, receiving interest and security on its respective shares. The correspondent bank is also compensated by demand balances held on deposit by the rural bank. When the participation originates with the correspondent bank, the correspondent bank shares its lending risks with other banks.

The correspondent banking system has experienced considerable changes in recent years. These changes reflect three factors: (1) higher interest rates, which encourage economizing on compensating balances with correspondent banks; (2) changing technology, which has significantly lowered the real cost of many banking services; and (3) changes in regulation initiated under 1980 legislation, which required Federal Reserve Banks to explicitly price their services and offer them to all banks. These developments have altered the competitive position of different suppliers of correspondent banking services as well as the nature of the demand for those services.

As correspondent banking and other forms of interbank relationships have become more competitive, the banks have developed new products. Nontraditional services such as financial planning, discount brokerage, interest-rate swaps, asset – liability management consulting, and marketing support are examples. Clearly, correspondent banking has become more competitive, and its customer base has become more volatile. It is based less on long-standing business relationships and more on competitive market factors.

Correspondent and rural bank loan participations may encounter some difficulties as well. Many correspondent banks are not staffed with agriculturally trained and oriented personnel. They may lack the ability to evaluate and control agricultural credits, especially if accounting information and loan documentation are limited. Distance between urban and rural banks often precludes the correspondent banker from participating directly in the credit evaluation. Finally, some rural banks hesitate to call on correspondents because they are apprehensive about losing customers to the larger banks.

Agricultural Credit Corporations

Agricultural credit corporations (ACCs) are commercial corporations organized primarily to provide credit services to farmers. They are profit-oriented businesses that provide farm loans at interest rates that cover expected costs of funds, operation and administration, and risk-bearing. Through a subsidiary arrangement, agricultural credit corporations offer ways for commercial banks or groups of banks, or other interested parties or firms, to fund farm lending from non-deposit sources. The participating parties, usually banks, must provide initial equity capital for the agricultural credit corporation, which is then levered with other funds, resulting in debt-to-equity ratios that range from 6 to 1 and 10 to 1.

Once equity capital and ACC management are available, the main problem is to develop the source of funds. Three sources are usually possible: (1) selling money market instruments; (2) discounting loans with Farm Credit Banks — called lending to "other financial institutions" (OFIs); and (3) borrowing from other financial intermediaries. Some agricultural credit corporations use all three sources simultaneously.

For smaller banks, the primary benefit of forming an agricultural credit corporation is to operate as another lending institution, thereby gaining access to funds from the Farm Credit Bank. Larger banks that are closely associated to money center banks can better utilize money market instruments to fund their ACC activities. Agricultural credit corporations may appeal to some banks because they provide greater flexibility for expanding into dynamic new geographic areas than is available through new bank charters or mergers.

FARM CREDIT SYSTEM

The Farm Credit System (FCS) is a system of federally chartered but privately owned banks, lending associations, and service units organized as cooperatives for the purpose of providing credit and related services to agricultural producers and their cooperatives in the United States. By virtue of the cooperative organization, the agricultural borrowers become the system's owners and holders of equity capital, and they are represented by elected boards of directors. The institutions of the Farm Credit System are regulated and examined by the Farm Credit Administration, an independent agency in the Executive Branch of the U.S. government.

Overview

During the 1980s and early 1990s, the Farm Credit System experienced significant structural reorganization. As of January 1, 1994, the Farm Credit System was composed of 8 district Farm Credit Banks, 3 Banks for Cooperatives, and 238 local lending associations. This structure is likely to undergo continuing transition. Other FCS service units and entities include the Federal Farm Credit Banks Funding Corporation, Farm Credit Leasing Services Corporation, the Farm Credit System Insurance Corporation, the Farm Credit Council, and the Farm Credit System Financial Assistance Corporation (see Figure 17.2). The objectives and organization of these FCS institutions will be described in later sections.

Congressional authority for the Farm Credit System, contained in the Farm Credit Act of 1971, as amended, specifies that the system work to improve the income and well-being of U.S. farmers and ranchers by furnishing sound, adequate, and constructive credit and closely related services to creditworthy borrowers and that it provide these services through favorable and unfavorable economic times. This mandate to be a reliable agricultural lender creates the need for a reliable source of funds and a heavy emphasis on risk management by the system's institutions.

Compared to other financial intermediaries, the Farm Credit System and its various units have a less complex overall financial structure. The assets of the FCS institutions are dominated by loans made by the Farm Credit Banks to the lending associations and by loans made by the lending associations and the Banks for Cooperatives to eligible agricultural borrowers. Financial reserves and physical facilities generally make up about 5 to 15 percent of the institutions' total assets. The Farm Credit Banks and Banks for Cooperatives obtain most of the FCS loan funds through the sale of system-wide securities in the national financial markets. The banks are subject to joint and several liability for these securities, and they participate in security sales according to their projected financing needs.

In broad terms, asset – liability management by the FCS institutions primarily involves the selection of maturity structures, the timing of security sales to minimize the costs-of-funds acquisition, the efficient allocation of these loan funds to individual borrowers according to their credit worthiness, and the management of interest rate risk, credit risk, and other important sources of risk. The total volume of FCS lending largely is determined by the borrower's loan demands and credit quality and by lender competition.

Also important is effective loan pricing. In the past, most of the loans to borrowers were priced with variable interest rates that were adjusted periodically in response to changes in the average costs of funds, capital or reserve

Major Components of the Farm Credit System

System Banks, Associations, and Borrowers

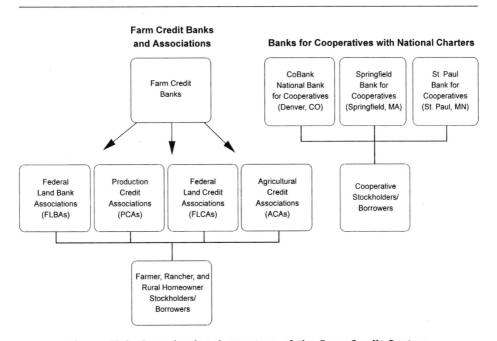

Figure 17.2. Organizational structure of the Farm Credit System.

requirements, and other intermediation costs of the banks or associations making the loans. Interest rates on FCS loans are exempt from state usury laws.

Average-cost pricing of loans was relatively simple to administer and, especially for long-term real estate loans, provided the Farm Credit System with a significant competitive advantage during the expansion period of the

1970s, when market interest rates were trending steadily upward. As a result, interest rates on FCS real estate loans were consistently below, sometimes substantially below, current market interest rates during this period. However, as interest rates declined during the 1980s, average cost pricing placed FCS real estate loans at a competitive disadvantage relative to interest rates on loans from other lenders. Consequently, the Farm Credit Banks and associations have shifted toward marginal cost pricing of loans in order to be more rate competitive and to offer a choice of either fixed-rate or variable-rate loans to borrowers.

The Farm Credit System experienced steady, yet moderate growth in loan volume during the 1950s and 1960s. Then, growth in loan volume accelerated significantly during the credit expansion and agricultural boom times of the 1970s. By 1980, many FCS institutions were experiencing loan quality problems, but because of heavy loan carryovers from year to year, they were not able to reduce loan volume. As total farm loan demand dropped off during the mid-1980s, loans held by the FCS institutions declined more than proportionately, and the Farm Credit System began to deal with significant deterioration of its loan portfolio. Substantial restructuring of FCS institutions and acquiring federal financial assistance were needed to put the system on a path toward financial recovery (see "Structure of Banks and Associations" for a discussion about the Agricultural Credit Act of 1987 and related developments).

Development of the Farm Credit System

The Farm Credit System dates back to 1916 when the Federal Farm Loan Act was passed authorizing the creation of a cooperative system of 12 Federal Land Banks (FLBs). The act also created the National Farm Loan Associations, subsequently called Federal Land Bank Associations (FLBAs), to act as the local lending agents of the Federal Land Banks. The 12 Federal Land Banks were originally capitalized with $750,000 each, primarily through stock purchases by the U.S. Treasury. As with the development of later FCS institutions funded with federal assistance, the original government capital has long since been repaid.

The Federal Land Banks were to provide farmers with long-term real estate mortgage loans on terms compatible with the unique characteristics of agriculture and more reasonable than those available from other lending sources at that time. Although the Federal Land Banks maintained a conservative posture through the 1920s, they were a major source of refinancing for many debt-stricken farmers during the depression years of the 1930s. Since then, they have grown to become the major farm real estate lender.

The Federal Intermediate Credit Banks (FICBs) were the second major type of lending institution within the Farm Credit System. Established by the Farm Credit Act of 1923, the 12 banks were to provide loan funds to agricultural cooperatives and to discount the short- and intermediate-term notes from various institutions, such as commercial banks or finance corporations. Thus, the Federal Intermediate Credit Banks were, in part, intended to be an additional source of funds for other agricultural lenders. During their early years, they experienced relatively slow growth because of conservative lending policies, a narrow spread between the FICB loan rate and the rate client banks were allowed to charge their customers, and the difficulties farmers had in establishing and operating their own finance corporations.

An enlarged credit program for agriculture was launched in 1933 when Congress authorized the establishment of local Production Credit Associations (PCAs), which could discount loans with, or borrow directly from, the Federal Intermediate Credit Banks and make short- and intermediate-term loans to eligible agricultural borrowers. In addition, the Banks for Cooperatives (BCs) also were established in 1933 as another key part of the Farm Credit System. Each of the 12 districts contained a Bank for Cooperatives, and a Central Bank for Cooperatives was established as well. The purpose of the Banks for Cooperatives is to provide seasonal and term loans to farmer-owned marketing, supply, and service cooperatives. All these cooperative organizations — Federal Land Banks, Federal Land Bank Associations, Federal Intermediate Credit Banks, Production Credit Associations, and Banks for Cooperatives — made up the Farm Credit System as it existed prior to 1988.

The Farm Credit Act of 1971 and the Farm Credit Amendments Act of 1980 provided the major legislative authority by which the FCS structure, services, and programs were modified and adapted during the 1970s and early 1980s. These acts helped to expand FCS services to selected farm-related businesses, to provide leasing services, and to finance limited international activities by the Banks for Cooperatives. Especially important for farm real estate lending was the provision of the 1971 act that increased the lending limit on a tract of farm land from 65 percent of the "normal agricultural value" to 85 percent of the current market value. This increase coincided with the substantial growth in farm real estate lending that occurred during the 1970s and with the explosive growth in the market price of farm land.

The rapid growth of farm debt during the 1970s, followed by the severe agricultural depression of the 1980s, placed many farmers and farm lenders in dire financial difficulty. The Farm Credit Amendments Act of 1985 was formulated to strengthen the Farm Credit System and to strengthen its regulatory agency, the Farm Credit Administration (FCA), in providing adequate examination, supervision, and regulations to help insure the continued viability of the system's institutions. The 1985 act was followed by the Farm Credit Amendments Act of 1986, which further responded to the system's mounting loan loss problems, and by the Agricultural Credit Act of 1987, which is described later.

Federal Regulation

The banks, associations, and related institutions making up the Farm Credit System are chartered, regulated, and examined by the Farm Credit Administration, which is managed by a full-time three-member board, with one member appointed as chair. In addition, other FCA executive and staff personnel are employed to carry out the agency's activities. The FCA Board is appointed by the president of the United States with the advice and consent of the U.S. Senate. According to the FCA Annual Report for 1990, the statutory powers of the board include:

- Approving the rules and regulations for the implementation of the Farm Credit Act of 1971, as amended.

- Providing for the examination of the condition of, and general regulation of the performance of all the powers, functions, and duties vested in, each FCS institution.

- Providing for the performance of all powers and duties vested in the Farm Credit Administration.

- Requiring such reports as it deems necessary from the FCS institutions.

In general, the Farm Credit Administration now functions as an arms-length regulator with substantial powers for regulating and evaluating the safety and soundness of the FCS institutions. Thus, it is structured similarly to federal regulators for commercial banks and other types of federally chartered or insured financial institutions. In contrast, prior to the mid-1980s, the Farm Credit Administration was thought of as a part of the Farm Credit System and as a system advocate rather than as an independent government regulator.

Ownership Structure

The basic ownership structure of the Farm Credit System reflects cooperative principles and exhibits a tiered approach to ownership and management. The farmer/borrower/patron of a lending association acquires an ownership interest in the local association and participates in the selection of management by voting for the association's board of directors. In turn, the association either obtains its funds from or serves as a lending agent of the district Farm Credit Bank. As a part of the funding and ownership arrangement, the local association then acquires an ownership interest in the district Farm Credit Bank and participates in the selection of its management by voting for the bank's board of directors. Similarly, the members of cooperatives who hold stock in the Banks for Cooperatives elect the boards of directors of their respective banks. In general, then, the ownership interests in the FCS institutions originate from the bottom up, in contrast to the ownership structure of a multi – bank holding company, for example.

Structure of Banks and Associations

The organizational structure of the FCS banks and associations was substantially altered by the Agricultural Credit Act of 1987. Prior to passage of the act, the FCS institutions had already made significant progress toward consolidating the management and geographic boundaries of the Production Credit Associations and the Federal Land Bank Associations in many of the farm credit districts. These consolidations were consistent with the unification of management and organizational structure that had occurred at the district level among the Federal Land Banks, the Federal Intermediate Credit Banks, and the Banks for Cooperatives. In several districts, most or all of the existing Production Credit Associations had been restructured into a single district-wide association with branches operating throughout the district having the same offices, personnel, and geographic boundaries as the Federal Land Bank Associations. Even the terminology had changed so that the traditional PCA and FLBA names were replaced by Farm Credit Services, or something similar. These changes were intended to bring about efficiencies in lending operations, improve risk management and control, respond to the decline in loan volume and the loss of high-quality borrowers in the early 1980s, and offer borrowers a complete and coordinated set of short-, intermediate-, and long-term credit services.

The 1987 act accelerated the pace of structural change and either required or offered new organizational alternatives. It also contained provisions affecting the capitalization of the banks and associations, making available federal financial assistance, and creating several new features that included an insurance fund, borrower rights and loan restructuring, a new secondary market for farm mortgage loans, and numerous others as well.

In restructuring the Farm Credit System, the 1987 act required a merger of the Federal Land Banks and Federal Intermediate Credit Banks in each district into a consolidated Farm Credit Bank (FCB).[1] The 1987 act also required the Farm Credit System to develop a proposal to consolidate the Farm Credit Banks in the 12 districts into no less than 6 financially viable Farm Credit Banks through inter-district mergers. Because Congress left decisions about inter-district consolidation to the Farm Credit System itself, FCB consolidations were discussed, but substantive progress did not materialize until the 1992 merger of the Farm Credit Banks of St. Louis and St. Paul into a new Farm Credit Bank called AgriBank. In 1994, the Louisville Farm Credit Bank also joined AgriBank, and the Farm Credit Banks of Omaha and Spokane merged to form Ag-America. Additional consolidations among the remaining farm credit districts could occur in the future.

At the association level, two new types of associations were created by the 1987 act. The Federal Lank Bank Associations and the Production Credit Associations were given the opportunity, with encouragement, to merge into consolidated Agricultural Credit Associations (ACAs), which would function as direct lenders, be subject to income taxation, offer a full line of credit services, and obtain their loan funds from their district Farm Credit Banks. Similarly, the Federal Land Bank Associations could be converted to Federal Land Credit Association (FLCAs), which would function as direct lenders for farm real estate loans and obtain loan funds from their district Farm Credit Banks. Thus, except for the Federal Land Bank Associations, the Agricultural Credit Associations and the Federal Land Credit Associations are now chartered as direct lenders, holding the loans they originate in their own portfolio, and obtaining their loan funds by borrowing from the district Farm Credit Banks. The Federal Land Bank Associations continue to originate and service long-term real estate loans for district Farm Credit Banks.

In general, then, because of consolidations and restructurings, the lending

---

[1]These mergers occurred in 11 of the farm credit districts, but not in the Jackson district. The Federal Land Bank of Jackson was placed in receivership on May 20, 1988, with its real estate lending territory now covered by the Farm Credit Bank of Texas. The Federal Intermediate Credit Bank of Jackson merged with the Columbia Farm Credit Bank.

associations of the Farm Credit System are much fewer in number, larger in size, and diverse in structure. For example, the number of associations declined from 915 in 1980 to 238 in 1994. Moreover, there are substantial differences in the configuration of associations within the farm credit districts. The organization of the Farm Credit Banks and associations on a national basis is shown in Figure 17.3 and summarized as follows:

Farm Credit System Institutions and Terms[2]

Farm Credit System (FCS) — System of privately owned lending institutions and other entities that are chartered under federal authority contained in the Farm Credit Act of 1971, as amended.

Farm Credit Administration (FCA) — The federal government agency with regulatory responsibility for the Farm Credit System.

Farm Credit Banks (FCBs) — FCS banks that make direct long-term agricultural real estate loans and / or provide short-, intermediate-, and long-term loan funds to Federal Land Credit Associations (FLCAs), Production Credit Associations (PCAs), and Agricultural Credit Associations (ACAs).

Farm Credit Districts — Territories served by Farm Credit Banks.

Federal Land Bank Associations (FLBAs) — FCS associations that take applications for and service long-term real estate loans for Farm Credit Banks.

Federal Land Credit Associations (FLCAs) — Federal Land Bank Associations that have been given direct long-term real estate lending authority and make such loans with funds obtained from Farm Credit Banks.

Production Credit Associations (PCAs) — FCS associations that make direct short- and intermediate-term loans with funds obtained from Farm Credit Banks.

Agricultural Credit Associations (ACAs) — FCS associations that make direct short-, intermediate-, and long-term loans with funds obtained from Farm Credit Banks.

---

[2]Farm Credit Administration Annual Report, 1990.

Banks for Cooperatives (BCs) — FCS banks that make loans of all kinds to agricultural, aquatic, and rural utility cooperatives.

Farm Credit System Insurance Corporation (FCSIC) — A federal government corporation established chiefly to insure the timely payment of principal and interest on notes, bonds, debentures, and other obligations of the Farm Credit Banks.

Federal Farm Credit Banks Funding Corporation (FCBFC) — An entity owned by the Farm Credit Banks that markets the securities sold to raise loan funds.

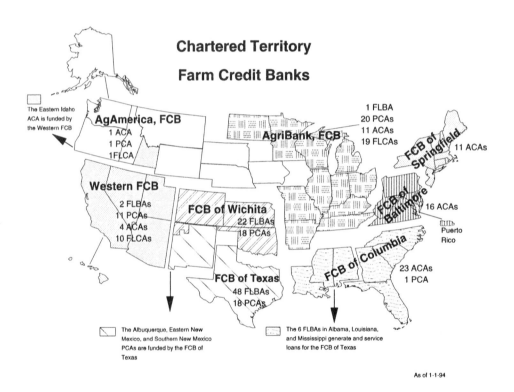

Figure 17.3. Geographic structure of the Farm Credit System.

Finally, the 1987 act required the 12 district Banks for Cooperatives to decide by stockholder vote whether to merge with the Central Bank for Cooperatives or remain separate. Ten of the Banks for Cooperatives elected to consolidate, thus forming a new National Bank for Cooperatives, called CoBank. The Bank for Cooperatives in the Springfield, Massachusetts, district and the one in the St. Paul district elected to remain separate. Thus, the CoBank and the Springfield and St. Paul Banks for Cooperatives now have national charters and may operate on a national basis with competing territories, if they so choose.

Funding the Farm Credit System

The Farm Credit Banks acquire loanable funds by issuing three types of securities in the national financial markets: Federal Farm Credit Banks Consolidated System – wide Bonds, Medium-Term Notes, and Discount Notes. Bonds are issued in book-entry form 16 times a year. Those with six-month maturities (and with three- and nine-month maturities as needed) are sold only in multiples of $5,000. Bonds with maturities of 13 months or longer are available in multiples of $1,000.

Discount notes are designed to provide flexibility in obtaining funds during the time interval between bond sales. They are comparable to commercial paper issued by corporate firms. These notes are issued daily with maturities of 5 to 270 days and are sold in denominations of $50,000 or multiples thereof.

The responsibility for selling securities rests with the Federal Farm Credit Banks Funding Corporation, located in Jersey City, New Jersey. After each Farm Credit Bank has determined and reported its financing needs, the Funding Corporation will consult with bond dealers and with the FCS marketing committee to determine the amount and terms of the issues to be offered. Bonds and notes are then sold through a nationwide chain of securities dealers. The dealers sell the bonds to private and public investors, mostly commercial banks, state and local governments, thrift institutions, and corporations. Included among the bondholders are investors from foreign countries. The bonds are priced to sell at par in their original issue. They are subsequently traded in an active secondary market at values based on current market yield.

Prior to 1979, the three types of Farm Credit Banks (Federal Land Banks, Federal Intermediate Credit Banks, and Banks for Cooperatives) issued their own consolidated securities. Now the issuances are all system-wide. The Farm Credit Banks also maintain credit lines with large commercial banks for added liquidity, and they may borrow from other banks within the Farm Credit System.

The farm credit bonds and notes are issued with joint and several liability among the banks, but not the lending associations. Joint and several liability means that investors in the bonds and notes can look to the performance of the Farm Credit Banks in aggregate, rather than individually, in judging the safety of these investments. The joint and several liability concept was untested until the agricultural finance crisis of the 1980s generated considerable, yet substantially different, degrees of financial adversity within the 12 farm credit districts, calling for intra-system assistance and federal assistance in some cases. Serious questions were then raised about the equitable treatment of banks that were at once joint and severally liable for each other's debt obligations to security holders, yet without the powers or inclination for joint performance monitoring and business control.

In response to these problems, the 1987 act authorized the creation of the Farm Credit System Insurance Corporation (FCSIC) within the system to become operational by 1993. The Farm Credit System Insurance Corporation is intended to (1) furnish a safety reserve for farm credit securities in case of financial distress experienced by the FCS institutions, (2) provide an alternative and prior risk control mechanism to the joint and several liability concept of the Farm Credit Banks, and (3) reduce the likelihood of system needs for public financial assistance in the future. The insurance fund is funded by premiums paid by the Farm Credit Banks. The Farm Credit System Insurance Corporation functions similar to the Federal Deposit Insurance Corporation for depositors of commercial banks.

Farm credit bonds and notes are treated as "government agency securities" in the financial market, even though the federal government does not guarantee them against default, which is explicitly stated. Agency status results from a set of regulatory exemptions and preferences of these securities, as they are traded in the financial markets, and the perception of implied government backing if the system experiences severe financial difficulty. Agency status is a significant factor in the ability of the Farm Credit Banks to market large volumes of securities at relatively favorable costs. Agency securities generally trade at yields that fall between the yields on U.S. Treasury securities and the yields on prime corporate bonds of comparable maturities — e.g., usually a 10- to 50-basis point spread over U.S. Treasury securities.

Equity capital of the various banks and associations is obtained by the sale of stock to borrowers and by earnings retained from interest, fees, and rental revenue paid by members. Prior to the 1987 act, agricultural borrowers were required to purchase stock in their local association at rates that fell between 5 percent and 10 percent of their loan amount. The value of the stock was fixed and largely considered risk-free. As a result of the act, the required minimum investment of each borrower was substantially reduced (to $1,000 or 2 percent of the loan amount, whichever is less), with most of the equity capital base of the lending associations now considered to be "at-risk" capital that is largely generated by retained earnings. Many associations, however, maintain borrower stock requirements at levels above the minimum. In practice, loaning additional amounts to borrowers to finance their stock purchases has created equity capital from bond sales. The stock requirement generally poses no cash flow hardship for borrowers, although it increases their effective rate of interest (see Chapter 14).

On occasion, other sources of funding arise to support FCS activities. An example is the federal assistance provided by the Agricultural Credit Act of 1987 in response to the financial problems experienced by several of the Farm Credit Districts during the farm financial crisis of the 1980s. The 1987 act created a special class of 15-year bonds to be sold under the auspices of a Farm Credit System Assistance Board and Farm Credit System Financial Assistance Corporation. The board was given a five-year life (until December 31, 1992) and a funding limit of $2.8 billion (subject to increase to $4 billion).

The assistance bonds are guaranteed by the U.S. government. The proceeds have been used to infuse money into financially troubled FCS institutions that needed outside help to redeem borrower stock at par value, pay off debt securities as they mature, or continue lending at competitive rates. The U.S. Treasury is providing assistance on interest payments to bondholders, although the Farm Credit System will ultimately be responsible for repaying all government assistance. By 1994, approximately $1.3 billion of securities had been issued by the Financial Assistance Corporation.

Income Tax Status

The banks and associations differ in their tax obligations, which in turn affects some of the system's financial operations. The Farm Credit Banks and Federal Land Bank Associations are not subject to taxation, except for taxes on the real estate they own. Production Credit Associations, Federal Land Credit Associations, Agricultural Credit Associations, and Banks for Cooperatives are subject to income taxes similar to other corporations operating on a cooperative basis. The tax exemption of some system institutions has offered a degree of tax management to the system that in the past allowed some credit services to be offered to borrowers at lower cost. However, the shift to Agricultural Credit Associations and Federal Land Credit Associations resulting from the 1987 act reduced this degree of flexibility in tax management.

Eligible Borrowers

The Farm Credit Banks and associations are authorized to make loans and provide leasing services to bona fide farmers and ranchers, rural residents for housing purposes, aquatic producers, and persons or businesses furnishing on-farm services to agricultural producers. Any person who owns agricultural land or is engaged in producing agricultural or aquatic products is eligible. Partnerships, corporations, trusts, estates, and other business forms may also qualify. Cooperatives engaged in the processing, handling, or marketing of farm or aquatic products, or in the acquisition of supplies or services for producers of such products, are eligible to borrow from the Banks for Cooperatives, provided they meet the minimum specified criteria. Of course, whether eligible entities who wish to borrow from an FCS institution actually become borrowers depends on their credit worthiness.

Real Estate Lending

Farm real estate lending is the largest and most significant lending program of the FCS institutions. Loans can be made for maturities of from 5 to 40 years, although most maturities currently fall in the 20- to 30-year range. Repayment programs usually are based on either equally amortized annual payments or on constant payments on loan principal plus interest on the remaining loan balance. However, payment size may vary if the loans are priced with variable or adjustable interest rates. While the loan proceeds may be used for any purpose, most real estate loans are used for purchasing new real estate, improving land, acquiring buildings, or refinancing previous real estate or non – real estate loans. The principal lending requirement is that all real estate loans be secured by a first lien on real estate.

The loan amount legally may not exceed 85 percent of the appraised value of the real estate pledged as security, although the Farm Credit Administration may by regulation require that loans not exceed 75 percent of the appraised value of the real estate security. The loan limit increases to 97 percent of the appraised value if a government guarantee by the Farmers Home Administration is provided. The actual percentage loaned, however, is generally much less than 85 percent and depends on the borrower's repayment capacity and other elements of credit worthiness. In the early 1990s, a typical loan-to-asset value is 65 percent, leaving the downpayment or equity requirement at 35 percent.

Prior to the late 1980s, the interest rates on farm real estate loans were variable (floating) rates in which the rates could change periodically (e.g., monthly) in response to changes in the average costs of funds of the associations or banks. In recent years, however, the associations also have offered adjustable-rate loans and fixed-rate loans. In an adjustable-rate loan, the rate is fixed for a specified period of time (1, 3, 5, or 10 years) while the loan repayment period may reflect amortization over a longer period of time. The interest rate then is adjusted at the end of the time interval to a level that is consistent with market interest rates at the adjustment time. The fixed rate during the adjustment period is possible because the Farm Credit Bank is participating in bond sales with the same or similar maturity as the length of the adjustment period. Thus, the bank is managing the interest rate risk for the association by matching maturities or durations on the loans and on the bonds. This matching results in a form of marginal cost pricing of loans by the bank and the association.

In a fixed-rate loan, the interest rate is fixed over the entire maturity of the loan. However, when the loan originates, the level of the fixed rate generally exceeds that of a variable or adjustable rate in order to compensate the lender for bearing the interest rate risk.

Advanced payments or total loan repayments may occur prior to maturity without penalty. In fact, few loans actually continue to maturity because they may be repaid early, they may be refinanced, they may have financing added, or they may be renewed by the borrowers.

Intermediate- and Short-Term Lending

Farm Credit Associations make intermediate- and short-term loans to farmers and other eligible borrowers for almost any purpose associated with farming, ranching, farm-related businesses, on-farm services, and rural life. The maximum loan maturity is 10 years, although most intermediate-term loans for machinery, equipment, and breeding livestock are in the three- to five-year range. Most loans are collateralized by the asset being financed; unsecured loans generally involve shorter maturities, high equity positions, and strong credit worthiness by the borrowers.

The terms and repayment patterns of most loans are based on the borrowers' projected cash flows. Repayment of intermediate-term loans may involve equally amortized loans or constant payment on principal loans. Operating loan funds are advanced to meet projected outlays as needed, and loans are repaid as income becomes available. This procedure keeps interest costs to a minimum. Service fees are usually charged to cover loan origination and servicing costs, and stock or fee requirements of the association's capitalization policy must be met. Smaller loans (e.g., less than $50,000) may be handled on a more expeditious or streamlined basis in which fees are a more significant part of loan pricing.

Variable or adjustable rates generally are charged, set at levels that cover the costs of loan funds borrowed from the district Farm Credit Bank, the costs of operating the association or service center, possible loan losses, and retention of earnings to build equity capital. Each association monitors its costs of funds on a monthly basis for possible changes in loan rates. Many associations have developed a system of multiple or tiered rates to borrowers based on differences in the borrowers' credit risk. Patronage refunds seldom are paid in cash; rather, they are retained by the association.

The legal lending limit to any one association borrower is 50 percent of the association's capital and surplus. This percentage increases to 100 percent if an approved loss-sharing agreement is in force with a loan-sharing participant. When a Farm Credit Bank participates in a loan made by an association, the bank's loan limit is 20 percent of the bank's capital and surplus. As with real estate loans, these high loan limits, along with the participation arrangements, make the legal loan limit inconsequential to most individual borrowers. Associations may also provide leasing or rental services to their members. Capital leasing is an additional method of intermediate-term financing for farmers.

Farm Credit Associations may also participate with commercial banks or other lenders in short- and intermediate-term agricultural lending. Participating commercial banks must fulfill one of the following:

1. Retain at least 50 percent of the total of each participation loan.

2. Retain at least 10 percent of the total of each participation loan, provided the commercial bank does not materially reduce its ratio of agricultural loans to total loans from the ratio maintained during the three preceding years.

3. Retain the maximum amount of the participation loan permitted by banking regulations to which the bank is subject. Banks must also purchase non-voting participation certificates from the association, which are analogous to the stock requirement of the association's customers.

Other financing institutions eligible to borrow from the Farm Credit Banks include commercial banks, trust companies, thrift institutions, credit unions, agricultural credit corporations, and any other agricultural producer associations engaged in extending credit to farmers. Other financial institutions typically have made only limited use of FCB funds, and this use declined substantially during the financial stress times of the 1980s. Nearly all of the OFI activity occurs with commercial banks, their affiliated agricultural credit corporations, and nonbank agricultural credit corporations.

For commercial banks or their agricultural credit corporations to become eligible for FCB funding, they must have:

1. A significant involvement in lending for agricultural or aquatic purposes.

2. A continuing need for funds for agricultural or aquatic purposes.

3. A limited access to national or regional capital markets and full utilization of local funds to finance local needs.

4. No intention of using FCB funds to expand financing for parties and purposes ineligible for financing by the Farm Credit Banks. Provisions also are specified to assure that the eligible other financial institutions utilize FCB funds on a continuing basis and to account for bona fide exceptions to the eligibility conditions cited previously.

Banks for Cooperatives

The Banks for Cooperatives form a credit system devoted solely to meeting the financing needs of cooperatives engaged in processing, handling, or marketing farm or aquatic products; furnishing products or services to farmers; or furnishing services to eligible cooperatives. The BC component of the Farm Credit System began in 1933 with 13 Banks for Cooperatives. As indicated earlier, restructuring in response to the 1987 act resulted in three BCs each having a national charter and authority to operate nationwide in competing territories if they elect to do so. These banks provide a major part of the credit used by many farm cooperatives in most states. Thus, they are an important part of the total credit system available to agriculture.

Banks for Cooperatives are authorized to make both seasonal and term loans. Seasonal loans are normally for up to 18 months. Term loans are used to finance structures, equipment, machinery, and other fixed assets. The Farm Credit Amendments Act of 1980 provided the authority for Banks for Cooperatives to offer basic financial services to cooperatives for export and import transactions and to engage in international banking activities that facilitate agricultural exports and imports by agricultural cooperatives.

Eligible cooperative borrowers from the Banks for Cooperatives must have 80 percent of their membership made up of farmers, although this requirement is lowered to 60 percent of voting members for rural electric, telephone, and other utility service and for certain local supply cooperatives. Banks for Cooperatives may also finance domestic leasing, including leveraged leasing transactions involving farm or aquatic cooperatives.

LIFE INSURANCE COMPANIES

In the early 1900s, life insurance companies and commercial banks were the major institutions making farm real estate loans. While the Farm Credit System and the Farmers Home Administration are now also involved, life insurance companies are still major suppliers of farm real estate debt. Their involvement varies substantially by region, by company, and with changes in financial market conditions.

Life insurance companies are profit-oriented institutions whose main financial activities are the sale of insurance policies and payments of claims to policy holders. The long-term nature of these activities requires the accumulation and holding of large financial reserves. These reserves are generally invested in a diversified portfolio of low-risk investments that may include farm real estate loans.

In total, investments in farm mortgages typically represent about 2 to 3 percent of the total assets of life insurance companies. Thus, farm mortgage lending is minor compared to other life insurance company investments, although some companies are much more involved in farm lending than others.

In the early 1990s, six companies dominated the farm real estate lending by life insurance companies. The market share of farm real estate debt held by life insurance companies was nearly 25 percent during the mid-1960s. It then declined steadily for nearly two decades because of competing uses for the investment funds, increased demand for loans from policy holders, and usury limits on interest rates in many states that became effective during periods of tight credit and rising market rates. More liberal lending by Federal Land Banks also depressed life insurance companies' market shares. By 1984, life insurance companies held only about 11 percent of the total farm real estate debt. However, because of the competitive problems of the Farm Credit Banks and because of the relative attractiveness of returns on farm loans compared to other investments by life insurance companies, this figure has increased slightly in recent years.

Life insurance companies prefer lending to larger, low-risk farming operations. The average size of a life insurance company loan is typically greater than a real estate loan from the Farm Credit System or a commercial bank. Some life insurance companies have concentrated their farm lending in selected areas such as the Corn Belt or irrigated farming areas where production is more stable and weather risks are fewer. However, during the 1980s, farm real estate loans held by life insurance companies shifted away from the Corn Belt and other major agricultural production regions to states in the Southeast and Pacific Coast regions.

Laws in some states limit the size of a life insurance company's loan to a designated proportion (e.g., 75 percent) of the appraised value of the farm land or other security. Considerable flexibility occurs in the choice of appraisal method. No legal limits are placed on individual loans, although companies may impose their own limits as a matter of policy.

The maturities and repayment schemes for life insurance company loans have changed as these companies sought interest rate protection from the effects of high, volatile inflation. These terms differ considerably, but a typical arrangement in recent years has been a loan maturity of from 10 to 20 years, a longer amortization schedule (e.g., 30 years), and a provision for interest rate adjustment every 5 years. Because the loan maturity is shorter than the amortization period, a substantial balloon payment is due at maturity. The borrower either meets the balloon payment with cash or refinances it with the life insurance company or another lender.

In recent times, life insurance companies have also adopted other innovative methods of responding to borrowers' liquidity problems caused by changes in inflation rates and swings in farm income. Mortgages with graduated payments, adjustable rates, and shared appreciation are examples. In a shared appreciation mortgage, the borrower agrees to give the lender a percentage of the capital gains on the land in return for a lower interest rate. The lower rate reduces the debt-servicing obligation, while the share in prospective capital gains makes up part of the lender's compensation. These arrangements also bring greater lending risks as well, since capital gains are subject to considerable uncertainty.

Life insurance company loans are made to farmers either directly from company offices or indirectly through correspondents, such as mortgage brokers, mortgage companies, real estate companies, and some commercial banks. Coordination between lenders helps both the borrowers and the lenders. For example, life insurance companies may arrange to purchase real estate loans which commercial banks have originated. This arrangement enables banks to offer a full range of loans to their customers and provides low-cost contact and servicing for the life insurance company.

TRADE CREDIT

Trade credit consists of financing provided to agricultural producers by merchants, dealers, and other agribusiness firms. Aside from direct personal loans, trade credit is perhaps the oldest form of credit known. It has survived competition from specialized financial institutions in widely differing environments around the world. It is a significant source of non – real estate financing for farmers. USDA data indicate that nearly $14.2 billion, or 21.7 percent, of the total non – real estate farm debt was held by "individuals and others" in 1994, mostly involving trade credit.

Trade credit generally arises from merchandising transactions. The originators typically are the sellers of farm inputs: seed, feed, fuel, fertilizer, machinery, storage facilities, or buildings. However, firms that buy crop and livestock products from farmers through procurement contracts may also include financing as part of the contract package. Trade credit provides a merchant or dealer with a means for facilitating sales, thereby maintaining or increasing sales volume.

Types of Trade Credit

Trade firms use several types of financing arrangements. The open account is especially convenient for the buyer and seller when frequent sales occur. The buyer receives deliveries without requiring immediate payment. An account receivable is generated and held by the seller, based on delivery slips. Billing usually occurs monthly, with prompt payment expected to bring the customer's account up to date.

The extended open account extends the open account, with or without interest, until a recognized source of payment becomes available. Again, the only security held by the seller is an account receivable based on delivery slips. Seasonality in farming makes this arrangement attractive to farmers and explains its wide use in acquiring crop and livestock inputs.

Bank credit cards are offered by some trade firms as an alternative to open account and extended open account credit. The trade firm and bank establish a line of credit for the farm customer and prepare appropriate identification cards and sales forms. The farmer may then charge purchases from the trade firm up to the credit limit, with the bank's having responsibility for billing and collection.

While open accounts and credit cards are convenient, the limited negotiability of the debt instrument presents trade firms with special liquidity problems. More formal evidence of indebtedness improves the marketability of the debt instrument. Three financing methods commonly used are the **unsecured promissory note,** the **secured promissory note,** and the **conditional sales contract.** The last two are important in capital transactions involving sales of buildings, machinery, and equipment to farmers. In addition, if an open account or an extended open account is not paid on schedule, the trade firm might have the farmer sign a promissory note and pledge collateral as security.

In installment lending, the trade firm or dealer may transfer the customer's note to a finance company, a subsidiary corporation of a manufacturer, or a local financial institution, such as a commercial bank or a Farm Credit Association, perhaps under point-of-sale financing. In these cases, the merchant or dealer shifts the financing responsibility to more specialized lenders. Loan documentation is usually more complete, and the borrower's credit worthiness is examined more fully.

The convenience which trade credit offers is an integral part of the merchandising system. Many manufacturers provide credit to their distributors, dealers, and retailers. This credit can be passed along to agricultural customers. However, as credit terms begin to exceed 30 days, the trade credit begins to compete with financing from other lenders. Thus, incentives must exist for trade firms to offer financing and for agricultural producers to accept it.

At the beginning of the 1990s, the credit subsidiaries of some large agribusiness firms began offering general lines of credit and operating loans. These firms believe they can serve as competitive, general-purpose lenders to agricultural borrowers, fill gaps created by other financial institutions during the 1980s, benefit from their established reputations and network of dealers, and sustain profitable lending operations. The growth and degree of future success achieved by these ventures will be interesting to observe.

The Agricultural Customer's Viewpoint

From a borrower's standpoint, trade credit could offer access to larger total credit, although split financing is discouraged by many agricultural lenders unless it is well planned and justified. Trade credit is also convenient for farmers, and on occasion it may be a cheaper means of financing.

In determining the cost of trade credit, the agricultural borrower must first be sure that the financed product is priced competitively. Free trade financing from a dealer who is non-competitive in price or product quality can be a costly source of funds. If a cash discount is not available, and if the quality and product price are acceptable, the farmer can benefit from trade credit if the interest cost is lower than from alternative sources and if the farm's primary lender does not respond adversely to the use of trade credit.

When a cash discount is available, the rate of interest paid by foregoing the cash discount is based on (1) the size of the cash discount, (2) the maximum time after billing before payment must occur in order for the borrower to qualify for a cash discount, and (3) the length of time before an unpaid account becomes delinquent. Most cash discounts range from 1 to 3 percent of the purchase price. Many suppliers allow a cash discount if payment occurs within 10 days of billing and impose penalties if an account is unpaid 30 days after billing. These terms vary among firms. Thus, the buyer should become familiar with the trade firm's credit policy.

For operating inputs, a credit policy might be represented by 3/10, n/30, which is interpreted as a 3 percent discount if paid within 10 days, with the bill due at face value within 30 days, after which it becomes delinquent.

Thus, the actuarial interest rate (i) the buyer pays to finance the account for the 20 days between the tenth and thirtieth day after billing is:

$$0 = -97 + \frac{100}{1 + i}$$

where i = 3.09 percent per 20 days. The annual percentage rate (i) and effective annual rates (i_e) are:

$$i = (0.0309)\left(\frac{365}{20}\right) = 56.40\%$$

and

$$i_e = (1 + 0.0309)^{356/20} - 1 = 74.26\% / \text{ yr.}$$

These are the buyer's costs for using the trade credit if a cash discount is available. A credit policy of 1/16 n/30 costs the buyer an annual percentage rate of 18.43 percent and an annual effective rate of 20.13 percent, compared to the cash discount.

The Trade Firm's Viewpoint

Consider now the issues involved in providing trade credit from the trade firm's point of view. The relative importance of these issues depends on the size, scope, and competitive position of the trade firm relative to other trade firms and to more traditional agricultural lenders.

- *Relationship to merchandising*

  The most basic function of trade credit is to serve as a tool to aid in merchandising the firm's major products. That is, trade firms often attempt to differentiate their products and increase their competitive position by offering credit services. However, when pricing of credit services and merchandising transactions are combined, it is difficult to distinguish between their separate effects.

- *Funding costs*

  Many large trade firms have credit subsidiaries that offer customer financing through their network of dealers and retail outlets. Funding sources for the credit subsidiaries include funds allocated by the parent company, lines of credit from commercial banks, sales of long-term securities, and issuances of commercial paper (the firm's own promissory notes) into the financial market. *Commercial paper financing* is often called **asset securitization.** It is an efficient, cost-effective form of financing, if the credit program has reached a sufficient size (e.g., $50 million per year). Even then, some credit enhancement (e.g., a standby letter of credit) might be needed for market acceptance. In any case, the funding costs of large trade firms generally are based on market rates of interest and are competitive with those of other lenders.

- *Credit evaluation procedures*

  In recent years, trade firms have become more formal in their credit evaluation procedures, especially firms that finance machinery, equipment, buildings, and storage facilities. These latter firms generally develop credit-scoring procedures (see Chapter 7) that can be broadly implemented throughout the firm's retail outlets. Thus, uniformity in the methods and documents used in credit evaluation tends to reduce evaluation costs and aids significantly in making credit decisions and pricing loans.

- *Geographic dispersion*

    The agribusiness firms that provide trade credit to farmers often are regional, national, or even international entities that conduct their lending activities over broad geographic areas. Such geographic diversification reduces total lending risks, enhances risk-bearing, adds to the marketability of debt securities, and facilitates the use of uniform credit evaluation procedures across their loan portfolios. In turn, these effects contribute to the efficiency of the lending operations and heighten the competitiveness of trade credit.

- *Customer relationships and information*

    Customer relationships are important in lending. For commercial banks, the Farm Credit System and other financial institutions, the lender – borrower relationship is the primary focus. In contrast, the customer relationship in agribusiness firms primarily involves the merchandising activity, in which the provision of credit may be a part of the total package. In its own way, the merchandising relationship between an agribusiness firm and its customers may be just as strong as the loan – deposit relationship in banking. Only the orientation differs.

    A long-standing relationship between trade firms and customers means that the trade firm's information base about its customers may be extensive. A seed company, for example, may be quite familiar with the crop production practices of its farm customers. Similarly, machinery and equipment dealers may be familiar with the investment and maintenance activities of their customers. Some agribusiness firms provide management information systems and other types of decision aids to their customers. In general, trade firms may have extensive information about the management ability, business practices, and financial performance of their agricultural customers; this information contributes importantly to evaluations of credit worthiness.

- *Regulation*

In contrast to specialized financial institutions, trade credit is subject to little government regulation. Financing provided to agricultural borrowers by agribusinesses is not subject to examination by regulators, legal lending limits, portfolio allocation limits, capital adequacy, liquidity requirements, or most other types of regulations affecting specialized lenders, especially depository institutions. Self-regulation and discipline imposed by the marketplace tend to prevail.

The absence of significant government regulations provides greater flexibility in the design and availability of loan programs and generates significant cost savings relative to the costs of compliance experienced by more regulated lenders. These factors also enhance the competitiveness of trade credit in the financial markets for agriculture.

- *Costs of credit delivery*

The credit evaluation procedures cited previously, together with specialized agricultural lending and sizable volume, allow for significant efficiencies in lending. In turn, efficiencies in lending contribute to competitive interest rates and a strong presence by agribusiness firms in the credit markets for agricultural borrowers.

INDIVIDUALS AND SELLER FINANCING

Individuals hold about 21 percent of the total farm real estate debt in the United States. This figure is down from the 30 to 35 percent range in the 1970s. Most of this financing comes from sellers of farm land who finance buyers' purchases either through contract arrangements or through notes secured by mortgages. The majority of these sales are financed with **land purchase contracts,** also known as **contracts of sale, conditional sales contracts, land contracts,** or **installment contracts.** These contracts are agreements to transfer control of land permanently to the buyers, while land title remains with the sellers until the contracted repayment conditions are met.

Historically, these contracts usually required relatively low downpayments, compared to downpayments required by commercial lenders. The

terms and rates of interest vary substantially, especially in transactions between close relatives. The term is often rather short — usually no more than
10 to 15 years. Partial amortization may occur, with requirements for a large
balloon payment (e.g., 30 to 40 percent of the sale price) at maturity. The
repayment schedule may be established through negotiation to meet the needs
of the buyer and the seller. Most contracts also require the buyer to maintain
the property to a specified standard until the title is transferred.

The Seller's Viewpoint

The seller of farm land is interested in contract sales for several reasons.
First, these sales can provide a steady, easily manageable flow of returns from
an agricultural investment as payments of principal and interest are made.
This feature may be important to retiring farmers or other land owners who
prefer to withdraw from active management.

Second, tax regulations in the United States permit the gains from the sale
of an asset (land in this case) to be spread over the years of repayment,
consistent with the pattern of cash flow in the loan arrangement. Sales of most
farm land usually involve substantial gains; thus, the installment sale may be
attractive to sellers even though significant reductions of tax rates during the
1980s have substantially lowered the degree of tax savings from this feature.

Third, the land purchase contract gives the seller a greater degree of control
over the asset, in case the buyer defaults, than is provided when the buyer receives
the title and pledges a real estate mortgage. Finally, by reducing the downpayment
requirement, the seller may increase the number of potential buyers.

The seller may also incur some disadvantages with a land purchase contract. The buyer's low equity makes the contract a high-risk loan. This financial
risk is offset somewhat by the simplified foreclosure procedures on many
contracts. The contract also has relatively low liquidity, since no formal secondary markets are available. Thus, the seller may have difficulty selling his or
her interest in the contract without accepting a heavy discount.

The Buyer's Viewpoint

For the buyer, a purchase contract provides the benefits and responsibilities of land ownership with a relatively small downpayment. Hence, the buyer
can achieve a high leverage position. If credit is limited from commercial

lenders, then seller financing may be the only available source. Moreover, the land contract allows a form of ownership that overcomes the tenure uncertainties of leasing.

Buying on contract also has several disadvantages. The risk of losing the farm by default in payment is typically greater with a purchase contract than with a note secured by a mortgage. The loss could include the downpayment and any equity built up from debt payments and land improvements. In recent years, the courts have tended to support the equity position of the contract buyer during default, although this situation varies among states.

The potential loss of the farm land may hamper short- and intermediate-term financing available from commercial lenders. First, the farmer's balance sheet position deteriorates under a purchase contract, primarily because lenders may not include the accumulation of equity in the land, even though they might include the current year's payment as a liability. Second, if the annual payments on purchase contracts are large, then less cash is available to service non – real estate loans. These factors may make non – real estate lenders more reluctant to finance farmers with contracts compared to financing those with mortgages or those who are renting. Before buying land on contract, the buyer should check carefully with other lenders to determine the potential impact on non – real estate credit. If operating credit were curtailed, the total cost of the purchase contract, including loss of liquidity, could be very high.

FARMERS HOME ADMINISTRATION

The Farmers Home Administration (FmHA) is a government lending agency operating under the authority of the U.S. Department of Agriculture. The functions now performed by this agency began, for the most part, in the 1930s, though emergency crop and livestock loans had been made earlier. The Farm Security Administration (FSA) was established in 1937 to concentrate on two programs: (1) farm production and subsistence loans and (2) farm establishment loans. In 1946, the Farmers Home Administration was formally designated to replace the Farm Security Administration with the purpose of continuing supervised farm credit to U.S. agriculture. Much of this lending has been to young, beginning farmers or to others with limited resources. However, the Farmers Home Administration also acquired authority to make various types of emergency loans to farmers who experience the effects of natural disasters and, on occasion, severe financial problems. FmHA farm borrowers have success potential but lack access to sufficient financing from commercial sources. As these farmers become better established, they usually graduate to commercial sources of financing.

FmHA farm lending occurs under four broad farm loan categories: farm ownership (including enlargement), farm operating, emergency, and others such as soil conservation, water conservation, and rural housing. Much of the rapid increase in FmHA lending during the 1970s and early 1980s took place under the economic emergency loan program, which was scaled back heavily in 1984. Farm loan assistance under the aforementioned programs can be as direct loans or as loan guarantees, although the guarantee approach was emphasized beginning in the mid-1980s.

Direct farm loans are funded from the Agricultural Credit Insurance Fund (ACIF), which is a revolving account funded through repayment of FmHA loans, congressional appropriations, and the sale of Certificates of Beneficial Ownership. These certificates are backed by FmHA – held mortgages and sold to the Federal Financing Bank (FFB). The Federal Financing Bank, in turn, uses these certificates as collateral for loans from the U.S. Treasury.

The current guaranteed loan program began in 1973 but was not heavily used until the late 1980s. Under this program, the loans are made and serviced by private lenders, with the Farmers Home Administration guaranteeing up to 90 percent of each loan if the borrower defaults. The guarantee may result in more favorable terms for the borrower, since much of the risk can be shifted to the Farmers Home Administration. The program enables agricultural banks and other lenders to originate and service the loans and to accommodate both large and small farm loans having high risk, since the guaranteed portion does not count against a bank's legal lending limit, and it may be sold through brokers in secondary markets.

FmHA farm borrowers are required to accept supervision and guidance in their farming operations. This supervision is provided through about 1,700 local offices in the United States, which serve all rural counties. Each county program is administered by a county supervisor and a committee partially composed of local farmers.

FmHA loans are subsidized in several ways. First, borrowers have access to credit that is not available from commercial sources. Second, interest rates are generally below market rates, although less so now than in the past. Rates are largely based on the federal government's costs of funds. Third, the management guidance program is greater than could be economically undertaken by a commercial lender. Fourth, government loan guarantees provide a lower cost source of insurance than would be available from commercial sources.

Farm ownership loans are mostly to finance the purchase of farm land, although the funds allocated to these loans were reduced substantially in the late 1980s. Each farm ownership loan is based on a farm plan designed to yield revenue sufficient for loan repayment and a reasonable standard of living. Loans are extended for up to 40 years, with a legal limit imposed per borrower to assure a wide distribution of loan funds. This limit has been adjusted periodically to reflect changes in borrowers' credit needs.

Farm operating loans are available to borrowers to buy livestock and equipment, pay operating expenses, refinance debts, and make other needed adjustments in farm operations. Loan repayments on operating loans generally occur when products are sold. Intermediate-term loans are also available for durable resources. As with farm ownership loans, a legal limit per borrower is specified.

The Farmers Home Administration has played an important role in financing high-risk borrowers. Moreover, it has often benefited other lenders by assuming the financing of deteriorating loan situations. Commercial lenders also benefit from the managerial skills acquired by borrowers who graduate from the Farmers Home Administration.

On occasion, the FmHA's purpose has been broadened to establish loan programs designed to improve rural communities and to alleviate rural poverty. These programs have included loans to individuals, groups of individuals, local businesses, and rural communities for establishing housing developments, community water handling systems, and recreational facilities, etc., and for providing loans to low-income rural families and the local cooperatives that serve them.

FEDERAL AGRICULTURAL MORTGAGE CORPORATION (FARMER MAC)

The Agricultural Credit Act of 1987 authorized the creation of a secondary market in farm mortgage loans that would operate under the jurisdiction of the Federal Agricultural Mortgage Corporation, or Farmer Mac. Farmer Mac operates as an independent entity within the Farm Credit System. It is supervised and regulated by the Farm Credit Administration. The purposes of the secondary market are to:

- Increase the availability of long-term credit to farmers and ranchers at stable interest rates.

- Provide greater liquidity and lending capacity in extending credit to farmers and ranchers.

- Facilitate capital market investments in providing long-term agricultural lending, including funds at fixed rates of interest.

- Improve the availability of credit for rural housing.

At the time of its creation, Farmer Mac represented a major innovation in agricultural finance. The design and development of Farmer Mac's operations are described in the following sections.

Basic Idea

The basic idea of the secondary market is to achieve a separation of the functions of loan origination, servicing, funding, and risk-bearing. As originally set forth in the 1987 act, Farmer Mac will oversee the purchases by agricultural mortgage marketing facilities (e.g., poolers organized by commercial banks, FCS institutions, insurance companies, and others) of farm mortgages that are originated and perhaps serviced by primary lenders. These marketing facilities will pool the individual loans into aggregate portfolios and then sell pooled participation securities to investors based on a pass-through of principal and interest payments by borrowers, or based on bonds sold to investors who are backed by the pools of loans. In turn, Farmer Mac will provide guarantees on these securities to assure their safety for financial market investors.

Basic Components

Six basic components make up the new secondary market for farm mortgage loans. They are:

1. *Agricultural borrowers,* who need long-term loans in which farm real estate is pledged as security for the loans.

2. *Loan originators,* who are the primary lenders to agricultural borrowers.

3. *Poolers,* who purchase part or all of the farm mortgage loans from the originators, with continued loan servicing provided either by the

originator, the pooler, or some other party. In turn, poolers issue mortgage-backed securities or pooled participation certificates based on the pass-through of principal and interest payments by borrowers.

4. *Investors,* who purchase in the financial markets the securities issued by poolers, thus, providing the ultimate sources of funds for the farm mortgage loans.

5. *Farmer Mac,* which oversees the operations of the poolers and originators, and provides a guarantee (i.e., a credit enhancement) on the securities issued by poolers to assure their safety for investors.

6. *Farm Credit Administration,* which supervises and examines the secondary market operations.

Conditions of Eligibility

Loans eligible to be included in the loan pools must meet Farmer Mac's standards and be documented in an acceptable manner. Only farm real estate loans or rural housing loans meet the standards. A farm real estate loan must be secured by a first mortgage and made to an agricultural operation. Loan size is initially limited to a maximum of $2.5 million per loan, although this limit is adjusted over time for inflation; loans may occasionally exceed that limit, depending on the collateral pledged.

Agricultural real estate loans must also meet Farmer Mac underwriting standards, which are formulated along the lines recommended by the Farm Financial Standards Task Force Report of 1991 (see Chapter 3). As examples, financial ratios are used to demonstrate that the borrower's anticipated cash flow will be sufficient to repay the loan; the loan-to-land value ratio may not exceed 0.75; the debt-to-asset ratio may not exceed 0.50; and other ratio and non-ratio standards apply as well.

Originators, Poolers, and Safety Mechanisms

A loan originator may be any financial institution, insurance company, business or industrial development company, agricultural association or cooperative, or any other entity that makes and services eligible loans. Originators must purchase a prescribed amount of stock from Farmer Mac. Based on initial purchases of stock, potential loan originators number about 1,700, with this total dominated by commercial banks and FCS institutions.

Originators may then sell loans to poolers. Each pool must be backed by a cash reserve equal to at least 10 percent of the value of the loans pooled or by a subordinated participation interest of the same amount in the loan pools. In the latter case, the originator essentially is selling only 90 percent of the loan and would incur the first 10 percent of the losses on the loan. Originators may retain the rights to service the loans and receive service fees as compensation.

Certified poolers must satisfy several requirements specified by Farmer Mac. Included are capital requirements, demonstrated management expertise, ability to issue securities in the financial markets, and ownership of at least 5,000 shares of Farmer Mac stock. Based on initial stock sales, 46 poolers were certified. Loan pools must achieve diversification of loans by geography, farm commodity, and loan size, and each pool must contain at least 50 loans. Loans of related borrowers cannot be included in the same pools. Once the pools are approved, the securities may be issued.

Several safety mechanisms are built into the Farmer Mac structure to deal with loan delinquencies and losses and thus protect investors in Farmer Mac securities. Included among these mechanisms according to their sequence of use are: (1) the cash reserve or subordinated participation requirements for originators and poolers; (2) a Farmer Mac reserve funded through the fees charged to participating financial institutions; (3) Farmer Mac's own equity capital; and (4) a line of credit established with the U.S. Treasury, in an amount of up to $1.5 billion.

Development of Farmer Mac Operations

While Farmer Mac was created by the Agricultural Credit Act of 1987, the first pooling operation did not begin until late 1991. Moreover, by late 1991, only five poolers had applied for and received certification, and by year-end 1992, only four loan pools had been established. The slow development of operations has been attributed to several factors: (1) weak loan demand, (2) strong liquidity of agricultural banks, (3) stringent capitalization requirements, (4) uncertain loan volume; and (5) questionable interest rate competitiveness. These factors are explained as follows.

The conditions leading to the creation of Farmer Mac included the strong loan demand by farmers in the 1970s and early 1980s and the desire by commercial banks for an effective way to fund long-term loans to agricultural borrowers. However, the financial stress conditions of the 1980s caused a significant downturn in farm loan demand and tightening of credit standards that continued into the 1990s. The slump in loan demand, together with moderate growth in deposits, created relatively high liquidity for agricultural banks and thus a diminished incentive for these banks to sell loans into a secondary market.

The incentive for loan sales also was reduced when regulatory agencies required banks and FCS institutions to meet equity capitalization requirements (i.e., maintain equity reserves) on the full amount of the sold loan rather than on the 10 percent portion of the loan held by the originating lender. This requirement reflected the regulator's view that the originating lender was essentially carrying all of the loan's credit risk.

Finally, potential poolers were reluctant to participate in the Farmer Mac program because they were concerned that the volume of loans might be insufficient to justify their commitment to the program or to support a viable Farmer Mac – sponsored secondary market. Part of the concern about loan volume reflected the view that interest rates and terms available to borrowers on loans originated for sale through the Farmer Mac program would not be competitive with loan terms from other lenders. The accumulation of reserves needed to provide the various safety mechanisms cited previously mostly comes from interest paid by borrowers; thus, safety for securities investors could come at the cost of less than competitive rates for borrowers. Since long-term fixed-rate loans were not widely available on farm real estate loans from 1987 to the early 1990s, it is difficult to directly compare estimates of Farmer Mac rates with interest rates and terms on loans currently offered to farmers.

Responses to Slow
Farmer Mac Development

Beginning in 1990, two major changes occurred in Farmer Mac authorizations that were intended to stimulate and expand the development of the secondary market. The first change was Farmer Mac's involvement in secondary loan sales of loans guaranteed by the Farmers Home Administration. The popular name for the activity is Farmer Mac II; the basic program described previously is now called Farmer Mac I.

Under the Farmer Mac II program, originating lenders can sell in the secondary market the FmHA guaranteed portion of operating loans or farm ownership loans with maturities of at least one year. The loans are sold for cash or are swapped for marketable securities. Farmer Mac functions as the pooler for this market. The sale terms are structured to provide lenders with the current market yield on the funds invested in the loans as well as fee income for continuing to service the loans. The loans may have fixed or variable interest rates. For variable rates, the loan rate is adjusted periodically in response to changes in a specified costs of funds index. Similar to Farmer Mac I, Farmer Mac II offers lenders the opportunity to increase liquidity and lending capacity while keeping interest rate risks to a minimum.

The second change created in late 1991 authorized Farmer Mac to fund loan pools by issuing its own unsecured debt securities, with these securities having agency status, similar to that of the farm credit bonds and discount notes. This program is called the **linked portfolio strategy (LPS).** The proceeds of the security sales are used to buy and hold other securities backed by qualifying pools of farm mortgage loans that are issued by certified Farmer Mac poolers and guaranteed by Farmer Mac. Interest rate risks are managed by closely matching terms and rates on Farmer Mac debt obligations with the terms and rates on the mortgage-backed securities purchased from poolers and by other methods of handling prepayment risk.

These LPS arrangements allow Farmer Mac to function more like the government-sponsored secondary markets (Fannie Mae and Freddie Mac) for residential housing mortgages. Thus, while both the Farmer Mac I and LPS programs are in place, the nature of Farmer Mac's role in the two programs differs substantially. Under the original Farmer Mac I, Farmer Mac is not directly involved in the intermediation process; however, under the LPS program, the agency does directly take a position in the flow of funds and securities.

COMMODITY CREDIT CORPORATION

Price and income support programs administered by the Commodity Credit Corporation (CCC) in the U.S. Department of Agriculture provide a comprehensive financial program that combines buffering of price variability for many commodities with provision of inventory financing, added marketing flexibility for farmers, and maintenance of farm incomes. In addition, the Commodity Credit Corporation administers longer-term storage and commodity reserve programs that extend the inventory financing and shift some of the marketing control to the government. Loans are also available for the purchase, construction, and installation of storage and drying equipment on the farm.

The Commodity Credit Corporation was established by executive order in 1933 as part of a package of emergency farm programs designed to boost farm income and to foster an efficient and orderly marketing system. The Commodity Credit Corporation is a USDA agency operating through personnel and facilities of the Agricultural Stabilization and Conservation Service and other USDA agencies. The Commodity Credit Corporation is authorized by charter and statute to buy, sell, make loans, store, transfer, export, and otherwise engage in agricultural commodity operations.

A basic justification of CCC loans under the original charter was that they enabled farmers to hold their commodities after harvest until prices rose as a result of production control programs. Although a primary objective of the Commodity Credit Corporation is to minimize the effects of low commodity prices on farm income, the inventory financing of the marketing of several important farm products has been greatly influenced by CCC lending and other price support activities.

The primary lending feature of the commodity program involves the pledging by eligible farmers of stored crops to Commodity Credit Corporation as collateral for a loan. The loan amount is equal to the level of loan amount for the particular commodity multiplied by the quantity of crops in storage. The farmer can then sell the stored crop on the open market and pay off the loan plus interest at a rate based on the government's costs of funds. However, if the commodity price remains below the loan rate, at the end of the loan contract (usually nine months), the commodity pledged as collateral can be transferred to CCC ownership, thus serving as full payment of loan principal and interest. This arrangement is called a **non-recourse loan.** It *assures the participating farmer of receiving a minimum price for the commodity.*

The basic eligibility requirements for a CCC loan are that producers comply with any allotment or acreage set-aside programs authorized by the U.S. Department of Agriculture in a particular crop year and store their crops in CCC – approved facilities. Besides the regular non-recourse loan program described above, a second choice for non-recourse loans has been the farmer-owned grain reserve program for producers of eligible commodities. Under the reserve program, participants have the same minimum price protection as under the regular program, but they enter a longer (three-year) contract, agreeing to hold grain in the reserve until the contract matures or until the national average market price reaches predetermined release or call levels. Currently, this program is not in operation. Hence, there is uncertainty over the future of the farm-owned reserve program.

CCC loan participation tends to be directly related to the difference between CCC loan levels and commodity market prices. When market prices are near or below the loan level, CCC program loan participation increases. The opposite occurs when market prices rise above the loan level. Thus, producers make greater use of the program when farm prices are declining. However, CCC loan activity is also influenced by other variables, such as interest rates, producers' expectations about variability in seasonal crop prices, the methods of making price and income support payments to farmers, and the preferences of commercial lenders. Indeed, many commercial lenders prefer farm borrowers to shift inventory financing to the Commodity Credit Corporation, using the proceeds of price support loans to pay off production and operating loans. Moreover, the occasional use by the government of "commodity certificates" as a means of making income transfers to participating farmers has also had downward effects on the volume and length of price support loans.

Summary

The major sources of debt capital for agricultural procedures in the United States include commercial banks, the Farm Credit System, life insurance companies, trade firms, individuals, the Farmers Home Administration, and the Commodity Credit Corporation. In addition, the Federal Agricultural Mortgage Corporation (Farmer Mac) oversees the operation of a secondary market for farm real estate lenders. These sources of debt capital differ significantly in organizational structure, operating characteristics, degrees of specialization, and sources of funds. The greatest volumes of loans are held by commercial banks and the Farm Credit System.

Other highlights are as follows:

- Smaller rural banks heavily dependent on local markets for sources of funds and lending opportunities remain a major source of bank financing for agricultural producers.

- Deregulation in banking has deinsulated rural financial markets from national market conditions and is leading to larger holding company and branching systems often operating on an interstate basis.

- High degrees of bank regulation influence a bank's geographic structure, reserves, capital position, lending allocations, and pricing practices.

- The Farm Credit System is a system of federally chartered but privately owned and cooperatively organized banks, lending associations, and service units that makes short-, intermediate- and long-term loans to eligible agricultural producers with funds obtained from the sale of farm credit securities in the financial markets.

- Life insurance companies concentrate on larger long-term loans in selected regions of the United States to agricultural producers and agribusinesses primarily to finance real estate and capital facilities.

- Trade credit provided to producers by input suppliers and other agribusiness firms is a convenient and increasingly cost-competitive source of financing for agricultural producers and usually is directly tied to a merchandising transaction.

- Individuals hold about 21 percent of the total farm real estate debt in the United States, primarily involving sellers of farm land.

- The Federal Agricultural Mortgage Corporation (Farmer Mac) oversees the operations of a secondary market in farm mortgage loans.

- U.S. government credit programs for agricultural producers are operated by the Farmers Home Administration and the Commodity Credit Corporation.

Topics for Discussion

1. Identify some of the unique features of the involvement of agricultural banks in farm lending. Distinguish between unit and branching systems of banking and their relationships to bank holding companies.

2. Identify some of the major regulations affecting commercial banks. How have interest rate controls on deposits and loans influenced banking over time? What effects have the removal of these ceilings had on bank lending to agriculture?

3. Evaluate the various sources and uses of bank funds, and contrast these with the FCS sources and uses of funds. What are the implications for portfolio management of these farm lenders?

4. Characterize the loan-to-deposit feedback relationship in banking. How does it arise? What are its implications for bank lending and investment activities?

5. What role do agricultural credit corporations play in agricultural finance? Why are they used by commercial banks? What are their major sources of funds?

6. Explain the role of correspondent relationships in banking, including the arrangements used in loan participation among banks. How and why are correspondent relationships changing?

7. Explain the organizational structure, ownership, control, and supervision of the cooperative Farm Credit System. Why did the system develop? What problems is the system currently facing?

8. What are the sources of loan funds for the Farm Credit System? What factors affect the system's access to the national financial markets? What are the important areas of the system's portfolio management?

9. How does the Farm Credit System generate its equity capital? What procedures are followed? What are the effects on farmers' costs of borrowing? How do income tax obligations of the Farm Credit Banks and associations affect farmers' costs of borrowing?

10. Give a brief synopsis of the lending programs of the Farm Credit Banks, Lending Associations, and Banks for Cooperatives.

11. Why has the involvement by life insurance companies in farm real estate lending declined relative to that of other major farm lenders?

12. Why does trade credit persist as a major source of farm lending even though trade companies do not specialize in farm lending? What types of trade credit are commonly used? How costly is trade credit?

13. What are the advantages and disadvantages of seller financing for farm land — from the seller's viewpoint? From the buyer's viewpoint?

14. Contrast and critique the role of government credit for farmers as administered through the Farmers Home Administration and the Commodity Credit Corporation. What purposes do these programs serve? For what types of borrowers? What are the credit terms? Sources of funds? Costs of funds? Do these programs compete with or complement lending programs of commercial lenders? Explain.

References

1. American Bankers Association, *FmHA Guaranteed Lending Manual*, Washington, D.C., 1986.

2. Barry, P. J., "Impacts of Regulatory Change on Financial Markets

for Agriculture," *American Journal of Agricultural Economics,* 63(1981):905 – 912.

3. Barry, P. J., "Needed Changes in Farmers Home Administration Lending Program," *American Journal of Agricultural Economics,* 67(1985):341 – 344.

4. Barry, P. J., "Prospective Trends in Farm Credit and Fund Availability: Implications for Agricultural Banking," *Futures Sources of Loanable Funds for Agricultural Banks,* Federal Reserve Bank of Kansas City (Missouri), 1981.

5. Barry, P. J., and W. Lee, "Financial Stress in Agriculture: Implications for Agricultural Lenders," *American Journal of Agricultural Economics,* 65(1983):945 – 952.

6. Board of Governors of the Federal Reserve System, *Improved Fund Availability for Rural Banks,* Washington, D.C., 1975.

7. Boehlje, M. D., "Non-institutional Lenders in the Agricultural Credit Market," *Agricultural Finance Review,* 41(1981):50 – 57.

8. Brake, J. R., and E. O. Melichar, "Agricultural Finance and Capital Markets," *A Survey of Agricultural Economics Literature,* University of Minnesota Press, Minneapolis, 1977.

9. Crane, D., R. Kimball, and W. Gregor, "The Effects of Banking Deregulations," Association of Reserve City Bankers, Washington, D.C., 1983.

10. Economic Research Service, USDA, *Preliminary Analysis of the Financial Condition of the Farm Credit System,* ERS – USDA, Agricultural Information Bulletin No. 490, March 1985.

11. Farm Credit System, *Project 1995,* FarmBank Services, Denver, 1984.

12. Fraser, D., "Structural and Competitive Implications of Interstate Banking," *Journal of Corporation Law,* Summer 1982, pp. 80 – 95.

13. Frisbee, O., "Bankers' Banks: An Institution Whose Time Has Come," *Federal Reserve Bank of Atlanta Economic Review,* 69(April 1984):31 – 35.

14. *Future Sources of Loanable Funds for Agricultural Banks,* Federal Reserve Bank of Kansas City (Missouri), 1981.

15. Hopkin and Associates, *Agriculture in Transition: A Perspective on Agriculture and Banking,* American Bankers Association, Washington, D.C., November 1986.

16. Koenig, S. R., and J. M. Stam, *Life Insurance Company Farm Lending During the 1980s: Evolution or Revolution*, unpublished paper, Economic Research Service, U.S. Department of Agriculture, 1992.

17. Lee, J. E., S. C. Gabriel, and M. D. Boehlje, "Public Policy Toward Agricultural Credit," *Future Sources of Loanable Funds for Agricultural Banks*, Federal Reserve Bank of Kansas City (Missouri), 1981.

18. Lee, W. F., et al., *Agricultural Finance*, 8th ed., Iowa State University Press, Ames, 1988.

19. Lins, D. A., "Life Insurance Company Lending to Agriculture," *Agricultural Finance Review*, 41(1981):41 – 49.

20. Melichar, E. O., "Farm Financial Experience and Agricultural Banking Experience: Banking Data Through the Third Quarter, 1985," presented at the Conference on Agricultural Finance, National Governors' Association, Chicago, January 21, 1986.

21. Norris, K., "The Farmers Home Administration: Where Is It Headed?," *Economic Review*, Federal Reserve Bank of Kansas City (Missouri), November 1986, pp. 21 – 32.

18

Policy Issues Affecting Financial Markets for Agriculture

The functions, management, and institutions involved in financial intermediation for agriculture were examined in Chapters 15, 16, and 17. In this chapter we broaden the setting to consider policy issues affecting financial markets for agriculture in both domestic and international arenas. The functions of the Federal Reserve System are summarized, followed by a discussion of how monetary and fiscal policies influence financial intermediaries in agriculture. Then, international factors are considered in light of the increasing globalization of the world's financial markets. The chapter concludes by evaluating the effectiveness of financial intermediaries for agriculture and by addressing future policy issues affecting agricultural credit markets.

MONETARY AND FISCAL POLICIES

Monetary policies refer to those actions of the Federal Reserve System that influence the cost and availability of money and credit in the economy. Fiscal policies are those powers of a government that allow it to tax and spend. Both sets of policies affect the behavior of financial intermediaries. Understanding them can assist managers in anticipating possible policy actions and in developing appropriate responses.

The Federal Reserve

The Federal Reserve System, often called the "Fed," is composed of 12 district banks and a Board of Governors located in Washington, D.C. Its control over money supplies — defined primarily as the total of currency and bank deposits — is achieved mainly through commercial banks, although other depository institutions also are subject to Federal Reserve requirements and other aspects of monetary control. Depository institutions are the focus of monetary control because of their capacity to "create money." These institutions expand the money supply by making new loans, which, in turn, creates new deposits. This linkage between lending and deposits gives new reserves a multiplying effect. Similarly, the withdrawal of reserves from commercial banks and other depository institutions can have a multiple contracting effect on the money supply.

The Fed contributes to national economic goals through its influence on the availability of money and the cost of borrowing. In the longer run, it attempts to insure sufficient growth in money and credit to achieve a rising standard of living. In the short run, its policies may combat inflationary and deflationary pressures and forestall liquidity crises and financial panics.

The Fed's actions on domestic problems frequently interact with policies affecting interest rates, exchange rates, and international trade. Similarly, the responses to international finance pressures also influence conditions in U.S. financial markets. The importance of these interactions has grown markedly during and since the 1970s with the growth of U.S. exports and U.S. bank lending in international markets.

The Fed's Monetary Tools

The Fed's primary tools for influencing the money supply and the cost of credit are (1) reserve requirements, (2) open market operations, and (3) the discount rate.

Reserve Requirements

All depository institutions in the United States are required to hold a specified percentage of their deposits on reserve (see Chapter 17). In turn, these reserve requirements strongly influence the financial system's capacity to expand the money supply. To illustrate, suppose a bank receives a new deposit for $1 million generated from a depositor's large sale of wheat. If this bank's reserve requirement is 20 percent, its required reserves increase by $200,000, leaving excess reserves of $800,000 for loans or investment in securities.

Assume the bank makes a new loan of $800,000 by depositing this amount in the new borrower's checking account. Deposits would immediately increase by $800,000, and required reserves would increase by $160,000 = (0.20)(800,000). Excess reserves would now equal $640,000 — which could be loaned — and so on. The proceeds of subsequent loans might be deposited in other banks, but the aggregate effects on money supplies would be the same.

The limit of this expansion through the banking system is the reciprocal of the reserve requirement. For a 20 percent reserve requirement, the new $1 million deposit could be expanded to $5 million of deposits. Of the $5 million in deposits, $1 million would be held in required reserves, and $4 million would be used for new loans.

The range of reserve requirements for various types of deposits is specified by law, and the Board of Governors must operate within these limits. Reserve requirements are changed infrequently, usually when the Fed wants to highlight an emerging financial or economic problem. The disadvantages of such changes are that they produce overly pronounced effects on financial markets and they rely greatly on loan demand, which is beyond monetary control.

Open Market Operations

The Fed's most frequently used and flexible tool is its "open market" operations. By selling or buying securities — primarily U.S. government securities — in the open money market, the Fed can reduce or increase the reserves of financial institutions and thereby influence the supply of loanable funds. Most transactions occur through dealers who have contacts with larger commercial banks, life insurance companies, thrift institutions (savings and loan associations and mutual savings banks), and other businesses with large financial reserves. The Federal Reserve Open-Market Committee (FOMC), made up of the seven members of the Board of Governors and presidents of five of the district Federal Reserve Banks, is responsible for implementing monetary policy. It conducts transactions continually and maintains confidentiality about its strategies.

The initial impact of FOMC action is on the availability and cost of reserves in depository institutions. However, the impact of FOMC actions will soon spread throughout the economy. Suppose the Federal Reserve Open-Market Committee is concerned about the inflationary effects of a rapidly expanding economy, and its recommended action is to sell a large volume of securities. This action would reduce bank reserves and deposits because of the funds needed to purchase the securities from the Fed. The value of bonds and other securities would decline, thus increasing interest rates on loans and other financial assets. The higher loan rates would discourage investment, alter savings patterns, and cause other secondary effects. The Fed's purchase of securities would have the opposite effects.

The Fed uses one or more indicators and targets to guide its open market operations. These targets include the level of bank reserves and, more recently, the rates of change in one or more of the money supply measures. Interest rates are considered less effective indicators because they respond to a broader range of factors, foreign as well as domestic. As an example, interest rates may increase because there is a greater demand for loanable funds or there is slower growth of the money supply. In any case, appropriate monetary policies are needed when the targets are not achieved.

Discount Rate

The discount rate is the rate charged depository institutions that borrow from the Fed to meet temporary needs. Control over this rate directly influences the short-term interest rates on loans, whereas the other monetary tools only indirectly affect interest rates. The Fed's influence on longer-term interest rates is indirect and is strongly influenced by the expectations of participants in financial markets.

When its policies change market interest rates, the Fed frequently must change the discount rate. In response to heavy open market sales, for example, short-term interest rates will rise. Banks and other borrowers will then find the Fed, which has not yet changed its discount rate, to be a cheap source of credit. To counter overuse of borrowing from the Fed, the Fed may raise the discount rate. *Thus, the discount rate generally follows rather than leads the market.*

The discount rate is not considered an effective monetary tool because it has overly pronounced effects and because the borrowing initiative lies with the depository institutions.

Minor Policy Tools

To achieve its objectives, the Fed occasionally uses other policy tools, such as control over the margin and a minimum downpayment required for stock purchases. The margin is raised to restrict speculative stock purchases and lowered to revive a sluggish stock market.

In addition, the Fed exerts substantial influence over the financial community through "moral suasion" or "jawbone control." Through its power to limit lending to potential borrowers and its possible influence on the activities of bank examiners, the Fed can strongly "encourage" financial institutions to support its policies and programs.

Fiscal Policies

Fiscal policies refer to the taxation and expenditure policies of government. Although both functions are essential to any government, there is sufficient flexibility in the magnitude and timing of expenditures and in the magnitude and structure of taxation that fiscal policies can significantly affect economic activity. In recessions, for example, expansionary monetary policies of the Fed may fail to generate sufficient demand to increase employment and stimulate the economy. In response, the federal government could introduce public expenditures, tax incentives, or both to stimulate capital investment and employment, and thereby augment monetary policies.

The relationships between fiscal and monetary policies are complex. If the federal government initiates a tax cut but does not reduce its expenditures, then it must finance the expenditures by borrowing in the financial markets. Such borrowing could create substantial federal deficits. Moreover, it could conflict with the Fed's monetary policies to cope with recession or inflation.

A number of stabilizers are built into the fiscal system. A graduated income tax, for example, immediately reduces tax withholdings at the start of a recession as personal and business incomes fall; this reduction tends to restrain decreases in disposable income. Tax surcharges are sometimes assessed to help curb inflation, and tax rebates or tax cuts may stimulate spending. If the tax cuts successfully stimulate the economy, the resulting higher incomes provide extra tax revenues.

On the expenditures side, unemployment compensation and welfare programs add stability through income effects that are contracyclical, just as public works and other government expenditures are designed to reduce unemployment. Farmers and rural areas are directly influenced through programs such as subsidized FmHA loans, rural development programs, and emergency relief programs for officially designated disaster areas resulting from drought, frost, or other forms of adverse weather. Payments to farmers under government agricultural programs are also important.

Besides the fiscal policies of the federal government, state and local governments frequently influence local economies through tax and spending programs. In turn, these tax and spending programs may influence the demand for and supply of loanable funds in the community and the behavior of local financial intermediaries.

Implications for Financial Intermediaries and Agriculture

Both the fiscal policies of government and the monetary policies of the Fed reflect the responses of these units to their perceptions of particular economic conditions. Ideally, these perceptions should be similar at any given time so that fiscal and monetary policies may be coordinated.

However, with the relative independence of the Federal Reserve System, the Board of Governors sometimes perceives economic problems to be different from those of Congress and the president, who largely determine fiscal policies. Moreover, congressional leaders do not always agree with the president on the seriousness of the fiscal problem or the appropriate courses of action. Consequently, fiscal policies are not always clearly defined nor are they always consistent with monetary policies. Nonetheless, the effects of both fiscal and monetary policies will influence the availability and cost of loan funds to agricultural producers who borrow from banks, the Farm Credit System, or other intermediaries.

INTERNATIONAL FINANCIAL MARKETS

The international dimension of financial markets has grown in importance for the total U.S. economy as well as for the agricultural sector and its sources of financial capital. Capital flows to the United States in exchange for exports, securities sold to non – U.S. buyers, and investments in the United States by people in foreign countries. These foreign purchases and investments must be paid for with U.S. dollars. Capital flows from the United States in exchange for imports, securities bought from non – U.S. sellers, and foreign investments made by U.S. firms and citizens. These foreign transactions by the United States must be paid for in the sellers' currency. Thus, international transactions involve the exchange of currencies as well as goods, services, and securities.

The exchange rate of a national currency is its value in terms of another national currency or of a collection of currencies, usually expressed as a trade-weighted index. Exchange rates are significant in trade and trade-related issues. For example, to buy soybeans from the United States, a Japanese buyer must exchange yen for U.S. dollars with which to pay for the soybeans. An increase in the yen price of a U.S. dollar means an increase in the yen price of U.S. soybeans in the absence of a compensating decrease in the price of the soybeans expressed in U.S dollars. A change in exchange rates immediately changes the relative prices between alternative sources of tradeable goods and the relative prices between tradeable and nontradeable goods within the countries with the altered exchange rates.

Of the various national currencies, the U.S. dollar is unique in that it has widespread use in international contracts that are related to the trading of commodities and financial assets. The U.S. dollar is also used widely in other countries as reserve currency. Some important historical bases for the prominent role of the U.S. dollar are treated in the following section.

The International Monetary Fund (IMF): Bretton Woods and Beyond

Economic history between World War I and II was marked by considerable chaos in international trade. The volume of world trade preceding World War II was less than that preceding World War I, despite a near doubling of world production. Near the close of World War II, an agreement that was designed to prevent a return to the economic chaos that occurred before the war was negotiated at Bretton Woods, New Hampshire. This agreement brought into existence exchange rates that fixed the values of major national currencies in terms of the British pound and the U.S. dollar and the value of these currencies in terms of gold. This agreement also authorized the formation of multi-national financial institutions headed by the World Bank and the International Monetary Fund (IMF).

However, during the Suez Crisis of the late 1950s, the exchange rate regime of Bretton Woods began to crumble with the exit of the British pound as a part of the support structure, leaving the U.S. dollar as the gold-based currency to anchor the fixed-rate currency exchange system. With the Vietnam war came the U.S. government's decision to finance its participation with fiscal budget deficits. The U.S. dollar was devalued relative to the price of gold, and then, in the early 1970s, it was decoupled from gold. This was the so-called "breakdown of Bretton Woods," which led to the current system of floating and managed exchange rates.

The events of the 1970s did not bring about a complete breakdown — the multi-national institutions, the International Monetary Fund and the World Bank, remain. Indeed, subsequent events may even have strengthened these institutions. Today, there are other multi-national financial institutions as well.

The presence of an international financial market is not new. For centuries, transactions between firms and governments of different nations have required insurance and financing with bonds or bank loans, often either on a government-to-government basis or with government guarantees. A novel feature of the modern financial market is the Euro dollar — a U.S. dollar held outside the United States (the U.S. dollar is not the only currency held outside its country of origin). Following World War II, countries with non-convertible currencies needed a medium of exchange to settle international accounts. The most important of these currencies were those of the Centrally Planned Countries (CPCs) (communist or highly socialized countries). The first significant demand is attributed to the People's Republic of China in 1949. Augmented by demand from former Warsaw Pact countries, the market in Euro dollars grew steadily through the 1950s and 1960s. The growth became explosive in the 1970s, fueled on the supply side by "Petro Dollars" from OPEC countries and on the demand side by loans to finance development in developing countries (LDCs) as well as in the Centrally Planned Countries.

In 1991, the value of goods and services in world trade had grown to more than $5 trillion (International Monetary Fund). In contrast, international financial transactions were nearing $200 trillion. The international financial market had grown to a huge presence, connecting central banks; large city banks with their international branches; the Bretton Woods multi-nationals, the multi-national financial institutions of Japan and Europe, notably the Bank for International Settlements, in Switzerland; and arbitrators throughout the world, from curb-side currency exchanges to sophisticated dealers in international securities.

The international financial market continues to grow and develop, thus increasing its integration with domestic financial markets in much of the world. As the market grows, the role of the International Monetary Fund is strengthened, although that role is still severely limited by the sovereign claims of nations that make up its membership. The international financial market remains largely unregulated, which is contrary to the relatively strong regulations that characterize many domestic financial markets.

Current Exchange Rate Arrangements

Current exchange rate arrangements are frequently described as being flexible. But, the flexibility is limited. Of the 160 national currencies of the members of the International Monetary Fund on December 31, 1990, only 20 could be described as being independently floating. However, these 20 include major currencies such as those of Japan, the United Kingdom, and the United States. Another 33 countries had currencies that were "pegged" (systemically related) to the U.S. dollar.

From the European Community (EC) has evolved the European Currency Unit (ECU), a basket of currencies that include the currencies of France and Germany as well as six other EC countries. Although the individual currencies do not float independently, the European Currency Unit does. Moreover, the currencies of 14 more countries have been pegged to the French franc alone.

The International Monetary Fund itself issues a currency known as the Special Drawing Right (SDR), which is also a basket of currencies. The basket is based upon five national currencies — the U.S. dollar, deutsche mark, French franc, British pound, and Japanese yen. The currencies of seven countries are pegged to the Special Drawing Right. Thus, 54 of the 160 currencies of IMF members are pegged or otherwise closely related to the floating currencies of the United States, European Community, and Japan. Much of the world trade in both commodities and financial assets is transacted in these currencies.

Exchange rate arrangements in the early 1990s were turbulent. New currencies were being initiated (e.g., among nations of the former USSR). The European Community was trying to develop a common currency for its members. However, not all its member countries were committed. Thus, the changes begun in the early 1990s are far from settled. Their future courses will have far-reaching consequences for commodities and financial assets in international markets.

Implications for Agricultural Finance

The globalization of international financial markets and the resulting linkages with domestic financial and commodity markets have made the financial performance of agricultural businesses and their lenders increasingly sensitive to world economic conditions. Pressures and shocks can spread quickly throughout the world. For example, the debt-servicing problems experienced by many developing countries that stemmed from substantial international borrowing in the late 1970s and early 1980s resulted in the subsequent problems of economic recession, high inflation, and then fluctuating real interest rates.

These countries avoided massive loan defaults only through debt rescheduling and/or refinancing by either commercial lenders or the International Monetary Fund, thus creating pressures on interest rates and credit availability that were felt throughout the financial markets of the world and illustrating how the performance of financial markets for agriculture in the United States and elsewhere is significantly influenced by international trade and monetary conditions.

EVALUATING FINANCIAL INTERMEDIATION IN AGRICULTURE

In general, financial intermediaries have exhibited strong performances in providing credit and related services to agriculture in the United States, although the specialized lending of some intermediaries (e.g., the Farm Credit System and small rural banks) has made them vulnerable to the swings in financial conditions in the agricultural sector. Funds for loans can be made available to farmers in a timely fashion, for various purposes and in amounts, costs, and maturities that compare favorably with other economic sectors.

These features of financial intermediaries have strengthened the linkage between agricultural and non-agricultural sectors over time and have increased the agricultural sector's sensitivity to changing conditions in national and international financial markets. However, relatively few studies have comprehensively evaluated the effectiveness of financial intermediation for agriculture in the United States. Most studies have focused on selected intermediaries and specific types of performance. In this section we will explore various performance criteria and consider some of the findings and judgments.

Efficient Allocation of Loan Funds

Efficiency criteria in economics specify that an optimal allocation of capital is characterized by equal marginal productivity of capital for all users in all geographic areas. Under perfect financial intermediation, two agricultural businesses producing the same products in different regions but having similar operations and risk characteristics should have similar access to loan funds and similar interest rates. However, imperfections in capital markets along with differences in lending risks can cause differences in financing terms and can account for contrasts in regional growth.

The nationally federated structure of the Farm Credit System, along with its member banks' sharing in the sale of consolidated Farm Credit Securities, helps to assure that FCS borrowers nationwide have high uniformity in interest rates and other credit terms. Differences in loan rates among FCS units primarily reflect diversities in these lenders' operating costs, risk reserves, and capitalization policies. On occasion, these differences may be significant, as occurred in the mid-1980s because of regional differences in the effects of financial stress in agriculture on the loan quality, loss positions, and performance of the district banks and local associations of the Farm Credit System.

For agricultural borrowers from commercial banks, variations in bank structure, size, diversity, and funding sources may cause differences in the availability and cost of borrowing for farmers. In addition, studies during the late 1960s found significant contrasts between various regions in interest rates for farm loans for similar purposes among farms of similar size and those for loans of similar risk and secured by similar assets. However, these differences have declined over time in response to deregulation and easing of geographic restrictions on banking, growing competition in financial markets, and a closer tie between rural financial markets and the national and international markets.

Operational Efficiency

Operational efficiency refers to the cost effectiveness of producing loans and other financial products and services offered by financial intermediaries.

While financial institutions divide these costs into detailed categories, they can be broadly considered as the interest costs of acquiring loanable funds, the non-interest operating costs of administering the lending programs and other activities, the costs of risk-bearing, and the costs of equity capital.

The costs of risk-bearing are represented by the provision for loan losses allowed for financial institutions, and the costs of equity capital are represented by the net earnings or profits earned by the institutions on various activities. Thus, using average cost concepts, one finds that higher efficiency is represented by lower average costs per unit of output for the various components that make up the total costs of operations.

Costs of funds may vary according to the size and type of the financial institution, depending on the specific sources of funds and the characteristics of the markets in which the institution lends and borrows. In the past, for example, smaller banks located in rural communities had relatively low costs of funds because of their strong reliance on non – interest-bearing demand deposits. However, the deregulation of deposit rates has substantially reduced the differences in costs of funds among depository institutions. Costs of funds may also be slightly higher for intermediaries making longer-term loans, due to the positive yield curve on longer-term sources of funds. Similarly, the cost of funds generally is higher as the intermediary's risk position increases.

Minimizing operating costs per dollar of loan volume is a desirable criterion. Some of the variables affecting operational efficiency are loan volume and economies of size, the availability and use of new technology by financial institutions, and the quality of management and personnel. For example, studies indicate that small banks may experience economies of size as bank growth occurs, although these gains diminish rapidly as total assets increase beyond the $50 million to $100 million size range, and costs vary according to the banks' organizational structure and product lines. Moreover, the potential economies of size differ, depending on whether large size results from the addition of branch offices or from the growth of existing offices. Similar analyses in the Farm Credit System have indicated that many lending associations are operating at a size that is less than the size that minimized average cost [14]. This finding in part explains the recent trend toward consolidation and growth in the size of lending associations in the Farm Credit System.

In general, the costs of financial intermediation relative to loan rates have declined over time for the major agricultural lenders. The primary forces causing this decline are (1) the development and use of new technology, such as electronic data processing; (2) economies of size achieved in a larger volume of agricultural lending; (3) improved management skills of lenders; and (4) increased competition between lenders. In contrast, the costs of funds generally fluctuate with changes in the levels of inflation rates and in the costs of risk-bearing.

Interest Rates on Agricultural Loans

A comparison of interest rates charged on agricultural loans by commercial banks and FCS lenders over time is shown in Figure 18.1. Several features are clearly evident. One is the sharp increase in rate levels in the early 1980s; the second is the high volatility, especially during the early 1980's; the third is the difference in the timing and magnitude of rate changes from these lenders; and the fourth is the decline in interest rates in the 1990s. In general, these rate characteristics reflect the continued deinsulation of farm credit from forces in national and international financial markets, attributed largely to financial deregulation, competitive responses of lenders, and macro-economic events and policies, including a shift in late 1979 by the Federal Reserve toward greater reliance on monetary aggregates as a policy target.

For agricultural banks, the rapid increases and sharp moves in agricultural

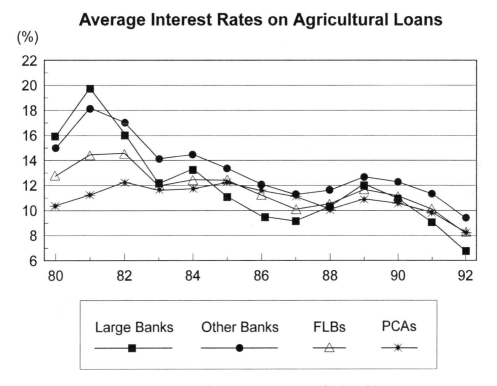

Figure 18.1. Average interest rates on agricultural loans.

loan rates were due to the strong reliance of these banks on rate-sensitive sources of funds. As market interest rates change, so will the banks' costs of funds. As Figure 18.1 shows, the rate swings are stronger at larger banks but still substantial at smaller banks.

The rate spread that favored the FCS rates during the early 1980s in part reflected the FCS practice of pricing loans with floating rates that were adjusted periodically for changes in PCA and FLB average costs of funds, the average maturity of which exceeded those of commercial banks. This practice caused the interest rates on FLB loans during the 1980 – 1982 period to be as much as 300 basis points below the rates on Treasury securities. The practice gave the Farm Credit System a competitive advantage over other lenders during periods of generally rising interest rates. However, as the general level of nominal interest rates declined, beginning in 1982, the rate advantage of the Production Credit Associations and the Federal Land Banks was transformed into a rate disadvantage.

Comparable data from other agricultural lenders are not available. However, rates on all types of agricultural loans now follow market interest rates more closely than in the past. In the future, these rates should continue to follow market rates, with their magnitude and volatility largely determined by the effects of macro-economic events and policies, inflation conditions, and structural characteristics of financial markets and financial intermediaries.

REGULATORY ISSUES
AND CREDIT POLICIES

The development of financial intermediation in agriculture has been a dynamic process characterized by institutional changes largely in response to regulations, innovations, and improvements in efficiency, many made possible by new technologies in information management. Continued changes will take place in the future as well.

The beginning of the 1980s, in particular, was a landmark in U.S. financial markets because of the regulatory changes that marked this period. Especially important was the passage of the Depository Institutions Deregulation and Monetary Control Act of 1980 (DIDMCA) and the Depository Institutions Act of 1982 (DIA). Other legislation specific to agricultural finance includes the Farm Credit Act Amendments of 1980, 1985, and 1986, the Agricultural Credit Act of 1987, which reflected an expressed shift toward a more conservative role for government credit programs. In this section we discuss regulatory and policy issues and consider their implications for agricultural finance.

Financial Deregulation

The changes contained in the 1980 DIDMCA and the 1982 DIA represented a significant deregulation of financial markets in the United States that profoundly affected structural characteristics, levels of performance, and competitive relationships throughout these markets. The 1980 DIDMCA established uniform Federal Reserve requirements at all depository institutions, provided for the orderly phase-out of interest rate ceilings on deposits, authorized interest-bearing transactions accounts, preempted state usury laws on certain types of loans (including most agricultural loans), increased federal insurance coverage on deposits and accounts of depository institutions, and required the Federal Reserve to competitively price its services and make them available to all depository institutions.

The 1982 DIA focused on the problems of ailing thrift institutions, although its provisions affected commercial banks and other institutions as well. Regulatory agencies were granted expanded authority to assist troubled depository institutions and to arrange acquisitions of such institutions across state lines and between banks and thrifts if conditions warranted. Federally chartered thrifts were granted expanded authority to make consumer and commercial loans. In addition, the phase-out of rate ceilings on deposits was accelerated, and many other provisions were also included.

For agricultural finance, the legislation significantly affected smaller banks that had fund availability problems and less flexibility in balance sheet management, compared to regional and money center banks. The changes in reserve requirements released some additional bank funds to support lending in rural areas. Elimination of interest rate controls on deposits and loans improved efficiency in local flows of funds, allowed more equitable competition by banks for funds, decreased differences in interest rates among lenders, and brought more market-oriented pricing at the bank, department, and customer levels. However, deregulation has also increased the risk positions of financial intermediaries, leading to greater concerns about their credit worthiness as well as greater merger and bankruptcy activity.

During the agricultural stress times of the 1980s, the profitability of agricultural banks came under pressure from mounting loan losses and loan delinquencies. Bank foreclosures accelerated and terms of lending became more stringent. However, the generally short-term nature of loans and other financial assets and the relatively strong liquidity and capital positions allowed most banks to weather the stress times and to restore acceptable levels of bank profitability. In contrast, the financial plight of thrift institutions was much more severe. Many thrifts experienced excessive lending risks, especially involving real estate loans, and foreclosures and institutional bankruptcies were widespread. Substantial assistance from the federal government was needed to protect depositors and to permit reasonable adjustments in the thrift industry.

Structural Changes in Banking

The structure of agricultural banking in the United States is changing significantly. Multiple offices, branch banks, and holding company – affiliated banks have become major suppliers of bank farm loans. These changes occurred even before late 1994 when Congress passed legislation that would permit interstate banking on a national basis, unless individual states choose not to participate.

Many states, however, have significantly reduced or eliminated restrictions on branch banking and multi – bank holding company expansion. Numerous states have enacted laws to permit some form of interstate banking, usually based on reciprocal arrangements with other states.

The consequences of the changes in the structure of banking are significant for agricultural finance. Multi-office banking, particularly branch banking, provides rural areas with easier access to money and capital markets, helps spread loan risks over a broader geographic area, reduces legal lending limit problems, allows banks to diversify into new markets and services, and promotes economies of size for functions such as centralized data processing, telecommunications and other electronic facilities, as well as providing highly specialized services such as estate management and market counseling.

The trend toward multi – bank holding companies and more liberal branching will likely continue. This will not preclude the existence of smaller, localized independent banks, although they will become fewer in number, but it does mean that these banks will need strong management, success niches in the market place, and effective relationships with larger banking systems.

Farm Credit System Issues

An important issue of the farm policy debates in the 1980s was the seriousness of the economic problems of the Farm Credit System and whether those problems were inherent in its specialized mission, decentralized structure, and operating policies. Given the general instability of agriculture, combined with deregulation of financial markets, broad questions were debated to determine what policy and structural changes were needed in the Farm Credit System and the Farm Credit Administration to enable them to continue their missions and whether or not the Farm Credit System needed federal assistance to cope with its financial problems resulting from the stress times in agriculture.

The Farm Credit Amendments Act of 1985 replaced the 13-member Farm Credit Board, elected mostly by the 12 farm credit districts, with a 3-member board appointed by the president of the United States. It created the Farm Credit System Capital Corporation as a temporary mechanism for administering "self-help" to the various units of the system. The act strengthened the Farm Credit Administration as an arm's-length regulator of the Farm Credit System, granting it powers similar to those of the regulators of other financial institutions, such as commercial banks and savings and loan associations. The Farm Credit Administration was given responsibility to set credit and accounting standards and was provided with cease-and-desist powers to insure that established policies and standards are maintained.

The concept of federal support to the system in times of economic crisis was clearly established by the Farm Credit Amendments Act of 1985. The support process was also defined. This process largely resolved, for the time being, the issue of "agency" status for the Farm Credit System and continued the set of regulatory exemptions and preferences that facilitate the sale of farm credit bonds, medium-term notes, and discount notes in the national and international financial markets.

Still another important policy issue for the Farm Credit System involved its responses to the significant lending problems and financial stresses of the second half of the 1980s. Loan volume and loan losses were declining rapidly. The system's "capital and surplus" declined by over $3.4 billion during 1985 primarily because of bad loans, with an additional $2.75 billion drop during 1986. Part of the problem involved mounting loan losses and a reduction in loan quality resulting from farmers' financial problems.

The problem was further aggravated by a large volume of long-term securities outstanding at high interest rates relative to current market rates. The Farm Credit Amendments Act of 1986 authorized the Farm Credit Banks to amortize a portion of their loan losses and bond premiums over up to a 20-year period. This action was intended to provide near-term help for the financial problems of the Farm Credit System and buy time so that further restructuring could occur.

As the loan losses, declining loan volume, and other financial problems affecting the Farm Credit System continued to mount, it became clear that additional legislative and financial assistance were needed. As a result, in December 1987, the U.S. Congress passed the Agricultural Credit Act of 1987, which was signed into law by the president. This act essentially culminated the series of Farm Credit Amendments Acts that had been passed in the preceding two years. Included among the provisions of the 1987 act were the restructuring of the Farm Credit System, federal financial assistance, greater flexibility in capitalizing the Farm Credit System, creation of an insurance fund, rights of borrowers and restructuring of loans, and a new secondary market for farm mortgage loans (see Chapter 17 for a more detailed description of the 1987 act).

These provisions allowed for a substantial restructuring of the Farm Credit System as it entered the 1990s. Moreover, the burden for the system remained of repaying the losses incurred from high-cost banks and loan losses, rebuilding loan volume, and meeting repayment obligations for federal financial assistance, while still rebuilding equity capital and maintaining competitive rates to borrower. In the early 1990s, most of the FCS institutions had restored the soundness of their operations, improved institutional profitability, streamlined operations, and were functioning competitively in the agricultural financial markets. However, restructuring of the FCS banks and lending associations was continuing, as these institutions sought greater efficiency and risk-bearing capacity in their lending operations. The Farm Credit System has also sought broader lending authorities in the 1990s, but the scope of the system's eligible borrowers is a sensitive policy issue.

Government Credit Programs

The growth of the U.S. government's agricultural lending through the Farmers Home Administration and the Commodity Credit Corporation in the late 1970s and 1980s clearly softened the impacts of volatile credit markets and variable agricultural incomes on the agricultural sector. However, much concern has surfaced about special credit treatment in agriculture, the proper balance between public and private lending, the degree of subsidy involved, the distribution aspects of various FmHA programs, and the resulting tax burdens. Some observers have suggested that the past structure of farm credit markets, including the prominence of government programs, resulted in over-financing of the agricultural sector, thus restraining the needs for resource adjustment, shifting too much risk-bearing to the government, and capitalizing the effects of concessionary financing terms into higher values of land and other agricultural assets. Considerable interest has centered on curtailing FmHA lending and restoring the agency's last resort role.

Besides the high taxpayer costs of government programs, these changes have also been based on the following conditions: (1) more moderate growth in agricultural debt than occurred in the 1970s; (2) expanded use of new federal crop insurance programs for disaster protection; (3) more effective coordination of public and private credit programs, tailored to beginning farmers or other target groups; and (4) greater innovation by commercial lenders in acquiring loan funds and in countering business and financial risks in agriculture. The objective was an intended pullback of the public safety net for agricultural lending.

Important changes have also been instituted in FmHA loan programs. There is increased realization that chronic financial problems in agriculture cannot be resolved by credit programs. Hence, there has been a strong shift from direct FmHA lending toward loan guarantees to agricultural producers who can demonstrate loan repayability but who, for various reasons, cannot qualify for financing with commercial lenders. The loan guarantee approach relies more heavily on funding from commercial lenders and utilizes the private lender's expertise and resources in loan administration, analysis, monitoring, and control. It also provides greater flexibility in loan pricing, reduces the emphasis on concessionary interest rates as a form of subsidizing FmHA borrowers, and improves the prospects for FmHA borrowers to become commercial credit borrowers. Further changes are in store for the future as well. For example, the agricultural loan program in the Farmers Home Administration will be transferred to a new farm services agency in the U.S. Department of Agriculture in order to streamline operations and to provide improved service for farmers.

Summary

Significant policy issues affecting agricultural finance have been considered in this chapter. Major points of interest include the following:

- Monetary policy conducted by the Federal Reserve System, especially through open market operations, may significantly affect the rate of growth of the money supply and the cost and availability of credit.

- Fiscal policy, which refers to the taxation and expenditure policies of government, may interact with monetary policy to influence aggregate economic performance and monetary conditions.

- The globalization of financial markets has closely integrated local financial market conditions with those in national and international financial markets. Changes in exchange rates on major currencies also have substantial, often unanticipated effects on financial market conditions.

- Efficiencies of funds allocations among financial markets have been increasing because of improved technologies in funds flows, economies of size, and more timely, comprehensive information about financial market conditions and the safety and soundness of financial institutions. As a result, geographic differences in interest rates have diminished over time.

- Deregulation in financial markets has contributed to market efficiency and closely integrated local, regional, national, and international financial markets.

- Continuing structural changes in banking are leading to larger banking systems, although market niches will remain for smaller, well-managed, and soundly capitalized individual banks, especially many such banks located in rural areas.

- The specialized loan functions of the Farm Credit System make its institutions vulnerable to swings in the financial conditions of agriculture, and these institutions may require federal assistance when these financial conditions are severe.

- Government credit programs, primarily through the Farmers Home Administration, have shifted significantly toward guaranteed loans in order to utilize commercial lenders' loan funds and expertise and to improve the borrowers' prospects for graduating to complete commercial lending.

Topics for Discussion

1. What are the primary objectives of the Federal Reserve System? In what ways might these objectives conflict?

2. What tools does the Federal Reserve System use to achieve its monetary objectives? Explain how each one works, and evaluate its relative effectiveness.

3. Distinguish between monetary policy and fiscal policy. Where does each originate? How do these policies, individually and jointly, impact on the ability of agricultural borrowers to obtain loans?

4. Compare FCS lenders, commercial banks in a unit banking state, and commercial banks with statewide branching in terms of:

   a. Their response to monetary policy.

   b. Their capacity to attract loanable funds to agriculture from nonlocal sources.

5. What factors explain the closer integration between conditions in rural financial markets and national financial markets in the 1980s and 1990s compared to earlier times?

6. Interest rates on agricultural loans from major agricultural lenders have become more volatile in recent times, compared to pre-1980 conditions. Why have these interest rate conditions occurred?

7. Identify the major components of interest rates on agricultural loans from commercial banks and FCS lenders. How do differences in these components affect the level and responsiveness of rates to changes in financial market conditions? Why do the earnings rates differ between these lenders?

8. Summarize the forms of financial deregulation and the major reasons for them in the 1980s. What are their major effects on agricultural lending by agricultural banks? What impacts will deregulation have on the level and volatility of agricultural loan rates?

9. What conditions have made liberalization of geographic restrictions on banking a relevant policy issue? What will be the effects of liberalization on credit services and bank performance in rural areas? What disadvantages may arise from geographic liberalization?

10. Explain some current policy issues affecting the Farm Credit System, its relationships to commercial banks, and its sale of securities in the national financial markets.

11. Identify and evaluate the major issues associated with government credit programs for agriculture. What is the proper balance between public and private sectors in meeting farmers' credit needs? What instruments and programs besides credit programs are possible?

References

1. Baker, C. B., "Agricultural Effects of Changes in Financial Markets," *American Journal of Agricultural Economics*, 66(1984):541–548.

2. Baker, C. B., *Current Financial Stress: Sources and Implications for U.S. Agriculture*, W. I. Myers Memorial Lecture, Cornell University, AE Res – 1, January 1987.

3. Barry, P. J., "The Farmers Home Administration: Current Issues and Policy Directions," *Looking Ahead*, The National Planning Association, September 1985:4 – 11.

4. Barry, P. J., "Financial Markets for Agriculture: An Overview of Performance and Policy Issues," *Financial Policies and Future Directions for Agriculture*, Department of Agricultural Economics, University of Illinois at Urbana – Champaign, 1985.

5. Barry, P. J., *Impacts of Financial Stress and Regulatory Forces on Financial Markets in Agriculture*, The National Planning Association, Washington, D.C., 1984.

6. Barry, P. J., "Impacts of Regulatory Change on Financial Markets for Agriculture," *American Journal of Agricultural Economics*, 63(1981):905 – 912.

7. Barry, P. J., "Needed Changes in Farmers Home Administration Lending Program," *American Journal of Agricultural Economics*, 67(1985):341 – 344.

8. Bennet, V., "Consumer Demand for Product Deregulation," *Federal Reserve Bank of Atlanta Economic Review*, May 1984.

9. Board of Governors, *The Federal Reserve System Purposes and Functions*, Federal Reserve System, Washington, D.C., 1974.

10. Cooper, R. N., "Macroeconomics in an Open Economy," *Science*, 123 (September 12, 1986):1115 – 1159.

11. Curry Foundation, *Confrontation or Negotiations: U.S. Policy and European Agriculture*, Assoc. Faculty Press, 1985.

12. Drucker, P. F., "The Changed World Economy," *Foreign Affairs*, 64:4(Spring 1986)768 – 791.

13. Farm Credit System, *Project 1995*, FarmBank Services, Denver, 1984.

14. Fraser, D., "Structural and Competitive Implications of Interstate Banking," *Journal of Corporate Law*, Summer 1984:642 – 654.

15. Hopkin and Associates, *Transition in Agriculture: A Perspective on Agriculture and Banking*, American Bankers Association, Washington, D.C., 1986.

16. Hughes, D. W., S. C. Gabriel, P. J. Barry, and M. D. Boehlje, *Financing the Agricultural Sector: Future Changes and Policy Alternatives*, Westview Press, Boulder, Colorado, 1986.

17. International Monetary Fund, *Exchange Arrangements and Exchange Restrictions*, Annual Report, 1990, pp. 576 – 581.

18. International Monetary Fund, *World Economic Outlook*, May 1992.

19. Khan, M., and J. S. Lizondo, "Devaluation, Fiscal Deficits, and the Real Exchange Rate," *The World Bank Economic Review*, 1:2(January 8, 1987):357 – 374.

20. Mendlesohn, M. S., *Money on the Move: The Modern International Capital Market*, McGraw-Hill Book Company, New York, 1980.

21. Osborn, D. D., and J. R. Hurst, "Economies of Size Among Production Credit Associations," *FCS Research Journal*, 4(November 1981):31 – 34.

22. Swackhammer, G. L., and R. J. Doll, *Financing Modern Agriculture: Banking Problems and Challenges*, Federal Reserve Bank of Kansas City (Missouri), 1969.

SECTION SIX

Other Topics

19

Legal Aspects of Agricultural Finance

This chapter summarizes many legal aspects of agricultural finance that are important to credit relationships and financial management involving transactions in both real property and personal property. Included are discussions of contracts, notes, the Uniform Commercial Code (UCC) involving personal property, real estate transactions, foreclosure and bankruptcy, co-ownership of property, other negotiable instruments, and other legal instruments and terminology important to financing agriculture.

CONTRACTS

A **contract** is *a legally binding agreement between two or more parties*. The contract may be written (preferably) or oral, and it underlies virtually all business transactions, including lending and borrowing, leasing, and buying and selling of goods and services.

A valid contract has four essential components:

1. The parties involved in the contract must be legally competent. They must be capable of fully understanding the implications of their actions. For example, contracts involving children who are minors and people who are mentally disabled are generally excluded.

2. The subject matter of the contract must be legal and proper.

3. All the parties to a contract must willingly consent to the agreement, as evidenced by an offer and an acceptance.

4. The transaction embodied in the contract must involve consideration; that is, the parties must receive and/or give up something of value.

NOTES

The basic legal document in most loan contracts is the note, or promissory note. It is the written promise of a borrower to repay a loan at a certain time, including any and all interest that has accumulated during the term of the loan. The note also may stipulate the actions the lender will follow if the borrower fails to meet the terms of the loan contract, as well as any additional obligations incurred by the borrower (e.g., late fees, interest penalties, costs of collection).

Notes may be unsecured or secured by real property, personal property, or both. A secured note is accompanied by a pledge of property (i.e., collateral) to the lender in the event the borrower defaults on the loan. The lender may, under specified circumstances, take possession of the property, sell it, and apply the proceeds to the debt obligation. Security reduces the riskiness of the loan and enhances its marketability if the loan is to be sold in a secondary market transaction.

An unsecured note has no specific pledge of property associated with it. Nonetheless, an unsecured lender still can share in the proceeds of the forced sale of a borrower's property, although secured or senior lenders receive compensation first with unsecured or junior creditors receiving a pro rata share of any remaining proceeds. Thus, unsecured lenders are in a higher-risk position than secured lenders.

REAL VERSUS PERSONAL PROPERTY

Financing and other types of transactions involve two types of property: *real property* and *personal property*. Real property consists of real estate (land), buildings, houses, and various types of improvements to real estate. Personal property, basically, is everything else, although many types of personal prop-

No. 2300-05 Endless Plains, Texas June 1 19___ $ 10,000

FOR VALUE RECEIVED, I, we, or either of us, as principals, promise to pay to the order of

LAST STATE BANK, ENDLESS PLAINS, TEXAS

Cr. Lfe. _____ ; ____ H & A ____ ; Interest _____ $300.00

at said address, the sum of $ 10,000 , payable in 4 monthly installments of $ 2,575.00 ,
the first installment being due and payable on or before the 1st day of July 19___ and the remaining installments being
due and payable on or before the same day of each succeeding month thereafter until said note is fully paid, with all remaining principal and interest maturing on
October 1 , 19___.

BORROWER shall pay interest after maturity at the highest legal rate and if this note is collected through judicial proceedings, BORROWER shall pay the amounts actually incurred by HOLDER as court costs and attorney's fees assessed by a court. BORROWER shall have the right at any time during regular business hours to prepay this note if full and shall receive refunds in accordance to the rule of 78's. BORROWER shall pay additional interest for default at $.05 per $1.00 of any installment due and remaining unpaid for 10 days or more. Deferred interest shall be charged as permitted by the Texas Consumer Credit Code. A default in payment of any amount when due will, at holder's option, mature the entire unpaid balance of this note, less unearned charges. BORROWER and each surety, endorser, or guarantor of this note hereby waive presentment, protest, notice of protest, and diligence in collection, and each hereby consents to any extension or extensions of this note which holder may make. HOLDER is expressly granted the right to offset against this note and all other liabilities of the BORROWER to HOLDER, all money or other property in its possession held for or owed to the borrower (including without limitation all deposits and accounts). To secure the payment of the foregoing obligation or any renewal of some and to likewise secure payment of any other indebtedness now or hereafter owed HOLDER by BORROWER, BORROWER hereby grants to HOLDER a security interest, subject to the terms and conditions stated on the reverse side hereof, in the personal property hereinafter described, together with any and all additions and accessions thereto and all proceeds therefrom, all hereinafter collectively called "Collateral".

Model 7000 A TRACTOR
WHEAT CROP   COTTON CROP

BORROWER hereby acknowledges that this combined Note, Truth in Lending Disclosure, and Security Agreement was completed as to all essential provisions and disclosures before it was signed by BORROWER and a copy thereof was delivered to BORROWER at time of signing. NOTICE: See other side for important information.

INS. _____

ADDRESS Star Route, Box 18   Joseph Farmer, Julie Farmer

PHONE _____   Address and Signature of BORROWER(S)

NOTE: Truth in Lending Disclosure, Security Agreement, Monthly Installments

---

DISCLOSURE COLUMN

1. Amount of Credit $ _____ 10,000

   $ ____ ; ____ Prepaid Finance Charge

   $ ____ ; ____ Required Deposit Balance

2. $ ____ ; ____ Total prepaid Finance Charge and Required Deposit Balance

3. Other Charges, Itemized

   Credit Life $ ____ ; ____
   Insurance
   Credit A/H $ ____ ; ____

   $ _____

4. Amount Financed $ 10,000.00

5. FINANCE CHARGE $ 300.00

6. Total of Payment $ 10,300.00

   ANNUAL PERCENTAGE RATE 9 %

Credit life or credit life accident and health insurance is voluntary and not required for this loan. This insurance is available for term or loan at cost shown below.

I desire (✓ if applicable) ☐ Credit life at

$ _____

☐ Credit life accident and health at

$ _____

☒ I do not desire insurance coverage.

DATED this the 1st day of June , 19___.

Joseph Farmer, Julie Farmer
[Signature of BORROWERS]

Figure 19.1. An example of a promissory note.

erty exist. Real property is considered permanent, while personal property is mobile and often has a limited life.

The laws and instruments involving buying, selling, and financing of real and personal property also differ. Real property utilizes a deed as an instrument of ownership and a mortgage (or deed of trust in some states) to perfect a lender's security interests in the financing of real property owned by a borrower. Transactions involving tangible and intangible personal property are governed by the laws reflecting the Uniform Commercial Code (UCC), which is ascribed to by all states in the United States except Louisiana. Thus, the discussion in the sections to follow maintains this distinction between real and personal property.

TRANSACTIONS SECURED BY REAL PROPERTY

An ownership interest in land and other real property usually is identified as an estate. An **estate** may be a *lease hold, which endures for a limited time* (whether 1 year or 99 years), or a *free hold, which exists for an indefinite length of time.* A free hold estate may be a life estate, in which title is held only during the holder's lifetime, or it may be a fee-simple estate. This distinction is important because only the latter is inheritable.

The legal instrument used to convey title to real estate is a **deed.** The most important types of deeds in the United States are the *general warranty deed,* the *quitclaim deed,* and the *deed of trust.*

Under a **general warranty deed,** *the grantor or seller essentially is promising a clear, fee-simple title to the land, except as noted on the deed.* If this promise is false, the buyer can sue for damages resulting from reliance on this promise. If a transaction is initiated under the promise of a general warranty deed and the seller cannot provide such a clear title, the buyer usually has grounds for nullifying the contract and recovering costs.

With a **quitclaim deed,** *the buyer receives only the seller's interest in the property.* That is, the quitclaim deed only claims that the seller's rights and interests in the property are being conveyed to the buyer. The seller is not guaranteeing that he or she has such an interest in the property or that it is free of claims by others.

Clearly, the buyer (and the buyer's lender) must thoroughly check the title, especially if a quitclaim deed is used to transfer title. To be sure of acquiring a clear title to the property, the buyer should secure either an abstract of title or title insurance on the property. An **abstract** is *a written, chronological summary of all deeds, mortgages, foreclosures, and other transactions affecting the title to a tract of land.* Title insurance protects a purchaser or a mortgage lender against losses arising from a defect in title to real estate, other than defects that have specifically been excluded.

A mortgage conveys a security interest in real property to the lender as an element of assurance about loan repayment if foreclosure or bankruptcy occurs. The mortgage becomes void when the debt is repaid. In most states, the mortgage creates a lien on the property, the title to which is owned by the owner-borrower.

In some states, a **deed of trust** or trust deed mortgage is used instead of a conventional mortgage. The trust deed *conveys title of a property to a trustee who holds it as security for the payment of the debt* (see Figure 19.2). When the debt is paid, the trust deed becomes void and title is reconveyed to the owner. However, if the debtor defaults, the trustee is authorized to sell the property to pay the debt.

When mortgaged property is sold and the previous mortgage is not immediately repaid, the buyer may acquire the property subject to the mortgage or the buyer may assume the mortgage. **Subject to the mortgage** means that *the buyer is obligated to pay off the real estate debt.* If foreclosure were to occur, the lender only has recourse to the mortgaged property and not to any other property owned by the buyer.

In contrast, assuming the mortgage means that the buyer is acknowledging the original mortgage, and in case of foreclosure, the lender has recourse to the property in question and to other property held by the buyer in order to satisfy the outstanding debt obligation.

Both deeds and mortgages are placed on record in county courthouses to signify current ownership of property and the existence of mortgage claims on the property by a lender. Subsequent claims by other lenders then represent junior mortgages, in order of claim, to the primary mortgage.

Finally, a substantial portion of the farm real estate transactions each year involve seller financing in which the seller (e.g., a retiring farmer) receives a small downpayment from the buyer and finances the rest of the purchase price using an installment land contract or simply a land contract. Generally, title is retained by the seller or by a trustee until the debt is repaid or until other conditions have been met. The advantages and disadvantages of seller financing using land contracts were discussed in Chapter 17.

Illustration Copy

DEED OF TRUST

THE STATE OF TEXAS   }   KNOW ALL PEOPLE BY THESE PRESENT:
COUNTY OF

That   Joseph and Julie Farmer

of _____ Brook _____ County, Texas, hereinafter called Grantors (whether one or more) for the purpose of securing the indebtedness hereinafter described, and in consideration of the sum of TEN DOLLARS ($10.00) to us in hand paid by the Trustees hereinafter named, the receipt of which is hereby acknowledged, and for the further consideration of the uses, purposes and trusts hereinafter set forth, have granted, sold and conveyed, and by these presents do grant, sell and convey unto _____ Robert and Regina Rancher _____ Trustees, of _____ Brook _____ County, Texas, and their substitutes or successors, all of the following described property situated in _____ Brook _____ County, Texas, to-wit:

> all that certain tract or parcel of land lying and being situated in Brook County, Texas, and being more particularly described on Exhibit "A" attached hereto and made a part hereof.

TO HAVE AND TO HOLD the above described property, together with the rights, privileges and appurtenances thereto belonging unto the said Trustees, and to their substitutes or successors forever. And Grantors do hereby bind themselves, their heirs, executors, administrators and assigns to warrant and forever defend the said premises unto the said Trustees, their substitutes or successors and assigns forever, against the claims, or claims, of all persons claiming or to claim the same or any part thereof.

This conveyance, however, is made in TRUST to secure payment of _____ one _____ promissory note _____ of even date herewith in the principal sum of __ Two hundred _____ thousand and no/100 --- Dollars ($200,000.00) executed by Grantors, payable to the order of _ LAST STATE BANK, ENDLESS PLAINS, _____ TEXAS _____, in the City of _____ ENDLESS PLAINS _____, _____ BROOK _____ County, Texas, as follows, to-wit:.

Figure 19.2. An example of a deed of trust.

TRANSACTIONS SECURED BY PERSONAL PROPERTY

Historically, security arrangements involving personal property could take a variety of forms. However, Article 9 of the Uniform Commercial Code brings all these devices for securing personal property under one law and presents a comprehensive, unified system for the regulation of security interests in personal property.

Under the Uniform Commercial Code, personal property is classified as being intangible and tangible. Intangible property consists of six types:

1. Accounts receivable.

2. Instruments (negotiable instruments such as notes, stocks, and bonds).

3. Chattel paper (a document that gives evidence of both a debt obligation and a security interest in specified property).

4. Documents of title (warehouse receipts, dock receipts, etc.).

5. Contract rights.

6. General intangibles.

Tangible personal property consists of:

1. Consumer goods (items held primarily for family or household use).

2. Equipment (items bought or used primarily for business use).

3. Inventory.

4. Farm products (crops, livestock, and supplies used or produced in farming operations and products of crops or livestock in their unprocessed state, such as wool, eggs, and milk; goods are farm products only if they are held by a debtor engaged in farming or ranching).

Under the Uniform Commercial Code, a **security interest** is *a claim, or a lien, which a secured party (lender) has on the personal property (i.e., the collateral) of the debtor.* The secured party creates the security interest by entering into a security agreement with the debtor. The **security agreement** is *a legal, written document that describes the assets pledged as security and is signed by the borrower.*

A security interest may arise in two ways. First, it may originate through the purchase of property. A purchase money security interest is created when the seller retains a security interest in an asset in order to secure a loan made

E-1462–SECURITY AGREEMENT–CLASS 4   FED. RES. c-10   CLARKE & COURTS, INC.

SECURITY AGREEMENT

Date _____ June 1, 19-- _____

A. PARTIES

1. Debtor _____ Joseph and Julie Farmer
   Check one: ☒ individual ☐partnership ☐corporation ☐other

2. Address: _____ Star Rt., Box 18 _____ Brook _____ Texas _____ 10000
   street or RFD   county   state   zip
   Address shown in ☐ place of business ☐ chief executive office (if more than one place of business) ☐ residence

3. Bank: _____ Last State Bank

4. Address: _____ Endless Plains, Texas 10000

(Information concerning this security interest may be obtained at the office of the bank shown above.)

B. AGREEMENT

Subject to the applicable terms of this security agreement, debtor grants to bank a security interest in the collateral to secure the payment of the obligation. A carbon, photographic, or other reproduction of this security agreement may be filed as a financing statement.

C. OBLIGATION

1. The following is the obligation secured by this agreement:
   a. All past, present, and future advances, of whatever type, by bank to debtor, and extensions and renewals thereof.
   b. All existing and future liabilities, of whatever type, of debtor to bank, and including (but not limited to) liability for overdrafts and as indorser and surety.
   c. All costs incurred by bank to obtain, preserve, and enforce this security interest, collect the obligation, and maintain and preserve the collateral, and including (but not limited to) taxes, assessments, insurance premiums, repairs, reasonable attorney's fees and legal expenses, feed, rent, storage costs, and expenses of sale.
   d. Interest on the above amounts, as agreed between bank and debtor, or if no such agreement, at the maximum rate permitted by law.
2. List notes included in the obligation as of the date of this agreement (show date and amount):

   June 1, 19-- Farm Line of Credit for 1994 in the amount of $80,000,
           withdraws of various amounts and dates under Bank's
           commitment granted December 9, 1994.

D. COLLATERAL

1. The security interest is granted in the following collateral:
   a. Describe the collateral and, as applicable, check boxes and provide information indicated in Item D.1.b. (If debtor's residence is outside the state: give location of consumer goods, farm products, and farm equipment, and if collateral includes accounts arising from the sale of farm products, give location and products sold.)

   MODEL 7000 A TRACTOR
   WHEAT   COTTON

   b. 1. ☒ The above described crops are growing on or are to be gown on:
   _____ Farm is located in Brook County, Texas, ten miles west of Endless Plains on Hwy. 38. _____
   (describe real estate)
   2. ☐ The above goods are to become fixtures on:

   _____
   (describe real estate, attach additional sheet, if needed)
   3. ☐ The above timber is standing on: _____
   (describe real estate: attach additional sheets, if needed)
   4. ☐ The above minerals or the like (including oil and gas) or accounts will be financed at the well head or mine head of the well or mine located on: _____

   _____
   (describe real estate, attach additional sheet, if needed)
   c. If b.2, b.3, or b.4, above, is checked, this security agreement is to be filed for record in the real estate records. (The description of the real estate must be sufficiently specific as to give constructive notice of a mortgage on the realty.)
   ☐ The debtor does not have an interest of record; the name of a record owner is _____
   d. All substitutes and replacements for, accessions, attachments, and other additions to, and tools, parts, and equipment used in connection with, the above property; and the increase and unborn young of animals and poultry.
   e. All property similar to the above hereafter acquired by debtor.
2. Classify goods under one or more of the following Uniform Commercial Code categories:
   ☐ Consumer goods   ☐ Equipment (farm use)   ☐ Inventory
   ☐ Equipment (business use)   ☐ Farm products
3. ☐   If this block is checked, this is a purchase money security interest, and debtor will use funds advanced to purchase the collateral, or bank may disburse funds direct to the seller of the collateral, and to purchase insurance on the collateral.
4. If any of the collateral is accounts, give the location of the office where the records concerning them are kept (if other than debtor's address in Item A.2.)
5. If this security agreement is to be filed as a financing statement, check this block ☐ if products are covered for financing statement purposes. Coverage of products for financing statement purposes is not to be construed as giving debtor any additional rights with respect to the collateral, and debtor is not authorized to sell lease, otherwise transfer, furnish under contracts of service, manufacture, process, or assemble the collateral except in accordance with the provisions on the back of this security agreement.

BANK, BY: LAST STATE BANK, ENDLESS PLAINS   DEBTOR, By: JOSEPH AND JULIE FARMER

Signature _Jerold Wayne_   Signature _Joseph Farmer, Julie Farmer_

Typed   Typed
Name and Title   JEROLD WAYNE, PRESIDENT   Name and Title   JOSEPH AND JULIE FARMER, OWNER

If this Security Agreement is to be filed as a financing statement, Bank must sign.

Figure 19.3. An example of a security agreement.

by the seller to the buyer. Similarly, the lender obtains a purchase money security interest in an asset when he or she loans money to the buyer to purchase an asset that is pledged as collateral to secure the loan.

A purchase money security interest has a high priority against third party claims, provided the security interest is perfected through proper filing (see filing discussion). For example, a dealer sells Joseph and Julie Farmer a tractor, requiring a 25 percent downpayment and the rest financed by an installment note secured by a purchase money security interest created by a security agreement. At about the same time, a local lender finances the Farmers and receives a security interest in their crops and machinery. If the Farmers later experienced bankruptcy, the dealer with a purchase money security interest in the tractor would prevail over the local lender's claim on the tractor, provided the dealer has perfected his or her security interest by proper filing.

The second way a security interest may arise is through a loan or an obligation on property already owned by a debtor. Three events must occur, in any order, for the security interest to be attached to the collateral and be enforceable against the debtor or third parties with respect to the collateral. These events are:

1. The secured party must give value (perform some service, advance a loan, or otherwise incur a legal obligation from the debtor).

2. The debtor must sign a security agreement that contains a description of the collateral.

3. The debtor must have rights in the property (e.g., a farmer who has ownership rights of growing crops and livestock being produced).

A security interest in personal property is perfected either by taking possession of the collateral (which may occur in the case of stocks, bonds, or other financial assets) or by filing the appropriate public notice of the secured party's interest in the collateral. Giving public notice is accomplished by filing a financing statement in either the county courthouse, or the state capital, or both, depending on applicable state laws. The **financing statement** is *a summary or an abstract of the security agreement.* It should contain a description of the collateral by item and type, the date of filing, and the signatures and addresses of both parties. The date of filing is important in establishing protection against third party claims. A properly filed financing statement is effective for up to five years. Arrangements may be made to extend the statement, to terminate it when the debtor's obligation is fulfilled, and to cover assignments or transfers of interest in the collateral.

This Financing Statement is presented to a Filing Officer pursuant to the Uniform Commercial Code

1. Debtor(s) Name and Mailing Address: (Do not abbreviate)

JOSEPH AND JULIE FARMER
STAR ROUTE, BOX 18
ENDLESS PLAINS, TEXAS

2. Secured Party(ies) Name and Address:

LAST STATE BANK
ENDLESS PLAINS, TEXAS

3. For Filing Officer (Date, Time, Number and Filing Office):

June 1, 19--
2:30 p.m.
File No. 25100

4. This Financing Statement covers the following types (or items) of property.
(**Warning:** If collateral is crops, fixtures, timber, or minerals, read instructions on back.)

5. Name and Address of Assignee of Secured Party: (Use this space to describe collateral, if needed)

All farm supplies, farm products, farm crops; and all natural increase thereof now or at any time hereafter located or growing or to be grown on farm located in Brook County, and all such crops after they have been severed and removed from said real property, and all farm equipment of debtor wherever located.

Check only if applicable
☐ This Financing Statement is to be filed for record in the real estate records.

Number of additional sheets presented ______
☒ Products of collateral are also covered.

6. This Statement is signed by the Secured Party instead of the Debtor to perfect a security interest in collateral.
(Please check appropriate box)
☐ already subject to a security interest in another jurisdiction when it was brought into this state or when the debtor's location was changed to this state, or
☐ already subject to a financing statement filed in another county
☐ which is proceeds of the original collateral described above in which a security interest was perfected, or
☐ as to which the filing has lapsed, or
☐ acquired after a change of name, identity, or corporate structure of the debtor.

JOSEPH AND JULIE FARMER — LAST STATE BANK, ENDLESS PLAINS

Use whichever signature line is applicable

By *Joseph Farmer, Julie Farmer*
Signature(s) of Debtor(s)

By *Jerold Wayne*, President
Signature(s) of Secured Party(ies)

(1) Filing Officer Copy—Numerical

STANDARD FORM—FORM UCC-1 (REV. 6-19-75) APPROVED BY THE SECRETARY OF STATE OF TEXAS—FORM E-1383—CLARKE & COURTS, INC.

Figure 19.4. An example of a financing statement.

Initially, Article 9 of the Uniform Commercial Code distinguished farm products from other personal property collateral by giving agricultural lenders special consideration for farm products as they passed beyond the farm gate. When purchasing farm products, the buyer was responsible for proving that a commodity was free of encumbrances, which was not the case for consumer goods or other commodities. This favorable consideration to farmers placed a heavy burden on the buyers of farm products.

This problem was resolved when Congress passed the "clear title" legislation in the Food Security Act of 1985. The act gives buyers clear title to farm products unless they are notified of an existing lien. Direct notification by agricultural lenders to buyers may occur after the farmer-borrower provides a list of potential buyers. The farmer may face criminal penalties if he or she sells "off the list" and does not use the proceeds to repay the loan. The federal legislation also permits states to set up central notification systems in which registered buyers automatically receive financing statement information from the central source (i.e., the secretary of state's office).

FORECLOSURE AND BANKRUPTCY

If loan delinquency or default has become serious enough, lenders can initiate foreclosure proceedings. In doing so, lenders are exercising their rights to have collateral pledged as security sold, with the sale proceeds applied to the outstanding loan balance. If the sale proceeds are sufficient to cover the debt, then the proceedings are closed. If, however, debt coverage is insufficient, the borrower remains liable for the remaining indebtedness. The lender then may seek a deficiency judgment in which other property owned by the buyer is sold until the debt obligation is fulfilled.

With personal property, the foreclosure procedure on secured loans is found in the Uniform Commercial Code (Part 5 of Article 9). Basically, the secured party may utilize the provisions set forth in the security agreement. Included may be repossession, sale, acceptance of collateral, or other available judicial procedures. Various alternatives may be followed in disposing of collateral, including possible redemption (i.e., repurchase) by the borrower before other disposition occurs.

Foreclosure proceedings for real estate generally involve initiation of foreclosure by the lender, a court hearing, and a judicial decision. The typical result is a judgment against the defaulting borrower for the outstanding debt, accumulated interest, and related costs. A redemption period following a foreclosure sale of real property may allow reacquisition of the property by the debtor. Interpretations vary among states regarding the status of the mortgage holder. In some cases, title is presumed to be held by the lender, while in others the lender only has an interest in the property. In either case, both parties are subject to the results of the court process.

Bankruptcy is *a legal process of either reorganizing an insolvent business or liquidating the business and distributing the sale proceeds to creditors.* Bankruptcy can be undertaken voluntarily by a borrower or initiated by creditors, although farmers are excluded from involuntary bankruptcy if at least 80 percent of the gross income during the preceding tax year came from farming. The general goal of bankruptcy proceedings is to alleviate the economic impact of insolvency to the debtor, while trying to diminish losses to legitimate creditors.

Several sections of the U.S. bankruptcy code are applicable to agricultural producers and lenders. Chapter 7 of the code is often used when a financially distressed farm business is entering into liquidation. When a farmer wishes to keep the farm property in agriculture under a debt restructuring and workout arrangement, the provisions of Chapter 11 and, for small operations, Chapter 13 of the bankruptcy code have been used in the past.

Chapter 12 became law in 1986. It is specifically designed for farmers whose debts do not exceed $1.5 million, at least 80 percent of which are related to the farming operation, and who derived more than 50 percent of their gross income from farming in the preceding tax year. Chapter 12 precludes foreclosure by lenders by providing a legislated workout formula for farmers whose land and other assets are indebted more than their market value. Under its provisions, the excess debt is placed into an unsecured category that the debtor will repay whenever profits exceeded living expenses. Payments on a new land mortgage will be based on cash rents and earnings.

Under Chapter 12, farmers have 90 days after filing to submit a reorganization plan, which the court must approve or disapprove within 45 days — approval by creditors is not needed. Upon completion of the plan, any remaining debts of the farmer are waived. Numerous other details and conditions are found in the bankruptcy code.

Bankruptcy is a legal tool available to debtors to escape the burden of debt and perhaps to obtain a fresh start in either the old business or a new business. Clearly, however, the bankruptcy becomes part of a borrower's financial history and adversely affects credit worthiness. The extent of the adverse effects depends on the circumstances leading to bankruptcy and on the experiences of the agricultural producer following bankruptcy.

Important income tax consequences also may be associated with the use of Chapter 12 rather than Chapter 7 or 13. Not all of the potential tax benefits available to the debtor under Chapters 7 and 13 are available under the 1986 version of Chapter 12. Hence, it is advisable for a financially troubled farmer contemplating relief through bankruptcy to seek legal counsel with expertise in both bankruptcy and taxation.

CO-OWNERSHIP OF AGRICULTURAL PROPERTY

Co-ownership of agricultural property (both personal and real) is quite common. Co-ownership (referred to as *concurrent interests)* may exist with any type of land ownership, although it is more prevalent in fee-simple estates. Most states have three types of co-ownership: **tenancy by entireties, joint tenancy with survivorship,** and **tenancy in common.** With all three types, the co-owners have an undivided interest in the property, although the rights as a co-owner are significantly influenced by the type of concurrent interest.

A **tenancy by entireties** is permitted in some states. It can be held only by a husband and wife. Under this form of co-ownership, *when one spouse dies, the surviving spouse automatically becomes the full and sole owner of property held in joint ownership.* Consequently, such property need not go through probate. However, during the lifetime of both spouses, the undivided interest of one spouse cannot be conveyed without the consent of the other.

Joint tenancy with survivorship may be held by any two or more co-owners. Again, the survivors automatically receive the interest of a deceased joint tenant so that the administrative expenses associated with probate of the jointly held property may be avoided. In contrast to a tenancy by entireties, a joint tenancy *enables an owner to convey interest in a lifetime deed without the consent of the co-owners.*

Tenancy in common also may exist between any two or more persons. However, *the interest of each co-owner is inheritable rather than passing to the survivor on the death of a co-owner.* Furthermore, with tenancy in common, co-owners may convey their respective interests to others without regard for the wishes of the other co-owners.

Some states retain the community property system as the basic property ownership law between a husband and wife. With community ownership, each spouse owns in his or her own right an equal, undivided portion of all marital property. Unless otherwise specified, all property belonging to a married couple — except property owned by one spouse before the marriage or acquired during marriage by gift or inheritance — is considered community property. As such, each party's share is inheritable in much the same manner as tenancy in common.

OTHER NEGOTIABLE INSTRUMENTS

A **draft,** or **bill of exchange,** is *an unconditional order in writing requiring the person to whom it is addressed (the drawee) to pay on demand, or in some specified future time, a specified sum of money to the person drawing the draft (the drawer) or to someone whom the drawer designates.* It is, in effect, an order by one person upon a second person to pay a sum of money under a specified set of circumstances to a third person. The drawee is not bound to pay the money unless the terms of the order are accepted. Once the transactions associated with the order have been accepted, the drawer becomes the acceptor.

There are several types of drafts. A **sight draft** is payable when the holder presents it to the drawee for payment, or "on sight." The **time draft** is payable at a specified future time, such as 60 days after acceptance. One type of bill of exchange important in agriculture is the **trade acceptance.** The bill of exchange is sent by a seller of goods to a purchaser with the understanding that if the goods are approved on arrival, the purchaser will accept the instrument immediately and abide by its terms.

A check is a bill of exchange drawn on a bank and payable on demand. To be legal, the drawer must have deposits in the bank to cover the check. A **cashier's check** *is drawn by a bank on itself directing that it pay the specified sum of money either to the purchaser of the check or to a designated person.* A **bank money order** is similar. It is *an order by a bank upon itself to pay a specified sum to a third party designated by the purchaser.* A **bank draft** is *a check drawn by one bank on another bank.*

These instruments serve essentially the same purpose — securing payment to fulfill a contract at a specified time and location. The purchaser of these money instruments must pay to the bank the amount specified on the instrument plus a service charge.

DOCUMENTS OF TITLE FOR GOODS IN STORAGE OR TRANSIT

The two most important documents of title for goods in storage or transit are (1) *bills of lading,* issued by carriers that accept the merchandise for shipments, and (2) *warehouse receipts,* issued by firms engaged in storing goods for hire. Other familiar documents are air bills, weight bills, air consignment notes, and any others that in the regular course of business or financing indicate that the holders are entitled to receive, hold, or dispose of the documents and the goods they cover.

Such titles may be either negotiable or non-negotiable. With negotiable documents, the **bailee** *(the individual receiving the property)* must deliver the goods only when a negotiable document of title is surrendered. With non-negotiable documents of title, the bailee may deliver goods upon a separate written authority and without a surrender of the document of title.

Warehouse receipts are important in agricultural financing. Warehouse receipts, certified by a bonded warehouse supervisor, can serve as evidence of ownership of property, with the bonded warehouse certifying the quantity and quality of the property. Such an instrument, therefore, can serve as collateral for the individual who is trying to secure a loan. The use of field warehouse receipts has extended far beyond the traditional "storable" commodities, such as corn, wheat, oil, and flour, to include dressed meats in packing houses, cattle in commercial feedlots, and cattle on pasture. This service of third party collateral control and certification reduces the lender's uncertainty about the existence and value of collateral and thus augments financing of such items. An extension of the concept of third party collateral control is accounts receivable certification, in which a bonded and third party assumes control over and certifies the value of the firm's accounts receivable. With such certification, as in the case of an agribusiness firm, the lender can advance funds against the accounts receivable.

A warehouse receipt should contain the location of the warehouse where the property is stored, the date of the receipt, the consecutive numbering of receipts, the name of the person who is to receive the goods — the bearer or a person specified on the order, the rate of storage charges, a description of the goods, a statement of charges subject to lien by the warehouse, and the signature of the warehouse supervisor.

To demonstrate how a warehouse receipt or a bill of lading might be used in a transaction, consider Henry and Louise Sellers who consign a carload of corn to the Union Pacific Railroad in North Platt, Nebraska, to be shipped to the buyer, Ms. B, in Los Angeles. The Sellers, on delivering the corn to the railroad, get a bill of lading stating the total amount and grade of corn being shipped. They then draw a draft on Ms. B. They attach the draft to the bill of lading and endorse both of them to their bank in North Platt.

At that time, their banker discounts the draft, giving the Sellers the amount of the draft less a fee for servicing and for financing until total collection occurs. The North Platt bank then forwards the draft and bill of lading to its correspondent bank, perhaps in Los Angeles. The correspondent then presents the draft to Ms. B, the buyer, who is the drawee of the draft. On paying the draft, Ms. B is given possession of the bill of lading. When the shipment arrives in Los Angeles, the railroad will notify Ms. B, who will then surrender her negotiable bill of lading to the shipper in order to receive the corn. The entire transaction is completed when the correspondent transfers to the North Platt bank the amount of the draft less its service charges.

OTHER LEGAL INSTRUMENTS
AND TERMINOLOGY

Several other legal instruments warrant consideration. In general, a security interest cannot be attached to livestock that are not conceived or to crops that are not yet planted. However, an **after-acquired property clause** attached to the security agreement permits the lender to create a valid security interest in livestock offspring that were not conceived when the agreement was executed. Similarly, the lender can advance money for crop production before the crop is planted, with the assurance that the security interest will be attached to the future crop.

One may **waive** rights under a contract by failing to demand performance by the other party or by failing to object when the other party fails to perform under the terms of the contract. For example, if a contract calls for Robert and Regina Rancher to pay the proceeds of all livestock sales from their farm operation to the bank, and the banker is aware that the Ranchers sold livestock from time to time without remitting the proceeds, the banker may have waived any rights under the contract to sue the Ranchers for violation of contract.

A **subordination agreement** frequently is used in agricultural financing. It is *an agreement subordinating one set of claims to another.* It might be used in financing a tenant farmer, for example, by requiring that the owner subordinate any claims on cash rent or on a share of the crop to the claims of the lender. Similarly, a lender may require that the commercial cattle feeder subordinate claims for any unpaid feed bills to the claims of the lender who is financing the cattle.

A **personal suretyship** is *a relationship and procedure by which one person becomes responsible for the debt or undertaking of another person.* In the usual parent – child farm arrangement, the parent may be called upon to sign a guaranty or suretyship, in which he or she assumes responsibility to repay a loan made to his or her child in case the child defaults. In financing partnerships or family corporations, the lender might require that one or more of the parties sign a guaranty for the indebtedness of the business.

An **acceleration clause** is a common provision of a mortgage or note *giving the lender the right to demand that the entire outstanding loan balance be immediately due and payable in the event of default.* An assignment is the transfer of title, property, rights, or other interest from one person or entity to another.

A **covenant** is *a legal promise in a note, loan agreement, security agreement, or mortgage to do or not to do specific acts;* or, it is *a legal promise that certain conditions* *do or do not exist.* A breech of a covenant can lead to the injured party's pursuing legal remedies and can be a basis for foreclosure.

An **encumbrance** is *a claim or interest that limits the rights of property.* Examples of encumbrances are liens, mortgages, leases, and easements.

A **lien** is *a claim by a creditor on property or assets of a debtor in which the property may be held as security or sold to satisfy a debt (full or partial).* A lien may arise through a borrowing transaction in which the lender is granted a lien on the borrower's property. Other examples of liens include a tax lien against real estate with delinquent taxes, a mechanic's lien against property on which work has been performed, and a land owner's lien against crops grown by a tenant.

A **loan conversion** provision is *an option provided by a lender to a borrower to change loan terms at a future date.* For example, at loan origination, a lender may provide a borrower with an option to convert from a variable to a fixed rate loan. Usually, the lender charges the borrower a fee for this option.

A **release** is *the cancellation of a claim on real estate or non – real estate assets.* It signifies complete fulfillment of the terms of the loan contract. A partial release is the release of a portion of collateral to the borrower for liquidation or other use.

Summary

Legal issues in agricultural finance depend significantly on whether transactions involve real property or personal property. Real property utilizes a deed as an instrument of ownership and a mortgage or deed of trust to provide loan security in debt-financed real property transactions. Personal property transactions are governed by the Uniform Commercial Code. The legal issues primarily are intended to facilitate transactions and to protect the financial interests of the parties to a transaction. For example, foreclosure and bankruptcy procedures aid lenders by allowing them to minimize loan losses in unsuccessful credit transactions. Bankruptcy procedures may also alleviate the economic impact of insolvency to borrowers. Other important legal issues in agricultural finance involve co-ownership of property, negotiable instruments, and title documents to goods in storage or transit.

Other highlights are as follows:

- The essential components of a contract are legally competent parties, legal and proper subject matter, an offer and an acceptance, and value consideration.

- A **note,** or promissory note, is *the basic legal document in a loan transaction.* A note may be secured or unsecured.

- A **security interest** is *a claim, or lien, that a secured party has on the personal property of a debtor.* The security interest is created by a written security agreement.

- A security interest is perfected by taking possession of the collateral or by filing a financing statement in the appropriate public office.

- The types of deeds to real property include the **general warranty deed,** the **quitclaim deed,** and the **deed of trust.**

- Lenders may initiate foreclosure proceedings in response to serious loan delinquency and default by borrowers.

- **Bankruptcy** is *a legal process whereby an insolvent business is reorganized or the business is liquidated and the proceeds from the sale are distributed to the creditors.*

- The major forms of co-ownership of property are **tenancy by entireties, joint tenancy with survivorship,** and **tenancy in common.**

- Other legal instruments and terminology important to agricultural finance include warehouse receipts, bills of lading, after-acquired property, waivers, subordination agreements, suretyship, acceleration clauses, covenants, assignments, liens, loan conversion, and releases.

Topics for Discussion

1. Identify the essential components of a contract between two or more parties.

2. What characteristics distinguish between real and personal property?

3. Identify and explain the major types of deeds associated with real property transactions.

4. In what ways are a security interest, a security agreement, and a financing statement associated with transactions involving personal property? How may a security interest arise?

5. Identify the steps involved in foreclosure proceedings.

6. Explain the concept of bankruptcy and describe the sections of the U.S. bankruptcy code that are applicable to agricultural producers and lenders.

7. Describe the types of co-ownership of personal and real agricultural property.

8. What are the major documents of title for goods in storage or transit?

9. Explain the meanings of *after-acquired property, waiver, subordination agreement, acceleration clause, covenant, encumbrance, lien, loan conversion,* and *release.*

References

1. Gustafson, C. R., D. M. Saxowsky, and J. Braaten, "Economic Impact of Laws That Permit Delayed and Partial Repayment of Agricultural Debt," *Agricultural Finance Review,* 47(1987):31 – 42.

2. Gustafson, C. R., J. Baltezore, and F. L. Leistritz, "Agricultural Credit Mediation: Borrower and Creditor Perspectives in North Dakota," *Agricultural Finance Review,* 51(1991):1 – 8.

3. Harl, N. E., *Agricultural Law: Estate, Tax and Business Planning,* Matthew Bender & Co., Inc., New York, 1992.

4. Harl, N. E., "Chapter 12 Bankruptcy: A Review and Evaluation," *Agricultural Finance Review,* 52(1992):1 – 11.

5. Lee, W. F., et al., *Agricultural Finance,* 8th ed., Iowa State University Press, Ames, 1988.

6. Looney, J., and D. L. Uchtmann, *Agricultural Law,* 2nd ed., McGraw Hill, Inc., New York, 1994.

20

Business Organization
in Agriculture

The ultimate financial success of agricultural businesses may be strongly influenced by their legal form of organization. Traditionally, agricultural businesses have been organized primarily as either sole proprietorships or family partnerships. However, other business organizations merit serious consideration. In this chapter we will outline and apply selected criteria for comparing alternative forms of business organization in agriculture. The consequences for estate management are also considered, with the tax implications developed in the appendix to this chapter.

EVALUATIVE CRITERIA

In earlier chapters, three criteria — profitability, liquidity, and risk — were used to evaluate management alternatives and to monitor and assess financial success. These criteria also serve to evaluate different business organizations, although expansion and modification are needed to compare their effects on the generation and conservation of the firm's equity capital.

Profitability

An important characteristic of modern agriculture is the high payoff for superior management. The farm business must attract and train effective management as well as maintain a high level of performance over time. The importance of business continuity is demonstrated in Figure 20.1. Line A reflects the changing performance of management for a single owner-operator family farm over three generations. Each generation, as a rule, passes through the four stages in its life cycle outlined in Chapter 2 — establishment, expansion, consolidation, and transfer. Thus, the first generation might begin at a relatively low level of management, lacking the necessary training, experience, and capital to achieve operational efficiency. As these are acquired, the level of management rises, reaching a peak before declining as the manager approaches retirement. Eventually, the business is transferred to the next generation.

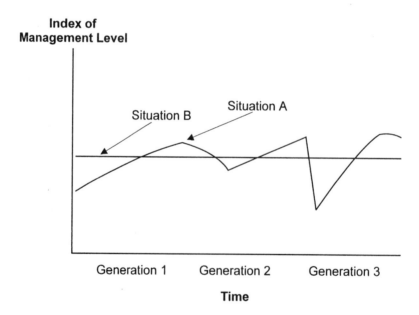

Figure 20.1. A comparison of management level over time under two hypothetical situations.

The second generation then repeats the process of gaining the necessary management experience and expanding the business. Let's assume the process ends abruptly with the untimely withdrawal of the manager. By the time new management is established and new financing arranged (and perhaps the estate transfer settled through the courts), much of the liquidity and momentum of the business and a sizable portion of the equity capital are lost. The next generation essentially starts over and completes its own life cycle of management effectiveness. Each generation, in turn, reaches a peak of efficiency during its life but maintains that level for a relatively brief span of its total responsibility.

Line B of Figure 20.1 represents the efficiency of similar resources when they are combined into a hypothetical firm (firm B) large enough to provide the necessary training and experience to assure continuity of competent management, labor, and finance over time — even during periods of intergeneration transfer. The business might be a large, industrial corporation, or it could be a family farm business large enough to involve two or more families and organized to provide management competence over time. Under these circumstances, firm B could bid resources away from firm A (the smaller firm) in an ongoing quest for survival, even if firm B never reached firm A's peak levels of efficiency.

Part of the problem in firm A is its small size of business. There are numerous important economies of size in agriculture, and those associated with intergeneration transfer are especially important. Farms large enough to provide full-time employment and income for two or more operators and their families also provide training experiences in business, financial planning, and decision making. Such experiences better prepare all the parties involved for eventual leadership.

Finally, some forms of business organization are more costly to organize than others. Differences in organization and administration costs should be considered when a business organization is being chosen.

Conserving Income

The usefulness of farm income depends on the earnings that remain *after taxes*. One aspect of conserving income is to provide as many employee benefits as possible with *before-tax* dollars. Such practices are legal with some types of business organizations and illegal with others.

Business organizations also differ in the total after-tax burden of social security taxes and in the social security payments allowed after retirement. Even more important are the differences in income tax rates experienced by different business organizations.

Generating Equity Capital

Historically, most new equity capital for agriculture has been provided by retained earnings and unrealized capital gains on farm assets. However, with increased capital needs, high risks, and narrow per unit profit margins on farm products, commercial agriculture is hard pressed to generate sufficient equity capital from retained earnings alone. As a result, an increasing number of farmers must find ways of tapping outside equity sources.

To make the investment convenient to prospective investors, business corporations that have ownership divided into minute shares of claims on the undifferentiated total net worth of the business have been developed. In contrast, the owners of a traditional family farm claim specific acres of land, head of livestock, kinds and numbers of machinery, etc. These differences in claims may significantly affect the firm's ability to generate outside equity capital. For example, a potential investor might be interested in owning 100 shares of stock in a successful agricultural venture rather than owning 1.4 of its tractors or 5.6 of its dairy cows.

Conserving Equity Capital

The choice of legal form of business organization can significantly influence the effectiveness in conserving equity capital. An agricultural business can lose its equity capital in several ways. When cash withdrawals from the business exceed farm income, farm equity is reduced. Equity capital losses also may arise from the liability of the firm and/or its owners for obligations, damage suits, and other contingent liabilities incurred by management and employees. The extent of this liability varies, depending on the type of firm organization and the effectiveness of risk management.

Equity capital is also lost through a decline in the value of farm assets. However, fluctuations in asset values are independent of the legal form of business organization and need not be considered here. It is worth noting, however, that acquiring capital with equity instruments instead of debt instruments reduces the firm's financial risk.

ORGANIZATIONAL ALTERNATIVES

An agricultural business may be organized as a sole proprietorship, a general partnership, a limited partnership, a "regular" or subchapter C corporation, or an S corporation. In some cases, a land trust may also be used.

Sole Proprietorship

Most farms in the United States are organized as sole proprietorships, wherein a single person or family has exclusive title and legal rights to the property of the business, has sole responsibility for business conduct, and is fully liable for obligations of the firm. The proprietorship offers a maximum of management control with a minimum of administrative costs. Its simplicity and flexibility make it an attractive form of business organization for a limited period of time. If, for example, the business is to be sold when the owner retires, the sole proprietorship is likely the most effective form of business organization. The fact that business performance and managerial control are limited only by the capacity and resources of the proprietor is both its strength and its weakness.

A sole proprietor is self-employed and pays social security taxes on "before-tax" net income. Thus, social security taxes are not a deductible expense for income tax purposes for a proprietorship.

The costs of most employee benefits to the proprietor and family are not tax-deductible. Exceptions include the cost of life insurance, when the lender requires insurance as a condition of a business loan, and a limited tax-sheltered retirement – investment program based on self-employment status.

A proprietorship incurs no income tax obligation. Instead, the proprietor is taxed on all profits whether or not the income is withdrawn from the business. Any salary paid to the proprietor as manager is also taxable to the proprietor. Consequently, both the business and the proprietor are indifferent as to whether the money is paid as a salary or as family withdrawals from profits.

The proprietorship has serious limitations in estate planning and inter-generation transfer. First, business continuity is linked directly to the life cycle of the proprietor. Consequently, the level of business performance tends to drop off significantly as the sole proprietor approaches age 60 or thereabout, due to limitations in energy and a tendency toward conservatism in planning and investment. Situation A in Figure 20.1 is representative of a sole proprietorship.

Second, the proprietor's wealth in the business is usually transferred to others as titles to specific agricultural assets — tractors, trucks, buildings, livestock acres of land. This process becomes so cumbersome that most proprietorships do not transfer either the title to or the control of business assets until the retirement or death of the proprietor. These procedures hamper the incoming generation and provide limited opportunity for minimizing estate taxes at the time of transfer. If the agricultural business is to continue into the next generation, other business organizations should be considered.

Partnership

A partnership is an association of two or more persons to carry on, as co-owners, a business for profit. Parent – child partnerships, for example, have become a popular form of business organization in U.S. agriculture. Two types of partnerships have a place in agriculture. They are general partnerships and limited partnerships.

In a **general partnership,** each partner is an integral part of both ownership and management, and each has unlimited liability not only for obligations incurred by the business but also for liabilities incurred by other partners while representing the business. A general partnership can be established and operated under a very informal agreement by the partners. It can be established to achieve any legal purpose. No written agreement is necessary, although a recorded agreement helps to avoid misunderstandings among general partners.

A **limited partnership,** on the other hand, is comprised of one or more *limited* partners and one or more *general* partners. A limited partnership differs from a general partnership in several important ways. First, a more formal legal arrangement is required for organizing and operating a limited partnership. This agreement must meet the requirements of the state in which it is organized and be filed with the appropriate legal body of that state, usually the secretary of state. Second, the liability of limited partners is limited to their investment in the business. This can be important to a prospective investor who wishes to avoid the liability of a general partnership. The burden of general and total liability in a limited partnership rests with the general partners, who are responsible for management.

Either type of partnership may have no more stability over time than a sole proprietorship, since both are generally dissolved on the death or withdrawal of a general partner. If desired, provisions that require the partnership, or the surviving partners, to purchase the interest of the deceased partner can be made for business continuity. However, the partnership relationship is usually a personal one and tends to dissolve when a general partner with-draws. This problem is usually less acute with parent – child partnerships, provided proper estate planning occurs. Many limited partnerships outline how limited partners may transfer their interest to other limited partners. Such transfers can usually occur without disrupting the business.

A general partnership must file an information return, but, like a proprie-torship, it pays no income taxes. Instead, all profits are allocated to the partners for income tax accounting. Salaries paid to partners are operating expenses to the partnership but are taxable to the partners. Thus, general partners are indifferent as to whether their income occurs as salary or distributed profits.

A limited partnership may be taxed either as a partnership or as a corpo-ration, depending on the election of the partnership (with concurrence of the Internal Revenue Service) or a ruling by the Internal Revenue Service. If the Internal Revenue Service determines that the partnership is more like a corpo-ration, it will be taxed as a corporation, and the "income pass-through" privi-lege will be lost. Since tax provisions are an important attraction of limited partnerships for many investors, those who draft limited partnership agree-ments should be sure these agreements resemble those for partnerships rather than for corporations. The primary criteria on which corporate resemblance is assessed are (1) continuity of life, (2) centralized management, (3) limited liability, and (4) free transferability of interests.

Limited partners do not receive a management salary because they cannot participate in management. Instead, they share directly in the profits or losses of the business. The typical limited partnership allocates a percentage of gross receipts to the general partners as a "cost" for management. Such costs are a deductible expense in computing profits for the limited partnership. Profits or losses are then allocated to limited partners, based on their percentage of ownership.

The social security tax for partnerships is similar to that for sole proprie-tors. It is paid by the partners and is not tax-deductible to either the partner-ships or the partners. Furthermore, there are restrictions on the amount of partnership income general partners may receive after they retire without disrupting social security benefits. Partnership income to limited partners does not affect social security payments, inasmuch as limited partners do not par-ticipate in management. The cost of employee fringe benefits for partners and their families is not tax-deductible for either general or limited partnerships.

New equity capital for partnerships can come from (1) partnership profits retained in the business, (2) additional investments by existing partners, or (3) new partners brought into the business. One advantage of limited partnerships is the attraction of outside equity capital from limited partners without management control being diluted by general partners.

There are also advantages to the limited partners. They can invest in a business without management responsibility or the unlimited liability of a general partnership or a proprietorship. At the same time, they incur any operating losses of the business during a given accounting year, thereby reducing taxes on their income. This tax shelter feature makes investment in a limited partnership attractive to persons with high incomes, such as doctors and lawyers (see Chapter 21).

A partnership is not very effective for estate planning and intergeneration transfer. If a partner dies, that partner's property is subject to probate. Moreover, when the partnership is dissolved, the transfer of title to the physical property tends to be cumbersome. Buy – sell agreements among the partners or between the partnership and a partner can be used to provide a market for the interest of a deceased or withdrawing partner. Funding of these agreements could come from life insurance so that the event that triggers the need also provides the necessary funds.

Problems arise in transferring an existing interest in a general partnership to minors. No partner can be forced to accept another person as a general partner, and assigning a partnership interest to a son or a daughter might suggest bringing in an additional partner. Such action could lead to the dissolution of the partnership. Some of these disadvantages do not exist in a parent – child partnership that is intended to achieve an intergeneration transfer over a period of time. However, there are obvious limitations on the ability of many minors to participate in management and assume its responsibilities.

Farm Leases

A business form that is in some ways unique to agriculture is the owner – tenant arrangement in a farm lease. Such an arrangement strongly resembles a partnership. Thus, it is considered here. Indeed, the lease terms must be properly fixed, lest the arrangement be legally interpreted as a partnership.

As indicated in Chapter 12, the more rapidly growing farm firms have their land resources partly or wholly leased. Such farms often depart dramatically from the popular image of the small-scale tenant farmer trying to become established on the "first rung of a tenure ladder," leading ultimately to full ownership of land resources. Instead, today's typical operation might combine the tracts of several land owners, who are each relatively small contributors to the farm's total resources.

The tenant may be a proprietor in the business (with all the attributes previously listed for sole proprietors) and may acquire the use of land resources without the capital commitments of ownership. In a livestock-share lease, the tenant may acquire the use of other resources as well. With all types of leases, land owners frequently participate in furnishing operating inputs, although participation is minimized in most cash leases. In most crop-share and livestock-share leases, the tenants' participation commonly is proportionate to their share of farm output and includes working with the owner in planning fertility, tillage, and crop enterprises.

Corporation

A **corporation** is *a legally created "person" or entity.* It can perform any economic function accomplished by a proprietorship: producing, marketing, financing, etc. It can sue or be sued independently of its human associates. It provides a means of separating ownership from management and of insulating both parties from the liabilities associated with the corporation. The liabilities of owners are limited to the owners' individual investments in the business. The liabilities of management, on the other hand, are defined only in terms of "prudence and reasonable competence in the performance of management."

Ownership of the corporation is held as small shares or claims on the net worth and profit stream of the business. However, several types of equity claims are possible. They are:

1. *Preferred stock: a stock having preference over other equity claims on the assets of the business in bankruptcy.* That is, the preferred stock is paid out as par value before any value is given to other equity claims. It also has preference over other equity claims on income. Most preferred stock carries a fixed-dividend obligation per share that must be paid whenever the company generates a profit. Only after the dividend claims of preferred stockholders have been met can dividends be paid on other equity claims.

   If the corporation does not generate profits in a given year, dividends need not be paid on preferred stock. However, if the stock is *accumulative preferred*, the fixed-dividend obligation accumulates from year to year when not paid and has a priority claim on income if and when the business shows a profit. Voting privileges may be attached to preferred stock but usually are not.

2. *Common stock: a stock that has no right or priority over any other stock for dividends or asset distribution, should the corporation be dissolved.* It is the ultimate residual claimant on both income and net worth. Each share of common stock usually entitles the holder to one vote on matters coming before the stockholders, such as electing the board of directors and establishing important policy matters. Income allocated to common stockholders is distributed equally per share of common stock. Market values tend to fluctuate more sharply for common stock than for preferred stock.

3. *Warrant: an assigned right to purchase a specific security at an established price.* Warrants are often used by established corporations in selling additional stock. The warrants are allocated to existing stockholders who can then either exercise their warrants themselves or sell them.

4. *Convertible debenture: a debt instrument carrying the right to be converted into equity claims, usually common stock, at a predetermined price at the option of the holder.* The "convertibility" attribute of the debenture usually lowers the interest rate required to make the instrument attractive to buyers. Once issued, however, it constitutes a potential dilution of existing common stock.

Two important types of corporations are the regular corporation and the tax option. These are called "subchapter C" and "subchapter S" corporations, respectively, because of the related sections of the Internal Revenue Code. The S corporation was authorized in 1958 to benefit small businesses, including those in agriculture. To qualify, a firm must meet all of the following conditions: (1) have only one class of stock outstanding;[1] (2) have no more than 35 initial stockholders, all of whom must be individuals or estates (i.e., another corporation could not be a stockholder); and (3) have the consent of all shareholders to the choice.

An S corporation may be terminated if the shareholders holding more than half of the shares voluntarily agree to termination and file such action appropriately with the Internal Revenue Service. Legislation in 1982 also changed the type of income that the business may earn. Any unincorporated

---

[1]Legislation passed in 1982 permits common stock in a subchapter S corporation to carry different voting rights. For example, this law enables company owners to give stock to family members, who are taxed at lower rates, without giving them voting powers. Moreover, an instrument treated as a straight debt instrument cannot be treated as a second class of stock.

business that becomes an S corporation or any business that has had an S status since it incorporated may now have unlimited "passive" income from sources such as interest, dividends, and rent. However, a regular corporation switching to S status may only earn up to 25 percent of its income from passive sources.

The S corporation differs from a regular corporation primarily on two counts. First, it is limited to only one class of stock and to 35 or fewer stockholders, as indicated earlier. Second, it is taxed as a partnership rather than as a corporation. Consequently, it pays no income tax.

Dividends from the regular corporation to stockholders are taxed at each stockholder's marginal tax rate. Thus, the payment of dividends leads to double taxation. Where possible, income to stockholders should be in the form of salaries, wages, and bonuses, which reduce corporate taxable income, rather than as dividends.

With the S corporation, *all profits* are allocated to the stockholders for income tax accounting, whether or not dividends are distributed; whereas, a C corporate stockholder pays taxes *only on the cash dividends received*. Moreover, long-term capital gains and losses are "passed through" (with tax implications) to the stockholders in an S corporation. In a regular corporation, capital gains taxes are added to the corporation's tax on ordinary income.

Most of the remaining discussion on corporations applies to both types. The corporation offers clear advantages over the sole proprietorship or partnership in generating equity capital. With a corporation, ownership is subdivided into minute fractional claims on the net worth of the business. This procedure makes it easier to redistribute existing ownership shares. It facilitates raising additional equity capital through selling stock. The fractionalization of shares also simplifies merging two or more businesses. Thus, a corporation provides an alternative to the limited partnership for selling equity claims having limited liability. A farmer who is a general partner would probably lose more management control by incorporating and selling stock, unless it was non-voting stock, than by selling limited partnership shares.

Various studies of farm debt show that corporations tend to carry heavier debt loads than other forms of business organizations. Several reasons besides differences in business size and levels of management might account for greater debt use by corporate firms. If, by incorporating, firms can attract outside equity capital, their credit would be strengthened. Moreover, the record keeping, business analysis, and financial planning required to make appropriate corporate reports may also strengthen the credit of corporate firms.

If, however, a business was incorporated primarily to limit the liability of owners, the financial strength of the corporation would probably be less impressive than that of the proprietorship or family partnership prior to incorporation. In this case, incorporation may reduce the firm's credit. Lenders usually respond by requiring the principal stockholders to personally guarantee any debt. This guarantee nullifies the limited liability condition for that particular debt but not for third-party claims.

The corporation also facilitates estate planning (see the appendix to this chapter). It is easier for an owner to give heirs shares of corporate stock than it is to give them title to tractors, buildings, or a small acreage of land. Existing tax laws in the United States permit both husband and wife to transfer to any individual up to $10,000 of value per year, tax-free. Few farm proprietorships take advantage of these gift privileges, partly because transferring titles to small tracts of land to several persons each year becomes very complicated. Corporations do not have this difficulty.

The subchapter C corporation may also have some distinct advantages over the S corporation in estate planning. The S corporation — similar to a proprietorship and a partnership — does not pay income tax, as indicated previously. However, this advantage also carries the disadvantage of inflexibility. With a regular corporation, money can flow from the business to the employee-owners either as salaries, which are tax-deductible to the business, or as dividends, which are not tax-deductible. With this flexibility, it is possible to achieve a tax-minimizing balance among salaries, corporate profits, and dividends.

Social security taxes are assessed to corporations at a specified corporate tax rate. This cost is tax-deductible to corporations. Employees pay an equal amount, but it is not tax-deductible to either the corporations or the employees. Thus, as indicated previously, the total cost of social security taxes, including both corporate and individual assessments, is higher for corporations than for proprietorships or partnerships. The corporations' payment of the social security assessment with "before-tax" dollars reduces the cost somewhat. But even on an "after-tax" basis, the total assessment is higher through corporations.

The costs of many employee benefits are deductible expenses for corporations, even though the employee-recipients are the owners or family members of owners. Tax-sheltered investment plans, medical and hospital insurance, group life insurance, health and accident insurance, sick-pay benefits, tax-free gifts on the death of an employee, and costs of certain employee recreational facilities might all be considered as "legitimate" and "reasonable" operating expenses of corporations. Most of these are subject to legal limits, however.

The costs of administration are likely higher for corporations than for other business organizations, although differences between corporations and limited partnerships are small in many states. The costs of establishing a corporation — including legal fees for drafting the articles of incorporation and for obtaining and registering the charter — are usually greater than for other types of business organizations. Furthermore, most states require chartered corporations to file an annual report and may levy various types of taxes on them.

Since the corporation is a legally created entity, it is eliminated only by legal action. Its continuity is not dependent on any person's life cycle. This continuity does not assure that management competence will continue at a high level over time. However, it does provide the legal and organizational setting for achieving such continuity. If a corporation is to be terminated, then the tax consequences of the alternatives for distributing assets, some of which may have experienced capital gains or losses, need careful consideration.

Table 20.1 summarizes each business organization (sole proprietorship, general partnership, limited partnership, regular corporation, and S corporation) with respect to several selected characteristics. Not all of the characteristics can be conveniently reduced to tabular form; however, the summary indicates the important features associated with each alternative. Farm leases are not included in this summary because they fall partly in the proprietor category and partly in the partnership category.

Land Trusts

By means of a trust, legal title to designated properties can be transferred to a trustee, who then becomes responsible for managing the beneficiary's properties. The objectives and constraints in management can be spelled out in the trust agreement. The **land trust** is *an organizational alternative by which ownership of a farm business can be — but does not have to be — separate from management.* Its uses tend to be restricted to agricultural businesses in which land ownership is an important business activity.

The land trust gives to the trustee the legal and equitable title to the real estate placed in trust and charges him or her with the responsibility of dealing with such title according to the trust agreement; however, the beneficiaries have the rights to (1) possession and control, (2) income generated, and (3) funds from the sale of the real estate. Even so, the interest of the beneficiaries is considered personal property rather than title to real estate.

The land trust is not available in all states. Most states have not established

Table 20.1. A Comparison of Selected Organizational Alternatives

Characteristics	Sole Proprietor	Partner-ship
		General
1. **Source of equity capital**	Limited to proprietor's equity; primarily new capital from retained earnings.	Limited to equity capital of partners; new capital from retained earnings and from equity of new partnerships.
2. **Liability to owners**	Liability extends to the total assets of the proprietor for obligations of the firm.	Liability extends to total assets of all partners for obligations of the firm and for actions of partners.
3. **Continuity of owners**	Linked directly to the life cycle of the proprietor.	Can be dissolved by any partner. Tends to be associated with life cycle of partners but does not have to be.
4. **Management and control**	Limited only by the capacity and resources of the proprietor.	Typically, partners share management equally. Each partner has power to bind partnership; thus, each partner must have high level of confidence.
5. **Income taxation**	All farm profits are taxable to the proprietor. Any salary paid to the proprietor as manager is taxable to the proprietor.	Partnerships pay no income taxes. All profits as well as losses are allocated directly to partners (both general and limited) for tax accounting. Salaries paid to general partners are taxable to the recipients and are a deductible expense to the partnership.
6. **Employee status**	Social security tax stipulated by law on earnings. Restrictions on amount of farm income proprietors can receive after their retirement without disrupting social security payments. Cost of employee benefits to proprietor and family not tax-deductible to business.	Partnership itself pays no social security tax. Restrictions on amount of partnership income a general partner can receive without disrupting social security benefits. Limitations do not apply to limited partners. Cost of employee benefits to partners and families not tax-deductible to business.
7. **Formalization and cost of organization**	No organizational cost. No minimum required records and reports except for income tax, employee withholdings, workers' compensation reports.	Can be organized with or without written agreement or contract. Records become more vital than for sole proprietorship.

for the Agricultural Business

ships	Corporations	
Limited	**Regular**	**S**
Specifically designed to attract equity capital from limited partners who do not have management responsibility.	From selling equity securities and from retained corporate earnings.	From selling preferred or common stock (not both) to 35 or fewer persons and from retained earnings.
Liability of limited partners limited to amount they invest in partnerships. General partners assume major liability.	Liability limited to the assets of the corporation. Liability of stockholders limited to their individual investments in the corporation.	
Can be dissolved by a general partner but is usually not affected by withdrawal of a limited partner.	Continuity through law is not associated with life cycle of any one person. Instead, the organization is created by law and is terminated by legal action.	
General partners share management, and each has power to bind the partnership. Limited partners cannot obligate the partnership.	Ultimate power rests with stockholders, who elect board of directors that appoints management.	With fewer stockholders there usually is a close link between stockholders, directors, and management.
Partnerships pay no income taxes. All profits as well as losses are allocated directly to partners (both general and limited) for tax accounting. Salaries paid to general partners are taxable to the recipients and are a deductible expense to the partnership.	Corporation pays tax on corporate profits. Salaries to employees are tax-deductible to the corporation and taxable to employees. Stockholders pay tax on dividends.	Taxed as a partnership, i.e., all profits are allocated to stockholders for tax reporting (stockholder could have taxable profits and no dividends with which to pay taxes).
Partnership itself pays no social security tax. Restrictions on amount of partnership income a general partner can receive without disrupting social security benefits. Limitations do not apply to limited partners. Cost of employee benefits to partners and families not tax-deductible to business.	Social security tax stipulated by law for salary to each employee, paid by corporation as tax-deductible expense. An equal amount paid by each employee is not tax-deductible. Only salaries and not dividends diminish social security payments to stockholders. Costs of many employee benefits to the owner-employees and families are tax-deductible as expenses to the corporation.	
Must be formally organized with written agreement. Requires official reports to limited partners.	Organization, legal, and filing fees more costly than for partnerships. Franchise tax varies by states.	

(Continued)

Table 20.1

| | | Partner- | |
Characteristics	Sole Proprietor	General
8. **Intergeneration transfer and estate planning**	Title to actual resources must be transferred after death of proprietor.	Usually difficult to plan for continuity of partnership business beyond first generation. Title to actual resources of partners must be transferred.

statutes specifically authorizing and regulating land trusts. In such states, some uncertainty exists concerning the legality of the land trust. In those states where the "Statute of Uses" is operative and where courts have given force to the concept that the "holder of the use" becomes the "holder of the legal estate," it is questionable whether land trusts are legally valid.

In states where the legality of land trusts is clearly established, land trusts appear to be an effective form for holding title to real estate properties, including farm and ranch holdings. In balance, when title to land is an important part of the farm business, the land trust seems to offer a promising organizational alternative. However, when considering the use of land trusts for the farm business, one should seek experienced legal counsel because the legal basis and constraints of the land trust are not clearly defined.

Business Organizations and Structural Characteristics

Data from the 1988 Agricultural Economics and Land Ownership survey, which is a supplement to the 1987 *Census of Agriculture*, indicate some of the structural characteristics of business organizations in agriculture. The form of farm business organization in the United States clearly is dominated by the proprietary or family type of operation — 1.648 million farms, or 87.7 percent of total farms. Partnerships comprise 9.1 percent of the total, family corporations 2.5 percent, other corporations 0.2 percent, and other forms of business organization 0.5 percent. The average size of partnerships (780.8 acres) is more than double that of individual or family units (368.9 acres), while family-held

(Continued)

ships	Corporations	
Limited	Regular	S
Usually difficult to plan for continuity of partnership business beyond first generation. Title to actual resources of partners must be transferred.	After death of stockholder, only the corporation stock owned by the decedent is subject to probate (not the underlying assets of that stock). Fractionalization of ownership in undifferentiated assets of the firm makes easier transfer possible by using tax-exempt gift privileges. Up to 49% of voting stock may be given away without losing control. Regular corporation can have two classes of stock, thus giving some advantages in estate planning.	

and other corporations average 1,937.6 acres and 3,370.1 acres, respectively. In general, the incidence of indebtedness and levels of leverage are higher in the corporate form of organization, especially in non – family-held corporations.

FINANCIAL IMPLICATIONS OF INTERFIRM COORDINATION

Commercial agriculture has experienced strong trends toward increasing specialization in production and a separation of the control of the various stages of production. New technology has created possibilities for significant economic gains from specialization — resulting in the creation of markets outside the farm for many products and services formerly produced on the farm. Increased specialization, then, has given rise to new economic stages of production.

A **stage of production** is *an operating process that starts with marketable inputs and ends with a salable product.* For example, most dairy farmers a few generations ago sold their products directly to consumers and produced essentially all of their inputs (feed, fertilizer, power, etc.) on the farm. Today, the milk industry is composed of many economic stages that include commercial fertilizer production and marketing, hay production and marketing, commercial feed manufacturing and distribution, equipment manufacturing and distribution, commercial production of dairy heifers, artificial breeding, milk production, milk transportation, dairy processing, and dairy retailing.

Changes in marketing have also increased the payoff from closer coordination of successive stages of production in order to assure that products meet market specifications at each successive stage. Products that do not meet such specifications are either sold at lower prices or, in extreme cases, denied a market at any price.

Successive stages of production are coordinated in several ways. In a market-oriented economy, coordination is largely guided by prices, open or spot markets, and other terms of exchange generated *external* to the firm. Coordination may be (1) **vertical** — *the coordination of successive stages in production and distribution* — or (2) **horizontal** — *the coordination of two or more units of production within the same economic stage.* The latter may achieve scale and/or scope economies in production or marketing or reduce business risks through geographic diversification.

Types of Vertical Coordination

Vertical coordination may range from forward contracting to total vertical integration. Forward contracting usually involves a relatively low level of interfirm coordination yet achieves some degree of interstage coordination. Various types of contracts have varying degrees of market coordination and management control:

1. The most prevalent is the **market-specification contract** often used with dairy farmers producing market milk or farmers producing chipper potatoes. Under such a contract, the farmer has substantial latitude in production decisions as long as the product meets the timing, quality, and quantity specifications of the processor or shipper. Forward contracts between farmers and grain buyers for the delivery at harvest of specified levels of production are another example.

2. The **production-management contract** reflects a higher level of interfirm involvement. Here, the buyer specifies certain resource inputs and management practices for producers to follow to assure that the product meets the specifications of the market.

3. The **resource-providing contract** also involves a higher level of interfirm involvement. Here, the coordinating firm supplies inputs of a particular specification (a patented variety of seed, a patented type of baby chick, a specified feed, etc.). These inputs are usually financed by the coordinating firm. Sometimes the coordinating firm becomes the farmer's major supplier of operating capital.

Varying arrangements of product pricing are used with these forward contracts. In some instances, the contract guarantees a particular price for specified amounts of products meeting the quality specifications of the contract. In other situations, the price applies to the total output from a specified acreage. In still other cases, the contract guarantees a market outlet for the product and outlines a procedure for price determinations.

At a still higher level of vertical coordination, the successive stages are combined into one decision-making unit by means of **total vertical integration.** It could represent the merger of a farm supply or farm marketing firm with a farm operation or a large farming operation integrating forward into processing or distribution. Vertical integration also occurs when groups of farms combine to perform their own cooperative processing and distribution functions.

There have been significant examples in recent years of cooperating grain growers entering into contracts involving commercial cattle feeding and hog feeding activities and dairy farmers forming cooperative milk processing and marketing businesses. Similarly, soybean growers have combined to provide their own crusher and oil mill cooperatives, cotton growers have provided their own cooperative cotton ginning and oil processing facilities, and many other types of producers have formed cooperative associations through which to process and/or market their products.

Some Possible Advantages of Vertical Coordination

There are several reasons why an individual farmer might consider some form of vertical coordination. They are:

1. *To ally closely with a particular market outlet to assure future market access.*

2. *To become tied more closely with a particular source of raw product,* such as a patented seed or special breeding stock that is not available on the open market but that might improve profitability.

3. *To acquire marketing and supervisory skills* in order to better coordinate production to market specifications.

4. *To more accurately meet changes in production specifications.* Changes in consumer demand attributed to nutrition and health

awareness, convenience, and a more diverse population have led to a widening variety of food products, more emphasis on targeting food products to market niches, and greater emphasis by retailers and processors on the production specifications of agricultural commodities. Accompanying and prospective biotechnology are helping to tailor food production to meet these consumer demands.

5. *To reduce or shift risks arising from production or price uncertainties.*

6. *To obtain better financing.* Farmers producing fruits or vegetables for the open market (either for the fresh market or for processing) face such a high degree of price uncertainty that lenders may refuse to finance them until they have a market contract. Better income prospects from contracting could also favorably affect financing terms. In addition, direct financing from a supply or market firm might be available.

Vertical coordination — either through a contractual arrangement or through total merger — may provide more alternatives in resource control and management. If so, a farmer should carefully consider the potential economic gains and compare the possible disadvantages — usually in the form of management constraints imposed by the coordinating firm. However, a producer who cannot obtain adequate financing because of (1) a relatively small, inefficient operation, (2) shortages in managerial skills, or (3) market uncertainties might sacrifice little real economic freedom for the benefits of vertical coordination.

The degree and type of vertical coordination vary greatly within agriculture. Nearly all dairy production, seed production, and most commercial fruit and vegetable production in the United States operate under some form of contract. The poultry industry generally and broiler production in particular have moved rapidly toward total contract production and vertical integration. The last few decades have witnessed substantial vertical coordination in the beef industry. The swine industry is also experiencing similar developments. In the past, significant government intervention through the target prices, price supports, and average restrictions forestalled substantial vertical coordination in the grain industry, although various types of forward contracting by grain farmers have been common.

Vertical coordination might affect different farmers to different degrees. In a multi-product firm where one product is "integrated," only part of the operator's business is affected. Whereas, in firms where one product dominates, the effect may be much greater.

Through vertical coordination, a farmer may shift price and yield uncertainties to the contracting firm; however, a new uncertainty associated with continuation of the contract may be incurred. Thus, the farmer's strategies will include efforts to preserve the new contractual relationship.

Summary

The choice of legal form of business organization may significantly affect the profitability, risk, and liquidity positions of agricultural businesses. Major organizational alternatives are proprietorships, partnerships, and corporations, as well as leasing arrangements and land trusts that are unique to real estate investments. Each of these organizational forms can be evaluated according to its effects on (1) source of equity capital, (2) liability to owners, (3) continuity of ownership, (4) management and control, (5) income taxation, (6) employee status, (7) formalization and cost of organization, and (8) intergeneration transfer and estate planning. Also important to the structure of agricultural businesses is the form of market coordination with other firms in the food and fiber system. These forms of vertical coordination range from open market transactions through various degrees of contracting and total vertical integration.

Other highlights are as follows:

- Sole proprietorships are the most prevalent form of business organization in the agricultural production sector — 87.7 percent of total farms in 1988 — although proprietorships on the average are smaller in size than other business forms.

- Partnerships may be general or limited. In limited partnerships, the limited partners do not participate in the management of the partnerships, and their financial liability extends to the limit of their investment.

- Corporations are characterized by limited liability of owners, marketable shares in the ownership of the firms, and a length of existence that is not tied to the life of any one individual. Smaller corporations meeting specified conditions may elect to be taxed as partnerships rather than as corporations.

- Income taxes, management responsibilities, and estate planning play a major role in the choice of business organization, especially as business size increases.

- The use of contractual arrangements between agricultural producers and input suppliers and food companies has increased considerably since the 1960s and is expected to continue in this trend.

Topics for Discussion

1. What attributes of the corporate form of business organization might explain the difference over time in the Index of Management Level — reflected in Figure 20.1?

2. Under what conditions is an S corporation more appropriate than a regular corporation?

3. Some have argued that incorporating an agricultural business is likely to reduce the amount of capital that can be attracted to the firm rather than increasing it. How can this be true?

4. What are the principal advantages and disadvantages of incorporating a farm business?

5. A firm can grow horizontally or vertically in terms of production stages and interstage markets. Which type of growth is induced by economies of size? By economies of complementarities in production?

6. Contrast and compare the corporate and limited partnership organizations of business.

7. The taxable estate of a farm business is $1 million, composed mostly of farm land. Identify and evaluate several ways in which the estate tax obligation could be reduced by prior planning.

References

1. Barkema, A., M. Drabenstott, and K. Welch, "The Quiet Revolution in the U.S. Food Market," *Economic Review,* Federal Reserve Bank of Kansas City, May – June 1991.

2. Coffman, G. W., *Corporations with Farming Operations,* Agricultural Economics Report 209, ERS – USDA, Washington, D.C., June 1971.

3. Garrett, W. B., *Land Trusts,* Chicago Title and Trust Company, Chicago, 1971.

4. Goldberg, R. A., *Agribusiness Coordination: A Systems Approach to the Wheat, Soybean, and Florida Orange Economies,* Graduate School of Business, Harvard University, Boston, 1968.

5. Harl, N. E., *Agricultural Law: Estate, Tax and Business Planning*, Matthew Bender & Company, Inc., New York, 1992.

6. Harl, N. E., *Farm Estate and Business Planning*, Century Communications, Inc., Skokie, Illinois, 1981.

7. Harris, M., and D. T. Massey, *Vertical Coordination via Contract Farming*, Miscellaneous Publication 1073, ERS – USDA, Washington, D.C., 1968.

8. Hopkin, J. A., "Conglomerate Growth in Agriculture," *Agricultural Finance Review*, Vol. 33, ERS – USDA, Washington, D.C., July 1972.

9. Kenoe, H. W., *Land Trust Practice*, 1974 Edition, Illinois Institute of Continuing Education, Chicago.

10. Krausz, N. G. P., *Corporations in the Farm Business*, Rev., Agricultural Extension Service Circular 797. University of Illinois at Urbana – Champaign, 1975.

11. Looney, J., and D. L. Uchtmann, *Agricultural Law: Principles and Cases*, McGraw-Hill, Inc., New York, 1994.

12. Matthews, S. F., and V. J. Rhodes, *The Use of Public Limited Partnership Financing in Agriculture for Income Tax Shelter*, North Central Regional Research Publication 223, July 1975.

13. Mighell, R. L., and L. A. Jones, *Vertical Coordination in Agriculture*, Agricultural Economics Report 19, USDA, Washington, D.C., 1963.

14. Scofield, W. H., and G. W. Coffman. *Corporations Having Agricultural Operations*, a preliminary report, Agricultural Economics Report 142, ERS – USDA, Washington, D.C., August 1968.

15. Thomas, K. H., and M. D. Boehlje, *Farm Business Arrangements: Which One for You?*, North Central Publication 50, College of Agriculture, Iowa State University, Ames, 1982.

16. Uchtmann, D. L., and C. A. Bock, *Agricultural Estate Planning and the Federal Estate and Gift Tax*, University of Illinois, Cooperative Extension Service, Urbana – Champaign, January 1986.

Appendix to Chapter 20

Estate Taxes: Implications for Farmers

In this appendix we review the effects of federal estate and gift taxes in the United States that are levied on transfers of property by donors at their death, or during their lifetime, and discuss some of the alternatives available to farmers in dealing with the tax obligations. The effects of these tax obligations on farmers vary over time, as farm estates vary in value. Farm estates are subject to potentially large transfer taxes and other costs but have limited capacity for generating easily the funds to pay such costs. In response to the high burden of transfer costs when the owners die, federal tax legislation passed in 1976 and in 1981 has eased these burdens and provided some special options for farm estates. Future changes will occur as well, so this discussion illustrates only some of the issues involved.

Estate management is a pervasive, long-term process. It includes all of the activities that go into building an estate, generating retirement income, planning an equitable distribution of property among heirs, and minimizing the cost of transferring assets. The costs involved in transferring assets are numerous. They include transfer taxes levied on the estate by the U.S. government, inheritance taxes levied on heirs by the respective state governments, transactions costs in liquidating assets, legal fees, costs of funeral and burial, payments of debts, and others.

If not well-planned, these costs may be very high and may substantially reduce the value of an estate. Moreover, inadequate planning can substantially lengthen the time involved in transferring a decedent's estate to heirs. These effects may adversely influence the operation and ownership of an ongoing farm business whose owner has died as well as affecting the financial position and management choices of heirs for many years.

Among the many transfer costs, the federal estate and gift tax obligation may be substantial. It is levied on transfers of property as gifts during the owner's life or as bequests after the owner has died. For transfers at death, the tax is basically levied on the decedent's net worth. However, special rules apply in determining the level of net worth. In order for a federal estate tax return to be filed, the taxable estate must be known. Deducting the allowable *deductions* from the decedent's *gross estate* will yield the *taxable estate*. The *gross estate* includes the fair market value of all property owned outright by the deceased, all property in joint tenancy (unless the surviving joint tenant can prove his or her original contribution), plus half the value of property owned by the decedent and another person as equal tenants in common, plus the proceeds of life insurance and life estates in which the deceased retained any incidence of ownership.

Recent taxable gifts are also included in the gross estate, although they are offset in the deductions when the taxable estate is determined.

The taxable estate is essentially found by reducing the gross estate for (1) claims against the estate, including those of creditors, (2) funeral expenses and other administrative costs, (3) charitable deductions, and (4) an unlimited marital deduction for bequests or gifts to a spouse. Federal estate taxes are then figured on the taxable estate, based on the unified transfer tax rate schedule, and an allowable tax credit is applied to reduce the tax liability to yield the amount of tax actually owed.

Table 20A.1 indicates the Unified Transfer Tax Rate Schedule in effect during 1988 and beyond. Determining the level of taxes actually paid involves subtracting a unified credit from the taxpayer's estate or gift tax liability. As of 1988, the unified credit is $192,800, which completely exempts taxable estates of up to $600,000 from taxation.

The unified credit protects many estates from federal taxation. However, the estates of many farmers may still exceed these levels because of the importance of farm land and other non–real estate assets. Thus, the tax obligation on estates could be significant.

To illustrate, consider the estate tax obligation on Farmer X who might die under current conditions. Farmer X is a surviving spouse who owns 800 acres of farm land valued at $1,500 per acre and has $140,000 of nonfarm assets. Thus, Farmer X's gross

Table 20A.1. Unified Transfer Tax Rate Schedule, 1988

If the Amount Is		Tentative Tax Is		
Over	But Not Over	Tax +	%	on Excess Over
$ 0	$ 10,000	$ 0	18	$ 0
10,000	20,000	1,800	20	10,000
20,000	40,000	3,800	22	20,000
40,000	60,000	8,200	24	40,000
60,000	80,000	13,000	26	60,000
80,000	100,000	18,200	28	80,000
100,000	150,000	23,800	30	100,000
150,000	250,000	38,800	32	150,000
250,000	500,000	70,800	34	250,000
500,000	750,000	155,800	37	500,000
750,000	1,000,000	248,300	39	750,000
1,000,000	1,250,000	345,800	41	1,000,000
1,250,000	1,500,000	448,300	43	1,250,000
1,500,000	2,000,000	155,800	45	1,500,000
2,000,000	2,500,000	780,800	49	2,000,000
2,500,000 and over	. . .	1,025,800	50	2,500,000

estate is projected to be $1,340,000. Assuming deductions of $40,000 and no previous gift transfers, the taxable estate is $1,300,000. The gross estate tax is projected as $469,800 (see Table 20A.1). Subtracting the unified credit of $192,800 leaves a net estate tax of $277,000 — a sizable amount. Because Farmer X's estate is dominated by farm land, the funds to meet the tax obligation must come from partial liquidation of the estate, from borrowing sources, or from special valuation or installment payment privileges allowed by the tax laws if the estate qualifies. In addition, Farmer X might consider other alternatives in estate management in order to reduce the tax burden or increase the estate's liquidity position. Some of these alternatives are considered as follows.

Gifts

Tax-free gifts that are made prior to the donor's death are one way to reduce the size of an estate and assure that continued growth in asset values will rest with the recipients of the gifts. Federal estate tax laws, in general, allow a 100 percent marital deduction on qualifying property passing to a donor's spouse. In addition, a donor may exclude up to $10,000 annually per donee from the value of total gifts subject to tax. For a husband and wife making gifts, the annual exclusion becomes, in effect, $20,000 per donee per year. Gifts above these levels are subject to the transfer tax although the unified credit applies.

Trusts and Life Estates

The **trust** is a *contractual instrument by which title to property is transferred to a trustee to hold and manage for the benefit of other persons designated as beneficiaries.* The trust agreement specifies the objectives and terms of the contract.

A trust created and operated during the lifetime of the grantor is called a **living trust.** If the grantor retains the right to change or terminate the trust, it is revocable and thus subject to estate taxes at the grantor's death. If the power to control the property or the income from it is given up, the trust is irrevocable and essentially a gift, and it largely eliminates the time-consuming and costly process of probating a will upon an individual's death.

A **testamentary trust** is created by a will at the grantor's death. The property in such a trust is part of the grantor's estate for tax purposes. The trust can be used to (1) relieve a spouse or other beneficiary of management burdens, while assuring the beneficiary of income benefits from the property, (2) provide income and care for an incompetent person, (3) prevent a beneficiary from unwisely spending an inheritance, (4) provide interim management for minors or temporarily incapacitated beneficiaries, and (5) provide a lifelong income to a spouse or children, with the remainder going to the children, grandchildren, or others.

The life estate arrangement provides some flexibility in protecting estates from taxation, in which immediate descendants may hold a current interest; however, limits are imposed on this technique, and referral to legal counsel is important to determine the merits of this approach.

Life Insurance

Life insurance on the grantor's life is one way to provide the liquidity needed by the estate or by the beneficiaries in transfer at the grantor's death. Either the estate or some persons may be designated as beneficiaries. If the grantor assigns all benefits to a beneficiary and relinquishes control rights over the policy, the life insurance proceeds likely will be excluded from the estate, with obvious tax benefits. If the grantor does not want the benefits paid directly to the beneficiary, they could be paid to a trust set up for the beneficiary, as designated by the grantor. In any case, life insurance provides an effective source of liquidity to meet an estate's financial obligations.

Incorporation

Corporations have distinct advantages over proprietorships and partnerships in transferring assets. Thus, they offer a high degree of flexibility in estate management and may aid in reducing transfer costs. Moreover, corporations can effectively maintain business control while still transferring assets to heirs or other parties.

One of the advantages of corporations in estate management is that of having more than one class of stock. A farmer can maintain managerial control over the business while transferring substantial ownership to heirs by transferring non-voting shares and maintaining voting shares. For example, a relatively young but successful farmer, with farm assets valued at, say, $1 million, could incorporate with 499,000 shares of common stock at par value of $1 and 501,000 shares of preferred stock at par value of $1 and pay a fixed dividend per share. Both classes could be made voting stock.

The common stock could be transferred to heirs either directly or through a trust, depending on the circumstances and preferences. It could be transferred all at once, subject to the transfer tax, or it could be transferred over several years to take advantage of the annual gift exclusions.

Alternatively, the stock could be sold to heirs in exchange for their debt, thus fixing the value of the wealth being transferred. The parent could then make gifts to the heirs over the years by canceling a portion of the debt each year. As the firm continues to expand, the increased net worth would accrue to the common stock held by the heirs. By holding the preferred stock, the parent maintains control of the business and assures a flow of income as long as the business is profitable. In this fashion, the growth in wealth accrues directly to the holders of common stock — the heirs — without costs of estate transfer.

Of course, numerous other factors associated with incorporation need to be considered. However, incorporation does offer additional flexibility in estate management.

Use Valuation of Farm Land

The Tax Reform Act of 1976 provides that property in a gross estate be valued at fair market value. However, the act, along with subsequent amendments, allows qualifying farm land to be valued on an "actual use" basis at the election of the executor or administrator of an estate. On an actual use basis, the value is determined by capitalizing the fair rental value of the property. Where relevant information is available, net cash rent on land, averaged for the past five years, is capitalized at the interest rate charged on new Federal Land Bank loans (now Farm Credit System Real Estate loans), also averaged over the past five years. Thus, for net cash rent of $80 per acre and interest of 10 percent, the actual use value would be $800. In contrast, the land's market value might be in the $1,500 to $2,000 per acre range. In times of rapidly inflating land values, this procedure will produce a value substantially less than could be obtained by selling the property. Thus, it is an attractive tax alternative for farmers.

However, using actual use values is based on several conditions. They are (1) qualifying conditions, (2) limiting conditions (the "maximum benefit rule"), and (3) recapture provisions. To qualify, at least half the value of the gross estate, net of indebtedness, must be made up of assets used in farming and must pass to one or more family members; at least 25 percent of the value of the gross estate must be composed of real property used in farming; and the deceased, or a member of the deceased's family, must have owned the real property for five of the eight years preceding the owner's death and, in like period, must have participated materially in the farming operation involving the real property.

The maximum benefit rule limits to $750,000 the amount by which the value of the estate can be reduced by actual use valuation. Finally, if within 10 years (15 years for deaths prior to 1982) after the death of the owner, the family recipient disposes of the qualifying real property to a nonfarm buyer, or converts it to a nonfarming use, any savings in estate taxes are subject to recapture. Thus, while the savings may be substantial, restrictions on subsequent use or disposition may also have substantial financial consequences.

Installment Payment of Federal Estate Tax

Farm estates frequently are plagued by relatively low liquidity, creating disruptive cash flow problems for transfer recipients who are confronted with large tax obligations. Tax laws have allowed for this problem by providing for a deferment of tax payments attributable to a farm, or other closely held business, if the interest in the closely held business or farm exceeds 35 percent (65 percent before 1982) of the decedent's adjusted gross estate. Installment payments may occur over a 15-year period at a favorable interest rate.

21

Outside Equity Capital
in Agriculture

This chapter focuses on the use of outside equity capital in agriculture. Included are the role of outside equity, the methods of obtaining it, and the agricultural operations that can be used with these methods. The impacts of tax obligations are a crucial factor affecting the availability of new equity capital. The chapter concludes with discussions of private versus public investment offerings and the process of going public.

THE ROLE OF
OUTSIDE EQUITY CAPITAL

Financial management in agriculture has centered on the management of debt capital under the assumption that farmers could generate sufficient equity capital from retained earnings, capital gains, inheritances, and gifts. This assumption has been consistent with the structural characteristics of agriculture, in which farming operations are largely controlled by individual farmers and farm families. However, these features are subject to change, as illustrated by the substantial leasing of farm land, which separates land ownership and control, and by the production of some commodities by nonproprietorship business organizations relying heavily on outside equity capital. Even growth-oriented proprietorship firms or partnerships may be under pressure to expand faster than they can generate equity capital from within their businesses. Alternatively, some highly leveraged operations may wish to attract additional equity in order to reduce leverage and thus high financial risks.

Within agriculture, farmers and ranchers often pool resources for a larger, more efficient operation. Some of these arrangements involve leasing; others are joint ventures, partnerships, mergers, or family corporations. Intergeneration arrangements are especially common within families. Pooling of family equity has often resulted from estate planning to minimize intergeneration transfer costs and to maintain farms as integral units. While these consolidations of resource control cause structural changes within farms, the total equity position of the participating parties remains largely unaffected.

Outside equity capital enters agriculture through both informal and formal market arrangements. Informal market transactions occur when outside investors buy land directly from farmers, for example. The investors generally arrange for labor, management, and other capital services through leasing contracts or direct hiring. The result is part-owner tenancy, although often with professional management services playing an important role.

Outside equity entering through formal market arrangements represents the specialized efforts of agricultural entrepreneurs to attract equity through limited partnerships, common stock, and agency services. These ventures have been commonly used in financing growth of large cattle feeding operations, as well as cattle breeding, citrus, and vineyard operations. Newer channels are associated with pension funds, life insurance companies, foreign investors, and equity participation loans considered by some lenders.

Outside equity may come from many sources, including individuals, associations, trusts, nonfarm corporations, and other organized groups with investable funds. Considerable interest is also developing in packaging outside equity capital for farm investments from people in nearby communities who have the savings and know the local farming industry. Most outside investors prefer not to participate in management. They generally have strong financial incentives for making the investments and prefer limited liability, reasonable liquidity, and undifferentiated ownership shares in the businesses. Proprietorships and general partnerships are poorly adapted to these investors.

Many prospective outside investors are strongly motivated to shelter high nonfarm incomes from tax obligations. In the past, tax shelters provided by farming have generally rested on the availability of cash accounting, current deductions of some capital expenditures for developing orchards and ranches, potentials for converting ordinary income to capital gains, and the pass-through of income losses to the investor, as with limited partnerships or S corporations. However, changes in tax laws over time have modified the benefits for nonfarm and farm investors from these tax provisions. An example is the Tax Reform Act of 1976, which, among its many provisions, stipulated the following: (1) large corporate businesses were required to use accrual accounting to figure their tax liabilities from farm operations; (2) loss deductions from farm operations were limited to the amount the investor actually has at risk in the operation; (3) cash basis taxpayers were prohibited from deducting prepaid interest; and (4) syndicates (including partnerships) were required to capitalize certain expenses. Tax legislation in 1986 further restricted the use of tax shelters in agriculture and other sectors as well.

Because changes in tax laws are difficult to predict and may occur relatively frequently, it is both difficult and inappropriate to establish definitive guidelines about their use in attracting outside equity into agriculture. Moreover, the impacts of changes in tax laws are hard to project. One expert, for example, observed, in response to the 1976 act, that "the incentives which have motivated outside investors in the past still exist, and although the means by which they can be tapped have changed, they will continue to attract risk capital to agriculture" [11].

EQUITY CAPITAL AND LIMITED PARTNERSHIPS

With limited partnerships, the financial liability of each partner is limited to his or her investment in the partnership, limited partners do not participate in management, and profits and losses are passed directly to the partners for income tax purposes.

Investment Characteristics and Limited Partnerships

Raising equity capital in U.S. agriculture through limited partnerships has been used in cattle feeding, as well as in cow – calf operations and orchard development. The development of large-scale commercial cattle feeding illustrates this process. Federal income tax policies clearly interacted with economies of size, new technologies, and profit potential to bring about the development of very large-scale, capital-intensive cattle feeding operations and a shift from small farm feedlots in the Midwest to the southern high plains of Texas, Oklahoma, Colorado, and Kansas. When rapid expansion of cattle feedlots occurred in the mid- to late 1960s, an important source of equity capital was the sale of limited partnerships in the national financial markets. Millions of dollars were raised in this manner between 1965 and the early 1970s. Much of this flow, in turn, was attributed to tax savings for their investors. These funds, when leveraged with debt capital at rates of up to 4 to 1, furnished the capital needed to buy millions of head of cattle and the necessary feed for this industry each year during this period.

The use of outside equity capital in cattle feeding also added to financial instability. Declining beef prices in 1972 to 1973 caused capital to be withdrawn from existing cattle limited partnerships, and few new partnerships were formed. In turn, cow-calf operators experienced short-term cash flow problems because they had to finance the placement of their calves in commercial feedlots rather than through sales to other custom feeders. The tax changes cited previously also dampened the attraction of limited partnerships in cattle operations.

Nonetheless, several characteristics of large-scale cattle feeding made limited partnerships attractive for raising equity capital. New technologies provided significant economies of size in cattle feeding and high capital needs to meet the related financing requirements. Innovative methods of attracting capital were needed. The deferral of near-term tax obligations was attractive to high-income nonfarm investors. Cattle feeding is also a relatively high-risk, high-return venture. High risk itself does not make a limited partnership attractive, but this type of business organization generally does offer some risk-reducing advantages.

A risk to the investor, in addition to the usual business risks, is the limited opportunity to resell limited partnership interests. Two courses of action may help reduce this risk. First, limited partnerships may provide investors with opportunities to liquidate their investments within a specified number of years, with provision for earlier withdrawal under a penalty. Second, some limited partnership offerings are designed to appeal to high-income investors, for whom short-term liquidity is not a problem. Frequently, the prospectus will stipulate a minimum salary and net worth for prospective investors.

Steps in Creating a Limited Partnership

One basic requirement for creating and selling limited partnership interests is the presence of at least one general partner who is responsible for management and liability. Since limited partners have limited liability, the general partner must assume unlimited liability for the partnership. Hence, the general partner must demonstrate financial responsibility to the limited partners.

The general partner may be an individual, a general partnership, an informal association, or a corporation. Having a corporation as the general partner introduces the limited-liability provision of the corporation as well. However, providing limited liability to the general partner increases the possibility of having the business classified by the U.S. Internal Revenue Service as a corporation for tax purposes.

The limited partnership must also be registered in its home state. If interstate offerings of partnership shares are anticipated, the registration must be filed and cleared with the Securities and Exchange Commission (SEC) in the U.S. Department of Commerce. The general partner(s) may then offer limited partnership shares to the public.

EQUITY CAPITAL AND CORPORATIONS

Incorporation for attracting outside equity capital has not been widespread in agriculture. Those firms that have publicly sold common stock generally cite high administrative costs and lengthy review of their prospectus by state and federal regulatory agencies. Liquidity of corporate shares can be high, but variability of earnings and difficulty in attracting high-quality management dampens income prospects for investors. The more successful corporate arrangements often arise when well-managed agricultural enterprises are combined with other enterprises in vertical sequence (input supplier, processor) or in conglomerate arrangements.

Nonetheless, the corporation offers several advantages over a proprietorship and a general partnership in tapping outside sources of equity capital. The regular corporation permits the use of two classes of stock. This provides unique advantages over the S corporation and the limited partnership in long-term tax planning and flexibility for channeling current returns and capital gains to investors.

The process of forming a corporation is relatively simple, although it varies among states. The articles of incorporation must be cleared by and registered with the appropriate state agency. An experienced attorney can help in determining in which state to register and in obtaining clearance and registration. One argument offered against incorporating a farm business without public stock offerings is the limited marketability of the stock. This creates liquidity problems for stockholders who desire to withdraw from the business. This problem is not unique to corporations however. It can be even more severe for proprietorships and partnerships with multiple owners. A market must be found for an undivided interest in the business, or some assets must be sold.

The marketability problem often depends on provisions in the articles of incorporation that control the trading of ownership interests. If the business is large enough to warrant selling stock to outsiders, the issues and suggestions that follow are applicable. If the business is not large enough to establish a market price, the stockholders can meet to establish the price for redeeming a member's stock during the following year.

In the following discussion, the central focus will be on the problems and processes of obtaining new equity capital by selling corporate stock. However, the discussion is also generally applicable to limited partnerships. Moreover, the issues and techniques apply to farm firms and to many farm-related businesses that are family-oriented.

Why Sell Corporate Stock?

One of the first questions the owner-manager should answer is: "Why sell stock in a successful business to outside investors?"

- *To increase the corporation's equity capital*

    One obvious reason is to obtain outside equity capital. Once the business is leveraged to a level that the owner-manager and lender feel is optimal, further growth is limited by the rate of equity accumulation. Unless new stock is sold, additional equity is limited to retained earnings and unrealized capital gains.

    Equity capital may be needed for other reasons too. If large debt obligations are severely straining the firm's cash position, the owner-manager might seek relief through a stock offering, using the proceeds to pay off debts. Substituting equity for debt reduces the heavy drain on cash to service principal and interest. However, prospective investors would probably need to be convinced that the adverse cash flow problem did not result from operational inefficiencies and poor management or from a declining economic outlook for the commodities involved.

- *To reduce private holdings of corporate stock*

A successful farmer with a sizable estate, completely tied up in a farming corporation, may prefer to sell corporate stock in order to diversify a personal investment. It is important to distinguish between stock offered by a corporate officer as an individual and stock offered by the firm itself. With many family corporations, the major investment holding of one parent is his or her stock in the corporation.

There are several reasons why the parent might diversify under these circumstances. One reason is to protect the business against potential sale for payment of estate and inheritance taxes. Although life insurance reduces this risk, there may be merit in selling corporate stock and investing the proceeds in noncorporate assets so that future income will not be solely dependent on the farm business.

Second, outside sales could establish a stock price for purposes of estate planning, settling estates, or making income and holding adjustments within a family. Sales of stock are always difficult in a closely held family corporation, and a sale to outside buyers might establish a base price for settlements.

Another reason for selling stock is that it allows an individual to withdraw from the business. The withdrawal of a valuable manager will clearly affect the business's income-generating capacity. Thus, if a parent who successfully built up the business is planning to withdraw, he or she should assure that strong management will continue. Otherwise, the stock's sale price could be adversely affected.

Several disadvantages of outside stock sales should also be considered. First, selling common stock dilutes the relative claims of existing owners on the business' stream of income and net worth. Management must choose between a total claim to a smaller cash flow and net worth and a partial claim to potentially larger sums.

Second, selling common stock extends voting rights and business control to new stockholders. Owner-managers of farm corporations are often reluctant to share control with outsiders. It is not uncommon to read of "outsiders" gaining control of a business, pushing the original management out, draining the firm's liquidity, and then selling for a "tax" profit. Maintaining majority control of the stock, combined with effective management, will insure against such a takeover.

Third, when a privately held corporation becomes a public corporation, its annual report (including balance sheet and income statement) is more readily available to outside parties and thus may be viewed as a disadvantage to the public holding.

Fourth, the transactions costs of obtaining outside equity capital are higher than those of debt capital — sometimes much higher. Examples are the costs of underwriting and distribution for common stock. Common stock typically must be sold at a higher expected yield than debt instruments because of its higher risk for the investor. Also, dividends are not tax-deductible, while interest paid on debt is tax-deductible.

Assume, for example, that a term loan is negotiated at 10 percent interest. For a corporation in the 34 percent tax bracket, this represents an after-tax equivalent of 6.6 percent. The sale of preferred stock to cover this investment likely would have to pay a dividend yielding at least 10 percent to the purchaser to make the returns on an equity investment comparable to the returns on a loan, including a risk premium on the equity investment to reflect its higher risk position. Thus, a $100 preferred stock would carry an annual dividend of $10. If it costs $10 per share to market the preferred stock, the selling firm would pay $10 per year on $90 of realized capital, or a minimum annual rate of 11.1 percent. Since stock dividends are not tax-deductible, the cost of preferred stock is considerably more than the after-tax cost of borrowed funds. The cost for common stock would be higher still, since it carries greater uncertainty for the investor.

A Private Placement Versus a Public Stock Offering

Once a farm corporation has decided to obtain new equity capital, it must make either a private placement or a public stock offering. For a corporation seeking less than, say, $500,000 in equity capital, the public sale would be too costly. Otherwise, for any but very large firms, this is an important decision.

A few close personal contacts may be interested in buying stock in the farm business. Where the capital needs are not too great, such a sale may be attractive to all parties. Even the small farm proprietorship might persuade a local lawyer, physician, or other person of means to invest in the farming operation.

Registration with the Securities and Exchange Commission (SEC) is not required for a private sale, thus saving important costs. The cost of an investment banker's services may even be saved. In most instances, however, a sizable private offering will be more successful (and certainly smoother) through the services of an investment banker. The investment banker has important contacts with individuals and firms seeking investments in private stock offerings.

In recent years, a number of mutual funds, insurance companies, and some pension funds have been attracted to private offerings of common stock or other forms of equity participation in successful small companies, including those in agriculture and related industries. The fees of an investment banker are generally less for arranging a private sale than for underwriting a public offering. Furthermore, if a farm corporation has unusual financial problems that are being resolved, or unrealized potential, these circumstances can often be more easily explained to, and better understood by, a few investors than by many small investors who must be reached through a prospectus.

However, private offerings have possible disadvantages. Unless the buyers know and have enthusiasm for the company and its management, they might insist on imposing tight control. They likely will follow management actions far more closely than public investors would and will probably respond more vigorously if they are not completely satisfied.

The regulations determining whether an issue can qualify as a private sale or must be registered with the Securities and Exchange Commission as a public sale are spelled out in the Securities Act of 1933. In broad terms, public offerings above a stipulated threshold level that will be offered in interstate sales must be registered. However, the number of offerees (and their intentions toward resale), the uniqueness of the offerings, and other factors must also be considered. Qualified legal counsel, along with economic analysis, is essential in determining whether or not to go public.

The best counselor for choosing between a private versus public sale is likely an investment banker who is experienced in placing agricultural securities. Whether to use a regional or a national firm depends on the ultimate objectives and characteristics of the business. If the farm corporation is primarily local in its resource base and market, a local company would probably best handle the underwriting.

Timing Problems

Two important timing questions must be resolved with a public issue. First, is the company ready for the market? If it has management, financial, or marketing problems that are being resolved or if it is on the threshold of a profit breakthrough, a public issue might be delayed until this situation is resolved. If delay is not possible, a private sale would presumably be preferred. A private sale can proceed immediately, whereas a public sale is subject to SEC approval and possible delays.

In determining whether or not the company is ready for the market, the owner-manager should honestly evaluate the business from the standpoint of both the underwriter (who might have to sell the stock) and the prospective investor. In doing so, he or she should consider the information in the prospectus. Realistically, the prospective investor might ask, "Given the alternatives available in the market, would I, as a detached outsider, be attracted to invest in this company? If so, why? Is it in a growth industry? Is it recession-proof? What is this firm's position within the industry in terms of earnings ratios, growth rates, liquidity, and risk?"

The second timing element relates to the market. Is the market ready for this type of new offering? Is the market "bearish," generally, and toward new issues, in particular? How will this market attitude affect price? Competent investment bankers are a good source of advice on timing.

If it seems wise to delay the public offering, interim sources for long-term funds will probably be needed. Perhaps a term loan with a bank or an insurance company is possible, with the agreement that the loan will be repaid when the stock is sold. The investment banker may be able to sell debentures. However, if short-term financing is already large, this lender may require that any long-term financing be subordinated to his or her debt.

BASIC STEPS IN GOING PUBLIC

After the decision to go public, the next step is to secure the services of three competent professionals — an investment banker, an attorney, and an accountant. The investment banker will advise on the marketability of a new stock offering, suggest the procedures to follow in preparing for the stock offering, and recommend the price. The banker's function in underwriting and syndicating the stock issues will depend on the method of underwriting, whether under a **fixed commitment basis** or a **best effort basis.**

Under the fixed commitment basis — the method most used by reputable investment banking firms — the underwriter agrees to purchase the stock offering at a mutually specified price, and a check for this amount, net of the underwriter's commission, is received immediately, once SEC clearance and all the registrations have been accomplished. Aside from certain *escape clauses,* the underwriter assumes the full risk of marketing the stock. The escape clauses should be studied carefully, for they outline the conditions under which the underwriter can cancel the agreement.

For more speculative issues, underwriters might prefer to enter agreements on a best effort basis, whereby they exert their best efforts to sell the stock offering but do not agree to purchase unsold stock. Most of the cattle feeding limited partnership subscriptions have been offered to the general public on a best effort basis. Some general partners also reserve the right to sell directly to the general public at the commission rate charged by underwriters. Under most circumstances, however, one should be wary of best effort contracts. They could reflect either a financially weak underwriter or a market situation judged weak or hazardous by the underwriter.

Investment bankers may also serve as agents in arranging private financing through private sale of stock, sale of debentures, purchase and leaseback, or mergers. As investment counselors, they have contact with many types of investors and other investment bankers. Consequently, they are in a strategic position to help evaluate the market for a particular company.

The selected attorney should be experienced in securities registration. The outcome success depends on a smooth and judicious process. If the registration proceedings are mishandled, the owner-manager can be subject to civil suit and penalties. It may be necessary to contact a regional financial center to find experienced legal counsel. Although the attorney must work closely with the underwriter's counsel, he or she should represent the client's interest in preparing the corporate documents, the prospectus, and the registration statement, and in negotiating the details of the underwriting agreement. The attorney is also responsible for obtaining SEC clearance and registration and for getting registration in each state where stock might be sold. An accountant will assist in preparing the prospectus, financial statements, and registration statement. Since these documents must be cleared by the Securities and Exchange Commission, an accountant familiar with SEC procedures can greatly facilitate the process.

The accountant, attorney, and underwriter must work jointly to cover the basic steps of (1) assembling the necessary information about the business to satisfy SEC regulations, (2) preparing the prospectus, (3) reorganizing the firm to its new capital and organizational structure after the stock offering, (4) taking care of all regulations, (5) printing stock certificates, and (6) arranging to sell the stock.

Costs of Going Public

The total costs of underwriting will vary substantially with the size of the issue, the strength of the firm selling the stock, the amount of legal and accounting detail required, the market conditions, and the way the underwriter views the marketability of the stock. Included in the total expenses, in addition to the underwriter's commission, are legal and accounting fees; state and federal taxes, fees, and registrations; costs of printing the preliminary and final prospecti, stock certificates, and other documents needed in volume; and travel and miscellaneous expenses. The total expenses, including the underwriter's commission, could range up to 20 percent or more of the gross revenue from the public sale. Naturally, these costs must be included in any budgeting analysis of the profitability of public stock offerings.

EQUITY CAPITAL
AND MERGERS

A **merger** is the *combination of two or more existing businesses into a single business entity.* The firms being merged may be proprietary, partnership, or corporate, and the merging process may be any of various methods. Similarly, the business established through merger may have any legal form. Three merger situations and the benefits stemming from them are presented as follows:

1. Many small businesses — including farm businesses — have solved pressing financial and marketing problems by merging with larger, established firms through some form of vertical or horizontal integration. In some instances, farmers have merged with a processing or marketing firm *(integrating forward)*, or they have merged with a supply firm *(integrating backward)*. Both pricing and tax advantages may arise in the exchange of equity (either proprietor's equity or corporate stock) in a privately held business for stock in a publicly held corporation. Moreover, the transaction is far less costly than a public offering. However, when a relatively small farm operation merges with a large corporation, management control may be lost to the large corporation.

2. A merger might be particularly attractive to a farmer approaching

retirement. An example is a producer of quality tree fruits, who is approaching retirement and whose children do not want to operate the orchard of 100+ acres. The food-processing firm that purchased the fruit for many years has offered to exchange corporate stock for the farmer's equity in the orchard. This firm's stock is listed on the New York Stock Exchange.

The processing firm is anxious to acquire control over orchards in the area to insure access to raw products. Moreover, acquiring the farm through stock exchange rather than cash would strengthen the firm's liquidity. Hence, it has offered the farmer a price somewhat above the going market price for quality orchards. The canner also has offered the farmer a two-year management contract to help get the farm's operation fully integrated into the corporate business.

By exchanging acres of land, irrigation equipment, sets of buildings, and machinery items for shares of stock in the food-processing corporation, the seller secures a flow of income that is more reliable than was the income from the orchard. The stock also provides considerably more flexibility in transferring the producer's wealth to heirs.

3. Merging might provide a way for two or more farmers to combine operations to achieve economies of size and specialization in management. However, the merging parties need not relinquish title to their real estate in order to achieve the desired benefits as long as the use of resources is pooled. A newly created firm might operate informally as a general partnership, or it might be incorporated.

To illustrate, consider an actual case of four young farmers, each of whom operated approximately 600 acres of corn and soybean land in central Illinois at an early stage in his or her farming career. In each case, the operator felt that his or her operation was too small to achieve efficient use of machinery, labor, and management. Moreover, none of them felt that they had sufficient equity capital and credit to expand. Hence, these four farmers decided to merge as a general partnership, providing a single operation of over 2,400 acres within an acceptable radius under an arrangement in which each retained control of his or her own land. These farmers also anticipated expansion of the land base by leasing additional farm land within 40 miles of the headquarters units.

The merger allowed each operator to specialize in one phase of management. One assumed responsibility for budgeting, financing, and fiscal control. Another took on the planning, operation, and maintenance of the production and harvesting equipment. As an extension of this specialization, this farmer supervised the machinery and crews in performing custom operations for other farmers. Another planned and supervised the cropping program, including the fertilizer and other farm chemical applications. Soon, this farmer was also selling fertilizers and farm chemicals, including applications. The remaining partner specialized in marketing, including programs for forward contracting and futures contracts.

While this operation is not indicative of what most average farmers are doing, it represents what can happen when farmers identify the nature of their problems, inventory their resources, and exercise imagination and initiative in resolving the problems.

Summary

Outside equity capital may enter the agricultural sector through both informal and formal market arrangements. Informal arrangements may reflect direct investments in farm real estate by nonfarm investors with leasing of the land to farm operators. Formal arrangements for outside equity represent public offerings of stock or limited partnership interests by agricultural production firms. Formal arrangements are not common in agriculture, mostly involving large cattle feeding operations, as well as cattle breeding, citrus, and vineyard operations. However, reliance on outside equity will undoubtedly increase in the future, at least for some types of operations that wish to (1) avoid the use of extensive financial leverage, (2) spread risk over more diverse set of investors, and (3) more effectively manage income tax obligations. Other highlights are as follows:

- Use of equity capital associated with various forms of sheltering income taxes may create financial instability because of potential changes in tax laws.

- Equity investments in limited partnerships and small-farm corporations may experience liquidity problems because of low marketability of the resulting financial assets.

- Alternative corporate forms provide for flexibility in long-term tax planning and for channeling current returns and capital gains to investors.

- Attracting outside equity capital can be costly because of the fees paid for accounting, legal, security, promotional, and investment banking functions, although the costs per dollar diminish as the equity investments increase in size.

- Timing of outside equity sales is important in terms of both financial market conditions and readiness of the issuing firm to obtain and successfully manage outside equity.

- Mergers often are a cost-effective way to combine two or more firms in order to achieve economies of scale and specialization in management.

Topics for Discussion

1. Discuss agriculture's need for outside equity capital.

2. Identify the types of financial criteria a prospective outside investor might use in evaluating a farm investment.

3. Why are some types of agricultural enterprises better suited than others for attracting outside equity capital?

4. Explain the concept of a tax shelter. What role does it play in attracting outside equity capital in agriculture?

5. Compare proprietorships, general partnerships, limited partnerships, subchapter S corporations, and regular corporations in terms of their adaptability for raising outside equity capital.

6. Why might a "private offering" be more advantageous to most agricultural firms than a "public ᵕale" as a means of raising equity capital? What disadvantages, if any, does it have?

7. Consider an agricultural business typical of your geographic area. In what ways might a merger help resolve some of the existing capital and management constraints?

References

1. Davenport, C., M. D. Boehlje, and D. Martin, *The Effects of Tax Policy on American Agriculture*, Agricultural Economics Report 480, ERS – USDA, Washington, D.C., 1982.

2. Fiske, J. R., M. T. Batte, and W. F. Lee, "Nonfarm Equity in Agriculture: Past, Present and Future," *American Journal of Agricultural Economics*, 8(1986):1319 – 1323.

3. Friend, I., *Investment Banking and the New Issues Market: Summary,*

Vol. 3, Wharton School of Finance and Commerce, University of Pennsylvania, Philadelphia, 1965.

4. Hopkin, J. A., *Cattle Feeding in California: A Study of Feedlot Finishing*, Bank of America, San Francisco, February 1957.

5. Hopkin, J. A., "Conglomerate Growth in Agriculture," *Agricultural Finance Review*, 33(1972): 15 – 21.

6. Krause, K. R., and H. Shapiro, "Tax-Induced Investment in Agriculture: Gaps in Research," *Agricultural Economics Research*, 26:1(1974):13 – 21.

7. Matthews, S. F., and D. H. Harrington, "Analysis of Nonfarm Equity Forms of Investment Applicable to Agriculture," *American Journal of Agricultural Ecoomics*, 68(1986):1324 – 1329.

8. Moore, C. V., "External Equity Capital in Production Agriculture," *Agricultural Finance Review*, 39(1979):72 – 82.

9. Scofield, W. H., "Nonfarm Equity Capital in Agriculture," *Agricultural Finance Review*, 33(1972):36 – 41.

10. Securities and Exchange Commission, Form S – 1 Registration Statement Under the Securities Act of 1933, U.S. Department of Commerce, Washington, D.C., 1958.

11. Sisson, C. A., "Tax Reform Act of 1976 and Its Effects on Farm Financial Structure," *Agricultural Finance Review*, 39(1979):83 – 90.

12. Solomon, M., Jr., *Investment Decisions in Small Business*, University of Kentucky, Lexington, 1963.

13. Thomas, K. H., and M. D. Boehlje, *Farm Business Arrangements: Which One for You?*, North Central Regional Extension Publication 50, College of Agriculture, Iowa State University, Ames, 1982.

14. Winter, E. L., *A Complete Guide to Making a Public Stock Offering*, Prentice-Hall, Inc., Englewood Cliffs, New Jersey, 1971.

15. Zwick, J., *A Handbook of Small Business Finance*, Small Business Administration, 15, Washington, D.C., 1965.

Appendix

Appendix Table I
Standard Normal Probabilities

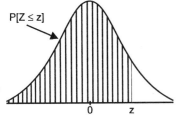

P[Z ≤ z]

Z	0.00	0.01	0.02	0.03	0.04	0.05	0.06	0.07	0.08	0.09
0.0	0.5000	0.5040	0.5080	0.5120	0.5160	0.5199	0.5239	0.5279	0.5319	0.5359
0.1	0.5398	0.5438	0.5478	0.5517	0.5557	0.5596	0.5636	0.5675	0.5714	0.5753
0.2	0.5793	0.5832	0.5871	0.5910	0.5948	0.5987	0.6026	0.6064	0.6103	0.6141
0.3	0.6179	0.6217	0.6255	0.6293	0.6331	0.6368	0.6406	0.6443	0.6480	0.6517
0.4	0.6554	0.6591	0.6628	0.6664	0.6700	0.6736	0.6772	0.6808	0.6844	0.6879
0.5	0.6915	0.6950	0.6985	0.7019	0.7054	0.7088	0.7123	0.7157	0.7190	0.7224
0.6	0.7257	0.7291	0.7324	0.7357	0.7389	0.7422	0.7454	0.7486	0.7517	0.7549
0.7	0.7580	0.7611	0.7642	0.7673	0.7704	0.7734	0.7764	0.7794	0.7823	0.7852
0.8	0.7881	0.7910	0.7939	0.7967	0.7995	0.8023	0.8051	0.8078	0.8106	0.8133
0.9	0.8159	0.8186	0.8212	0.8238	0.8264	0.8289	0.8315	0.8340	0.8365	0.8389
1.0	0.8413	0.8438	0.8461	0.8485	0.8508	0.8531	0.8554	0.8577	0.8599	0.8621
1.1	0.8643	0.8665	0.8686	0.8708	0.8729	0.8749	0.8770	0.8790	0.8810	0.8830
1.2	0.8849	0.8869	0.8888	0.8907	0.8925	0.8944	0.8962	0.8980	0.8997	0.9015
1.3	0.9032	0.9049	0.9066	0.9082	0.9099	0.9115	0.9131	0.9147	0.9162	0.9177
1.4	0.9192	0.9207	0.9222	0.9236	0.9251	0.9265	0.9279	0.9393	0.9306	0.9319
1.5	0.9332	0.9345	0.9357	0.9370	0.9382	0.9394	0.9406	0.9418	0.9429	0.9441
1.6	0.9452	0.9463	0.9474	0.9484	0.9495	0.9505	0.9515	0.9525	0.9535	0.9545
1.7	0.9554	0.9564	0.9573	0.9582	0.9591	0.9599	0.9608	0.9616	0.9625	0.9633
1.8	0.9641	0.9649	0.9656	0.9664	0.9671	0.9678	0.9686	0.9693	0.9699	0.9706
1.9	0.9713	0.9719	0.9726	0.9732	0.9738	0.9744	0.9750	0.9756	0.9761	0.9767
2.0	0.9772	0.9778	0.9783	0.9788	0.9793	0.9798	0.9803	0.9808	0.9812	0.9817
2.1	0.9821	0.9826	0.9830	0.9834	0.9838	0.9842	0.9846	0.9850	0.9854	0.9857
2.2	0.9861	0.9864	0.9868	0.9871	0.9875	0.9878	0.9881	0.9884	0.9887	0.9890
2.3	0.9893	0.9896	0.9898	0.9901	0.9904	0.9906	0.9909	0.9911	0.9913	0.9916
2.4	0.9918	0.9920	0.9922	0.9925	0.9927	0.9929	0.9931	0.9932	0.9934	0.9936
2.5	0.9938	0.9940	0.9941	0.9943	0.9945	0.9946	0.9948	0.9949	0.9951	0.9952
2.6	0.9953	0.9955	0.9956	0.9957	0.9959	0.9960	0.9961	0.9962	0.9963	0.9964
2.7	0.9965	0.9966	0.9967	0.9968	0.9969	0.9970	0.9971	0.9972	0.9973	0.9974
2.8	0.9974	0.9975	0.9976	0.9977	0.9977	0.9978	0.9979	0.9979	0.9980	0.9981
2.9	0.9981	0.9982	0.9982	0.9983	0.9984	0.9984	0.9985	0.9985	0.9986	0.9986
3.0	0.9987	0.9987	0.9987	0.9988	0.9988	0.9989	0.9989	0.9989	0.9990	0.9990
3.1	0.9990	0.9991	0.9991	0.9991	0.9992	0.9992	0.9992	0.9992	0.9993	0.9993
3.2	0.9993	0.9993	0.9994	0.9994	0.9994	0.9994	0.9994	0.9995	0.9995	0.9995
3.3	0.9995	0.9995	0.9995	0.9996	0.9996	0.9996	0.9996	0.9996	0.9996	0.9997
3.4	0.9997	0.9997	0.9997	0.9997	0.9997	0.9997	0.9997	0.9997	0.9997	0.9998
3.5	0.9998	0.9998	0.9998	0.9998	0.9998	0.9998	0.9998	0.9998	0.9998	0.9998

Appendix Table II

N	0.5%	0.75%	1%	1.5%	2%	3%	4%
1	1.0050	1.0075	1.0100	1.0150	1.0200	1.0300	1.0400
2	1.0100	1.0151	1.0201	1.0302	1.0404	1.0609	1.0816
3	1.0151	1.0227	1.0303	1.0457	1.0612	1.0927	1.1249
4	1.0202	1.0303	1.0406	1.0614	1.0824	1.1255	1.1699
5	1.0253	1.0381	1.0510	1.0773	1.1041	1.1593	1.2167
6	1.0304	1.0459	1.0615	1.0934	1.1262	1.1941	1.2653
7	1.0355	1.0537	1.0721	1.1098	1.1487	1.2299	1.3159
8	1.0407	1.0616	1.0829	1.1265	1.1717	1.2668	1.3686
9	1.0459	1.0696	1.0937	1.1434	1.1951	1.3048	1.4233
10	1.0511	1.0776	1.1046	1.1605	1.2190	1.3439	1.4802
11	1.0564	1.0857	1.1157	1.1779	1.2434	1.3842	1.5395
12	1.0617	1.0938	1.1268	1.1956	1.2682	1.4258	1.6010
13	1.0670	1.1020	1.1381	1.2136	1.2936	1.4685	1.6651
14	1.0723	1.1103	1.1495	1.2318	1.3195	1.5126	1.7317
15	1.0777	1.1186	1.1610	1.2502	1.3459	1.5580	1.8009
16	1.0831	1.1270	1.1726	1.2690	1.3728	1.6047	1.8730
17	1.0885	1.1354	1.1843	1.2880	1.4002	1.6528	1.9479
18	1.0939	1.1440	1.1961	1.3073	1.4282	1.7024	2.0258
19	1.0994	1.1525	1.2081	1.3270	1.4568	1.7535	2.1068
20	1.1049	1.1612	1.2202	1.3469	1.4859	1.8061	2.1911
24	1.1272	1.1964	1.2697	1.4295	1.6084	2.0328	2.5633
25	1.1328	1.2054	1.2824	1.4509	1.6406	2.0938	2.6658
30	1.1614	1.2513	1.3478	1.5631	1.8114	2.4273	3.2434
36	1.1967	1.3086	1.4308	1.7091	2.0399	2.8983	4.1039
40	1.2208	1.3483	1.4889	1.8140	2.2080	3.2620	4.8010
48	1.2705	1.4314	1.6122	2.0435	2.5871	4.1323	6.5705
50	1.2832	1.4530	1.6446	2.1052	2.6916	4.3839	7.1067
60	1.3489	1.5657	1.8167	2.4432	3.2810	5.8916	10.5196

Future Value of $1.00 $V_N = \$1(1 + i)^N$

5%	6%	7%	8%	9%	10%	11%
1.0500	1.0600	1.0700	1.0800	1.0900	1.1000	1.1100
1.1025	1.1236	1.1449	1.1664	1.1881	1.2100	1.2321
1.1576	1.1910	1.2250	1.2597	1.2950	1.3310	1.3676
1.2155	1.2625	1.3108	1.3605	1.4116	1.4641	1.5181
1.2763	1.3382	1.4026	1.4693	1.5386	1.6105	1.6851
1.3401	1.4185	1.5007	1.5869	1.6771	1.7716	1.8704
1.4071	1.5036	1.6058	1.7138	1.8280	1.9487	2.0762
1.4775	1.5938	1.7182	1.8509	1.9926	2.1436	2.3045
1.5513	1.6895	1.8385	1.9990	2.1719	2.3579	2.5580
1.6289	1.7908	1.9672	2.1589	2.3674	2.5937	2.8394
1.7103	1.8983	2.1049	2.3316	2.5804	2.8531	3.1518
1.7959	2.0122	2.2522	2.5182	2.8127	3.1384	3.4984
1.8856	2.1329	2.4098	2.7196	3.0658	3.4523	3.8833
1.9799	2.2609	2.5785	2.9372	3.3417	3.7975	4.3109
2.0789	2.3967	2.7590	3.1722	3.6425	4.1772	4.7846
2.1829	2.5404	2.9522	3.4259	3.9703	4.5950	5.3109
2.2920	2.6928	3.1588	3.7000	4.3276	5.0545	5.8951
2.4066	2.8543	3.3799	3.9960	4.7171	5.5599	6.5436
2.5269	3.0256	3.6165	4.3157	5.1417	6.1159	7.2633
2.6533	3.2071	3.8697	4.6610	5.6044	6.7275	8.0623
3.2251	4.0489	5.0724	6.3412	7.9111	9.8479	12.2391
3.3864	4.2919	5.4274	6.8485	8.6231	10.8347	13.5855
4.3219	5.7435	7.6123	10.0627	13.2677	17.4494	22.8923
5.7918	8.1473	11.4239	15.9682	22.2512	30.9127	42.8181
7.0400	10.2857	14.9745	21.7245	31.4094	45.2592	65.0008
10.4013	16.3939	25.7289	40.2106	62.5852	97.0172	149.7970
11.4674	18.4201	29.4570	46.9016	74.3575	117.3908	184.5645
18.6792	32.9877	57.9464	101.2570	176.0312	304.4812	524.0562

(Continued)

Appendix Table II

N	12%	13%	14%	15%	16%	17%	18%
1	1.1200	1.1300	1.1400	1.1500	1.1600	1.1700	1.1800
2	1.2544	1.2769	1.2996	1.3225	1.3456	1.3689	1.3924
3	1.4049	1.4429	1.4815	1.5209	1.5609	1.6016	1.6430
4	1.5735	1.6305	1.6890	1.7490	1.8106	1.8739	1.9388
5	1.7623	1.8424	1.9254	2.0114	2.1003	2.1920	2.2878
6	1.9738	2.0820	2.1950	2.3131	2.4364	2.5652	2.6996
7	2.2107	2.3526	2.5023	2.6600	2.8262	3.0012	3.1855
8	2.4760	2.6584	2.8526	3.0590	3.2784	3.5115	3.7589
9	2.7731	3.0040	3.2519	3.5179	3.8030	4.1084	4.4355
10	3.1058	3.3946	3.7072	4.0456	4.4114	4.8068	5.2338
11	3.4785	3.8359	4.2262	4.6524	5.1173	5.6240	6.1759
12	3.8960	4.3345	4.8179	5.3503	5.9360	6.5801	7.2876
13	4.3635	4.8980	5.4924	6.1528	6.8858	7.6987	8.5994
14	4.8871	5.5348	6.2613	7.0757	7.9875	9.0075	10.1472
15	5.4736	6.2543	7.1379	8.1371	9.2655	10.5387	11.9737
16	6.1304	7.0673	8.1372	9.3576	10.7480	12.3303	14.1290
17	6.8660	7.9861	9.2765	10.7613	12.4677	14.4265	16.6722
18	7.6900	9.0243	10.5752	12.3755	14.4625	16.8790	19.6733
19	8.6128	10.1974	12.0557	14.2318	16.7765	19.7484	23.2144
20	9.6463	11.5231	13.7435	16.3665	19.4608	23.1056	27.3930
24	15.1786	18.7881	23.2122	28.6252	35.2364	43.2973	53.1090
25	17.0000	21.2305	26.4619	32.9190	40.8742	50.6578	62.6686
30	29.9599	39.1159	50.9502	66.2118	85.8499	111.0647	143.3706
36	59.1356	81.4374	111.8342	153.1519	209.1643	284.8991	387.0368
40	93.0508	132.7816	188.8835	267.8635	378.7212	533.8687	750.3783
48	230.3908	352.7816	538.8065	819.4007	1,241.6051	1,874.6550	2,820.5665
50	289.0015	450.7359	700.2330	1,083.6574	1,670.7038	2,566.2153	3,927.3559
60	897.5950	1,530.0535	2,595.9187	4,383.9987	7,370.2014	12,335.3565	20,555.1400

(Continued)

19%	20%	21%	22%	23%	24%	25%
1.1900	1.2000	1.2100	1.2200	1.2300	1.2400	1.2500
1.4161	1.4400	1.4641	1.4884	1.5129	1.5376	1.5625
1.6852	1.7280	1.7716	1.8158	1.8609	1.9066	1.9531
2.0053	2.0736	2.1436	2.2153	2.2889	2.3642	2.4414
2.3864	2.4883	2.5937	2.7027	2.8153	2.9316	3.0518
2.8398	2.9860	3.1384	3.2973	3.4628	3.6352	3.8147
3.3793	3.5832	3.7975	4.0227	4.2593	4.5077	4.7684
4.0214	4.2998	4.5950	4.9077	5.2389	5.5895	5.9605
4.7854	5.1598	5.5599	5.9874	6.4439	6.9310	7.4506
5.6947	6.1917	6.7275	7.3046	7.9259	8.5944	9.3132
6.7767	7.4301	8.1403	8.9117	9.7489	10.6571	11.6415
8.0642	8.9161	9.8497	10.8772	11.9912	13.2148	14.5519
9.5964	10.6993	11.9182	13.2641	14.7491	16.3863	18.1899
11.4198	12.8392	14.4210	16.1822	18.1414	20.3191	22.7374
13.5895	15.4070	17.4494	19.7423	22.3140	25.1956	28.4217
16.1715	18.4884	21.1138	24.0856	27.4462	31.2426	35.5271
19.2441	22.1861	25.5477	29.3844	33.7588	38.7408	44.4089
22.9005	26.6233	30.9127	35.8490	41.5233	48.0386	55.5112
27.2516	31.9480	37.4043	43.7358	51.0737	59.5679	69.3889
32.4294	38.3376	45.2593	53.3576	62.8206	73.8641	86.7362
65.0320	79.4968	97.0172	118.2050	143.7880	174.6306	211.7582
77.3881	95.3962	117.3909	144.2101	176.8593	216.5420	264.6978
184.6753	273.3763	304.4816	389.7579	494.9129	634.8199	807.7936
524.4337	708.8019	955.5938	1,285.1502	1,724.1856	2,307.7070	3,081.4879
1,051.6675	1,469.7716	2,048.4002	2,847.0378	3,946.4305	5,455.9126	7,523.1638
4,229.1603	6,319.7487	9,412.3437	13,972.4277	20,674.9920	30,495.8602	44,841.5509
5,988.9139	9,100.4382	13,780.6123	20,796.5615	31,279.1953	46,890.4346	70,064.9232
23,104.9709	56,347.5144	92,709.0688	151,911.2161	247,917.2160	402,996.3472	652,530.4468

Appendix Table III

N	0.5%	0.75%	1%	1.5%	2%	3%	4%
1	0.9950	0.9926	0.9901	0.9852	0.9804	0.9709	0.9615
2	0.9901	0.9852	0.9803	0.9707	0.9612	0.9426	0.9246
3	0.9851	0.9778	0.9706	0.9563	0.9423	0.9151	0.8890
4	0.9802	0.9706	0.9610	0.9422	0.9238	0.8885	0.8548
5	0.9754	0.9633	0.9515	0.9283	0.9057	0.8626	0.8219
6	0.9705	0.9562	0.9420	0.9145	0.8880	0.8375	0.7903
7	0.9657	0.9490	0.9327	0.9010	0.8706	0.8131	0.7599
8	0.9609	0.9420	0.9235	0.8877	0.8535	0.7894	0.7307
9	0.9561	0.9350	0.9143	0.8746	0.8368	0.7664	0.7026
10	0.9513	0.9280	0.9053	0.8617	0.8203	0.7441	0.6756
11	0.9466	0.9211	0.8963	0.8489	0.8043	0.7224	0.6496
12	0.9419	0.9142	0.8874	0.8364	0.7885	0.7014	0.6246
13	0.9372	0.9074	0.8787	0.8240	0.7730	0.6810	0.6006
14	0.9326	0.9007	0.8700	0.8118	0.7579	0.6611	0.5775
15	0.9279	0.8940	0.8613	0.7999	0.7430	0.6419	0.5553
16	0.9233	0.8873	0.8528	0.7880	0.7284	0.6232	0.5339
17	0.9187	0.8807	0.8444	0.7764	0.7142	0.6050	0.5134
18	0.9141	0.8742	0.8360	0.7649	0.7002	0.5874	0.4936
19	0.9096	0.8676	0.8277	0.7536	0.6863	0.5703	0.4746
20	0.9051	0.8612	0.8195	0.7425	0.6730	0.5537	0.4564
24	0.8872	0.8358	0.7876	0.6995	0.6217	0.4919	0.3901
25	0.8828	0.8296	0.7798	0.6892	0.6095	0.4776	0.3751
30	0.8610	0.7992	0.7419	0.6398	0.5521	0.4120	0.3083
36	0.8356	0.7641	0.6989	0.5851	0.4902	0.3450	0.2437
40	0.8191	0.7416	0.6717	0.5513	0.4529	0.3066	0.2083
48	0.7871	0.6986	0.6203	0.4894	0.3865	0.2420	0.1522
50	0.7793	0.6883	0.6080	0.4750	0.3715	0.2281	0.1407
60	0.7414	0.6387	0.5504	0.4093	0.3048	0.1697	0.0951

Present Value of $1.00 $V_0 = \$1(1 + i)^{-N}$

5%	6%	7%	8%	9%	10%	11%
0.9524	0.9434	0.9346	0.9259	0.9174	0.9091	0.9009
0.9070	0.8900	0.8734	0.8573	0.8417	0.8264	0.8116
0.8638	0.8396	0.8163	0.7938	0.7722	0.7513	0.7312
0.8227	0.7921	0.7629	0.7350	0.7084	0.6830	0.6587
0.7835	0.7473	0.7130	0.6806	0.6499	0.6209	0.5935
0.7462	0.7050	0.6663	0.6302	0.5963	0.5645	0.5346
0.7107	0.6651	0.6227	0.5835	0.5470	0.5132	0.4817
0.6768	0.6274	0.5820	0.5403	0.5019	0.4665	0.4339
0.6446	0.5919	0.5439	0.5002	0.4604	0.4241	0.3909
0.6139	0.5584	0.5083	0.4632	0.4224	0.3855	0.3522
0.5847	0.5268	0.4751	0.4289	0.3875	0.3505	0.3173
0.5568	0.4970	0.4440	0.3971	0.3555	0.3186	0.2858
0.5303	0.4688	0.4150	0.3677	0.3262	0.2897	0.2575
0.5051	0.4423	0.3878	0.3405	0.2992	0.2633	0.2320
0.4810	0.4173	0.3624	0.3152	0.2745	0.2394	0.2090
0.4581	0.3936	0.3387	0.2919	0.2519	0.2176	0.1883
0.4363	0.3714	0.3166	0.2703	0.2311	0.1978	0.1696
0.4155	0.3503	0.2959	0.2502	0.2120	0.1799	0.1528
0.3957	0.3305	0.2765	0.2317	0.1945	0.1635	0.1377
0.3769	0.3118	0.2584	0.2145	0.1784	0.1486	0.1240
0.3101	0.2470	0.1971	0.1577	0.1264	0.1015	0.0817
0.2953	0.2330	0.1842	0.1460	0.1160	0.0923	0.0736
0.2314	0.1741	0.1314	0.0994	0.0754	0.0573	0.0437
0.1727	0.1227	0.0875	0.0626	0.0449	0.0323	0.0234
0.1420	0.0972	0.0668	0.0460	0.0318	0.0221	0.0154
0.0961	0.0610	0.0389	0.0249	0.0160	0.0103	0.0067
0.0872	0.0543	0.0339	0.0213	0.0134	0.0085	0.0054
0.0535	0.0303	0.0173	0.0099	0.0057	0.0033	0.0019

(Continued)

Appendix Table III

N	12%	13%	14%	15%	16%	17%	18%
1	0.8929	0.8850	0.8772	0.8696	0.8621	0.8547	0.8475
2	0.7972	0.7831	0.7695	0.7561	0.7432	0.7305	0.7182
3	0.7118	0.6931	0.6750	0.6575	0.6407	0.6244	0.6086
4	0.6355	0.6133	0.5921	0.5718	0.5523	0.5337	0.5158
5	0.5674	0.5428	0.5194	0.4972	0.4761	0.4561	0.4371
6	0.5066	0.4803	0.4556	0.4323	0.4104	0.3898	0.3704
7	0.4523	0.4251	0.3996	0.3759	0.3538	0.3332	0.3139
8	0.4039	0.3762	0.3506	0.3269	0.3050	0.2848	0.2660
9	0.3606	0.3329	0.3075	0.2843	0.2630	0.2434	0.2255
10	0.3220	0.2946	0.2697	0.2472	0.2267	0.2080	0.1911
11	0.2875	0.2607	0.2366	0.2149	0.1954	0.1778	0.1619
12	0.2567	0.2307	0.2076	0.1869	0.1685	0.1520	0.1372
13	0.2292	0.2042	0.1821	0.1625	0.1452	0.1299	0.1163
14	0.2046	0.1807	0.1597	0.1413	0.1252	0.1110	0.0985
15	0.1827	0.1599	0.1401	0.1229	0.1079	0.0949	0.0835
16	0.1631	0.1415	0.1229	0.1069	0.0930	0.0811	0.0708
17	0.1456	0.1252	0.1078	0.0929	0.0802	0.0693	0.0600
18	0.1300	0.1108	0.0946	0.0808	0.0691	0.0592	0.0508
19	0.1161	0.0981	0.0829	0.0703	0.0596	0.0506	0.0431
20	0.1037	0.0868	0.0728	0.0611	0.0514	0.0433	0.0365
24	0.0659	0.0532	0.0431	0.0349	0.0284	0.0231	0.0188
25	0.0588	0.0471	0.0378	0.0304	0.0245	0.0197	0.0160
30	0.0334	0.0256	0.0196	0.0151	0.0116	0.0090	0.0070
36	0.0169	0.0123	0.0089	0.0065	0.0048	0.0035	0.0026
40	0.0107	0.0075	0.0053	0.0037	0.0026	0.0019	0.0013
48	0.0043	0.0028	0.0019	0.0012	0.0008	0.0005	0.0004
50	0.0035	0.0022	0.0014	0.0009	0.0006	0.0004	0.0003
60	0.0011	0.0007	0.0004	0.0002	0.0001	0.0001	0.0000

(Continued)

19%	20%	21%	22%	23%	24%	25%
0.8403	0.8333	0.8264	0.8197	0.8130	0.8065	0.8000
0.7062	0.6944	0.6830	0.6719	0.6610	0.6504	0.6400
0.5934	0.5787	0.5645	0.5507	0.5374	0.5245	0.5120
0.4987	0.4823	0.4665	0.4514	0.4369	0.4230	0.4096
0.4190	0.4019	0.3855	0.3700	0.3552	0.3411	0.3277
0.3521	0.3349	0.3186	0.3033	0.2888	0.2751	0.2621
0.2959	0.2791	0.2633	0.2486	0.2348	0.2218	0.2097
0.2487	0.2326	0.2176	0.2039	0.1909	0.1789	0.1678
0.2090	0.1938	0.1799	0.1670	0.1552	0.1443	0.1342
0.1756	0.1615	0.1486	0.1369	0.1262	0.1164	0.1074
0.1476	0.1346	0.1228	0.1122	0.1026	0.0938	0.0859
0.1240	0.1122	0.1015	0.0920	0.0834	0.0757	0.0687
0.1042	0.0935	0.0839	0.0754	0.0678	0.0610	0.0550
0.0876	0.0779	0.0693	0.0618	0.0551	0.0492	0.0440
0.0736	0.0649	0.0573	0.0507	0.0448	0.0397	0.0352
0.0618	0.0541	0.0474	0.0415	0.0364	0.0320	0.0281
0.0520	0.0451	0.0391	0.0340	0.0296	0.0258	0.0225
0.0437	0.0376	0.0323	0.0279	0.0241	0.0208	0.0180
0.0367	0.0313	0.0267	0.0229	0.0196	0.0168	0.0144
0.0308	0.0261	0.0221	0.0187	0.0159	0.0135	0.0115
0.0154	0.0126	0.0103	0.0085	0.0070	0.0057	0.0047
0.0129	0.0105	0.0085	0.0069	0.0057	0.0046	0.0038
0.0054	0.0042	0.0033	0.0026	0.0020	0.0016	0.0012
0.0019	0.0014	0.0010	0.0008	0.0006	0.0004	0.0001
0.0010	0.0007	0.0005	0.0004	0.0003	0.0002	0.0000
0.0002	0.0002	0.0001	0.0001	0.0000	0.0000	0.0000
0.0002	0.0001	0.0001	0.0000	0.0000	0.0000	0.0000
0.0000	0.0000	0.0000	0.0000	0.0000	0.0000	0.0000

Appendix Table IV

N	0.5%	0.75%	1%	1.5%	2%	3%	4%
1	1.0000	1.0000	1.0000	1.0000	1.0000	1.0000	1.0000
2	2.0050	2.0075	2.0100	2.0150	2.0200	2.0300	2.0400
3	3.0150	3.0226	3.0301	3.0452	3.0604	3.0909	3.1216
4	4.0301	4.0452	4.0604	4.0909	4.1216	4.1836	4.2465
5	5.0503	5.0755	5.1010	5.1523	5.2040	5.3091	5.4163
6	6.0755	6.1136	6.1520	6.2295	6.3081	6.4684	6.6330
7	7.1059	7.1595	7.2135	7.3230	7.4343	7.6625	7.8983
8	8.1414	8.2132	8.2857	8.4328	8.5830	8.8923	9.2142
9	9.1821	9.2748	9.3605	9.5593	9.7546	10.1591	10.5828
10	10.2280	10.3443	10.4622	10.7027	10.9497	11.4639	12.0061
11	11.2792	11.4219	11.5688	11.8633	12.1687	12.8078	13.4864
12	12.3356	12.5075	12.6825	13.0412	13.4121	14.1920	15.0258
13	13.3972	13.6014	13.8093	14.2368	14.6803	15.6178	16.6268
14	14.4642	14.7034	14.9474	15.4504	15.9739	17.0863	18.2918
15	15.5365	15.8137	16.0969	16.6821	17.2934	18.5989	20.0236
16	16.6142	16.9323	17.2579	17.9324	18.6393	20.1569	21.8245
17	17.6973	18.0593	18.4304	19.2041	20.0121	21.7616	23.6975
18	18.7858	19.1947	19.6147	20.4894	21.4123	23.4144	25.6454
19	19.8797	20.3387	20.8109	21.7967	22.8406	25.1169	27.6712
20	20.9791	21.4912	22.0190	23.1237	24.2974	26.8704	29.7781
24	25.4320	26.1885	26.9735	28.6335	30.4219	34.4265	39.0826
25	26.5591	27.3848	28.2432	30.0630	32.0303	36.4593	41.6459
30	32.2800	33.5029	34.7849	37.5387	40.5681	47.5754	56.0849
36	39.3361	41.1527	43.0769	47.2760	51.9944	63.2759	77.5983
40	44.1588	46.4465	48.8864	54.2679	60.4020	75.4013	95.0255
48	54.0978	57.5207	61.2226	69.5652	79.3535	104.4084	139.2632
50	56.6452	60.3943	64.4632	73.6828	84.5794	112.7969	152.6671
60	69.7700	75.4241	81.6697	96.2147	114.0515	163.0534	237.9907

Future Value of a Uniform Series $V_N = \$1\{[(1+i)^N - 1] \div i\}$

5%	6%	7%	8%	9%	10%	11%
1.0000	1.0000	1.0000	1.0000	1.0000	1.0000	1.0000
2.0500	2.0600	2.0700	2.0080	2.0900	2.1000	2.1100
3.1525	3.1836	3.2149	3.2464	3.2781	3.3100	3.3421
4.3101	4.3746	4.4399	4.5061	4.5731	4.6410	4.7097
5.5276	5.6371	5.7507	5.8666	5.9847	6.1051	6.2278
6.8019	6.9753	7.1533	7.3359	7.5233	7.7156	7.9129
8.1420	8.3938	8.6540	8.9228	9.2004	9.4872	9.7833
9.5491	9.8975	10.2598	10.6366	11.0285	11.4359	11.8594
11.0266	11.4913	11.9780	12.4876	13.0210	13.5795	14.1640
12.5779	13.1808	13.8164	14.4866	15.1929	15.9374	16.7220
14.2068	14.9716	15.7836	16.6455	17.5603	18.5312	19.5614
15.9171	16.8699	17.8885	18.9771	20.1407	21.3843	22.7132
17.1730	18.8821	20.1406	21.4953	22.9534	24.5227	26.2116
19.5986	21.0151	22.5505	24.2149	26.0192	27.9750	30.0949
21.5786	23.2760	25.1290	27.1521	29.3609	31.7725	34.4054
23.6575	25.6725	27.8881	30.3243	33.0034	35.9497	39.1899
25.8404	28.2129	30.8402	33.7502	36.9737	40.5447	44.5008
28.1324	30.9057	33.9990	37.4502	41.3013	45.5992	50.3959
30.5390	33.7600	37.3790	41.4463	46.0185	51.1591	56.9395
33.0660	36.7856	40.9955	45.7620	51.1601	57.2750	64.2028
44.5020	50.8156	58.1767	66.7648	76.7898	88.4973	102.1742
47.7271	54.8645	63.2490	73.1059	84.7009	98.3471	114.4133
66.4388	79.0582	74.4608	113.2832	136.3075	164.4940	199.0209
95.8363	119.1209	148.9135	187.1021	236.1247	299.1268	380.1644
120.7998	154.7620	199.6351	259.0565	337.8824	442.5926	581.8261
188.0254	256.5645	353.2701	490.1322	684.2804	960.1723	1,352.6996
209.3480	290.3359	406.5289	573.7702	815.0836	1,163.9085	1,668.7712
353.5837	533.1282	613.5204	1,235.2133	1,944.7921	3,034.8164	4,755.0658

(Continued)

Appendix Table IV

N	12%	13%	14%	15%	16%	17%	18%
1	1.0000	1.0000	1.0000	1.0000	1.0000	1.0000	1.0000
2	2.1200	2.1300	2.1400	2.1500	2.1600	2.1700	2.1800
3	3.3744	3.4069	3.4396	3.4725	3.5056	3.5389	3.5724
4	4.7793	4.8498	4.9211	4.9934	5.0665	5.1405	5.2154
5	6.3528	6.4803	6.6101	6.7424	6.8771	7.0144	7.1542
6	8.1152	8.3227	8.5355	8.7537	8.9775	9.2068	9.4420
7	10.0890	10.4047	10.7305	11.0668	11.4139	11.7720	12.1415
8	12.2997	12.7573	13.2328	13.7268	14.2401	14.7733	15.3270
9	14.7757	15.4157	16.0853	16.7858	17.5185	18.2847	19.0859
10	17.5487	18.4197	19.3373	20.3037	21.3215	22.3931	23.5213
11	20.6546	21.8143	23.0445	24.3493	25.7329	27.1999	28.7551
12	24.1331	25.6502	27.2707	29.0017	30.8502	32.8239	34.9311
13	28.0291	29.9847	32.0887	34.3519	36.7562	39.4040	42.2187
14	32.3926	34.8827	37.5811	40.5047	43.6720	47.1027	50.8180
15	37.2797	40.4175	43.8424	47.5804	51.6595	56.1101	60.9653
16	42.7533	46.6717	50.9804	55.7175	60.9250	66.6488	72.9390
17	48.8837	53.7391	59.1176	65.0751	71.6730	78.9792	87.0680
18	55.7497	61.7251	68.3941	75.8364	84.1407	93.4056	103.7403
19	63.4397	70.7494	78.9692	88.2118	98.6032	110.2846	123.4135
20	72.0524	80.9468	91.0249	102.4436	115.3797	130.0329	146.6280
24	118.1552	136.8315	158.6586	184.1678	213.9776	248.8076	289.4945
25	133.3339	155.6196	181.8708	212.7930	249.2140	292.1049	342.6035
30	241.3327	293.1992	356.7868	434.7451	530.3117	647.4391	790.9480
36	484.4631	618.7493	791.6729	1,014.3457	1,301.0270	1,669.9945	2,144.6459
40	767.0914	1,013.7042	1,342.0251	1,779.0903	2,360.7572	3,134.5218	4,163.2130
48	1,911.8980	2,707.6334	3,841.4753	5,456.0047	7,753.7817	11,021.5002	15,664.2586
50	2,400.0182	3,459.5071	4,994.5213	7,217.7163	10,435.6488	15,089.5017	21,813.0937
60	7,471.6411	11,761.9498	18,535.1333	29,219.9916	46,057.5085	72,555.0381	114,189.6664

(Continued)

19%	20%	21%	22%	23%	24%	25%
1.0000	1.0000	1.0000	1.0000	1.0000	1.0000	1.0000
2.1900	2.2000	2.2100	2.2200	2.2300	2.2400	2.2500
3.6061	3.6400	3.6741	3.7084	3.7429	3.7776	3.8125
5.2913	5.3680	5.4457	5.5242	5.6038	5.6842	5.7656
7.2966	7.4416	7.5892	7.7396	7.8926	8.0484	8.2070
9.6830	9.9299	10.1830	10.4423	10.7079	10.9801	11.2588
12.5227	12.9159	13.3214	13.7396	14.1708	14.6153	15.0735
15.9020	16.4991	17.1189	17.7623	18.4300	19.1229	19.8419
19.9234	20.7989	21.7139	22.6700	23.6690	24.7125	25.8023
24.7089	25.9587	27.2738	28.6574	30.1128	31.6434	33.2529
30.4035	32.1504	34.0013	35.9620	38.0388	40.2379	42.5661
37.1802	39.5805	42.1416	44.8737	47.7877	50.8950	54.2077
45.2445	48.4966	51.9913	55.7459	59.7788	64.1097	68.7596
54.8409	59.1959	63.9095	69.0100	74.5280	80.4961	86.9495
66.2607	72.0351	78.3305	85.1922	92.6694	100.8151	109.6868
79.8502	87.4421	95.7799	104.9345	114.9834	126.0108	138.1085
96.0218	105.9306	116.8937	129.0201	142.4295	157.2534	173.6357
115.2659	128.1167	142.4413	158.4045	176.1883	195.9942	218.0446
138.1664	154.7400	173.3540	194.2535	217.7116	244.0328	273.5558
165.4180	186.6880	210.7584	237.9893	268.7853	303.6006	342.9447
337.0105	392.4842	457.2249	532.7501	620.8174	723.4610	843.0329
402.0425	471.9811	554.2422	650.9551	764.6054	898.0916	1,054.7912
966.7122	1,181.8816	1,445.1507	1,767.0813	2,160.4907	2,640.9164	3,227.1743
2,754.9143	3,539.0094	4,545.6648	5,837.0466	7,492.1113	9,611.2791	12,321.9516
5,529.8299	7,343.8578	9,749.5248	12,936.5353	17,154.0456	22,728.8026	30,088.6554
22,253.4753	31,593.7436	44,815.9222	63,506.4897	89,886.9215	127,061.9174	179,362.2034
31,515.3363	45,497.1908	65,617.2016	94,525.2793	135,992.1536	195,372.6442	280,255.6929
179,494.5838	281,732.5718	441,466.9943	690,500.9823	1,077,896.5910	1,679,147.2800	2,610,117.7870

Appendix Table V

N	0.5%	0.75%	1%	1.5%	2%	3%	4%
1	0.9950	0.9926	0.9901	0.9852	0.9804	0.9709	0.9615
2	1.9851	1.9777	1.9704	1.9559	1.9416	1.9135	1.8861
3	2.9702	2.9556	2.9410	2.9122	2.8839	2.8286	2.7751
4	3.9505	3.9261	3.9020	3.8544	3.8077	3.7171	3.6299
5	4.9259	4.8894	4.8534	4.7826	4.7135	4.5797	4.4518
6	5.8964	5.8456	5.7955	5.6972	5.6014	5.4172	5.2421
7	6.8621	6.7946	6.7282	6.5982	6.4720	6.2303	6.0021
8	7.8230	7.7366	7.6517	7.4859	7.3255	7.0197	6.7327
9	8.7791	8.6716	8.5660	8.3605	8.1622	7.7861	7.4353
10	9.7304	9.5996	9.4713	9.2222	8.8926	8.5302	8.1109
11	10.6770	10.5207	10.3676	10.0711	9.7868	9.2526	8.7605
12	11.6189	11.4349	11.2551	10.9075	10.5753	9.9540	9.3851
13	12.5562	12.3423	12.1337	11.7315	11.3484	10.6350	9.9856
14	13.4887	13.2430	13.0037	12.5434	12.1062	11.2961	10.5631
15	14.4166	14.1370	13.8651	13.3432	12.8493	11.9379	11.1184
16	15.3399	15.0423	14.7179	14.1313	13.5777	12.5611	11.6523
17	16.2586	15.9050	15.5623	14.9076	14.2919	13.1661	12.1657
18	17.1728	16.7792	16.3983	15.6726	14.9920	13.7535	12.6593
19	18.0824	17.6468	17.2260	16.4262	15.6785	14.3238	13.1330
20	18.9874	18.5080	18.0456	17.1686	16.3514	14.8775	13.5903
24	22.5629	21.8891	21.2434	20.0304	18.9139	16.9355	15.2470
25	23.4456	22.7188	22.0232	20.7196	19.5235	17.4131	15.6221
30	27.7941	26.7751	25.8077	24.0158	22.3965	19.6004	17.2920
36	32.8710	31.4468	30.1075	27.6607	25.4888	21.8323	18.9083
40	36.1722	34.4469	32.8347	29.9158	27.3555	23.1148	19.7928
48	42.5803	40.1848	37.9740	34.0426	30.6731	25.2667	21.1951
50	44.1428	41.5664	39.1961	35.9997	31.4236	25.7298	21.4822
60	51.7256	48.1734	44.9550	39.3803	34.7609	27.6756	22.6235

Present Value of a Uniform Series $V_0 = \$1\{[1 - (1 + i)^{-N}] \div i\}$

5%	6%	7%	8%	9%	10%	11%
0.9524	0.9434	0.9346	0.9259	0.9174	0.9091	0.9009
1.8594	1.8334	1.8080	1.7833	1.7591	1.7355	1.7125
2.7232	2.6730	2.6243	2.5771	2.5313	2.4869	2.4437
3.5460	3.4651	3.3872	3.3121	3.2397	3.1699	3.1024
4.3295	4.2124	4.1002	3.9927	3.8897	3.7908	3.6959
5.0757	4.9173	4.7665	4.6229	4.4859	4.3553	4.2305
5.7864	5.5824	5.3893	5.2064	5.0330	4.8684	4.7122
6.4632	6.2098	5.9713	5.7466	5.5348	5.3349	5.1461
7.1078	6.8017	6.5152	6.2469	5.9952	5.7590	5.5370
7.7217	7.3601	7.0236	6.7101	6.4177	6.1446	5.8892
8.3064	7.8869	7.4987	7.1390	6.8052	6.4951	6.2065
8.8633	8.3838	7.9427	7.5361	7.1607	6.8137	6.4924
9.3936	8.8527	8.3577	7.9038	7.4869	7.1034	6.7499
9.8986	9.2950	8.7455	8.2442	7.7862	7.3667	6.9819
10.3797	9.7122	9.1079	8.5595	8.0607	7.6061	7.1909
10.8378	10.1059	9.4466	8.8514	8.3126	7.8237	7.3792
11.2741	10.4773	9.7632	9.1216	8.5436	8.0216	7.5488
11.6986	10.8276	10.0591	9.3719	8.7556	8.2014	7.7016
12.0853	11.1581	10.3356	9.6036	8.9501	8.3649	7.8393
12.4622	11.4699	10.5940	9.8181	9.1285	8.5136	7.9633
13.7986	12.5504	11.4693	10.5288	9.7066	8.9847	8.3481
14.0939	12.7834	11.6536	10.6748	9.8226	9.0770	8.4217
15.3725	13.7648	12.4090	11.2578	10.2737	9.4269	8.6938
16.5469	14.6210	13.0352	11.7172	10.6118	9.6765	8.8786
17.1591	15.0463	13.3317	11.9246	10.7574	9.7791	8.9511
18.0772	15.6500	13.7305	12.1891	10.9336	9.8969	9.0302
18.2559	15.7619	13.8007	12.2335	10.9617	9.9148	9.0417
18.9293	16.1614	14.0392	12.3766	11.0480	9.9672	9.0736

(Continued)

Appendix Table V

N	12%	13%	14%	15%	16%	17%	18%
1	0.8929	0.8850	0.8772	0.8696	0.8621	0.8547	0.8475
2	1.6901	1.6681	1.6467	1.6257	1.6052	1.5852	1.5656
3	2.4018	2.3612	2.3216	2.2832	2.2459	2.2096	2.1743
4	3.0373	2.9745	2.9137	2.8550	2.7982	2.7432	2.6901
5	3.6048	3.5172	3.4331	3.3522	3.2743	3.1993	3.1272
6	4.1114	3.9975	3.8889	3.7845	3.6847	3.5892	3.4976
7	4.5638	4.4226	4.2883	4.1604	4.0386	3.9224	3.8115
8	4.9676	4.7988	4.6389	4.4873	4.3436	4.2072	4.0776
9	5.3282	5.1317	4.9464	4.7716	4.6065	4.4506	4.3030
10	5.6502	5.4262	5.2161	5.0188	4.8332	4.6586	4.4941
11	5.9377	5.6868	5.4527	5.2337	5.0286	4.8364	4.6560
12	6.1944	5.9176	5.6603	5.4206	5.1971	4.9884	4.7932
13	6.4235	6.1218	5.8424	5.5831	5.3423	5.1183	4.9095
14	6.6282	6.3025	6.0021	5.7245	5.4675	5.2293	5.0081
15	6.8109	6.4624	6.1422	5.8474	5.5755	5.3242	5.0916
16	6.9740	6.6039	6.2651	5.9542	5.6685	5.4053	5.1624
17	7.1196	6.7291	6.3729	6.0472	5.7487	5.4746	5.2223
18	7.2497	6.8399	6.4674	6.1280	5.8178	5.5339	5.2732
19	7.3658	6.9380	6.5504	6.1982	5.8775	5.5845	5.3162
20	7.4694	7.0248	6.6231	6.2593	5.9288	5.6278	5.3527
24	7.7843	7.2829	6.8351	7.4338	6.0726	5.7465	5.4509
25	7.8431	7.3300	6.8729	6.4641	6.0971	5.7662	5.4669
30	8.0552	7.4959	7.0027	6.5660	6.1772	5.8294	5.5168
36	8.1924	7.5979	7.0790	6.6231	6.2201	5.8617	5.5412
40	8.2438	7.6344	7.1050	6.6418	6.2335	5.8713	5.5482
48	8.2972	7.6705	7.1296	6.6585	6.2450	5.8792	5.5536
50	8.3045	7.6752	7.1327	6.6605	6.2463	5.8801	5.5541
60	8.3240	7.6873	7.1401	6.6651	6.2492	5.8819	5.5553

(Continued)

19%	20%	21%	22%	23%	24%	25%
0.8402	0.8333	0.8264	0.8197	0.8130	0.8065	0.8000
1.5465	1.5278	1.5095	1.4915	1.4740	1.4568	1.4400
2.1399	2.1065	2.0739	2.0422	2.0114	1.9813	1.9520
2.6386	2.5887	2.5404	2.4936	2.4483	2.4043	2.3616
3.0576	2.9906	2.9260	2.8636	2.8035	2.7454	2.6893
3.4098	3.3255	3.2446	3.1669	3.0923	3.0205	2.9514
3.7057	3.6046	3.5079	3.4155	3.3270	3.2423	3.1611
3.9544	3.8372	3.7256	3.6193	3.5179	3.4212	3.3289
4.1633	4.0310	3.9054	3.7863	3.6731	3.5655	3.4631
4.3389	4.1925	4.0541	3.9232	3.7993	3.6819	3.5705
4.4865	4.3271	4.1769	4.0354	3.9018	3.7757	3.6564
4.6105	4.4392	4.2784	4.1274	3.9852	3.8514	3.7251
4.7147	4.5327	4.3624	4.2028	4.0530	3.9124	3.7801
4.8023	4.6106	4.4317	4.2646	4.1082	3.9616	3.8241
4.8759	4.6755	4.4890	4.3152	4.1530	4.0013	3.8593
4.9377	4.7296	4.5364	4.3567	4.1894	4.0333	3.8874
4.9897	4.7746	4.5755	4.3908	4.2190	4.0591	3.9099
5.0333	4.8122	4.6079	4.4187	4.2431	4.0799	3.9279
5.0700	4.8435	4.6346	4.4415	4.2627	4.0967	3.9424
5.1009	4.8696	4.6567	4,4603	4.2786	4.1103	3.9539
5.1822	4.9371	4.7128	4.5070	4.3176	4.1428	3.9811
5.1951	4.9476	4.7213	4.5139	4.3232	4.1474	3.9849
5.2347	4.9789	4.7463	4.5338	4.3391	4.1601	3.9950
5.2531	4.9929	4.7569	4.5419	4.3453	4.1649	3.9987
5.2582	4.9966	4.7596	4.5439	4.3467	4.1659	3.9995
5.2619	4.9992	4.7614	4.5451	4.3476	5.1665	3.9999
5.2623	4.9995	4.7616	4.5452	4.3477	4.1666	3.9999
5.2630	4.9999	4.7619	5.5454	4.3478	4.1667	4.0000

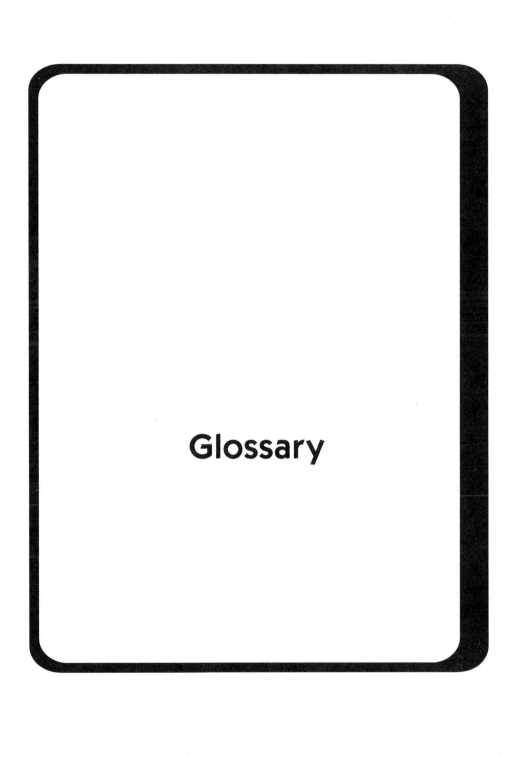

Glossary

TERMS USED IN AGRICULTURAL FINANCE

Abstract: A written, chronological summary of all deeds, mortgages, foreclosures, and other transactions affecting the title to a tract of land. Also called *abstract of title*.

Accelerated cost recovery system (ACRS): A method of determining depreciation charges for income tax reporting in the United States. The ACRS is based on (1) the 200 percent declining balance method with a switch to straight-line depreciation at a time to maximize the depreciation charge, and (2) substantially shortened lives of most qualified property and equipment.

Accrual accounting: A method of accounting in which income is reported in the year it is earned and expenses are reported in the year they are incurred. Changes in inventories, accounts receivable, prepaid expenses, accounts payable, accrued interest, and other accrual adjustments are accounted for under the accrual method.

Accrued interest: A balance sheet item that represents interest expense (income) that has been incurred (earned) but not paid (received). Accrued interest is generally reported with current liabilities.

Actuarial rate: The interest rate per conversion period that equates the present value of all cash inflows associated with a loan transaction to the present value of all cash outflows.

Add-on interest: A method of determining interest on a loan in which interest is calculated on the original loan balance for the entire period of the loan. The total interest due over the life of the loan is calculated by multiplying the original principal amount by the interest rate. This result is added to the original principal amount and divided into equal payments that are paid over the life of the loan.

Adjustable rate loan: A type of loan that has provisions to change the interest rate at pre-specified points in time based on changes in the market index, a lender's cost of funds, or other factors as determined by the lender.

Agent: A person (firm) authorized by another person (firm) to perform activities. The agent acts on behalf of his or her principal but does not have title to the property of his or her principal.

Agricultural bank: A type of bank that has a significant involvement in agricultural lending. Three methods are commonly used to identify an agricultural bank. They are: (1) location in an agricultural community; (2) a specified proportion of agricultural loans to total loans (e.g., above the average ratio for all banks or greater than 25 percent agricultural loans to total loans); and (3) a specified minimum amount of agricultural loans (e.g., $5 million).

Agricultural credit association: A lending association of the Farm Credit System that makes direct short-, intermediate-, and long-term loans with funds borrowed from a Farm Credit Bank.

Amortization: A process in which the level of debt is reduced when equal periodic debt payments that cover interest and principal are made. It can also be used in reference to depletion of intangible assets.

Annual percentage rate (APR): The term used in the Truth in Lending Act. It is an *actuarial* representation of the total financing cost of credit expressed as a percent per annum. The annual percentage rate (APR) is calculated similarly across different institutions.

Annuity equivalent: The determination of a series of equal periodic payments (i.e., an annuity) for the economic life of an investment, given the present value of the investment's cash flows.

Appraisal: The written summary by a qualified individual setting forth an estimated value of a specific asset or group of assets, most often used in reference to real estate.

Asset-generating loan: A loan whose proceeds are used to finance assets with high collateral value.

Asset replacement: The use of present value methods to determine the optimal time to replace an asset.

Assets: The items and property owned or controlled by an individual or a business that have commercial or exchange value. Items may also include claims against others. All assets are reported on a balance sheet at market or cost value.

Assignment: The transfer of title, property, rights, or other interests from one person or entity to another.

Average cost of funds: A method of determining the cost of funds at a lending institution. This method uses an average cost of existing funds. In contrast, the marginal cost of funds uses cost of new funds only.

Balance sheet: The financial statement that reflects the value of the assets of an individual or a business and the financial claims on these assets at a specific point in time.

Balloon payment: A lump sum final payment of a loan. It reflects the entire remaining balance of a shorter-term loan (e.g., five years), which is amortized over a longer term (e.g., 10 to 20 years).

Bank holding company (BHC): A corporation that owns or controls at least 25 percent of the voting stocks of one or more banks.

Bankruptcy: The federal court proceeding by which a debtor (individual or corporation) may obtain protection from creditors.

Banks for Cooperatives: Banks of the Farm Credit System that specialize in financing agricultural cooperatives.

Base rate: An interest rate used as a basis for pricing loans. A margin reflecting the riskiness of the individual or operation is added to or subtracted from the base rate to determine the loan rate. The bank's funding, operating costs, and required return are reflected in the base rate.

Basis point: Most often used in describing interest rate movements or interest costs. One basis point is 1/100 of 1 percent. For example, 50 basis points are 0.5 percent.

Bid price: The maximum price an investor can pay for real estate or other assets in order to earn a required rate-of-return.

Branch banking: A system of banking in which individual banks may have multiple offices dispersed over geographic areas.

Business risk: All risks that are independent of a unit's financial structure and sources of financing. Examples in agriculture include unanticipated variations in agricultural production, commodity markets, and real estate prices.

Cap: Term used with variable or adjustable rate loans that refers to the maximum allowable adjustment in interest rate.

Capital budgeting: A process of determining the profitability of an investment, usually based on present value methods.

Capital debt repayment capacity (CDRC): Capital debt repayment capacity is the projected amount of funds available to a borrower to repay principal and interest on term loans.

Capital gains (losses): The amount by which the proceeds from the sale of capital assets exceed their cost or book value.

Capital leasing: See *Financial leasing*.

Cash accounting: A method of accounting whereby income is reported in the year it is received and expenses are reported in the year they are paid.

Cash flow budget: A financial statement reflecting the projected sources and uses of cash during an accounting period. Items on the statement are usually categorized as business or nonbusiness with subdivisions for funds from business operations and funds from financing.

Cash flow coverage: A financial ratio defined as excess cash available divided by the sum of interest paid plus principal payments on intermediate- and long-term debt.

Cash lease: A leasing method for farm land in which the tenant makes a cash rental payment to the land owner for use of the land.

Certainty equivalent: The certain amount of cash return that gives the same utility as a risky amount of cash return.

Closing costs: The costs incurred by the borrower and the seller in completing a loan transaction. Included are origination fees, inspections, title insurance, appraisals, attorney's and realtor's fees, and other costs of closing a loan.

Coefficient of variation: A relative measure of variability defined as the standard deviation divided by the expected or average value of a risky enterprise.

Collateral: Property pledged to assure repayment of debt.

Commitment: A formal agreement between a lender and a borrower in which a specified amount of money will be loaned at a specified future date subject to specific performance criteria and repayment terms.

Commodity Credit Corporation (CCC): The Commodity Credit Corporation (CCC) is a wholly owned federal corporation within the U.S. Department of Agriculture that was formed to finance price support loans for agricultural commodities.

Common stock: A financial security (securities) that represents ownership in a corporation.

Community property: A system of property ownership in which each spouse owns in his or her own right an equal, undivided portion of all marital property.

Compensating deposit balance: A minimum deposit balance that is sometimes required by a bank from a borrower. The balance is usually expressed as a percentage of the total loan commitment and/or a stipulated percentage of the amount of commitment actually used by the customer.

Compound interest: The interest that is paid, added to, or compounded into the principal and thereafter also earns interest. Common compounding periods are daily, monthly, quarterly, annually, and continuously. The more frequent the compounding, the higher the effective rate of interest.

Constant payment loan: Repayment of a loan over several periods with equal payments of principal plus interest in each period.

Constant payment on principal: Repayment of a loan over time with equal payments of principal in each period plus interest paid on a declining loan balance.

Contingent tax liability: The tax liability that may arise if an asset is sold.

Contract interest rate: The interest rate stated on a note.

Contract sale: A sale of property directly between the buyer and the seller. An agreement is made between the buyer and the seller determining price, payment terms, and title transfer.

Cooperative: An organization that is owned by and operated for the benefit of its patrons. An example is the Farm Credit System in which member-borrowers are the owners of the lending associations.

Coordinated financial statements: Interrelated financial statements, including the balance sheet, income statement, and statement of cash flows.

Corporation: A legal form of business organization in which owners have shares of equity capital and their liability is limited to the extent of their individual investments. A corporation acts as a single person even though it is composed of one or more persons.

Correlation: A statistical concept describing the degree of association or interdependence between two random variables.

Correspondent bank: A bank that performs specific functions for another bank (respondent bank). Functions may include loan participation, check clearing, data processing, cash management, and consulting services.

Co-signer: An individual in addition to the borrower who signs a note and thus assumes responsibility and liability for repayment.

Cost of capital: The cost of using debt and/or equity capital, often measured as a weighted average of the cost of debt and equity capital.

Cost of debt: The interest plus noninterest cost of using debt capital.

Cost of equity: The cost of using equity capital, often called a required rate-of-return on equity capital.

Cost of funds: The interest and noninterest cost of obtaining equity and debt funds.

Covenant: A legal promise in a note, loan agreement, security agreement, or mortgage to do or not to do specific acts; or a legal promise that certain conditions do or do not exist. A breach of a covenant can lead to the "injured party" pursuing legal remedies and can be a basis for foreclosure.

Coverage ratios: Ratios depicting the returns to assets relative to financial claims by lenders on those returns. Several types of coverage ratios are commonly used.

Credit: The borrowing capacity of a firm or other economic unit, as determined by lenders based on an evaluation of the firm's credit worthiness.

Credit reserve: Unused borrowing capacity that is available for liquidity purposes.

Credit scoring: A quantitative approach used to measure and evaluate the credit worthiness of a loan applicant. Profitability, solvency, management ability, and liquidity are commonly included in a credit-scoring model.

Credit worthiness: The ability, willingness, and financial capability of a borrower to repay debt.

Current assets: Cash and near-cash items with values that will probably be realized in cash or used up during the year. Also assets whose conversion to cash would only minimally disrupt the normal business operation.

Current liabilities: Financial obligations that are due and payable within the next year.

Current ratio: A liquidity ratio calculated as current assets divided by current liabilities.

Custom hiring: A form of leasing or rental that combines the hiring of labor services and the use of a tangible asset.

Debt-to-asset ratio: A solvency ratio calculated as total liabilities divided by total assets.

Deed: The legal instrument used to convey title to real estate.

Deed of trust: A written instrument that conveys or transfers property to a trustee. Property is transferred by the borrower to a trustee, who holds it as security for the payment of debt, and upon full payment of the debt, returns it to the borrower. In some states, a deed of trust is used in place of a mortgage.

Default: The failure of a borrower to meet the financial obligations of a loan or a breach of any of the other terms or covenants of a loan.

Deferred taxes: The tax consequences attributable to income resulting from the sale of assets reported on the balance sheet. These amounts are separated into current and noncurrent liabilities.

Delinquency: The status of overdue principal and/or interest payments on a loan.

Demand loan: A loan with no specific maturity date. The lender may demand payment on the loan at any time.

Depreciation: A decrease in value of real property caused by age, use, obsolescence, and/or physical deterioration. A noncash accounting expense that reflects the allowable deduction in book value of assets such as machinery, buildings, and breeding livestock.

Discount interest: A method of computing interest on a loan in which interest is calculated on the original amount of a loan for its full period and then subtracted from the amount of the loan at the beginning.

Discount rate: The rate used in determining present values of series of payments. It may represent an opportunity cost, a cost of capital, or a required rate-of-return.

Disintermediation: The withdrawal of deposits in response to higher unregulated rates-of-return in money market mutual funds or for direct placement in other money market instruments.

Diversification: The process of holding multiple assets in order to reduce the total variability of returns from the assets, relative to holding individual assets.

Dividends: Payments made by corporations to stockholders.

Downpayment: The equity amount invested in an asset purchase. The downpayment plus the amount borrowed generally equals the total value of the asset purchased.

Draft: An order for the payment of money drawn by one person or bank on another. Often used in the dispersal of an operating loan to a borrower for payment of bills.

Duration: The weighted average maturity of an asset's cash flows, using the present value of each cash flow as weights.

Effective interest rate: The calculated interest rate that may take account of stock, fees, and compounding, in contrast to a quoted rate of interest.

Encumbrance: A claim or interest that limits the right of property. Liens, mortgages, leases, dower rights and easements are examples of encumbrances.

Equity capital: See *Net worth.*

Estate: The total value of all assets owned by an individual, often applicable when the individual dies or declares bankruptcy. Also refers to an ownership interest in land.

Exchange rate: The value of one country's currency expressed in terms of another country's currency.

Expected value: The mean or weighted average of a probability distribution of potential outcomes using the probabilities as weights.

Farm Credit Bank: An institution of the Farm Credit System that loans the proceeds of the system's bond sales to local lending associations and provides other financial services to these associations.

Farm Credit System (FCS): A system of federally chartered, but privately owned, banks and associations that lend primarily to agricultural producers and their cooperatives.

Farmer Mac: See *Federal Agricultural Mortgage Corporation.*

Farmers Home Administration (FmHA): A U.S. government agency operating under the authority of the U.S. Department of Agriculture. Programs principally provide or guarantee credit for agricultural and rural borrowers who show promise for financial viability but are unable to independently obtain financing from commercial sources.

Federal Agricultural Mortgage Corporation (FAMC): A federally chartered instrumentality of the United States that provides guarantees for the timely payment of principal and interest on securities representing interests in, or obligations by, pools of qualified loans. Oversees the secondary market in farm mortgage loans. Often called *Farmer Mac.*

Federal Land Bank Association (FLBA): An institution of the Farm Credit System that takes applications for and services long-term real estate loans for a Farm Credit Bank.

Federal Land Credit Association (FLCA): A Federal Land Bank Association that has been given direct long-term real estate lending authority and makes such loans with funds obtained from a Farm Credit Bank.

Federal Reserve System: The central banking system of the United States, consisting of a board of governors, 12 federal reserve banks and their branches, and all member banks. Often called the Fed, it is charged with control over the nation's money supply and monetary policies.

Fees: Fixed charges or payments for services associated with loan transactions.

Filing: Giving public disclosure of a lender's security interest or assignment in collateral.

Financial control: The process of measuring and monitoring business performances over time to maintain desired standards of performance.

Financial feasibility: The ability of a business plan or investment to satisfy the financing terms and performance criteria agreed to by a borrower and a lender.

Financial intermediation: The channeling of funds and securities between savers and investors by financial institutions.

Financial leasing: A long-term contractual arrangement in which the lessee acquires control of an asset in return for rental payments to the owner. Also called *capital leasing.*

Financial market: A market in which financial assets are bought and sold.

Financial risks: Risks associated with the use of borrowing and leasing. Uncertainties about the ability to meet financial obligations.

Financial statement: A written report of the financial condition of a firm. Balance sheets, income statements, statements of changes in net worth, and statements of cash flow are examples of financial statements.

Financing statement: A statement filed by a lender with a public official. The statement reports the security interest or lien on the borrower's assets.

First mortgage: A real estate mortgage that has priority over all other mortgages on a specified piece of real estate.

Fiscal policy: Taxation and expenditure policies of government.

Fixed-rate loan: A loan that bears the same interest rate until loan maturity.

Floating-rate loan: See *Variable-rate loan.*

Foreclosure: The legal process by which a lien against property is enforced through the taking and selling of the property.

Forward contracting: Contracting for the sale or purchase of a firm's assets at a stipulated price prior to the exchange.

Future value: The value that a payment, or a series of payments, will have at some time in the future, based on compounding the payments at a specified interest rate.

Futures contract: Standardized agreement made on an organized exchange representing a commitment to buy or sell a commodity at a future time.

Gap: The margin of difference between rate-sensitive assets and rate-sensitive liabilities held by a financial institution.

Gap ratio: The ratio whereby rate-sensitive assets are divided by rate-sensitive liabilities.

Hedging: Reducing price risk by taking a position in a commodity futures contract that is the opposite of a cash or spot position. An example is the sale of a futures contract on a growing crop.

Income statement: A summary of the revenue (receipts or income) and expenses (costs) of a business over a period of time to determine its profit position. Also called a *profit and loss statement,* an *earnings statement,* and an *operating statement.*

Inflation: A general increase in the price level of a set of goods and services.

Interest coverage: The financial ratio that expresses a firm's return to assets relative to the amount of interest charges due on debt obligations.

Interest expense ratio: An operating performance measure calculated as interest expense (cash and accrued) divided by value of farm production or gross revenue.

Interest rate risk: Unanticipated or random variations in interest rates on major sources of funds for a borrower.

Internal rate-of-return (IRR): A profitability criterion used in capital budgeting, which is defined as the discount rate that equates the net present value of a projected series of cash flow payments to zero.

Investment analysis: See *Capital budgeting.*

Joint tenancy: Co-ownership of property in which the joint tenants have undivided interests in property and the ability to convey these interests without the co-owners' consent, although a joint tenant assumes full ownership at the death of the other joint tenant.

Land contract: An installment contract for the purchase of land. The seller gives the buyer the title or deed upon payment of the last installment.

Land trust: Legal entity by which an interest in real property is converted to an interest in personal property. Owners deed real property to the trust and receive an interest in the trust in return.

Lease: A contract whereby an individual acquires the control of an asset (e.g., land, machinery) by renting for a specified period of time. A rental payment is made by the lessee (or tenant) to the lessor (or land owner) to cover the lessor's cost of ownership. Examples of types of leases are operating leases, financial leases, real estate leases, and custom hiring.

Lease-purchase: A financing arrangement in which an asset (e.g., a tractor) is leased for a period of time and then purchased at a price specified in the lease-purchase contract.

Legal lending limit: A legal limit on the total loans and commitments a financial institution can have outstanding to any one borrower. The limit usually is determined as a specified percentage of the financial institution's own net worth or equity capital. The purpose of the legal lending limit is to help the financial institution avoid excessive exposure to credit risk.

Lessee: A tenant or person renting an asset from another party.

Lessor: A land owner or person renting an asset to another party.

Leverage: The use of debt capital in combination with equity capital.

Leveraged leasing: A three-party leasing arrangement in which an asset owner finances the asset with funds borrowed from a lender and then leases the asset to a lessee.

Lien: A claim by a creditor (lender) on property or assets of a debtor (borrower) in which the property may be held as security or sold in satisfaction (full or partial) of a debt. A lien may arise through a borrowing transaction where the lender is granted a lien on the borrower's property. Other liens include tax liens against real estate with delinquent taxes, a mechanic's lien against property on which work has been performed, and a land owner's lien against crops grown by a tenant.

Life cycle: The relationship between a firm's performance and the age of its managers. Stages making up the life cycle are (1) start, (2) growth, (3) consolidation, and (4) transfer.

Line of credit: An arrangement by a lender to make an amount of credit available to a borrower for use over a specified period of time. It is generally characterized by a master note, cash flow budgets, and periodic and partial disbursements and repayments of loan funds. A formal agreement of similar characteristics is a credit commitment.

Liquidation: The sale of assets to generate cash needed to meet financial obligations, transactions, or investment opportunities.

Liquidity: The ability of a business to generate cash, with little risk of loss of principal value, to meet financial obligations, transactions, or investment opportunities.

Loan agreement: A written agreement between a lender and a borrower stipulating terms and conditions associated with a financing transaction and in addition to those included to accompanying note, security agreement, and other loan documents. The agreement may indicate the obligations of each party, reporting requirements, possible sanctions for lack of borrower performance, and any restrictions placed on a borrower.

Loan commitment: A formal agreement to loan up to a specified dollar amount during a specified period.

Loan committee: A group of loan officers, executive personnel, and/or directors of a financial institution who establish lending policies and/or approve loan requests that exceed the lending authority of individual loan officers.

Loan guarantee: An agreement by an individual, a unit of government, an insurance firm, or other party to repay all or part of a loan made by a lender in the event that the borrower is unable to repay.

Loan participation: A loan in which two or more lenders share in providing loan funds to a borrower. Generally, one of the participating lenders originates, services, and documents the loan.

Loan-to-asset value: The ratio of loan balance to the value of assets pledged as collateral to secure a loan.

Marginal cost of funds: A method of determining the cost of funds at a lending institution in which interest rates on new loans are based on the cost of new funds acquired in financial markets to fund the loans. This pricing policy contrasts with loan pricing based on the average cost of funds already acquired by the lending institution.

Master note: A note (promise to repay) often used in combination with line-of-credit financing to cover present and future borrowing needs through periodic disbursements and repayments of loan funds.

Maturity: Amount of time until the loan is fully due and payable. For example, a five-year term loan has a maturity of five years.

Modified internal rate-of-return: A modification of the internal rate-of-return that explicitly considers reinvestment rates for earnings and resolves potential inconsistencies between the IRR and the NPV methods of capital budgeting.

Monetary policy: Actions taken by the government to influence the size and rate of change of the money supply and the availability of credit in an economy.

Mortgage: A legal instrument that conveys a security interest in real estate property to the lender.

Net cash flow: The difference between all cash inflows and cash outflows during an accounting period.

Net interest margin: The difference between the interest earnings on assets and the interest costs of earning assets for a financial institution.

Net present value (NPV): The discounted sum of the projected series of net cash flows for an investment, including the initial investment requirements and the terminal value of the investment at the end of the planning horizon.

Net worth: The financial claim by owners on the total assets of a business, calculated as total assets minus total liabilities. Also called *equity capital* and *ownership equity.*

Nominal value: A money value reported in current or actual dollars; not adjusted for inflation.

Noncurrent assets: All assets not classified as current assets.

Noncurrent liabilities: All liabilities not classified as current liabilities.

Nonrevolving line of credit: A line of credit in which the maximum amount of a loan is the total of loan disbursements. Repayments do not make loan funds available again as in a revolving line of credit.

Note: A written document in which a borrower promises to repay a loan to a lender at a stipulated interest rate within a specified time period or upon demand. Also called a *promissory note.*

Off-farm income: Income earned by a farm operator or member of the operator's family from employment off the farm or from investments made in nonfarm activities or ventures.

Operating expenses: Outlays incurred or paid by a business for all inputs purchased or hired that are used up in production during the accounting period.

Operating loan: A short-term loan (i.e., less than one year) to finance crop production, livestock production, inventories, accounts receivable, and other operating or short-term liquidity needs of a business.

Operating receipts: Receipts from products produced and sold or held for sale during an accounting period.

Option: A contract granting the holder the right, but not the obligation, to buy or sell a commodity at a stipulated price.

Origination fee: A fee charged by a lender to a borrower at the time a loan is originated to cover the costs of administering the loan, evaluating credit, checking legal records, verifying collateral, and other administrative activities.

Partnership: An association of two or more persons to carry on, as co-owners, a business for profit.

Payback period: A capital budgeting method that estimates the length of time required for an investment to pay itself out.

Personal property: Any tangible or intangible property that is not designated by law as real property. Personal property is not fixed or immovable.

Planning horizon: The number of years over which the projected cash flows of an investment will be evaluated.

Points: A form of loan fee generally charged by long-term lenders at loan origination to cover a portion of the lender's administrative and funding costs. Points typically are expressed as a percentage of the total loan. For example, 3 points equal 3 percent of the loan amount.

Portfolio analysis: Analysis of the risk and return characteristics of a mix, or combination, of assets, enterprises, or investments.

Prepaid expenses: A balance sheet item that represents cash expenditures for items that will be incurred in future periods. Examples are prepaid rent, insurance, and taxes. These items are generally reported with current assets.

Prepayment penalty: An amount charged by a lender on a loan paid prior to its maturity.

Present value: The value that a future payment, or a series of payments, will have at present, based on discounting the payments at a specified discount rate.

Prime rate: A nationally quoted rate believed to represent the interest rate charged by U.S. money center banks to their most credit worthy corporate borrowers. Prime rate may also refer to an individual lender's interest rate charged to its most credit worthy borrowers, although the term *base rate* is more commonly used.

Principal: The dollar amount of a loan outstanding at a point in time, or the portion of a payment that represents a reduction in loan balance. Principal is distinguished from interest due on a loan or the interest portion of a loan payment.

Probability: The likelihood of possible outcomes for a future event.

Production Credit Association (PCA): A lending association of the Farm Credit System that makes short- and intermediate-term loans with funds borrowed from a Farm Credit Bank.

Profitability: The relative profit performance of a business, enterprise, or other operating unit. Profitability comparisons often occur over time, across peer groups, relative to projections, and relative to norms or standards.

Pro forma statements: A financial statement or presentation of data that represents financial performance based on projections of events and conditions. Examples are a pro forma balance sheet and a pro forma income statement.

Proprietorship: An unincorporated business owned and operated by a single individual.

Public offering: Sale of equity or debt securities by a business to investors in the financial markets.

Rate adjustment: A change in interest rate on an existing loan. Rate adjustments may occur on variable or adjustable rate loans.

Rate-of-return on assets (ROA): A profitability measure representing the rate-of-return on business assets during an accounting period. ROA is calculated by dividing the dollar return to assets during the accounting period by the value of assets at the beginning of the period or the average value of assets over the period.

Rate-of-return on equity (ROE): A profitability measure representing the rate-of-return on the equity capital that owners have invested in a business. ROE is calculated by dividing the dollar return to equity capital during an accounting period by the value of equity capital at the beginning of the period or the average value of equity capital over the period.

Rate-sensitive assets: Financial assets whose interest rate is subject to change within a specified period of time.

Rate-sensitive liabilities: Financial liabilities whose interest rate is subject to change within a specified period of time.

Real value: A money value reported in terms of constant purchasing power; adjusted for inflation.

Real property: Land, buildings, minerals, and other kinds of property that are legally classified as real.

Refinancing: A change in an existing loan designed to extend and/or restructure the repayment obligation or to achieve more favorable loan terms by transferring the financing arrangement to another lender or loan type.

Remaining balance interest: A method of determining interest obligations on a loan in which interest is calculated on the outstanding loan balance, net of past principal payments.

Renewal: A form of extending an unpaid loan in which the borrower's remaining unpaid loan balance is carried over (renewed) into a new loan at the beginning of the next financing period.

Repayment ability: The anticipated ability of a borrower to generate sufficient cash to repay a loan plus interest according to the terms established in the loan contract.

Retained earnings: The portion of net income that is retained within a business and added to net worth.

Revolving line of credit: A line of credit made available to a borrower in which the borrower can usually borrow, repay, and re-borrow funds at any time and in any amounts up to the credit limit, but not above, during a specified period of time.

Risk: Potential adversity, often expressed by unanticipated or random variation.

Risk assessment: The procedures a lender follows in evaluating a borrower's credit worthiness, repayment ability, and collateral position relative to the borrower's intended use of the loan proceeds. Risk assessment is similar to *credit scoring* and *risk rating*.

Risk aversion: An attitude toward risk in which the investor requires compensation for taking risk.

Risk dominance: One risky investment dominates another if it has the same expected return and less risk, or the same risk and a higher expected return. Also called *risk efficiency*.

Risk efficient set: A collection of risky assets or enterprises, none of which dominates another.

Risk premium: The adjustment of a lender's base interest rate in response to the anticipated level of a borrower's credit risk in a loan transaction. Higher risk loans may carry higher interest rates, with the rate differential representing the risk premium.

Risk rating: The relative amount of credit risk associated with a loan transaction. The lender may use credit-scoring or risk assessment procedures to evaluate loan requests and group borrowers into various risk classes for purposes of loan acceptance or rejection, loan pricing, loan control, degree of monitoring, and level of loan documentation.

Second mortgage: The use of two lenders in a real estate mortgage in which one lender holds a first mortgage on the real estate and another lender holds a second mortgage. The first mortgage holder has first claim on the borrower's mortgaged property and assets in the event of loan default and foreclosure or bankruptcy.

Secondary market: An organized market in which existing financial assets are bought and sold.

Secured loans: Loans in which specific assets have been pledged by the borrowers as collateral to secure the loans. Security agreements and mortgages serve as evidence of security in secured loans.

Security agreement: A legal instrument signed by a debtor granting a security interest to a lender in specified personal property pledged as collateral to secure a loan.

Security interest: A claim, or lien, which a secured party (i.e., a lender) has on the personal property of a debtor. A security interest is usually created by a security agreement.

Self-liquidating loan: A loan made for a purpose that generates sufficient income to repay the loan within the maturity period.

Seller financing: A loan provided by the seller of property to its buyer.

Share lease: A leasing arrangement for real estate in which the land owner receives a share of the production as rent and pays a share of the business's variable expenses. The land owner often shares in the management decisions with the tenant.

Shared appreciation mortgage: A financing arrangement for real estate in which the lender reduces the interest rate on the loan in return for a stipulated share of the appreciated value of the land being financed at a designated time in the future.

Simple interest: A method of calculating interest obligations in which no compounding of interest occurs. Interest charges are the product of the loan principal times the annual rate of interest, times the number of years or proportion of a year the principal has been outstanding.

Simple rate-of-return: A profitability measure calculated by dividing an investment's average annual profits by either the original investment value or the average investment value over the investment's expected life.

Solvency: A business condition of financial viability in which net worth is positive and the business is expected to meet its financial obligations as they come due.

Split line of credit: A financing situation in which a borrower obtains operating credit from two or more lenders.

Standard deviation: A statistical measure of the amount of dispersion or random variation of a projected outcome about its expected value; the square root of the variance.

Statement of cash flows: A financial statement presenting the cash receipts and cash payments over a specified period of time. The cash receipts and payments are separated into operating, investing, and financing activities.

Stock requirement: A method of capitalizing lending institutions such as the co-operative Farm Credit System. The borrower is required to purchase stock in the lending association to obtain a loan.

Tenancy by entireties: Co-ownership of property by spouses, in which no conveyance of an interest to a third party may occur without the consent of both spouses.

Tenancy in common: Co-ownership of property in which the tenants have undivided interests and may convey their interests to others during their lifetime or at their death, without the consent of both co-owners.

Time preference: The strength of an individual's preference for receiving payments now rather than in the future.

Title insurance: Insurance that protects a purchaser or mortgage lender against losses arising from a defect in title to real estate, other than defects that have been specifically excluded. A clear title is free of any claims, mortgages, liens, or other encumbrances and has no ownership interest other than that of the owner of record.

Trade credit: Credit arising from a merchandising transaction in which the seller finances a purchase by the buyer.

Truth in Lending: The federal Truth in Lending Act is intended to assure a meaningful disclosure of credit terms to borrowers, especially on consumer loans. Excluded transactions include loans for commercial or business purposes, including agricultural loans; loans to partnerships, corporations, cooperatives, and organizations; and loans greater than $25,000, except for owner-occupied, residential real estate mortgages where compliance is required regardless of the amount.

Turnover ratio: The ratio of the value of farm production to average total assets.

Uniform Commercial Code: The body of laws regulating exchanges of personal property.

Uniform series: A series in which the payments are equal in each conversion period.

Unsecured loans: Loans for which there are no guarantors or co-signors and no specific assets have been pledged by the borrowers as collateral to secure the loans.

Usury laws: Laws that establish legal ceilings on the interest rates charged for various types of loans. In states where usury laws exist, most usury limits are well above market interest rates and often are indexed to change with changes in market interest rates or other leading rate indicators.

Variable-rate loan: A loan transaction in which the interest rate may be changed within the period of the loan contract. Generally, rate changes occur in response to changes in the lender's cost of funds or a specified index.

Variance: A statistical measure of the amount of dispersion or random variation, based on squared deviations about a variable's expected value.

Vertical coordination: The coordination of successive stages in a production and distribution system ranging from spot or open market transactions through market contracting to complete ownership or vertical integration.

Vertical integration: Ownership of two or more successive stages of a production and distribution system by a single party.

Warehouse receipt: A receipt issued by a warehouse worker providing evidence of title to stored goods (especially commodities) and thus validating the existence of collateral that may be pledged to secure a loan.

Weighted average cost of capital: A combined measure of the costs of debt and equity capital, using the ratios of debt-to-assets and equity-to-assets as the weights.

Working capital: The difference between current assets and current liability. Often used as a measure of business liquidity.

Indexes

NAME INDEX

SUBJECT INDEX

I

L